# MATHEMATICAL MODELING
## FOR THE SCIENTIFIC METHOD

# The Jones & Bartlett Learning Series in Mathematics

## Geometry

*Geometry with an Introduction to Cosmic Topology*
Hitchman (978-0-7637-5457-0) © 2009

*Euclidean and Transformational Geometry: A Deductive Inquiry*
Libeskind (978-0-7637-4366-6) © 2008

*A Gateway to Modern Geometry: The Poincaré Half-Plane, Second Edition*
Stahl (978-0-7637-5381-8) © 2008

*Understanding Modern Mathematics*
Stahl (978-0-7637-3401-5) © 2007

*Lebesgue Integration on Euclidean Space, Revised Edition*
Jones (978-0-7637-1708-7) © 2001

## Precalculus

*Precalculus: A Functional Approach to Graphing and Problem Solving, Sixth Edition*
Smith (978-0-7637-5177-7) © 2012

*Precalculus with Calculus Previews (Expanded Volume), Fourth Edition*
Zill/Dewar (978-0-7637-6631-3) © 2010

*Precalculus with Calculus Previews (Essentials Version), Fifth Edition*
Zill/Dewar (978-1-4496-1497-3) © 2012

*Algebra and Trigonometry, Third Edition*
Zill/Dewar (978-0-7637-5461-7) © 2012

*College Algebra, Third Edition*
Zill/Dewar (978-1-4496-0602-2) © 2012

*Trigonometry, Third Edition*
Zill/Dewar (978-1-4496-0604-6) © 2012

## Calculus

*Calculus of a Single Variable: Early Transcendentals, Fourth Edition*
Zill/Wright (978-0-7637-4965-1) © 2011

*Multivariable Calculus, Fourth Edition*
Zill/Wright (978-0-7637-4966-8) © 2011

*Calculus: Early Transcendentals, Fourth Edition*
Zill/Wright (978-0-7637-5995-7) © 2011

*Multivariable Calculus*
Damiano/Freije (978-0-7637-8247-4) © 2012

*Calculus: The Language of Change*
Cohen/Henle (978-0-7637-2947-9) © 2005

*Applied Calculus for Scientists and Engineers*
Blume (978-0-7637-2877-9) © 2005

*Calculus: Labs for Mathematica*
O'Connor (978-0-7637-3425-1) © 2005

*Calculus: Labs for MATLAB*
O'Connor (978-0-7637-3426-8) © 2005

## Linear Algebra

*Linear Algebra with Applications, Seventh Edition*
Williams (978-0-7637-8248-1) © 2011

*Linear Algebra with Applications, Alternate Seventh Edition*
Williams (978-0-7637-8249-8) © 2011

*Linear Algebra: Theory and Applications*
Cheney/Kincaid, Second Edition
(978-1-4496-1352-5) © 2012

## Advanced Engineering Mathematics

*Advanced Engineering Mathematics, Fourth Edition*
Zill/Wright (978-0-7637-7966-5) © 2011

*An Elementary Course in Partial Differential Equations, Second Edition*
Amaranath (978-0-7637-6244-5) © 2009

## Complex Analysis

*A First Course in Complex Analysis with Applications, Second Edition*
Zill/Shanahan (978-0-7637-5772-4) © 2009

*Complex Analysis for Mathematics and Engineering, Sixth Edition*
Mathews/Howell (978-1-4496-0445-5) © 2012

*Classical Complex Analysis*
Hahn (978-0-8672-0494-0) © 1996

## Real Analysis

*Elements of Real Analysis*
Denlinger (978-0-7637-7947-4) © 2011

*An Introduction to Analysis, Second Edition*
Bilodeau/Thie/Keough (978-0-7637-7492-9) © 2010

*Basic Real Analysis*
Howland (978-0-7637-7318-2) © 2010

*Closer and Closer: Introducing Real Analysis*
Schumacher (978-0-7637-3593-7) © 2008

*The Way of Analysis, Revised Edition*
Strichartz (978-0-7637-1497-0) © 2000

## Topology

*Foundations of Topology, Second Edition*
Patty (978-0-7637-4234-8) © 2009

## Discrete Math and Logic

*Discrete Structures, Logic, and Computability, Third Edition*
Hein (978-0-7637-7206-2) © 2010

*Essentials of Discrete Mathematics, Second Edition*
Hunter (978-1-4496-0442-4) © 2012

*Logic, Sets, and Recursion, Second Edition*
Causey (978-0-7637-3784-9) © 2006

## Numerical Methods

*Numerical Mathematics*
Grasselli/Pelinovsky (978-0-7637-3767-2) © 2008

*Exploring Numerical Methods: An Introduction to Scientific Computing Using MATLAB*
Linz (978-0-7637-1499-4) © 2003

## Advanced Mathematics

*Mathematical Modeling with Excel*
Albright (978-0-7637-6566-8) © 2010

*Clinical Statistics: Introducing Clinical Trials, Survival Analysis, and Longitudinal Data Analysis*
Korosteleva (978-0-7637-5850-9) © 2009

*Harmonic Analysis: A Gentle Introduction*
DeVito (978-0-7637-3893-8) © 2007

*Beginning Number Theory, Second Edition*
Robbins (978-0-7637-3768-9) © 2006

*A Gateway to Higher Mathematics*
Goodfriend (978-0-7637-2733-8) © 2006

*For more information on this series and its titles, please visit us online at http://www.jblearning.com/math. Qualified instructors, contact your Publisher's Representative at 1-800-832-0034 or info@jblearning.com to request review copies for course consideration.*

# The Jones & Bartlett Learning International Series in Mathematics

*Mathematical Modeling with Excel*
Albright (978-0-7637-6566-8) © 2010

*An Introduction to Analysis, Second Edition*
Bilodeau/Thie/Keough (978-0-7637-7492-9) © 2010

*Basic Real Analysis*
Howland (978-0-7637-7318-2) © 2010

*Advanced Engineering Mathematics, Fourth Edition, International Version*
Zill/Wright (978-0-7637-7994-8) © 2011

*Calculus: Early Transcendentals, Fourth Edition, International Version*
Zill/Wright (978-0-7637-8652-6) © 2011

*Elements of Real Analysis*
Denlinger (979-0-7637-7947-4) © 2011

*Mathematical Modeling for the Scientific Method*
Pravica/Spurr (978-0-7637-7946-7) © 2011

*A Journey into Partial Differential Equations*
Bray (978-0-7637-7256-7) © 2012

*Multivariable Calculus*
Damiano/Freije (978-0-7637-8247-4) © 2012

*Complex Analysis for Mathematics and Engineering, Sixth Edition*
Mathews/Howell (978-1-4496-0445-5) © 2012

*Functions of Mathematics in Liberal Arts*
Johnson (978-0-7637-8116-3) © 2012


*For more information on this series and its titles, please visit us online at http://www.jblearning.com/math. Qualified instructors, contact your Publisher's Representative at 1-800-832-0034 or info@jblearning.com to request review copies for course consideration.*

# MATHEMATICAL MODELING
## FOR THE SCIENTIFIC METHOD

DAVID W. PRAVICA
East Carolina University

MICHAEL J. SPURR
East Carolina University

*World Headquarters*

Jones & Bartlett Learning
40 Tall Pine Drive
Sudbury, MA 01776
978-443-5000
info@jblearning.com
www.jblearning.com

Jones & Bartlett Learning
Canada
6339 Ormindale Way
Mississauga, Ontario L5V 1J2
Canada

Jones & Bartlett Learning
International
Barb House, Barb Mews
London W6 7PA
United Kingdom

Jones & Bartlett Learning books and products are available through most bookstores and online booksellers. To contact Jones & Bartlett Learning directly, call 800-832-0034, fax 978-443-8000, or visit our website, www.jblearning.com.

Substantial discounts on bulk quantities of Jones & Bartlett Learning publications are available to corporations, professional associations, and other qualified organizations. For details and specific discount information, contact the special sales department at Jones & Bartlett Learning via the above contact information or send an email to specialsales@jblearning.com.

Copyright © 2011 by Jones & Bartlett Learning, LLC

All rights reserved. No part of the material protected by this copyright may be reproduced or utilized in any form, electronic or mechanical, including photocopying, recording, or by any information storage and retrieval system, without written permission from the copyright owner.

**Production Credits**
Senior Acquisitions Editor: Timothy Anderson
Associate Editor: Melissa Potter
Production Director: Amy Rose
Senior Marketing Manager: Andrea DeFronzo
Composition: Northeast Compositors
Cover Design: Kristin E. Parker
Cover Image: © Comstock Images/age fotostock
Printing and Binding: Malloy, Inc.
Cover Printing: Malloy, Inc.

**Library of Congress Cataloging-in-Publication Data**
Pravica, David W.
 Mathematical modeling for the scientific method / David W. Pravica and Michael J. Spurr.
   p. cm.
 Includes bibliographical references and index.
 ISBN-13: 978-0-7637-7946-7 (casebound)
 ISBN-10: 0-7637-7946-6 (casebound)
 1. Science—Methodology—Mathematical models.  I. Spurr, Michael J. II. Title.
 Q175.32.M38P73 2011
 001.4'20151—dc22
                                        2010025297

6048
Printed in the United States of America
14 13 12 11 10   10 9 8 7 6 5 4 3 2 1

Dedicated to
Hippasus of Metapontum,
whose one critical question
changed the world.

The first author
would like to thank his wife,
Sandra J. Pravica,
for her continued support,
encouragement, and her
enlightening
ideas and suggestions.

The second author
would like to thank his wife,
Elaine M. F. Spurr,
for her patience and support
during the completion of this work.

# Contents

List of Tables   xvii

List of Figures   xix

Preface   xxi

Anxiety with Mathematics   xxv

Ethics in Science   xxvii

Acknowledgments   xxx

## Part I   Foundations   1

**Chapter 1   Review of Algebra   3**

    1.1   Logic and Sets   4
    1.2   General Algebraic Equations   7
        1.2.1   Polynomials and Zeros   9
        1.2.2   Quadratic Equations   9
        1.2.3   Cubic Equations   11
        1.2.4   Quartic Equations   14
        1.2.5   Higher Degree Equations   15
        1.2.6   Iterative Solutions to Algebraic Equations   16
    1.3   Extensions of Mathematical Spaces   19
        1.3.1   Introduction to Groups   20
        1.3.2   Rings and Fields   22
        1.3.3   Introduction to Algebraic Numbers   23
        1.3.4   Continued Fractions   24
    1.4   Valuations, Inequalities, and Distance   25
    1.5   The Real Numbers   29
    1.6   Fundamental Theorems   30
    1.7   Sequences and Series   33
    1.8   Complex Numbers   36

**Chapter 2   Functions and Relations   41**

    2.1   Sets   42

- 2.2 A Catalogue of Functions 47
  - 2.2.1 Constant Functions 49
  - 2.2.2 Characteristic Functions 49
  - 2.2.3 Linear Functions 51
  - 2.2.4 Power Functions 52
  - 2.2.5 Nonlinear Polynomials 52
  - 2.2.6 Exponentials 53
  - 2.2.7 Hyperbolic Trigonometric Functions 54
  - 2.2.8 Trigonometric Functions 54
- 2.3 Euler's Formula 58
- 2.4 Function Rings, Algebras, Ideals, and Fields 59
- 2.5 Reciprocal Functions 61
  - 2.5.1 Rational Functions 61
  - 2.5.2 Rational Exponentials and Hyperbolic Trigonometrics 61
  - 2.5.3 The Six Trigonometric Functions 62
- 2.6 A Catalogue of Inverse Functions 63
  - 2.6.1 Constant Functions 63
  - 2.6.2 Characteristic Functions 63
  - 2.6.3 Linear Functions 64
  - 2.6.4 Root Functions 64
  - 2.6.5 Nonlinear Polynomials 64
  - 2.6.6 Logarithms 64
- 2.7 Continuous Functions 66
- 2.8 Algebra of Functions 69
- 2.9 Power Series and Products 70
- 2.10 Linearly-Independent Spanning Sets of Functions 73
- 2.11 Probability Functions 74
  - 2.11.1 Combinations and Permutations 76
  - 2.11.2 Discrete Distributions 78
  - 2.11.3 Discrete Uniform Distributions 78
  - 2.11.4 Bernoulli and Binomial Distributions 79
  - 2.11.5 Poisson Distribution 80
  - 2.11.6 Continuous Distributions 81
  - 2.11.7 Continuous Uniform Distributions 81
  - 2.11.8 Arbitrary Continuous Distributions 81
  - 2.11.9 Exponential Distributions 82
  - 2.11.10 Gaussian (Normal) Distribution 83
  - 2.11.11 Bayes' Formula and Inference 84
- 2.12 Relations 86

## Chapter 3 Statistics 89

- 3.1 The Problem with Precision 89
- 3.2 Random Variables, Models, and Parameters 90
    - 3.2.1 Terminology 91
    - 3.2.2 States of a System 91
- 3.3 Descriptive Statistics 92
    - 3.3.1 Stem-Leaf Plots 93
    - 3.3.2 Dot and Bar Plots 95
    - 3.3.3 Five-Number Summary and Box Plots 95
- 3.4 Distributions 97
    - 3.4.1 Frequency Diagrams and Histograms 97
- 3.5 Introduction to Inference 101
    - 3.5.1 Bayes Rule for Uniformly Distributed States 101
    - 3.5.2 Bayes Rule for Nonuniform A Priori Distributions 102
- 3.6 A Catalogue of Distributions 103
    - 3.6.1 Gaussian (Normal) Distribution 103
    - 3.6.2 Uniform Distribution 105
    - 3.6.3 Exponential Distribution 106
    - 3.6.4 Bernoulli and Binomial Distributions 107
    - 3.6.5 Poisson Distribution 107
- 3.7 Parameter Estimation 109
    - 3.7.1 Exponential Distribution 109
    - 3.7.2 Normal Distribution 109
    - 3.7.3 Tests for Probability Models 110
- 3.8 Statistical Inference 113
    - 3.8.1 Confidence Intervals 114
    - 3.8.2 Hypotheses Testing 115
    - 3.8.3 Differences of States 119
- 3.9 Linear Regression 120
    - 3.9.1 Example of the Deaf Frog 121
    - 3.9.2 Linear Models 121
    - 3.9.3 Scatter Plots 122
    - 3.9.4 Control Charts 123
    - 3.9.5 Correlation Coefficient 124
    - 3.9.6 Significant Association 125
- 3.10 Main Theorems 128
    - 3.10.1 Law of Large Numbers and Central Limit Theorems 128
    - 3.10.2 Theoretical Details 129
- 3.11 Bivariate Analysis 133
    - 3.11.1 Confidence in a Linear Model 134

## Part II  Multidimensionality  137

**Chapter 4  Linear Algebra and Matrices  139**
- 4.1  Lists of Data  139
- 4.2  Vector Spaces  140
  - 4.2.1  The Dot Product  142
- 4.3  Bases, Dimension, Linear Independence, and Vector Sums  145
- 4.4  Projections  148
- 4.5  Vectors in Two- and Three-Dimensional Space  150
  - 4.5.1  Cross Product in $\mathbb{R}^2$ and $\mathbb{R}^3$  153
- 4.6  Linear Systems and Matrix Equations  154
  - 4.6.1  Matrix Multiplication  157
- 4.7  Vector, Matrix, and Linear Algebra  158
  - 4.7.1  Matrix Equations  159
  - 4.7.2  Gaussian Elimination  160
- 4.8  Matrix Inverse and the Gauss-Jordan Method  164
- 4.9  Valuations of a Matrix, Determinant, and Trace  166
- 4.10  Elementary Matrices  169
- 4.11  Eigenvalues and Eigenvectors  171
- 4.12  Sum and Product Matrix Representations  175
  - 4.12.1  Spectral and $QDQ^T$ Representations of a Symmetric Matrix  175
  - 4.12.2  Perm × LDU Decomposition  178
- 4.13  Linear Regression  180
  - 4.13.1  Discussion of the F-Distribution  184

**Chapter 5  Calculus  191**
- 5.1  The Advancing Operation  192
- 5.2  Continuity  192
- 5.3  Slope Function, Linear Approximation, and the Derivative  194
- 5.4  Derivatives of Basic Functions  200
- 5.5  The Chain and Quotient Rules  210
- 5.6  Derivatives of Trigonometric Functions  214
- 5.7  Derivatives of Exponential Functions  222
- 5.8  The Derivative of an Inverse Function  227
- 5.9  A Table of Derivatives  233
- 5.10  The Differential of a Function  233
- 5.11  Continuous Differentiability and Higher Derivatives  236
- 5.12  Optimization  238

5.13 The Mean Value Theorem and its Consequences   241
5.14 Newton's Method for the Zeros of a Function   245
5.15 Linearization and L'Hopital's Rule   247
5.16 Antiderivatives   249
5.17 Initial Value Problems and Euler's Method   254
5.18 Area and the Fundamental Theorem of Calculus   256
5.19 Applications of the Definite Integral   265
    5.19.1 Volumes by Cross-Sectional Area & Volumes of Revolution   266
    5.19.2 Mass and Probability   267
    5.19.3 The Center of Mass and the Mean of a Probability Density   269
5.20 Improper Integrals and L'Hopital's Rule   273
5.21 Integration Techniques   277
    5.21.1 Integration by Parts   277
    5.21.2 Integrals of Trigonometric Expressions   279
    5.21.3 Trigonometric Substitutions   284
    5.21.4 Examples Illustrating the Method of Partial Fractions   286
5.22 Series and Taylor Series   291

**Chapter 6  Vector Calculus   303**

6.1 Algebra and Geometry in $\mathbb{R}^n$   303
    6.1.1 The Algebra of $\mathbb{R}^n$ as a Vector Space   304
    6.1.2 The Dot Product and its Geometry in $\mathbb{R}^n$   305
    6.1.3 The Cross Product in $\mathbb{R}^3$   311
    6.1.4 Flat Sets in $\mathbb{R}^n$: Lines, Planes, and Hyperplanes   316
6.2 Examples of Surfaces in $\mathbb{R}^3$   319
6.3 Functions of One Variable: Curves in $\mathbb{R}^n$   321
6.4 Functions of Several Variables   324
6.5 The Chain Rule; Partial and Directional Derivatives   328
6.6 Vector-Valued Functions of Several Variables   334
6.7 Change of Variables   339
    6.7.1 Parameterized Surfaces   339
    6.7.2 Parameterized Regions   342
    6.7.3 The Boundary of a Region in $\mathbb{R}^n$   343
6.8 Integration, Green's Theorem, and the Divergence Theorem   344
    6.8.1 Double Integrals   344
    6.8.2 Triple Integrals   348

        6.8.3   Change of Variables; Reparameterization   351  
        6.8.4   Surface Integrals   357  
        6.8.5   Line Integrals   359  
        6.8.6   Green's Theorem; the Divergence Theorem   362  
6.9   Vector Fields, Divergence, and Curl   370  
6.10  Optimization and Linear Regression   374  
6.11  A Revisit to Gradient Flow   377  
6.12  The Multivariable Newton's Method for Finding Zeros of a Function   378  

# Part III   Applications   381

## Chapter 7   Mathematical Modeling   383

7.1   Objectives of Mathematical Modeling   384  
        7.1.1   Prediction   384  
        7.1.2   Cataloging   384  
        7.1.3   Postdiction   385  
7.2   Difference Equations   386  
        7.2.1   Change Versus Equilibrium   386  
7.3   First-Order Single-Variable Differential Equations   387  
7.4   First-Order Linear Differential Equations with Forcing   389  
        7.4.1   Forced Linear Equations   390  
        7.4.2   Variable Coefficients   391  
7.5   Constant-Coefficient Systems of Linear Differential Equations   392  
        7.5.1   Forced Constant-Coefficient Linear-Vector Equations   396  
        7.5.2   Linear Electrical Circuits   397  
7.6   Systems of Nonlinear Differential Equations   401  
        7.6.1   Existence and Uniqueness   401  
        7.6.2   Linear Approximations of Nonlinear Systems and Quasilinearity   403  
7.7   Stability Versus Instability   406  
        7.7.1   Equilibria of Autonomous Systems   406  
        7.7.2   Instability and Negative Feedback   407  
7.8   System Instability Due to Parameter Changes   410  
        7.8.1   Phase Plots and Bifurcations   411  
        7.8.2   Parameter Space Diagrams   412  
7.9   Models of Mass and Electric Charge   415

        7.10  Motions in Space   418
                7.10.1  Quantum and Electromagnetic Fields   419

**Chapter 8**  **The Scientific Method**   **423**
        8.1  Quarks and Groups   426
        8.2  Measuring Radioactivity   428
        8.3  Chemical Equilibrium and Reaction Rates   430
                8.3.1  Rates of Reaction   430
                8.3.2  Chemical Equilibrium   431
        8.4  A Small Ising Model of Magnetism   433
                8.4.1  High Temperature Limit   434
                8.4.2  Low Temperature Limit   434
        8.5  The Shape of DNA   435
        8.6  Cellular Homeostasis   436
        8.7  Epidemics   438
        8.8  PCP Diagnosis   440
        8.9  Single-Species Population Models   443
                8.9.1  Exponential Growth   443
                8.9.2  Faster-Than-Exponential Growth   444
                8.9.3  Faster-Than-Exponential Logistic Growth   446
                8.9.4  Oscillatory Logistic Example   446
                8.9.5  The Lemming Example   447
                8.9.6  Logistic Growth with Harvesting   449
        8.10  Interacting Species   451
                8.10.1  Equilibria   451
                8.10.2  Few Carnivores and Herbivores   452
                8.10.3  Few Carnivores and Many Herbivores   452
                8.10.4  Few Herbivores and Many Carnivores   452
        8.11  Thermodynamics of Oceans and Atmosphere   453
                8.11.1  Dissolved $CO_2$   455
        8.12  Climate and Tree Rings   457
        8.13  The Motion of the Planets—the Royal Science   460
        8.14  The Sky Is Blue, but the Universe Is Red   462
        8.15  Quantum Cosmology and the Big-Bang Theory   464

**References**   471

**Glossary**   475

**Index**   477

# List of Tables

1.1  The addition and multiplication tables for $\mathbb{Z}_3$. 22

3.1  Hand-widths statistics. 93
3.2  Deaf frog data. 122
3.3  Length of pendulum vs. period of oscillations. 126
3.4  LLN for proportions and means. 129
3.5  CLT for proportions and averages. 129
3.6  Random sample data. 133

5.9  Table of Derivatives. 233
5.16 Table of Integrals. 250–251

7.1  The Integrating Factor Method for $y' + p(t) \cdot y = q(t)$. 391
7.2  Table of different $2 \times 2$ matrix generators. 395

8.1  Two species tree-ring data. 459

# List of Figures

1.1   The simple and generalized Roy Baker triangles. 12
1.2   Pythagorean proof. 31
1.3   Naturally occurring irrational lengths. 32
1.4   Roots of unity. 38

2.1   The continuous hill functions $h_1(x)$ and $h_2(x)$. 47
2.2   The $Stairs(x)$ and $Sawtooth(x)$ functions. 51
2.3   Right angle triangles and trigonometric functions. 55
2.4   Hexagon in a circle. 55
2.5   Special right-angle triangles. 56
2.6   Left: $e^x$ and $\ln(x)$ functions; Right: $e^{-x}$ and $\ln(-x)$ functions. 65
2.7   A continuous function. 67
2.8   Probability distribution of tails $T$ for three fair coins. 75
2.9   A Lissajous relation. 87

3.1   Hand-widths for H, F, M: Dot plot. 95
3.2   Hand-widths for H, F, M: Box plot. 96
3.3   Probability distribution using $\mu$ and $\sigma$ for hand widths. 99
3.4   Cumulative hand-width distribution. 100
3.5   Exp Distribution. 106
3.6   Chi-Squared Distribution. 112
3.7   Scatter plot. 123
3.8   Time series. 124

4.1   Vector and vector addition. 152
4.2   The Fisher Distribution. 185

5.1   Zooming in on a smooth function. 192
5.2   Intermediate Value Theorem. 193
5.3   Linearization with HOT error. 195

5.4  The geometry of a key trigonometric estimate. 215
5.5  Cosine and Sine function and HOT. 215
5.6  Comparing the area of a circular sector with areas of triangles. 216
5.7  Mean Value Theorem. 242
5.8  Rectangular, triangular, and trapezoidal areas. 257
5.9  Error in area estimation bounded by the area of a rectangular stack. 259
5.10 Estimating an integral above and below by the areas of two rectangles. 262
5.11 Fundamental Theorem of Calculus II via linearization of an integral function. 264

6.1  Surface, tangent plane, and HOT. 318
6.2  Region with boundary flow. 369

7.1  Phase plots. 411
7.2  Bifurcation plots. 412
7.3  population parameter plot. 413

8.1  PCP disease distributions. 441

# Preface

A purpose of *Mathematical Modeling for the Scientific Method* is to clarify the connection between deductive and inductive reasoning as used in mathematics and science. Critical ideas, when first introduced, will be *italicized*, however many definitions will be implied before being formally stated. Students should inquire with the instructor to clarify the meaning of anything new, or concepts insufficiently explained in the text. The goal is to be introductory while covering a broad range of techniques and applications. Ideally, one should strive for the least number of steps required before a cognitive concluding statement can be made. However, this is an art, and often it is better to give more information than be too brief.

None of the data presented here should be considered rigorously obtained or confirmed by experiment. Hopefully the reader will be motivated to further explore issues that have been raised. Calculations should require only elementary operations available on typical calculators. The choice of presentation was made for brevity and clarity, allowing a class of students with a variety of backgrounds to experience the material in a manner that unites perspective by the end of the textbook.

This textbook is appropriate for the following courses:

| |
|---|
| Mathematical Modeling |
| Calculus for the Biological Sciences |
| Mathematics for the Chemical and Physical Sciences |
| Concise Differential Equations/Linear Algebra with Applications |
| Probability Functions |
| Statistical Methods in the Sciences |
| Reference Manual for basic Algebra, Statistics, Linear Algebra, Calculus, Vector Calculus, Differential Equations, and Modeling in the Sciences |
| Iteration Methods as applied to Science |

*Mathematical Modeling for the Scientific Method* can be used in several settings:

> Part I: The textbook begins with several basic and fundamental mathematical spaces. Manipulations using algebra are presented, along with their conclusions. The distinction between variable and parameter is emphasized. Chapter 2 begins the discussion of specific special functions, with an emphasis on those used in the study of probability. Only single-variable functions are considered. The first part then ends with an introduction to statistics, which technically requires only basic algebra, but philosophically explores the issue of confidence and likelihood. Parameters are determined from data, and the concept of a best model can be introduced.
>
> Part II: Chapter 4 continues the discussion of linear equations. When there is more than one variable and one equation to be studied, the equations become systems. Parameters are estimated using matrix equations. This requires the concept of a vector space. In Chapter 5, the calculus of continuous functions is developed. The idea that many functions behave linearly near a point of interest is explored. This allows the use of algebraic techniques to study continuous functions. Families of linearly-independent functions lead to vector spaces. This part ends with the merging of the previous two topics into vector calculus.
>
> Part III: Difference and differential equations are discussed almost interchangeably. In particular, Chapter 7 presents methods for solving differential equations, but also for finding approximations using iteration methods. There is an example of how to modify equations to gain stability. Then the equations that model gravity and electromagnetism are presented. Finally, a series of topics from quarks through population models to galaxies are briefly described, with problems given that are intended to inspire discussion and further exploration.
>
> Mathematical Modeling: The textbook is designed to cover the essential topics that every mathematics teacher should be familiar with. Chapter 1 discusses mathematics spaces, variables, parameters, relations, and transformations, with basic results and simple proofs. Chapter 2 reviews the main functions of science with particular reference on the probability functions. Statistics is introduced, in a minimally rigorous manner, in Chapter 3. Linear and vector algebra is introduced in Chapter 4. This ends the topics that merely require algebra. Chapters 5 and 6 require notions of limits and continuity, although an algebraic approach is used to introduce the concept of linearization. Chapter 7 reviews techniques for solving differential equations. A method of analysis that may find use in equation-modification is presented for the purpose of system stabilization. Also included is a list of elements needed to obtain an iterated solution rigorously.

Calculus for the Biological, Chemical, and Physical Sciences: The latter Chapters 5, 7, and 8 form the foundation for an applications-oriented mathematical modeling course with options to emphasize biology, physics, or chemistry. Sections 8.5–8.11 emphasize biological applications. Sections 8.1–8.2, 8.4, 8.13–8.15 are oriented to physics applications. Section 8.3 emphasizes applications to chemical reactions. Specifically, Chapter 5 can be used as the basis for a calculus for the biological sciences offering, which can be supplemented with subsections of 8.9.

Probability and Statistical Methods in the Sciences: Chapters 2 and 3 support a course in statistical methods for the sciences.

Reference Manual for basic Algebra, Statistics, Linear Algebra, Calculus, Vector Calculus, Differential Equations, and Modeling in the Sciences and Iteration Methods as applied to Science: Chapters 1, 2, and with selections from Chapter 3 form the basis for a general introductory course in mathematics for the sciences. Chapters 4 and 7 form the basis for a concise combined course in differential equations and linear algebra.

The book, as a whole, provides a broad spectrum mathematical reference for quantitative analysis in the sciences.

# Anxiety with Mathematics

Mathematics, more than any other field of study, provides a forum for creativity and discovery. However, some of the difficulty with learning topics in math is that every new lesson contains new definitions, properties, rules, applications, etc. It is the freedom and unboundedness of the subject that can often be the most overwhelming for students. We try to mediate this problem by beginning every new topic with an application. In this way, abstraction is arrived at naturally. Indeed, the origins of algebra begin with the idea that one tree and another tree indicates the presence of two trees. This holds for apples or sheep or stars. Thus,

$$x + x = 2x, \quad x = \text{anything}.$$

If $x$ is the number of cartons of eggs, and a farmer has 360 eggs ready for market, then the number of cartons that can be delivered is a solution to the equation

$$12x = 360, \quad x = \text{number of cartons}. \qquad (\text{P.1})$$

The formulation of an algebraic equation like this is often more connected to our individual goals than is the actual process of solving for $x$. Keeping an eye on our true goals, we may be more interested in a qualitative estimate, like

$$12 > 10 \text{ and } x \geq 0,$$

which are obviously true statements for this situation. These observations logically imply that

$$12x \geq 10x.$$

Now, from (P.1) we obtain

$$360 = 12x \geq 10x \text{ or just } 360 \geq 10x,$$

which implies $36 \geq x$. Thus, the farmer cannot deliver more than 36 cartons, and disappointment is now justified if the goal was to deliver 40 cartons. The abstractly introduced variable $x$ allows for a series of mechanically logical steps to be performed, at least by the initiated. These types of quick estimates are as much a part of mathematical thought as is the obtaining of the precise $x = 30$ answer. Hence, people think mathematically all the time. However, formalism is typically chosen by history and preference of the original developers in the different subareas. Consequently, mathematical ideas may appear random and imposed. Not only should students not feel excluded from the practice of symbolism, but they need to be allowed to take ownership in the creative process, in particular with the naming of quantities. This is how students can learn to be comfortable working in the abstract.

The goal of this textbook is to give a sense of scope to those areas in mathematics that have found application in science. The trade off is that many definitions will be inferred rather than presented explicitly and deeper results may just be stated, with only references in place of proofs. However, some techniques of proof will be developed so the reader will have an opportunity to obtain a sense of the foundations of mathematics. An objective of scientists, in particular theoretical physicists, is to state a set of *postulates* that then act as *axioms* on which *deductive reasoning* can be performed.[1] The postulates arise from many experiments, which upon reflection suggest certain laws of nature. Once these laws are *axiomized*, the philosopher can explore the consequences. This led Maxwell to create the first unified field theory of electro-magnetism, and from it estimate the speed of light [2]. It also led Dirac to suggest the existence of *anti-particles* many years before their experimental discovery [7]. For the most part, however, scientists look to confirm the postulates, rather than motivate new axioms to explain new phenomena. Thus the logical framework that mathematics provides sets up a *paradigm* of thought into which researchers get locked. This can be a blessing or a curse, a magnifying glass or blinders, liberation or confinement. This textbook will emphasize the notion of mathematical spaces, within which logical conclusions can be rigorously deduced. For the purpose of this epistle, these spaces should provide a mental rock on which to peer across the landscape of experiential truths.

---

[1] Newton postulated 3 laws of nature [52], thermodynamics is based on 4 laws [39], and quantum mechanics has 6 postulates on the evolution and observation of quantum states [9].

# Ethics in Science

One motivation for writing this book was to review topics in mathematics that commonly appear in science. The intent is to reveal a connection between apparently disparate topics in a way that is natural. To some extent, all computations can be reduced to various algebraic actions on variables. Hence, many problems can be formulated in a way that can be resolved using computer programs. This has led many to think that the need for mathematical training has faded away. This text was written with the belief that this will never be the case.

It should be clear that a rationalization of our experiences requires some intellectual structure. The subject of algebra forces the performer of computations to introduce unknown variables to assist in dealing with the abstract. This opens up a world of creativity if only because the selection of $y$, rather than $x$, for an unknown variable, is in a sense a personal choice. The significance is that one is faced with the ultimate question: "What do we know?" The only answer, according to Descartes, is that something must exist, even if our experiences are only illusions. Who knows something exists? You do! Thus you exist. Beyond that you can be certain of nothing.

Science intrinsically must deal with uncertainty. This conflicts with the goals of mathematics, which consist of obtaining pure knowledge via the rules of logic. Thus the development and understanding of mathematics requires deduction, whereas the natural sciences continue to expand according to the process of induction. It is statistics that works to bridge the two areas of intellectual inquiry. This is why we have decided to introduce a discussion of this subject early in the text.

Human actions are considered by psychologists to derive from two sources: intent and opportunity. The latter comes from our environment, and is most often out of our control. Intent can have two sources: social obligations and personal attitudes [64]. What others think are our obligations is part of the social norms of our society [74], and again this can be considered, to a large extent, out of our control. However, our attitudes are something we can work on all our lives. Attitudes come from measures that we call values. Good values are known as virtues, and bad values are vices. Assuming that we are driven toward positive values, then how we assign value is intimately connected to our intentions. Even if the ego is subordinate to social norms, our values, good or bad, will play out in actions that may be hard to explain to others, and ourselves. So there is always a need to review and refresh our values on a regular basis. It is important to continually develop our personal opinions using information and analysis.

Ethics can therefore be studied from the point of view of values, which are divided into three basics types: utilitarian values, moral values, and existential values [32]. The latter is the most important and most personal. There is no direct evidence that our lives have meaning, but it seems reasonable to just assume that it does. If you are studying mathematics, statistics, and science, then you are certainly motivated. You need to be considering what is good and bad, and also how you could justify a value judgment. From that, inner questioning derives what you will consider to be right and wrong, and thus your morality.

Excessive concentration on utilitarian values has become a major distraction for people in our society, and this has led to the problems of pollution, poverty, and the mismanagement of resources. Conservation efforts have struggled to make a case for preservation. Failure can sometimes be traced back to ill-fated attempts at appealing to a cost–benefit analysis of environmental issues. In fact, if the death of a species means profit for some group, then the price for delivery of a species increases as it becomes depleted, leading to increased harvesting. This works to snuff the species right out of existence. A heavy emphasis on utilitarian values is, by its very nature, unbalanced.

The program presented in this book is intended to take the anxiety out of learning mathematical concepts and tools. Problems should be worked to the point that it becomes clear to the reader that a new idea is now understood. Mastery of any specific topic is outside the scope of what this text can offer, and will require consulting works dedicated to such a goal. Most material presented is standard, albeit in an atypical order. Some proofs of important results are presented to reiterate the language of deductive reasoning, and the use of symbolic logic. However, in many cases, a sketch and example will be used to build the intuition needed later in the book.

Throughout the text the reader should be thinking critically. Indeed, there will certainly be places where the format fails, if only because we are all different people [30]. Be sure to imagine a better way to have presented a topic. This should instill confidence in your understanding and a willingness to create your own symbolisms to represent concepts. A well-rounded experience will involve four states of mind, according to educators. They are (1) Thinking-Watching, (2) Thinking-Doing, (3) Feeling-Watching, (4) Feeling-Doing. An emphasis on lectures must always begin with state (1). However, an enthusiastic lecturer will engage the students, creating a transition to state (2). If organized properly, lectures will end with the solution of a problem and a conclusion. Students can be asked about their suggested policies based on a computation. Furthermore, as is common with mathematical problems, formulating and solving in the abstract can have a very satisfying effect, resulting in state (3). At the end of the course, problems become projects, and students will have to do a series of steps to solve a single big problem. Once the topics of mathematical modeling and the scientific method are addressed, it is hoped that the reader will feel empowered. This will mean that state (4) has been reached. It may seem that feelings should play little or no part in any scientific investigation. However, it is now understood that feeling is connected with memory. The consequence is that feelings of achievement, which most consider to be good feelings, may result in a deeper understanding, and a greater motivation, in the student.[2] This, in itself, constitutes an advance in the fields of science and mathematics.

---

[2]Human motivation is an important area of study for which there are many theories [50].

# Acknowledgments

The authors would like to thank our students, whose insights and questions over the years helped make this book possible. We also thank our colleagues at East Carolina University, especially Robert Bernhardt, John Crammer, Robert Joyner, Njina Randriampiry, Heather Ries, Catherine Rigsby, and Zach Robinson, and our colleague at the University of Windsor, Gerry McPhail, for their invaluable advice, support, and philosophical discourse in this and many other endeavors. The second author would like to thank Shondell Jones for his expertise at physical therapy, which allowed for timely completion of our book. We also express our thanks and gratitude to Tim Anderson, Lindsey Jones, Melissa Potter, and Amy Rose of Jones & Bartlett Learning, along with their excellent proofreading staff, for helping improve our book and bringing it to fruition.

# I Foundations

In Part I, the foundations of logical thinking will be reviewed. Standard results from algebra, special function theory, probability and statistics will be formulated in preparation for connecting mathematics and science. The subtle difference between logical inference and statistical inference will be demonstrated.

Algebraic equations are the oldest problems involving rational processes. Statistical problems require a different sense of what constitutes a solution. Functional equations need sufficient conditions, like invertibility, for a solution to even exist.

# 1 Review of Algebra

**A** review of some basic mathematical ideas is presented starting with algebra and the sets on which it is performed. The process by which new sets are created from initially simple sets is explained, where the square root of 2 is used as a key example.

## Introduction

To express rational ideas, the following logical symbols will be used;[1]

$$\{\ \}\ \in\ |\ \ni:\ \forall\ \exists\ \exists!\ \implies\ \impliedby\ \iff\ \not\Longrightarrow\ \lor\ \land\ \setminus\ \times\ /\ \leq<>\geq\ ==\equiv\ |\cdot|$$

A glossary is provided at the end of the text, but each symbol will also be defined in this section when used for the first time.

Mathematics is the study of *mathematical spaces*. A mathematical space consists of *elements*, as well as *operations* and *relations* on or between the elements. A relation can have a truth value of either *true* or *false*. Not all statements that follow the rules of logic have an obvious *truth value*. For example, the statement

$$\textit{This sentence is false.} \tag{1.1}$$

cannot be true or false. The art of mathematics is motivated by the concern for establishing truth values for statements within mathematical spaces.[2]

The first use of abstract ideas appears to have involved the *counting of elements*. The elements used in this process are called the *natural numbers* or the *counting numbers*, which is a set that will be represented by the symbol $\mathbb{N}$. The relation of addition is binary, in that two elements from $\mathbb{N}$ result in a new number in $\mathbb{N}$. Let $\mathbb{N} \times \mathbb{N}$ be the set of ordered pairs $\langle a, b \rangle$ for $a, b \in \mathbb{N}$. Then addition is the mapping,

$$+ : \mathbb{N} \times \mathbb{N} \to \mathbb{N}, +\langle a, b \rangle = a + b \in \mathbb{N}.$$

---

[1] Standard symbols will be employed. These are used extensively in computer science [19].

[2] In 1931 Kurt Gödel provided insight into the limitations of the rules of logic [29].

The expression $a + b$ will be used in this text and corresponds to the *addition operation*. An operation creates new elements. Furthermore, the concept of addition will be considered sufficiently fundamental and will not be explored further, except to say that a rigorous definition of addition is a rather subtle issue.[3] The set $\mathbb{N}$ is closed under addition. When combining more than two numbers, the order of operation does not affect the final answer; for example,

***Associativity:*** $(a + b) + c = a + (b + c) = a + b + c$
***Commutativity:*** $a + b = b + a$

The properties of the space $\langle \mathbb{N}, + \rangle$ make it easy to work with, and it appears to have been the original framework used to express quantitative concepts and meaning. However, in the study of complicated systems, this space is of limited use when solving modern scientific problems. For example, there are many systems that do not have the commutative property. Consider trying to open a door by pulling the handle before turning the door knob. Furthermore, given $a$ and $b$ in $\mathbb{N}$, a *linear* equation $a \cdot x = b$ can only be solved for $x \in \mathbb{N}$ if $a$ *divides* $b$, written $a|b$. Otherwise, if $a \nmid b$, then a solution $x$ will be in $\mathbb{Q}^+$, the set of positive (nonzero) fractions. Three equations that can be expressed in $\langle \mathbb{N}, + \rangle$, but cannot be solved in this space are

$$x + x = x, \; y + 5 = 1, \; z + z + z = 1.$$

To solve these equations, one must expand the spaces one works in to include the extended numbers $x = 0$, $y = -4$, and $z = 1/3$. Quite often mathematics is advanced by the need to solve problems, which cannot be done by remaining in familiar spaces. There are unlimited ways to invent new number systems.

This chapter begins with a review of the history of algebraic equations. New spaces are the consequence of the search for solutions to equations.

## ■ 1.1 Logic and Sets

The natural numbers $\mathbb{N}$ is a *closed* set under the operation of addition $(+)$ and multiplication $(\cdot)$. The symbols $+, -, \times, \div$ should be known to the reader, but the symbol "$\cdot$" will be used for multiplication in this text rather than "$\times$". Multiplication represents the process of *collecting* numbers, so that

$$x + x = 2 \cdot x, \; y = 1 \cdot y, \; z + z + z = 3 \cdot z.$$

---

[3]In *Principia Mathematica* the proof that $1 + 1 = 2$ takes 300 pages of preparation [63].

Multiplication is considered to be the mapping,

$$\cdot : \mathbb{N} \times \mathbb{N} \to \mathbb{N}, \cdot \langle a, b \rangle = a \cdot b \in \mathbb{N}.$$

The process of counting and collecting natural numbers is known as *combinatorial analysis* and is the foundation of *probability theory* [61]. Proficiency with the space $\langle \mathbb{N}, +, -, \cdot, \div \rangle$ goes back to Mesopotamia between 4000 to 5000 BC.[4] The operation "$\div$", which will often be written as "/", and the operation "$-$" are usually suppressed to give the expression $\langle \mathbb{N}, +, \cdot \rangle$, since they derive from "$\cdot$" and "$+$" as respective inverse operations. Furthermore, the exponent notation $x^n$ should be familiar. For example,

$$x^1 = x, \ x^2 = x \cdot x, \ x^3 = x \cdot x \cdot x, \ x^{-1} = 1/x, \ x^{-2} = 1/x^2, \ x^{1/2} = \sqrt{x}.$$

The triple $\langle \mathbb{N}, +, \cdot \rangle$ will be referred to as the *space of natural numbers*.

**Definition 1.1**

- Natural numbers: $\mathbb{N} = \mathbb{N}_1 \equiv \{1, 2, 3, \ldots\}$,
- Whole numbers: $\mathbb{N}_0 \equiv \{0, 1, 2, 3, \ldots\}$,
- Numbers starting at $n$: $\mathbb{N}_n \equiv \{n, n+1, n+2, n+3, \ldots\}$,
- Integers: $\mathbb{Z} \equiv \{\ldots, -3, -2, -1, 0, 1, 2, 3, \ldots\}$,
- Integers mod $n$: $\mathbb{Z}_n \equiv \{0, 1, 2, 3, \ldots, (n-1)\} \equiv \{1, 2, 3, \ldots, n\} \equiv$ etc.,
- Rational numbers: $\mathbb{Q} \equiv \{n/m : m \in \mathbb{Z}, n \in \mathbb{N}\}$.

Mounds of clay tables have been found in Mesopotamia, mostly created by students as part of their training, which demonstrates a sophisticated understanding of computations within the *space of positive rational numbers* $\langle \mathbb{Q}^+, +, \cdot \rangle$.[5] Until about 400 BC this space was thought of as being sufficient to describe the measurement of all aspects of elements within the human experience. This was maintained by a society of Greek mathematicians, called the Pythagoreans. However, a contradiction by Hippasus of Metapontum showed that the *space of rational numbers* $\langle \mathbb{Q}, +, \cdot \rangle$, is not closed under solutions of nonlinear algebraic equations. This issue will be revisited with greater detail in Section 1.6.

**Definition 1.2** Prime numbers $\mathbb{P}$ are a subset of $\mathbb{N}_2$ so that for each $p \in \mathbb{P}$ the only divisors of $p$ in $\mathbb{N}$ are 1 and $p$. In other words,

---

[4] A discussion can be found in Chapter 2 of Eves [11].

[5] Some of these computations were incorrect. See Section 2.6 of [11]. To maintain mastery over mathematical techniques, practice is essential.

- $(p \in \mathbb{P}) \iff (p \neq n \cdot m) \; \forall n, m \in \mathbb{N}_2$.
  ($p$ is prime if and only if it is not the product of two natural numbers greater than 1.)

- Prime numbers: $\mathbb{P} \equiv \{2, 3, 5, 7, 11, 13, 17, 19, 23, \dots\}$.

As mathematical spaces, $\langle \mathbb{N}, +, \cdot \rangle$, $\langle \mathbb{N}_0, +, \cdot \rangle$, and $\langle \mathbb{Z}, +, \cdot \rangle$ are closed. However $\mathbb{P}$ is definitely not closed under products, and it is uncommon that two or more primes add to another prime. In fact, the sum of any two primes greater than 2 is divisible by 2, and thus is not a prime.

### EXAMPLE 1.1

**Twin-prime conjecture:** There is an infinite subsequence $\{p_k\}$ from $\mathbb{P}$ where $p_k + 2 \in \mathbb{P}$.

There are many twin primes, $\langle 3, 5 \rangle$, $\langle 5, 7 \rangle$, $\langle 11, 13 \rangle, \dots$, but there is no proof of the preceding conjecture. This is one of many classical unresolved questions of mathematics.[6]

**Definition 1.3** A set $\mathcal{S}$ is called *ordered* if there is a relation "$\leq$" on $\mathcal{S}$ so that it is

*Reflexive*: $(\forall a \in \mathcal{S}) \implies (a \leq a)$;
*Antisymmetric*: $(\forall a, b \in \mathcal{S})(a \leq b) \wedge (b \leq a) \implies (a = b)$; and
*Transitive*: $(\forall a, b, c \in \mathcal{S})(a \leq b) \wedge (b \leq c) \implies (a \leq c)$.

A set is called *totally ordered* if the following holds:

*Totality*: $(\forall a, b \in \mathcal{S}) \implies (a \leq b) \vee (b \leq a)$;

which is another way of saying that every element in $\mathcal{S}$ can be compared using the ordering relation "$\leq$". Within the space of natural numbers this corresponds to the *less-than-or-equal* relation. This allows for the process of ranking the elements of a set. The following will also be used:

*Less than*: $(\forall n, m \in \mathbb{N})(n < m) \iff ((n - m) \notin \mathbb{N}_0)$; and
*Less than or equal*: $(\forall n, m \in \mathbb{N})(n \leq m) \iff ((n - m) \notin \mathbb{N})$.

---

[6]An example is given in Section 1.2 of Strayer [70].

The symbol "∧" is used for *and*, "∨" for *or*, "∈" for *is an element of*, and "∀" for *for all*. Also, "$\iff$" means *if and only if*, and "$\alpha \notin A$" means that $\alpha$ is *not in* $A$.

**Remark 1.1** This text will strive to be clear about operations. One obscurity that can occur is with expressions like

$$2\frac{1}{2} \stackrel{?}{=} \begin{cases} 2 + \frac{1}{2} = \text{two and a half} \\ 2 \cdot \frac{1}{2} = \frac{1}{2} \cdot 2 = \text{a half of two} \end{cases} \quad (1.2)$$

Other potential ambiguities will be presented.

### ■ Exercises

1. What are the next three prime numbers after 23?

2. The number of primes is infinite, i.e., $\#(\mathbb{P}) = \infty$. By considering the integer

$$N \equiv 1 + p_1 \cdot p_2 \cdot p_3 \ldots p_{k-1} \cdot p_k \in \mathbb{N}, \quad (1.3)$$

   where $k < \infty$, and $\mathcal{S} \equiv \{p_1, p_2, p_3, \ldots, p_{k-1}, p_k\} \subset \mathbb{P}$ is a finite subsequence, show that there must be a prime that divides $N$ that is not in the set $\mathcal{S}$. (*Hint:* Suppose $N = p_i \cdot m$ for some $p_i \in \mathcal{S}$, $m \in \mathbb{N}$ and obtain a contradiction.)

3. Use the construction in Eq. (1.3) to find three primes greater than 23 using primes that are less than 23.

### ■ 1.2 General Algebraic Equations

In this text a clear distinction between *parameters* and *variables* will be made.

Parameters are considered to be fixed constants that characterize a system. They are often measured by scientists and, in reality, have a range of values depending, in part, on how carefully they were obtained. The letters $\{a, b, c, a_i, A, B, C, \ldots\}$ will typically be used to represent parameters.

Variables are unknown quantities of a system, but can sometimes be solved for by using mathematical techniques. There may be many solutions to a problem, which will occur as points or as a range of values. The letters $\{x, y, z, x_i, X, Y, Z, X_i, \ldots\}$ will often be used to represent variables, i.e., desired solutions to a problem.

To begin the study of equations, consider the case where the parameters are $a$ and $b$, with $a \neq 0$, and the variable is $x$, which is in $\mathbb{Z}$,

$$a \cdot x + b = 0. \tag{1.4}$$

This linear equation has only one solution $x = -b/a = -(b \div a)$. If $a, b \in \mathbb{Z}$ but $a \nmid b$, then there is no solution $x \in \mathbb{Z}$. An extended space was created long ago where such equations could be solved, and this is the set of rational numbers $\mathbb{Q}$. In the case that $a = 0$, there is no solution in $\mathbb{Q}$ if $b \neq 0$. However, rather than thinking of $x$ as being unknown or undefined, the solution will be expressed as $\pm\infty$ (i.e., not uniquely defined, just as the solution of $x^2 = 2$ is not uniquely defined). This allows the following identities to be consistently applied,

$$\frac{1}{0} = \pm\infty, \ \frac{1}{\infty} = 0, \ \infty + \infty = \infty, \ 2^{-\infty} = \frac{1}{2^\infty} = 0.$$

The quantities $\pm\infty$ are not thought of as rational numbers. Rather, they bound or provide limits to the range of all numbers, so that

$$\forall x \in \mathbb{Q} \implies -\infty < x < \infty, \ \forall x \in \mathbb{Q}\setminus\{0\} \implies -\infty < \frac{1}{x} < \infty.$$

Indeed, suppose that $b > 0$ is fixed, but the parameter $a > 0$ is allowed to get close to 0 while staying positive. Then the solution $x = b/a$ will increase, so that in the limit that $a = 0^+$, $x = +\infty$, which is unique.

Next consider the equation

$$0 \cdot x + 0 = 0, \text{ or simply } 0 \cdot x = 0.$$

Clearly any $x \in \mathbb{Q}$ provides a solution. In this case the solution, which may be written as $x = 0/0$ is *undefined*, because any number ($x \neq \pm\infty$) will do. Examples of undefined expressions that will be discussed in this text are

$$\frac{0}{0}, \ \frac{\infty}{\infty}, \ \infty - \infty, \ 0^0, \ 1^\infty.$$

To solve equations that lead to similar expressions, one needs to know more about the derivation of the equation. Thus, derivations of equations will be considered as important as their solutions.

## 1.2.1 Polynomials and Zeros

Equation (1.4) is of *degree* 1. A second degree equation is of the form $ax^2+bx+c = 0$. In general the equation $p(x) = 0$ is of degree $n \in \mathbb{N}$ if $x^n$ is the highest power term of $x$ in the polynomial $p(x)$. The *set of polynomials* is defined $\forall n \in \mathbb{N}$ as

$$\mathcal{P} \equiv \{p(x) \mid \exists \{a_j\}_{j=0}^n \implies p(x) = a_n \cdot x^n + \cdots + a_2 \cdot x^2 + a_1 \cdot x + a_0\}, \quad (1.5)$$

where the entities $a_j$ must come from a mathematical space where two operations "+" and "·" are defined.

**Definition 1.4** The *zeros* of a polynomial $p(x)$ are special values $x_*$ of the variable $x$ such that $p(x_*) = 0$. The quantities $x_*$ are also called *roots* of the polynomial $p(x)$, since, by the *Fundamental Theorem of Algebra*, a polynomial is characterized by its zeros.

If $x_*$ is a zero of $p(x)$, then $(x-x_*)$ is a factor of $p(x)$. Thus $q(x) \equiv p(x) \div (x-x_*)$ is a polynomial of degree one less than $p(x)$, and $p(x) = (x - x_*) \cdot q(x)$.

## Exercises

1. Write the following polynomials as a product of linear factors, by first finding their roots. (*Hint:* Some guessing may be required.)

   (a) $p_a(x) = x^2 - 3x + 2$
   (b) $p_b(x) = x^3 - 7x^2 + 10x$
   (c) $p_c(x) = x^3 + 2x^2 - 5x - 6$

## 1.2.2 Quadratic Equations

Equations of the form, for $a \neq 0$,

$$a \cdot x^2 + b \cdot x + c = 0, \quad (1.6)$$

can be solved using the *quadratic formula*.[7] The formula is derived by dividing both sides by $a$ to obtain a *monic* polynomial equation (where the highest power has a

---

[7] Solving this problem appears to be a very ancient ability of civilization [14].

coefficient of 1) then adding the square of one-half the linear term $b \cdot x$ to obtain

$$1 \cdot x^2 + (b/a) \cdot x + [b/(2a)]^2 = -(c/a) + [b/(2a)]^2.$$

This technique is called *completing the squares*. Factoring the left-hand side and finding a common denominator on the right-hand side gives

$$[x + b/(2a)]^2 = (b^2 - 4ac)/(4a^2).$$

As with Eq. (1.4) there are two possible solutions (unless $b^2 = 4ac$), since the square of a negative number is positive, so that

$$x = \frac{-b \pm \sqrt{b^2 - 4ac}}{2a}. \tag{1.7}$$

An alternative approach involves using a *transformation*. The variable $y$ is introduced, and the substitution $x = y - b/(2a)$ is used in Eq. (1.6). To make the substitution clear, write the quantities needed as

$$x = y - b/(2a),$$
$$x^2 = y^2 - b \cdot y/(a) + b^2/(4a^2).$$

After expanding and simplifying, the preceding quantities give

$$\left(a \cdot y^2 - b^2/(4a) + c = 0\right) \implies \left(y^2 = b^2/(4a^2) - c/a\right).$$

The quadratic formula is now obtained by solving for $y$, and then making the *back substitution* $y = x + b/(2a)$.

### EXAMPLE 1.2

**The Golden Ratio:** The quadratic equation $x^2 = x + 1$ can be rewritten and solved to give

$$x^2 - x - 1 = 0 \implies x = \frac{1 \pm \sqrt{5}}{2}. \tag{1.8}$$

The quantity $\phi = (1 + \sqrt{5})/2$ is called the Golden Ratio, and this number appears quite often in nature. For example, it is the diagonal of a regular unit pentagon [11]. Also, the algebraic formula in Eq. (1.8) appears often in biological systems.

### ■ Exercises

1. Show that the two solutions in Example 1.2 are actually negative reciprocals of each other.

2. Show that if $a \neq 0$ and $c \neq 0$ in the quadratic formula (1.7), then $x^{-1} = -(ax+b)/c$, unless $x = -b/a$.

### ■ 1.2.3 Cubic Equations

The Cardano method[8] for solving equations of the form, where $a \neq 0$,

$$a \cdot x^3 + b \cdot x^2 + c \cdot x + d = 0, \tag{1.9}$$

is achieved by applying an algebraic transformation to first remove the quadratic term. So let $x = y - b/(3a)$. The consequence is a sequence of equations

$$\begin{aligned} x &= y - b/(3a), \\ x^2 &= y^2 - 2b \cdot y/(3a) + b^2/(9a^2), \text{ and} \\ x^3 &= y^3 - b \cdot y^2/a + b^2 \cdot y/(3a^2) - b^3/(27a^3), \end{aligned} \tag{1.10}$$

so that substitution and division by $a$ gives the *monic* cubic equation,

$$y^3 + B \cdot y + C = 0, \tag{1.11}$$

where the new coefficients are

$$B \equiv \left(-b^2/(3a^2) + c/a\right), \ C \equiv \left(2b^3/(27a^3) - bc/(3a^2) + d/a\right). \tag{1.12}$$

Similarly, using the Vieta substitution [36] $y = z - B/(3z)$ gives the monic quadratic equation:

$$(z^3)^2 + C \cdot (z^3) - B^3/27 = 0. \tag{1.13}$$

This solution, found in the 16th century, is not usually memorized and will not be used extensively in this text. However it demonstrates how knowledge about a simpler equation, like the formula for quadratic equations, can extend to knowledge

---
[8]Some techniques in *Ars Magna* [6] used ideas of Tartaglia.

about more complicated equations. This example also shows how a transformation or substitution, as a mathematical technique, facilitates a connection between complicated equations and easily solvable equations.

### EXAMPLE 1.3

**The Roy Baker problem:**[9] A 10 cm strip of metal is to overlap an inverted right triangle $\triangle abc$. Just before the bending process $(1) \to (2) \to (3)$, the height above the triangle is $h = 5$ cm. In the first bend $(1) \to (2)$, the metal strip is exactly above the triangle. Find the dimensions of $\triangle abc$.

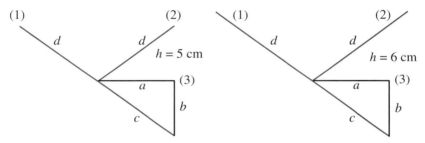

**FIGURE 1.1** ■ The simple and generalized Roy Baker triangles.

Figure 1.1 gives four equations for the four parameters $\{a, b, c, d\}$,

$$
\begin{aligned}
d + c &= 10 & (1) & \quad \text{total length of strip} = 10 \text{ cm} \\
a^2 + b^2 &= c^2 & (2) & \quad \text{Pythagorean's theorem for } \triangle abc \\
a + b &= d & (3) & \quad \text{perimeter excluding hypothenuse} \\
a^2 + 25 &= d^2 & (4) & \quad \text{height of bent strip} = 5 \text{ cm}
\end{aligned}
$$

---

[9]Private communication with Roy Baker, who was a tool and dye maker. These types of problems often appear in industrial applications.

Combining equations using the rules of addition and multiplication gives

$$
\begin{aligned}
(3) &\implies & [b = d - a] &\implies b^2 = d^2 - 2ad + a^2 & (5) \\
(2) \wedge (5) &\implies & c^2 &= d^2 - 2ad + 2a^2 & (6) \\
(1) &\implies & [c = 10 - d] &\implies c^2 = 100 - 20d + d^2 & (7) \\
(6) \wedge (7) &\implies & [-2ad &+ 2a^2 = 100 - 20d] & \\
&\implies & a^2 - 50 &= (a - 10) \cdot d & (8) \\
(8)^2 \wedge (4) &\implies & a^4 - 100a^2 + 2500 &= (a^2 - 20a + 100) \cdot (a^2 + 25) & \\
\textbf{Conclusion} &\implies & 5a \cdot (4a^2 - 45a + 100) &= 0 &
\end{aligned}
$$

The problem leads to a cubic equation that factors into a linear and quadratic equation. The three distinct solutions are

$$a_1 = 0, \ a_2 = \frac{45 + 5\sqrt{17}}{8}, \ a_3 = \frac{45 - 5\sqrt{17}}{8}.$$

Only $a_3 \simeq 3.05$ cm is the reasonable, nontrivial solution. The $a_2 \simeq 8.20$ cm solution with (2) and (1) gives $d < 2$, which cannot be reconciled with (4). Also, the $a_1 = 0$ solution corresponds to a simple bend in the middle of the strip.

### EXAMPLE 1.4

**The Generalized Roy Baker problem:** A 10 cm strip of metal is to overlap an inverted right triangle $\triangle abc$. Before the bending process $1 \to 2 \to 3$, the height above the triangle is $h = 6$ cm. In the first bend $(1) \to (2)$, the metal strip is exactly above the triangle. Find the dimensions of $\triangle abc$.

To modify the preceding solution, replace Eq. (4) with (4.h) defined by

$$a^2 + h^2 = d^2 \qquad (4.h) \qquad \text{height of bent strip above triangle} = h \text{ cm}$$

Incorporating this change into the previous system, with $h = 6$,

$$
\begin{aligned}
(8)^2 \wedge (4.h) &\implies & a^4 - 100a^2 + 2500 &= (a^2 - 20a + 100) \cdot (a^2 + 36) & \\
\textbf{Conclusion} &\implies & 20 \cdot a^3 - 236 \cdot a^2 + 720 \cdot a - 1100 &= 0 & (1.14)
\end{aligned}
$$

Thus, in general (i.e., $h \neq 5$), the cubic formula must be used. Dividing Eq. (1.14) by 20, then defining $a = y + 236/60$ gives, after substitution and simplification,

$$y^3 - \left(236^2/60 - 720\right) \cdot y - \left(2 \cdot (236)^3/(27 \cdot 400) + 1100\right) = 0.$$

Finally, using the transformation $y = z + 781/(225 \cdot z)$, results in the quadratic equation in $z^3$:

$$(z^3)^2 - (118,483/3,375) \cdot (z^3) - (781/225) = 0. \tag{1.15}$$

From the quadratic formula, taking the cube root and back substitution,

$$z^3 \simeq \begin{cases} 33.8713 \\ 1.2347 \end{cases} \implies z \simeq \begin{cases} 3.2355 \\ 1.0728 \end{cases} \implies y \simeq \begin{cases} 4.3083 \\ 4.3083 \end{cases}, \tag{1.16}$$

and so only one solution is obtained from this procedure. The value $a_1 \simeq 8.2417$ is *extraneous*. Two other solutions exist for this problem, but their discovery will have to wait until after the introduction of the set of complex numbers. This is done in Section 1.8.

### ■ 1.2.4 Quartic Equations

The Ferrari method to solving equations of the form, for $a \neq 0$,

$$a \cdot x^4 + b \cdot x^3 + c \cdot x^2 + d \cdot x + e = 0,$$

can be achieved[10] by applying the algebraic transformation $x = y - b/(4a)$ to remove the cubic term. Dividing the result by $a \neq 0$ gives a simpler quartic equation,

$$y^4 + B \cdot y^2 + C \cdot y + D = 0; \tag{1.17}$$

with the corresponding coefficients,

$$B = (8c - 3b^2)/(8a),$$
$$C = (8d - 4bc + b^3)/(8a),$$
$$D = (64e - 16b^2d - 3b^4)/(256a).$$

Now, by solving the *resolvent* cubic equation, $p(\beta_*) = 0$ where

$$p(\beta) \equiv \beta^3 - B \cdot \beta^2 - 4D \cdot \beta + (4BD - C^2),$$

---

[10] Obtained by Ferrari in the 16th century and published in [6], apparently without permission.

one can *complete the squares* in Eq. (1.17) and obtain

$$(y^2 + \beta_*/2)^2 - \left(y\sqrt{\beta_* - B} - C/(2\sqrt{\beta_* - B})\right)^2 = 0.$$

This gives two, typically independent, quadratic equations:

$$2y^2 - (2\sqrt{\beta_* - B})y + \left(\beta_* + C/\sqrt{\beta_* - B}\right) = 0,$$
$$2y^2 + (2\sqrt{\beta_* - B})y + \left(\beta_* - C/\sqrt{\beta_* - B}\right) = 0.$$

One lesson here is that knowledge of both cubic and quadratic equations is needed to resolve the quartic equation. In addition, the complexity can appear overwhelming for the general situation. This is mitigated by treating established knowledge of simpler equations as resolved and tangible. However, an ever-growing knowledge base of mathematical techniques and tricks is required if one desires to solve increasingly general equations perfectly. One must ask whether this is too great a request. There are many unsolved problems in mathematics where precise answers are still unknown.

**EXAMPLE 1.5**

It is known that $e, \pi \notin \mathbb{Q}$ [45]. However, it has not been proven that $e^\pi \notin \mathbb{Q}$. Although one might expect this to be true, one cannot assume it. In fact $e^{\ln(2)} = 2 \in \mathbb{Q}$, even though $e, \ln(2) \notin \mathbb{Q}$. Thus it may also be true that $e^\pi \in \mathbb{Q}$.

An unsolved mathematical problem may be interesting, while having no significant scientific consequence.

### ■ 1.2.5 Higher Degree Equations

A general polynomial of degree $n$, for $a_n \neq 0$, will be written as

$$p(x) = a_n \cdot x^n + a_{n-1} \cdot x^{n-1} + \cdots + a_1 \cdot x + a_0 \equiv \sum_{k=0}^{n} a_k \cdot x^k, \qquad (1.18)$$

where $\sum$ denotes a *sum of terms* and $k$ is the *index of the sum*. Also, for $x \neq 0$,

$$x^0 = x^{1-1} = x^1 \cdot x^{-1} = x/x = 1.$$

So using $x^0 = 1$ in Eq. (1.18) is consistent with both sides of the equation. If $p(x)$ is an $n$th degree polynomial, then for $n \geq 5$ there is no general procedure for finding the roots of Eq. (1.18) precisely, unlike in the $n = 1, 2, 3$, and 4 cases.[11] Instead, for applications, one must use approximation techniques to estimate the zeros of a function or the roots of a polynomial. A simple approach is to compute a sequence $\{x_k\}$ by starting with some nonzero *seed* $x_0 \in \mathbb{Q}$, close to a suspected zero of $p(x)$,

$$x_{k+1} = -\left(\frac{a_{n-1}}{a_n} + \frac{a_{n-2}}{a_n}\frac{1}{x_k} + \cdots + \frac{a_1}{a_n}\frac{1}{x_k^{n-2}} + \frac{a_0}{a_n}\frac{1}{x_k^{n-1}}\right). \quad (1.19)$$

The concern with this approach is that there is no guarantee that the sequence $\{x_k\}$ will contain numbers even close to a root of $p(x)$ in Eq. (1.18). To measure the accuracy of the approximation $x_{approx} = x_k$ to an exact value $x_{true} = x_*$, one can compute the

$$\begin{aligned} relative\ error &= \left|\frac{x_{true} - x_{approx}}{x_{true}}\right| \\ relative\ \%\ error &= relative\ error \times 100\%. \end{aligned} \quad (1.20)$$

Of course there is a problem with this measure if $x_{true} = 0$. In this case one would just compute the $error = |x_{true} - x_{approx}| = |0 - x_{approx}| = |x_{approx}|$.

### ■ Exercises

1. Consider the equation $x^2 - x - 1 = 0$ and write the iteration scheme suggested in Eq. (1.19). Then obtain $x_4$ starting with $x_0 = 1$. Find the relative % error between $x_4$ and the Golden Ratio, Eq. (1.8).
2. Consider the equation $x^2 - 2 = 0$ and write the iteration scheme suggested in Eq. (1.19). Then obtain $x_4$ starting with $x_0 = 1$. What is the value of $x_4$ if the seed is $x_0 = 5$ instead? Prove $\forall x_0 > 0 \implies \sqrt{2} \in [\min\{x_n\}, \max\{x_n\}]$.

### ■ 1.2.6 Iterative Solutions to Algebraic Equations

It is possible to be very efficient in finding approximations to solutions of algebraic equations. The derivation, known as Newton's method, will be considered in Chapter 5.

---

[11] This was proven by Abel and is the origin of the Galois theory [28].

## ■ Theorem 1.1

For each positive $A \in \mathbb{Q}$, the square root of $A$, written $\sqrt{A} = A^{1/2}$, is approximated by the ordered sequence $\langle x_k \rangle \subset \mathbb{Q}$, which is obtained by the iterative process,

$$x_{k+1} = \frac{x_k}{2} + \frac{A}{2x_k}, x_0 = 1. \tag{1.21}$$

In particular, the difference $(A - x_k^2)$ vanishes as $k \in \mathbb{N}$ increases.

---

This, the first proof of the text, can be considered to be optional at this point, and may be better understood after reading Section 1.6.

**Proof of Theorem 1.1:** In the following it will be assumed that $A > 1$, since the $0 < A < 1$ cases follow from the observation that $\sqrt{A} = A/\sqrt{A}$. The proof will be broken into the following five distinct steps:

**Step 1:** Verify that the statement of the theorem is reasonable. Replace $x_{k+1}$ and $x_k$ with $x_\infty$. Then multiplying both sides by $x_\infty$ gives the quadratic equation

$$x_\infty^2 = \frac{x_\infty^2}{2} + \frac{A}{2}, \tag{1.22}$$

which is the same as $x_\infty^2 = A$. Thus $x_\infty = \sqrt{A}$ for $A > 1$ is a *stationary value* for the iterative Eq. (1.21). In particular, if $x_0 = \sqrt{A}$ then $x_k = \sqrt{A}, \forall k \in \mathbb{N}$.

**Step 2:** Observe that for any seed $x_0$ between 1 and $A$, $x_1$ must also be between 1 and $A$, and so this will be the case for all $x_k$. Thus the sequence $\{x_k\}$ is *bounded* between 1 and $A$.

**Step 3:** Information is needed on the product $x_k \cdot x_{k-1}$, which can be obtained by multiplying Eq. (1.21) by $x_k$ to obtain the bounds, due to Step 2,

$$\frac{1+A}{2} \leq x_{k+1} \cdot x_k = \frac{x_k^2}{2} + \frac{A}{2} \leq \frac{A^2 + A}{2}. \tag{1.23}$$

These bounds are independent of $k$. Furthermore, Eq. (1.23) implies the reciprocal bound $(x_{k+1} \cdot x_k)^{-1} \leq 2/(1+A)$.

**Step 4:** It must be shown that the sequence $\{x_k\}$ *converges*. Consider subtracting Eq. (1.21) from the previous iterate $x_k = x_{k-1}/2 + A/(2x_{k-1})$ for $k \geq 1$,

$$\begin{aligned} x_{k+1} - x_k &= \frac{x_k - x_{k-1}}{2} + \frac{A}{2x_k} - \frac{A}{2x_{k-1}} \\ &= (x_k - x_{k-1})\left(\frac{1}{2} - \frac{A}{2x_k \cdot x_{k-1}}\right). \end{aligned} \tag{1.24}$$

The important observation here is that the last factor (on the right-hand side) is bounded above by $1/2$, which is less than 1, and bounded below by $-A/(1+A)$, which is greater than $-1$. Define the constant

$$L \equiv \frac{A}{1+A} = \frac{1}{1+1/A}. \tag{1.25}$$

Since $A > 1$, the bounds $1/2 < L < 1$ hold. Thus the differences between successive iterates is vanishing. Indeed, if $x_k = x_{k-1}$ then Eq. (1.24) and Step (1) imply that the root has been found. Otherwise,

$$-1 < -L \leq \left(\frac{x_{k+1} - x_k}{x_k - x_{k-1}}\right) \leq L < 1. \tag{1.26}$$

From Eq. (1.21), the difference $x_1 - x_0 = (A-1)/2 > 0$. Thus, from successive multiplications by ratios of differences (which are not zero),

$$(x_{k+1} - x_k) = \frac{x_{k+1} - x_k}{x_k - x_{k-1}} \cdot \frac{x_k - x_{k-1}}{x_{k-1} - x_{k-2}} \cdots \frac{x_2 - x_1}{x_1 - x_0} \cdot \frac{A-1}{2}, \tag{1.27}$$

which is bounded between $\pm L^k(A-1)/2$. As $k \to \infty$ (i.e., $k$ becomes infinitely large) the quantity in Eq. (1.27) vanishes since $0 < L < 1$.

**Step 5:** Finally, multiplying Eq. (1.21) by $2x_k$ and subtracting by $2x_k$ gives

$$2x_{k+1}x_k - 2x_k^2 = 2x_k \cdot (x_{k+1} - x_k) \tag{1.28}$$
$$= A - x_k^2. \tag{1.29}$$

The right-hand side vanishes as $k \to \infty$ due to Eq. (1.27) and the fact that $x_k$ is restricted between 1 and $A$ by Step 2. Thus the relative error $|1 - x_k^2/A|$ vanishes as $k \to \infty$ and so $x_k \to \sqrt{A}$ as $k \to \infty$. □

**Remark 1.2** The preceding analysis represents a central theme for this text, so each part should be understood separately, before one tries to grasp the entire proof. Also, the initial statement of the theorem should be reread until the goals of the result are clear. However, the connection between the details of the proof, and the actual statement can be hard to maintain. This is common, so one should be aware that two ways of thinking or mental states are often needed to grasp the significance of an analysis argument.

### ■ Theorem 1.2

For each positive $A \in \mathbb{Q}$ and $n \in \mathbb{N}$, the $n$th root of $A$, written $\sqrt[n]{A} = A^{1/n}$ is approximated by the sequence $\{x_k\}$ defined by,

$$x_{k+1} = \frac{(n-1) \cdot x_k}{n} + \frac{A}{n \cdot x_k{}^{n-1}}, x_0 = 1. \qquad (1.30)$$

The difference $(A - x_k^n)$ vanishes as $k \in \mathbb{N}$ increases.

The details for this more general case are left as a writing assignment. The iteration scheme is generated by *Newton's method*, which requires calculus to explain (see Section 5.5). Newton's method applies even to functions that are not polynomials, and converges for many more cases than the method suggested in Eq. (1.19). However, even Newton's method can fail to give convergent sequences to a zero of a function. In such cases the more dependable *Bisection method* is required (see Section 5.14).

### ■ Exercises

1. For $A = 2$ approximate $\sqrt{2}$ by using Theorem 1.1 to compute $x_3$ using $x_0 = 1$. At each step simplify to a fraction of the form $n/m$. Do the same for $A = 4$.
2. Consider the polynomial $p(x) = x^2 - 1$. Factor this polynomial. What are the zeros of $p(x)$? Write the iteration scheme that will solve $p(x) = 0$ using Theorem 1.1, and find $x_3$ with seed $x_0 = 2$.
3. Prove Step 2 in Theorem 1.1 using Eq. (1.21) and the restriction that $1 < x_0 < A$. This should consist of just showing that $(1 < x_0 < A) \implies (1 < x_{k+1} < A)$.
4. Use Theorem 1.2 to obtain the values in Eq. (1.16) from $z^3$. A table may help, and just a few iterations should be sufficient.
5. Write the proof of Theorem 1.2 for general $n \in \mathbb{N}_2$. (The proof should be 1–2 pages in length.)

### ■ 1.3 Extensions of Mathematical Spaces

Some spaces of mathematics are now examined for their structural properties.

### 1.3.1 Introduction to Groups

One of the most basic spaces, that has an interesting structure, consists of a set $G$ and a binary operation $* : G \times G \to G$. The pair $\langle G, * \rangle$ is called a *group* if the following four properties hold:

**Closure:** $(\forall a, b \in G) \implies a * b \in G$;
**Associativity:** $(\forall a, b, c \in G) \implies ((a * b) * c = a * (b * c))$;
**Identity (Neutral) Element:** $(\exists e \in G) \ni ((\forall a \in G) \implies (a * e = e * a = a))$;
**Inverse Element:** $(\forall a \in G) \implies ((\exists a^{-1} \in G) \ni (a * a^{-1} = a^{-1} * a = e))$.

Note that in the notation used above, one has $a = a^1$.

The pair $\langle \mathbb{N}, + \rangle$ satisfies the first two properties and $\langle \mathbb{N}_0, + \rangle$ satisfies the first three. However it is $\langle \mathbb{Z}, + \rangle$ that forms an *infinite* group. An example of a finite group is $\langle E_2, \cdot \rangle$ where $E_2 = \{1, -1\}$. This can be viewed as a *solution space* of the equation

$$(x^2 = 1) \implies x \in \{1, -1\} = E_2.$$

In general $E_n$ will denote the solution space of $x^n = 1$.

The *order* of a group is the number of elements in the corresponding set, and the notations $|G| = \#(G)$ will be used. This is also known as the *cardinality* of the set. Thus $|E_2| = 2$ but $|\mathbb{Z}| = \infty$. Groups can have extra properties:

- Abelian: $(\forall a, b \in G) \implies a * b = b * a$;
- Cyclic: $(\exists a \in G) \ni (\forall g \in G) \implies ((\exists n \in \mathbb{Z}) \ni (g = a^n))$.

Nonabelian groups appear in many applications and can have nonintuitive properties. A cyclic group $G$ is always abelian and has at least one special element $a \in G$, called a *generator*. In this case $G$ can be written as

$$G = (a) \equiv \{a^n | n \in \mathbb{Z}\} = \{\ldots, a^{-2}, a^{-1}, e, a, a^2, \ldots\}. \quad (1.31)$$

The *order* of an element $g \in G$ is written $o(g) \in \mathbb{N}$ and is the smallest natural number so that $g^{o(g)} = e$. Thus the order of a cyclic group generated by $a \in G$ is $o(a) = |(a)| = |G|$.

---

**Definition 1.5** Given $n \in \mathbb{N}$ and $q \in \mathbb{Z}$, then $n$ *divides* $q$ if $\exists s \in \mathbb{Z}$ so that $q = n \cdot s$. This is written $n \mid q$. When $n \nmid q$ then $\exists r \in \{1, 2, \ldots, (n-1)\} = \mathbb{Z}_n \backslash \{0\}$, called the *remainder* of $q \div n$, so that $q = n \cdot s + r$. With this, two natural binary operations, $+_n$ and $*_n$, can be defined on $\mathbb{Z}_n = \{0, 1, 2, \ldots, (n-1)\}$ as

$$q_1 +_n q_2 \equiv \text{the remainder of } (q_1 + q_2) \div n \equiv (q_1 + q_2) \mod (n),$$
$$q_1 *_n q_2 \equiv \text{the remainder of } (q_1 \cdot q_2) \div n \equiv q_1 \cdot q_2 \mod (n).$$

The expression $mod(n)$ stands for *modulus n*, or just *mod n*.

---

Some of the preceding statements contain facts that will not be shown here. Euler's algorithm provides the foundation for the ideas behind the division of integers, and the Fundamental Theorem of Algebra explains their products. As an application, a subset $S \subset G$ is a *subgroup* of $G$ if it is a group with the same operation as $G$. For example, $E_2$ is a subgroup of $E_4$, or any $E_{2 \cdot n}$, $\forall n \in \mathbb{N}$. See Fig 1.4 in Section 1.8.

### ■ Theorem 1.3

For each $a \in G$, where $\#(G) < \infty$ and $G$ is a group, then $(a) \subset G$ is a subgroup and $o(a)$ divides $\#(G)$.

---

**Proof** If $(a) \subsetneq G$ then choose $b_1 \in G \backslash (a)$ and define the left coset $b_1 \cdot (a)$, which is not a subgroup. Also, $(a) \cap b_1 \cdot (a) = \emptyset$, since otherwise $\implies a^n = b \cdot a^m$ or simply $a^{n-m} = b \implies b \in (a) \implies\!\!\Leftarrow$. Next define $b_2 \in G \backslash [(a) \cup b_1 \cdot (a)]$. Similarly, a finite sequence $\{b_i\}_{i=1}^k$ can be found where the cosets $\{b_i \cdot (a)\}_{i=1}^k$ partition $G$. Then $\#(G) = \sum_{i=1}^k |b_i \cdot (a)| = \sum_{i=1}^k |(a)| = k \cdot |(a)|$. □

There are a lot of subgroup structures in $\mathbb{Z}$.

### ■ Theorem 1.4

For each $n \in \mathbb{N}$ the pair $\langle \mathbb{Z}_n, +_n \rangle$ defines a cyclic group of order $n$. For each $p \in \mathbb{P}$ the pair $\langle \mathbb{Z}_p \backslash \{0\}, *_p \rangle$ is a cyclic group of order $(p-1)$.

---

Consequently, $\langle \mathbb{Z}_p, +_p \rangle$ and $\langle \mathbb{Z}_p \backslash \{0\}, *_p \rangle$ are both groups $\forall p \in \mathbb{P}$; note that $\langle \mathbb{Z}_3 \backslash \{0\}, *_3 \rangle$ and $\langle E_2, \cdot \rangle$ have the same operation tables. Two such groups are said to be *isomorphic*. See Table 1.1.

| $\langle \mathbb{Z}_3, +_3 \rangle$ | | | | $\langle \mathbb{Z}_3 \setminus \{0\}, *_3 \rangle$ | | | $\langle E_2, \cdot \rangle$ | | |
|---|---|---|---|---|---|---|---|---|---|
| $+_3$ | 0 | 1 | 2 | $*_3$ | 1 | 2 | $\cdot$ | 1 | $-1$ |
| 0 | 0 | 1 | 2 | 1 | 1 | 2 | 1 | 1 | $-1$ |
| 1 | 1 | 2 | 0 | 2 | 2 | 1 | $-1$ | $-1$ | 1 |
| 2 | 2 | 0 | 1 | | | | | | |

**TABLE 1.1** ■ The addition and multiplication tables for $\mathbb{Z}_3$.

## ■ Exercises

1. Construct the $\mathbb{Z}_5$ addition table, and the $\mathbb{Z}_5 \setminus \{0\}$ multiplication tables.
2. Define $E_2 \times E_2 \equiv \{(1,1), (1,-1), (-1,1), (-1,-1)\}$, where the group operation is defined by $(a,b) * (c,d) \equiv (a \cdot c, b \cdot d)$.

   (a) Show that this group is abelian.
   (b) Show that this group is not cyclic.
   (c) Define all subgroups, including the trivial ones.

3. Define $E_2 \times E_2 \times E_2 \equiv \{(\pm 1, \pm 1, \pm 1)\}$, with operation $(a,b,c) * (d,e,f) \equiv (a \cdot d, b \cdot e, c \cdot f)$.

   (a) Show that this group is abelian using the same for $E_2$.
   (b) Show that this group is not cyclic.
   (c) Why is there no cyclic subgroup of order 4?
   (d) Find a subgroup of order 2 and one of order 4.

### ■ 1.3.2 Rings and Fields

The space $\langle \mathbb{Z}, +, \cdot \rangle$ is called a *ring* because it satisfies three properties:

- Group over +: The pair $\langle \mathbb{Z}, + \rangle$ is an abelian group;
- Closed over $\cdot$: $\mathbb{Z}$ is closed under multiplication; $(\forall a, b \in \mathbb{Z}) \implies (a \cdot b \in \mathbb{Z})$;
- Distributive Laws: $(\forall a, b, c \in \mathbb{N}) \implies$

$$(a \cdot (b+c) = a \cdot b + a \cdot c) \quad \text{and} \quad ((b+c) \cdot a = b \cdot a + c \cdot a). \quad (1.32)$$

Many problems that can be posed in a ring have solutions that cannot be found in that ring. In particular, questions involving linear equations in the ring $\langle \mathbb{Z}, +, \cdot \rangle$ directly lead to the space $\langle \mathbb{Q}, +, \cdot \rangle$, which is a special ring called a *field*. A field has the following three properties:

- Group over $+$: The pair $\langle \mathbb{Q}, + \rangle$ is an abelian group;
- Group over $\cdot$: The pair $\langle \mathbb{Q}^\times, \cdot \rangle$ is an abelian group, where $\mathbb{Q}^\times \equiv \mathbb{Q} \backslash \{0\}$;
- Distributive Laws: $(\forall a, b, c \in \mathbb{Q}) \implies$ Eqs. (1.32) hold.

Many equations can be formulated and solved in this field. However, it has sufficient structure to pose new problems that cannot be solved. For example, it was discussed that $x^2 = 4$ is solvable in $\mathbb{Q}$ but $x^2 = 2$ is not. A new space can be created by *adjoining* the objects $\sqrt{2}$ and $-\sqrt{2}$ to $\mathbb{Q}$ in the following manner:

$$\mathbb{Q}[\sqrt{2}] \equiv \{a + \sqrt{2}b \mid a, b \in \mathbb{Q}\}.$$

It should be clear that this space satisfies the properties of a ring. In fact, $\mathbb{Q}[\sqrt{2}]$ is a field, and the only issue is with the multiplicative property. It is sufficient to find a multiplicative inverse for a general, nonzero element $x = (a + \sqrt{2}b)$. This is just the reciprocal, and the technique needed here is multiplication by the *radical conjugate* of $x$ that is $(a - \sqrt{2}b)$. For $(a \neq 0) \vee (b \neq 0)$,

$$\begin{aligned}
(a + \sqrt{2}b)^{-1} &= \frac{1}{a + \sqrt{2}b} = \frac{1}{a + \sqrt{2}b}\left(\frac{a - \sqrt{2}b}{a - \sqrt{2}b}\right) \\
&= \frac{a - \sqrt{2}b}{a^2 - 2b^2} \\
&= \frac{a}{a^2 - 2b^2} + \sqrt{2}\frac{-b}{a^2 - 2b^2},
\end{aligned}$$

which is in $\mathbb{Q}[\sqrt{2}]$ because $a^2 - 2b^2 \neq 0$ for all $a, b \in \mathbb{Q}$. This is due to the Hippasus Corollary shown in Section 1.6. To acknowledge that the space is in fact a field, one now writes $\mathbb{Q}(\sqrt{2})$. Similarly it can be shown that $\mathbb{Q}(\sqrt{2}, \sqrt{3})$ is a field and this contains the *subfield* $\mathbb{Q}(\sqrt{6})$. By adjoining new elements to existing spaces, greater flexibility is sometimes attained to solve equations. For example, the simple union $\mathbb{Q}(\sqrt{2}) \cup \mathbb{Q}(\sqrt{3})$ is not a field or subfield, and does not contain a solution to $x^2 = 6$. Adjoining new elements and creating new spaces must be done in a manner that preserves the structure of the original space, while being logically consistent.

### ■ 1.3.3 Introduction to Algebraic Numbers

By expanding the types of equations that one wants to solve, new spaces need to be created. The field of *algebraic numbers* $\mathbb{A}$ is defined to contain all the roots of polynomials with rational coefficients. Clearly $\mathbb{Q} \subset \mathbb{A}$, and in fact

$$\mathbb{Q}(\sqrt{\mathbb{P}}) \equiv \{Q(\sqrt{p_1}, \sqrt{p_2}, \sqrt{p_3}, \dots) \mid p_i \in \mathbb{P}\} \subset \mathbb{A}.$$

Furthermore, $\mathbb{A}$ must contain all $\sqrt[n]{p}$ for all $n \in \mathbb{N}_2$ and $p \in \mathbb{P}$, none of which are in $\mathbb{Q}$. However, $\mathbb{A}$ is not an ordered field because solutions to

$$x^n + 1 = 0, \ \forall n \in \mathbb{N}_2, \tag{1.33}$$

also need to be adjoined to any set containing all the various roots of primes. Here the solutions to $x^2 = -1$ will be denoted $i \equiv \sqrt{-1}$ and $-i = -\sqrt{-1}$. Square roots of negative numbers are called *imaginary* numbers. It may seem reasonable to expect that a new symbol should be created to write the solutions of $x^4 = -1$, however, one can show that they are all in $\mathbb{Q}(\sqrt{2}, i)$. Furthermore, it can be shown that $\mathbb{A}$ is larger than even $\mathbb{Q}(\sqrt[\mathbb{N}_2]{\mathbb{P}}, i)$. However, the algebraic numbers are still not sufficient to describe nature. Indeed, $\pi \notin \mathbb{A}$, and this alone is motivation to explore deeper extensions.

### EXAMPLE 1.6

To approximate $\sqrt{2}$ using fractions, consider the number $\alpha = 1 - \sqrt{2}$. This is negative because $1 < 2$ implies that $\sqrt{1} < \sqrt{2}$. Furthermore,

$$2 = (1-\alpha)^2 = \alpha^2 - 2\alpha + 1 \implies 1 \geq 1 - \alpha^2 = -2\alpha \geq 0 \implies -1/2 \geq \alpha \geq 0.$$

Finally, note that $\alpha \in \mathbb{Q}(\sqrt{2})$. Then by multiplying $\alpha$ by itself repeatedly, a shrinking value is obtained in $\mathbb{Q}(\sqrt{2})$ that allows for an improved rational approximation of $\sqrt{2}$;

$$\frac{-1}{2^n} \leq \alpha^n \leq \frac{1}{2^n} \text{ and } \alpha^n = a_n - \sqrt{2}b_n \implies \sqrt{2} \simeq \frac{a_n}{b_n}.$$

Here $a_n, b_n \in \mathbb{Z}$ because $\mathbb{Z}[\sqrt{2}]$ is a ring (i.e., $\mathbb{Z}[\sqrt{2}]$ is closed under addition and multiplication).

To compute a sequence of approximations one can use previous results to speed up convergence as follows:

$$\alpha^2 = -3 + 2\sqrt{2}, \quad (\alpha^2)^2 = 17 - 12\sqrt{2}, \quad ((\alpha^2)^2)^2 = -577 + 408\sqrt{2},$$
$$\sqrt{2} \simeq 3/2 = 1.5, \quad \sqrt{2} \simeq 17/12 = 1.417, \quad \sqrt{2} \simeq 577/408 = 1.414216.$$

### ∎ 1.3.4 Continued Fractions

Another way of extending from $\mathbb{Q}^+$ is by using the expression

$$[A; a_1, a_2, \dots] \equiv A + \cfrac{1}{a_1 + \cfrac{1}{a_2 + 1/\dots}}, \ A, a_1, a_2, \dots \in \mathbb{N}_0. \tag{1.34}$$

Every fraction is expressible as a sequence involving only a finite number of integers. Thus one can use an ordered sequence to write

$$\sqrt{2} = [1; 2, 2, 2, \ldots] \simeq [1; 2, 2, 2, 1, 0, 0, \ldots] \simeq 1 + \cfrac{1}{2 + \cfrac{1}{2 + 1/2}} = \frac{17}{12} \simeq 1.417.$$

This has advantages over the iterative formulation in Eq. (1.21) because one can easily truncate the continued fraction expression to obtain the approximation, rather than having to continually compute iterates. This text will emphasize the qualitative differences between solutions to problems, when more than one is available.

### ■ Exercises

1. Let $\mathbb{Z}^2 \equiv \{\langle n, m \rangle | n, m \in \mathbb{Z}\}$ with operation $*$ where

$$\langle n_1, m_1 \rangle * \langle n_2, m_2 \rangle \equiv \langle n_1 + n_2, m_1 + m_2 \rangle$$

   Show that $\langle \mathbb{Z}^2, * \rangle$ is a group.

2. Rationalize the denominator using the radical conjugate.

   (a) $\frac{1}{1+\sqrt{3}}$.
   (b) $\frac{1}{\sqrt{2}+\sqrt{3}}$.
   (c) $\frac{1}{1+\sqrt{2}+\sqrt{3}}$ (*Hint:* Choose one radical and remove it first.)
   (d) $\frac{1}{1+\sqrt{2}+\sqrt{3}+\sqrt{6}}$ (*Hint:* Use a radical conjugate twice.)

3. Compute and simplify $(1+i)^4$ using $i^2 = -1$. Use the simplification to find a solution of $z^4 + 1 = 0$.

4. Use Eq. (1.8) to find an equation for $\phi$ in terms of $1/\phi$. From the definition in Eq. (1.34) show that $\phi = [1; 1, 1, 1, \ldots]$. Use this and Eq. (1.34) to find an approximation to $\sqrt{5}$ to 2 decimal places.

### ■ 1.4 Valuations, Inequalities, and Distance

When dealing with a large set, like $\mathbb{Q}$, a measure is needed to assess quantities. Here the *absolute value* is defined to be

$$|a| \equiv max\{a, -a\} = \sqrt{a^2}, \forall a \in \mathbb{Q}.$$

This notation will be used later when different spaces are discussed.[12] More generally, consider the positive part of an ordered field $\mathbb{O}^\times$ (like $\mathbb{Q}$ or $\mathbb{Q}(\sqrt{2})$ or $\mathbb{Q}(\sqrt[N_2]{\mathbb{P}})$, etc.) as a measuring space.

**Definition 1.6** An *archimedean valuation* "$v(*)$" on a field $\mathbb{F}$ to an ordered field $\mathbb{O}^\times$, has the following properties:

- Unique zero: $v : \mathbb{F} \to \mathbb{O}^\times \cup \{0\}$, where $v(x) = 0 \iff x = 0$;
- Product Homomorphism: $(\forall a, b \in \mathbb{F}) \implies (v(a \cdot b) = v(a) \cdot v(b))$;
- Triangle Inequality: $(\forall a, b \in \mathbb{F}) \implies (v(a + b) \leq v(a) + v(b))$.

A valuation becomes *nonarchimedean* if it has the stronger property that

- $(\forall a, b \in \mathbb{F}) \implies (v(a + b) \leq \max\{v(a), v(b)\})$.

The absolute value on $\mathbb{Q}$ is an archimedean valuation. A nonarchimedean example is the $p$-adic valuation $v_p : \mathbb{Q} \to \mathbb{Q}^+$, obtained for each $p \in \mathbb{P}$, by

$$v_p(m/n) \equiv p^{k_n - k_m}, \qquad p^{k_m} | m \wedge p^{k_n} | n; \tag{1.35}$$

where the integers $m, n, k_m, k_n$ are defined by

$$m = (r/s) \cdot p^{k_m}, \qquad n = (t/u) \cdot p^{k_n}, \qquad p \nmid r, s, t, u.$$

For example, if $m = 18/5$ and $n = 14/15$, then

$$v_3(m) = 1/9, \qquad v_3(n) = 3,$$
$$v_3(m \cdot n) = v_3(84/25) = 1/3,$$
$$v_3(m + n) = v_3(68/15) = 3.$$

**Remark 1.3** The quantity $\sqrt{4}$ is simply 2. This is different than saying that the solution(s) of $x^2 = 4$, (of which there are two) are $x = \pm 2$. Since an alternative of the absolute value is $|x| = \sqrt{x^2}$, an alternative to $x^2 = 4$ is the equation $|x| = 2$.

■ **Lemma 1.1**
An equation of the form $|x| = a$ for $a > 0$, has the solutions $x = \pm a$.

---

[12]The absolute-value notation will be used in different contexts. One must be aware of what space the notation $|*|$ is meant to apply.

The *distance* between two points $x, y \in \mathbb{F}$ can be defined as

$$dist(x, y) \equiv d_v(x, y) = v(x - y).$$

This gives a measurable sense of the separation between points. For example, the numbers 1 and $-1$ are the same distance from 0 when using the absolute value. Furthermore, the distance between 1 and $-1$ is $d(1, -1) = |1 - (-1)| = |2| = 2$.

**Definition 1.7** Let $\mathcal{S} = \{x_n\}$ be a sequence from a field $\mathbb{F}$ with valuation $v(*)$. Then $\mathcal{S}$ is called a *Cauchy sequence* if for each $M \in \mathbb{O}^+$ there is an $N_M \in \mathbb{N}$ and $X_M \in \mathbb{F}$ so that

$$d_v(x_n, X_M) \leq M^{-1}, \quad \forall n \geq N_M. \tag{1.36}$$

For $\mathbb{F} = \mathbb{Q}$ and $\mathbb{O}^+ = \mathbb{Q}^+$, which are ordered fields, this is the same as writing

$$x_n \in [X_M - 1/M, X_M + 1/M] \iff X_M - M^{-1} \leq x_n \leq X_M + M^{-1}$$
$$\iff |x_n - X_M| \leq M^{-1}.$$

**Remark 1.4** The quantities $X_M$ in the definition of a Cauchy sequence are approximations in $\mathbb{F}$ of the *limit* of the sequence, written $x_\infty$. The limit may not be in $\mathbb{F}$, however it is an entity such that

$$|x_\infty - X_M| \leq 2/M \text{ and } |x_\infty - X_n| \leq 2/M, \quad \forall n, M \geq N_M.$$

■ **Theorem 1.5**

The sequence $\{x_n\}$ where each element is defined as

$$x_n = [1; 2, 2, \cdots, 2, 1, 0, 0, \ldots] \quad \text{with } n \text{ 2's,}$$

is a Cauchy sequence.

**Proof** Only the approximations $\{X_M\}$ will be found and studied here. The following argument will constitute a *proof by induction*. There are two cases:
Case $M = 1$: Compute

$$X_1 = [1; 2, 1, 0, 0, \ldots] = 1 + \frac{1}{2+1} = \frac{3}{2} = 1.5,$$

and define $N_1 \equiv 2$. It is clear that, as required in the theorem,

$$1 < \sqrt{2} < \sqrt{4} = 2 \implies 1 = 1.5 - 1/2 < \sqrt{2} < 1.5 + 1/2 = 2.$$

Case $M+1$: By the definition of continued fractions, one should see that

$$X_{M+1} = 1 + \frac{1}{1 + X_M} = \frac{X_M + 2}{X_M + 1}.$$

Now assume that $0 \leq X_M - \sqrt{2} \leq 2^{-M}$. Then

$$\sqrt{2} + 1 \leq X_M + 1 \leq \sqrt{2} + 1 + \frac{1}{2^M} \implies$$

$$X_{M+1} - \sqrt{2} = \frac{X_M + 2 - \sqrt{2}X_M - \sqrt{2}}{X_M + 1} \leq \frac{(1-\sqrt{2})(X_M - \sqrt{2})}{\sqrt{2}+1} \leq 0,$$

which suggests that the sequence $\{X_M\}$ alternates with approximations above and below $\sqrt{2}$. Alternatively, using the same assumptions,

$$X_{M+1} - \sqrt{2} = \frac{-(\sqrt{2}-1)(X_M - \sqrt{2})}{X_M + 1} \geq \frac{-2^{-1}2^{-M}}{\sqrt{2}+1-2^{-M}} \geq \frac{-1}{2^{M+1}(\sqrt{2}+1)-2},$$

which is greater than $-2^{-M-1}$ when $M \geq 1$. A similar argument holds under the assumption $-2^{-M} \leq X_M - \sqrt{2} \leq 0$. Thus it is shown that

$$d(X_M, \sqrt{2}) = |X_M - \sqrt{2}| \leq 2^{-M} \implies d(X_{M+1}, \sqrt{2}) = |X_{M+1} - \sqrt{2}| \leq 2^{-(M+1)},$$

and so the result holds by induction $\forall M \in \mathbb{N}$. $\square$

### ■ Exercises

1. Rewrite the following as intervals. (*Hint:* Solve the case of equality first.):
    (a) $|x-5| < 2$, for $x \in \mathbb{R}$
    (b) $|y^2 - 5| < 4$, for $y \in \mathbb{R}$

2. Verify the triangle inequality $|x+y| \leq |x| + |y|$ by squaring both sides, simplifying, and squaring again. (*Hint:* Use that $|w|^2 = |w^2| = w^2$, $\forall w \in \mathbb{R}$.)

3. Suppose $a = 2$ and $b = -3$.
    (a) Compute $|a+b|$
    (b) Compute $|a| + |b|$

4. Suppose $p = 3$. Find the following $p$-adic valuations:

   (a) Compute $v_3(25/54)$
   (b) Compute $v_3(12/45)$
   (c) Verify the non-archimedean property for the case $v_3(25/54 + 12/45)$.

5. Show that $\{\phi_n\}$ defined as $\phi_n \equiv [1; 1, 1, 1, \ldots, 1, 0, 0, \ldots]$ with $n$ ones after the ";" is a Cauchy sequence. (*Hint:* Write $\phi_{n+1}$ in terms of $\phi_n$ and show that $\phi_{n+1} \in [3/2, 2]$.)

## 1.5 The Real Numbers

To describe the real numbers, first choose any finite sequence $\{a_j\}_{j=0}^N$ from $\mathbb{Z}_{10}$ and exponent $A \in \mathbb{Z}$. Then construct the fraction

$$r = a_0 10^A + a_1 10^{A-1} + \ldots a_{N-1} 10^{A-N+1} + a_N 10^{A-N} \in \mathbb{Q}. \tag{1.37}$$

The sequence $\{a_j\}$ can be thought of as being infinite, with only a finite number of $a_j$ terms as being nonzero. This expression for $r$ is called a *decimal* expansion because a *base* of 10 is being used. There is no way to express fractions like $1/3$ using a finite decimal expansion; however this can be done with a base that is divisible by 3. Rather than changing base, infinite sequences can be allowed, so that

$$\frac{1}{3} = \frac{3}{10} + \frac{3}{100} + \frac{3}{1000} \cdots = \sum_{k=1}^{\infty} \frac{3}{10^k}. \tag{1.38}$$

In this case, $A = -1$, $N = \infty$, and $a_j = 3, \forall j \in \mathbb{N}_0$. The set of *real numbers* (in base 10) is defined to be

$$\mathbb{R} \equiv \left\{ \sigma \cdot 10^A \cdot \sum_{j=1}^{\infty} \frac{a_j}{10^j} \,\middle|\, \sigma \in E_2, A \in \mathbb{Z}, (\forall j \in \mathbb{N}) \implies (a_j \in \mathbb{Z}_{10}) \right\}. \tag{1.39}$$

The common method used to express real numbers as in Eqs. (1.37) or (1.39), is

$$\sigma a_0 a_1 \ldots a_{A-1} a_A \cdot a_{A+1} a_{A+2} \ldots \quad \iff \quad r \in \mathbb{R},$$

for $A > 0$ and $\sigma \in E_2 = \{-1, 1\}$. For $A \leq 0$,

$$\sigma 0.00 \ldots 0 a_0 a_1 \ldots \quad \iff \quad r \in \mathbb{R},$$

where the number of zeros appearing $0.00\ldots 0$ is $|A|$. In either case one can write

$$r = \sigma a_0.a_1 a_2 \cdots \times 10^A = sign \times significant \times base^{exponent},$$

which is called the *scientific notation* for $r$. Such expressions have an intrinsic ambiguity, but they provide convenient shorthand notations for presentation and computation. Every real number can be approximated by a fraction, to any degree of accuracy. A fraction can always be represented as a decimal that eventually repeats. To verify this requires a study of *Geometric series*, where Eq. (1.38) is an example.

### ■ Exercises

1. Rewrite the following in the form of Eq. (1.38):

   (a) 32.105
   (b) 17/12 (only express 6 terms)

## ■ 1.6 Fundamental Theorems

Mathematics is founded on clear natural definitions, and rigorous true statements. Understanding the key elements in any topic allows a meaningful perspective to any problem posed in the area. Possibly the first and most famous abstract mathematical result is Pythagorean's Theorem.

### ■ Theorem 1.6

**Pythagorean's Theorem**
Let $a, b$, and $c \in \mathbb{Q}$ be positive numbers that are the lengths of a right triangle, with $c$ the longest side (hypotenuse). Then $a^2 + b^2 = c^2$.

**Proof** The idea behind the proof follows from a simple diagram. Consider a square with sides of length $a + b$. Then a smaller square can be inscribed in the original square with unknown side length $c$. By conservation of area

$$\begin{aligned}(a+b)^2 &= c^2 + 4 \cdot (ab/2) \\ a^2 + 2ab + b^2 &= c^2 + 2ab \\ a^2 + b^2 &= c^2,\end{aligned}$$

where the area of the right triangle in Fig. 1.2 is $ab/2$. This completes the result. □

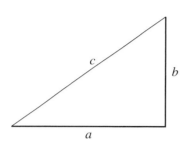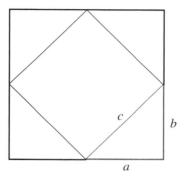

**FIGURE 1.2** ■ The proof of the Pythagorean Theorem, with aid of a diagram.

Although not difficult to see, when this result was discovered, the significance was recognized. It marked a leap in human mental endeavors. However, the principle was known to be used by farmers for surveying plots of land. Indeed, by tying 13 knots in a rope, evenly spaced, one can create a $\langle 3, 4, 5 \rangle = \langle a, b, c \rangle$ triangle by meeting the first and last knot. Since $\triangle(a, b, c)$ is a right triangle, this allows one to easily make a 90° angle.

Another well-known fact of numbers is the Fundamental Theorem of Arithmetic.

■ **Theorem 1.7**

**Fundamental Theorem of Arithmetic**
For each $n \in \mathbb{N}_2$ there is a finite sequence of prime numbers $\{p_j\}_{j=1}^k \subset \mathbb{P}$ and powers $\{a_j\}_{j=1}^k \subset \mathbb{N}$, so that

$$n = p_1^{a_1} \cdot p_2^{a_2} \cdot \cdots \cdot p_k^{a_k} = \prod_{j=1}^k p_j^{a_j}. \qquad (1.40)$$

If the primes are listed in strictly increasing order, i.e., $p_j < p_{j+1}$, then this representation is unique.

Here $\prod$ denotes a *product of factors* and $j$ is the *index of the product*.

**Proof** If $n = 2$ then the result holds with $k = 1$, $p_1 = 2$, and $a_1 = 1$. For $n \geq 3$, if $2|n$ then set $p = 2$ and $a = 1$. Now do the same for $(n/2) \in \mathbb{N}$. Else, if $3|n$ set $p = 3$ and $a = 1$. Proceeding in this manner will exhaust all possible primes less than the original $n$ in no more than $n$ iterations. □

Combining the preceding two fundamental ideas gives the Hippasus corollary.

**Corollary 1.1  The Hippasus Corollary**

For each $p \in \mathbb{P}$ the square root $\sqrt{p}$ is not a fraction.

**Proof** To prove by contradiction, suppose that $\sqrt{p} \in \mathbb{Q}^+$. Then $\exists m, n \in \mathbb{N}$ so that

$$\sqrt{p} = n/m \implies p \cdot m^2 = n^2. \tag{1.41}$$

What is left is a matter of counting primes, i.e., $p \cdot m^2$ has an odd number of primes, and $n^2$ has an even number. Thus Eq. (1.41) is absurd. Consequently, $\sqrt{p} \notin \mathbb{Q}^+$. □

$\mathbb{Q}$ is not closed under solutions of polynomial equations. In particular $\sqrt{2} \notin \mathbb{Q}$. This was quite disturbing to the Pythagoreans since $\sqrt{2}$ is a geometrically natural number, as is $\sqrt{3}$ and $\pi$. See Fig. 1.3.

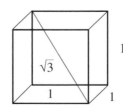

**FIGURE 1.3** ■ Irrational lengths $\{\sqrt{2}, \sqrt{3}, \pi\}$ defined geometrically by a unit square, unit cube, and circle with unit diameter, respectively.

**Definition 1.8** A *zero* of a polynomial $p(x)$ that is not in $\mathbb{Q}$ is called *algebraic*. The set of algebraic numbers is denoted $\mathbb{A}$. A number that is not in $\mathbb{Q}$ is called *irrational*, and a number that is not in $\mathbb{A}$ is called *transcendental*.

**Remark 1.5** The Golden Ratio (or Golden Mean or Golden Section) $\phi \equiv (\sqrt{5}+1)/2$ is an algebraic number. $\pi$ is transcendental, however it is quite difficult even to show that it is not rational.[13] Thus, it should be clear that rational numbers are insufficient to encompass many of the basic geometrical entities.

### EXAMPLE 1.7

Inscribing a hexagon in a circle of radius 1, shows that the perimeter is $2\pi > 6$. By fitting a circle of radius 1 inside a hexagon, one can see that $2\pi < 4\sqrt{3} < 7$.

### ■ Exercises

1. Although $6 \notin \mathbb{P}$, why is $\sqrt{6}$ irrational? Why does this argument not work for $\sqrt{4}$, which is rational?
2. How many prime divisors does 720 have, including repeated primes?

### ■ 1.7 Sequences and Series

An ordered list of numbers, indexed by $\mathbb{N}_n$ for some $n \in \mathbb{Z}$, is called a *sequence*. This has already been discussed in the context of approximating square roots. Geometrically, $\sqrt{2}$ is the diagonal of a unit square. However, algebraically, $\sqrt{2}$ can be thought of as a sequence of converging rational numbers.

**Query 1.1** If $\mathbb{Q}$ is the set of truly rational numbers, then is $\sqrt{2}$ a number, or is it an algorithm that generates a convergent sequence?

---

[13]Herstein presents a proof based on relatively elementary concepts [28].

**Remark 1.6** Every *algorithm* must terminate in a finite number of steps in order to be useful to applications. Thus the concept of *tolerance*, or being sufficiently close, needs to be a part of any problem-solving process.

---

**Definition 1.9** For any two quantities in a field $a, b \in \mathbb{F}$ and $n \in \mathbb{N}$, the identity,

$$(a+b)^n = (a+b) \cdot (a+b) \cdots (a+b) = \sum_{j=0}^{n} \binom{n}{j} \cdot a^j \cdot b^{n-j}, \qquad (1.42)$$

is called a *binomial series*, and the quantity

$$\binom{n}{j} \equiv \frac{n \cdot (n-1) \cdots (j+2) \cdot (j+1)}{(n-j) \cdot (n-j-1) \cdots (2) \cdot (1)} = \frac{n \cdot (n-1) \cdots (n-j+2) \cdot (n-j+1)}{j \cdot (j-1) \cdots (2) \cdot (1)} \qquad (1.43)$$

is called the *binomial coefficient*. The *factorial* of a natural number $n \in \mathbb{N}$ is defined recursively by $n! \equiv n \cdot (n-1)!$ with $1! = 1$ and $0! = 1$.

---

The binomial coefficients can be derived using Pascal's triangle. In particular, from Eq. (1.43) one has the iterative identity

$$\binom{n+1}{j+1} = \binom{n}{j} + \binom{n}{j+1}, \quad \binom{n}{-1} = \binom{n}{n+1} = 0, \qquad (1.44)$$

for $0 \leq j \leq n$, which is sufficient for a proof of Eq. (1.42). The binomial series also holds for fractional powers $n \in \mathbb{Q}$ if the last expression in Eq. (1.43) is used.

---

**Definition 1.10** For each positive fraction $m/n \in \mathbb{Q}^+$ with $0 < m/n < 1$, the infinite sum of fractions

$$\sum_{j=0}^{\infty} \left(\frac{m}{n}\right)^j = \frac{1}{1-(m/n)} = \frac{n}{n-m}, \qquad (1.45)$$

is also a fraction. The infinite sum in Eq. (1.45) is called a *Geometric series*, and Eq. (1.45) is the Geometric series formula.

---

Conversely, every fraction $m/n$ between 0 and 1 can be expressed as a Geometric series. This is seen by replacing $n$ with $n+m$ to get

$$\frac{m}{n} = \sum_{j=0}^{\infty} \left(\frac{m}{n+m}\right)^j, \qquad (1.46)$$

and all other fractions can be obtained from multiplication of Eq. (1.46) by an appropriate integer. In this sense the set $\mathbb{Q}$ is closed under Geometric series.

### ■ Theorem 1.8

Every real number can be approximated by a fraction.

**Proof** For each $N \in \mathbb{N}_0$, and any positive real number $r > 0$, an approximating fraction $q \simeq r$ is

$$q = 10^A \cdot \sum_{j=0}^{N} \frac{a_j}{10^j}, \qquad a_j \in \mathbb{Z}_{10} = \{0, 1, 2, \ldots, 9\}.$$

A measure of the accuracy of the approximation is obtained from Eq. (1.45) by observing

$$0 \leq r - q < 10^A \cdot \frac{1}{10^{N+1}} \sum_{j=0}^{\infty} \frac{9}{10^j} = 10^A \cdot \frac{9}{10^{N+1}} \cdot \frac{10}{10-1} = 10^A \frac{1}{10^N} = \frac{1}{10^{N-A}}.$$

Thus, for $N$ sufficiently large, the approximation $q$ is as close to $r$ as desired. □

The space $\mathbb{Q}$ is not closed under infinite decimal expansion.

**Definition 1.11** For each $n \in \mathbb{N}_0$ the factorial of $n$, written as $n!$, is defined to be

$$0! = 1, \ 1! = 1, \ 2! = 2, \ 3! = 6, \ (n+1)! = (n+1) \cdot n!.$$

Napier's number, written as $e$, is defined to be the real number

$$e \equiv \sum_{n=0}^{\infty} \frac{1}{n!} = 1 + 1 + \frac{1}{2} + \frac{1}{6} + \cdots \simeq 2.718281828. \tag{1.47}$$

It also has the expression $e = [2; 1, 2, 1, 1, 4, 1, 1, 6, 1, 1, 8, \ldots]$. The symbol $e$ is used in memory of Euler. The number $e$ is known to be transcendental.

Napier's number is irrational. Using an argument presented in Ross [59], assume that $e = p/m$ for some $p, m \in \mathbb{N}_2$. Then the product $e \cdot m! \geq p \geq 2$ must be an integer and so, using Eq. (1.47), the same must be true for the sum,

$$1 < \sum_{n=m}^{\infty} \frac{m!}{n!} \leq 1 + \frac{1}{m+1} + \sum_{j=2}^{\infty} \frac{1}{(m+1)^j} = \frac{m+1}{m} = 1 + \frac{1}{m} < 2. \qquad (1.48)$$

But this cannot be the case for $m \in \mathbb{N}_2$. To show that $e$ is not algebraic is difficult, but can be found in the classical work of Mahler [45].

### EXAMPLE 1.8

For all real numbers $x \in \mathbb{R}$, the quantity $1/(1+x^2)$ is finite and can be computed. For $x = 1$ this quantity is $1/2$, however the corresponding Geometric series $\sum_{j=0}^{\infty}(-1)^j$, obtained from Eq. (1.45) with $m = -1$ and $n = 1$, cannot be determined, i.e., it is undefined.

## ■ Exercises

1. Compute the fraction $\sum_{j=2}^{\infty} 3^{-j}$ using Eq. (1.45).
2. Compute the fraction $\sum_{j=2}^{\infty} (2/3)^j$.
3. Prove Eq. (1.44) using the factorial expression. (*Hint:* First consider the $n = 0$ case. Then consider the right-hand side and get a common denominator.)
4. Prove Eq. (1.42) by induction using Eq. (1.44).
5. For each $k \in \mathbb{N}$ it can be shown that $e^{1/k} = \sum_{n=0}^{\infty} (k^n \cdot n!)^{-1}$. Modifying the argument in Eq. (1.48) to show that $\sqrt[k]{e} = e^{1/k} \notin \mathbb{Q}$, extend this fact to negative integer roots.

## ■ 1.8 Complex Numbers

The final extended space of this chapter includes both algebraic numbers $\mathbb{A}$ and real numbers $\mathbb{R}$ to obtain a consistent, closed set, under solutions of algebraic equations and Cauchy sequences. This set is called the *complex numbers* $\mathbb{C}$ and can be written as a linear combination of real and imaginary numbers,

$$\mathbb{C} \equiv \{a + ib \mid a, b \in \mathbb{R}\}, a = \text{the real part}, b = \text{the imaginary part}, i = \sqrt{-1}.$$

This is algebraically similar to the field $\mathbb{Q}(\sqrt{2})$ and it is quite meaningful to write $\mathbb{C} = \mathbb{R}(i)$. For $a, b \in \mathbb{R}$ and $p \in \mathbb{P}$, the radical conjugate of $a + \sqrt{p}b$ is $a - \sqrt{p}b$ and the identity

$$(a + \sqrt{p} \cdot b) \cdot (a - \sqrt{p} \cdot b) = a^2 - p \cdot b^2 \in \mathbb{R}.$$

Conversely, the *complex conjugate* of $(a + ib)$ is $(a - ib)$ and is commonly expressed in two different ways

$$a - ib \equiv (a + ib)^* \equiv \overline{a + ib} \in \mathbb{C}.$$

The important identity that holds is

$$|a + ib|^2 \equiv (a + ib) \cdot (a + ib)^* = (a + ib) \cdot (a - ib) = a^2 + b^2 \in \mathbb{R}_0^+,$$

where $\mathbb{R}_0^+ \equiv \mathbb{R}^+ \cup \{0\}$. The *modulus* (not to be confused with $mod\ (n)$) is defined to be the archimedean valuation on $\mathbb{C}$

$$|a + ib| \equiv \sqrt{a^2 + b^2},\ |\cdot| : \mathbb{C} \to \mathbb{R}_0^+,$$

and is an extension of the absolute value on $\mathbb{Q}$. The unit circle in the complex plane is the set of points

$$\mathcal{S}^1 \equiv \{a + ib \mid a^2 + b^2 = 1, a, b \in \mathbb{R}\} = \{z \in \mathbb{C} : |z| = 1\},$$

and clearly $\mathcal{S}^1 \subset \mathbb{C}\backslash\{0\}$, which is a subgroup under multiplication.

■ **Theorem 1.9**

**Fundamental Theorem of Algebra**
Each polynomial $p(x)$ of degree $n \in \mathbb{N}$, with complex coefficients, has at least one zero in $\mathbb{C}$. Thus every polynomial can be factored into a product of linear quantities $(a \cdot x + b)$ for some $a, b \in \mathbb{C}$.[14]

For example, $x^2 + 1 = (x + i) \cdot (x - i)$. Furthermore, $x^4 + 1 = (x^2 + 1) \cdot (x^2 - 1)$ thus one obtains the solution space for

$$(x^4 = 1) \implies x \in \{1, -1, i, -i\} = E_4.$$

---

[14] A proof will be possible after Chapter 6, using techniques of vector calculus.

The elements of $E_4$ are called the 4th *roots of unity*. The $n$th roots of unity solve $x^n = 1$ and the set of solutions is denoted as $E_n$. Note that $\langle E_n, \cdot \rangle$ is a cyclic group[15] of order $n$. Also, $E_n \subset \mathcal{S}^1$ is a subgroup under multiplication.

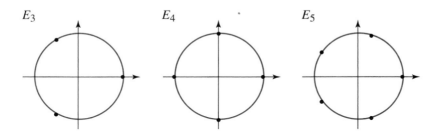

**FIGURE 1.4** ■ $E_3$, $E_4$, and $E_5$ inscribed on a unit circle.

**Remark 1.7** Example 1.4 can now be revisited and the third solution to the problem found. This is done by defining the cubic roots of unity

$$(x^3 = 1) \implies x \in \{1, \omega, \omega^2\} = E_3,$$

where the complex numbers $\omega$ and $\omega^2$ are

$$\omega = (1 + i\sqrt{3})/2, \ \omega^* = \overline{\omega} = \omega^2 = \omega^{-1} = (1 - i\sqrt{3})/2.$$

Then, from quadratic Eq. (1.15), and choosing the solution $z^3 = 1.2347$, there are three complex solutions

$$z \simeq \begin{cases} 1.0728 \\ 1.0728\omega \\ 1.0728\omega^2 \end{cases} \implies y \simeq \begin{cases} 4.3084 \\ 2.1542 + i1.0814\sqrt{3} \\ 2.1542 - i1.0814\sqrt{3} \end{cases} \implies a \simeq \begin{cases} 8.2417 \\ 1.7792 + i1.8730 \\ 1.7792 - i1.8730 \end{cases}.$$

Note that the complex solutions are conjugates of each other.

Based on this analysis, the generalized Roy Baker problem has no realistic solution for $h = 6$ cm.

---

[15]The notation of Herstein [28] will be employed.

## Exercises

1. Draw the following inequalities as discs in $\mathbb{C}$:

    (a) $|x - 1| < 2$
    (b) $|x - i| < 2$

2. Verify the triangle inequality $|z_1 + z_2| \leq |z_1| + |z_2|$ for $z_1, z_2 \in \mathbb{C}$ by squaring, simplifying, and squaring again. (*Hint:* Use the fact that $|z|^2 = x^2 + y^2$ for $z = x + i \cdot y$ and $x, y \in \mathbb{R}$.)

3. Suppose $a = 2$ and $b = -3i$.

    (a) Compute $a^2 + b^2$ and $a^2 - b^2$
    (b) Compute $|a + b|$ and $|a - b|$

4. Suppose $a + ib, c + id \in \mathcal{S}^1$. Show that $(a + ib) \cdot (c + id) \in \mathcal{S}^1$. Also, with $1 + i0 \in \mathcal{S}^1$ as identity, find the reciprocal of $a + ib \in \mathcal{S}^1$, showing that $\langle \mathcal{S}^1, \cdot \rangle$ is a group.

# 2 Functions and Relations

The common functions of mathematics and science are reviewed. Applications to geometry and probability theory are used to motivate the introduction of many of the special functions.

## Introduction

Single actions, transformations, or computations are identified with the use of symbols like $f, g, h$ or $A, B, C$ or $T, S, R$. They indicate the act of *converting* a variable from one value to another value. Consider the following list of expressions indicating the *from-to* relationship:

| | |
|---:|:---|
| *abscissa* | *ordinate* |
| *input* | *output* |
| *factor* | *response* |
| *cause* | *effect* |
| *domain* | *range* |
| *independent* | *dependent* |
| *predictor* | *prediction* |

Different areas of mathematics and science express the triple $\langle function, from, to \rangle$ in different ways, but most problems can be formulated as:

$$\text{given equation: } f(x) = y \qquad \text{solve for: } x$$

These problems are resolved by finding the inverse function $f^{-1}$ to a function $f$. Science often generates equations like $f(x) = y$, and it is for mathematics to solve for $x$, if only to a limited degree of accuracy.

**Remark 2.1** The notation for the inverse function can be confusing:

$$f^{-1}(y) \quad \overset{?}{=} \quad \begin{cases} (f(y))^{-1} = 1/f(y) & = \text{ the reciprocal of } f(y) \\ \{x | f(x) = y\} & = \text{ preimage of } y \text{ under action of } f \end{cases}$$

The confusion is exacerbated by the common use of the notation $f^n(x) = (f(x))^n$ for all $n \neq -1$. However, this text will continue this tradition, while striving to be clear with the inverse function notation.

## 2.1 Sets

There is an important connection between *statements* and *sets*. Given the symbols and rules of logic, consider the space of statements $\mathcal{S}$ that are syntactically correct. Within $\mathcal{S}$ there is a subset of those sentences that are true $\mathcal{T}$, those that are false $\mathcal{F}$, and those in $\mathcal{S}\backslash(\mathcal{T} \cup \mathcal{F})$ that are nonsense, as in Eq. (1.1). Two sentences $\phi, \psi \in \mathcal{S}$ are *mutually exclusive* or *disjoint* if $(\phi \in \mathcal{T}) \iff (\psi \in \mathcal{F})$. In particular, the sets $\mathcal{T}$ and $\mathcal{F}$ are disjoint. An individual's space of sentences $\mathcal{S}_{me}$ will be a very small subset of $\mathcal{S}$ as will be $\mathcal{T}_{me} \subset \mathcal{T}$ and $\mathcal{F}_{me} \subset \mathcal{F}$.[1] There will be sentences $\phi$ like

$$\phi \in \mathcal{U}_{me} \equiv (\mathcal{T} \cup \mathcal{F}) \backslash (\mathcal{T}_{me} \cup \mathcal{F}_{me}),$$

whose truth value can be estimated with a personal *belief* function, usually denoted as $P(\phi)$ for the *probability of $\phi$ being true*. The function $P : \mathcal{U}_{me} \to [0,1]$ goes from the space of personal *unknowns* $\mathcal{U}_{me}$ to the closed interval $[0,1]$, where $P = 1$ means true and $P = 0$ means false. To improve outcomes of personal situations, which are consequences of actions based on beliefs where $0 < P < 1$, a deep consideration of how a belief system can be founded on rationality and logic may help.[2]

Before a discussion of functions can properly begin, a review of basic set theory is in order. Three types of sets will be repeatedly used:

**(1)** *Finite sets*, which are all equivalent to $\mathbb{Z}_n$ for some $n \in \mathbb{N}$;
**(2)** *Countable sets*, where the elements can be indexed by $\mathbb{N}$; and
**(3)** *Borel sets* $\mathcal{B}$, are all intersections and countable-unions of intervals from $\mathbb{R}$.

The set $\mathbb{Q}$ provides a dilemma because it is countable, yet has metric properties akin to $\mathbb{R}$. In fact, the Borel sets can be defined as the minimal extension of

---

[1] These subset relations are actually quite optimistic.
[2] An in-depth formulation with further references is given by Ginsberg [21].

countable intersections and countable unions of finite intervals with end points from $\mathbb{Q}$.

The set of no elements, written $\{\} \equiv \emptyset$, is called the *empty set*. Two sets $A$ and $B$ are called *disjoint* if $A \cap B = \emptyset$, where $\cap$ stands for *intersection*. Two sets $A$ and $B$ are called a *cover* for a set $C$ if $C \subset A \cup B$, where $\cup$ stands for *union*. Two sets $A$ and $B$ *exactly cover* a set $C$ if $C = A \cup B$. If $C$ is a set with two disjoint subsets $A, B \subset C$ where $A \cup B = C$, then $\{A, B\}$ is a called a *partition* of $C$. In particular, a *disjoint-exact* cover defines a partition. A common method of obtaining a partition on $C$ is by defining an equivalence relation "$\sim$" on $C$ so that it is

*Reflexive:* $(\forall a \in C) \Longrightarrow (a \sim a)$;
*Symmetric:* $(\forall a, b \in C) \Longrightarrow ((a \sim b) \iff (b \sim a))$; and
*Transitive:* $(\forall a, b, c, \in C)\, (a \sim b) \wedge (b \sim c) \Longrightarrow (a \sim c)$.

The set $C$ decomposes into equivalence classes $\{C_i\}$, which gives a disjoint-exact cover of $C$.

The *complement* of a set $A$ is $A^c \equiv \{x : x \notin A\}$. This definition is ambiguous. For example $\{0\}^c$ is different in $\mathbb{Z}, \mathbb{Q}, \mathbb{R}, \mathbb{C}$, etc. An alternative is to define *set subtraction* "$-$", and the *symmetric difference* "$\Delta$" [62],

$$\begin{aligned} B - A &\equiv \{x \in B | x \notin A\}, and \\ A \Delta B &\equiv (A \cup B) - (A \cap B) = \{(x \in A \wedge x \notin B) \vee (x \notin A \wedge x \in B)\}, \end{aligned} \qquad (2.1)$$

so that if $A \subset X$, and computations are restricted to $X$, then $A^c = X - A$. The *power set* of $X$ is $\mathcal{P}(X) \equiv \{A \subseteq X\}$. The proof of the following is left as an exercise.

■ **Theorem 2.1**

The space $\langle \mathcal{P}(X), \Delta \rangle$ is an abelian group, and the space $\langle \mathcal{P}(X), \Delta, \cup \rangle$ is a ring.

**Definition 2.1** The *cardinality* or *size* of $\mathbb{Z}_n$ is $n$, written $\#(\mathbb{Z}_n) = n$. The cardinalities of $\mathbb{N}, \mathbb{Z}$, and $\mathbb{Q}$ are infinite, but countable since they can be enumerated by $\mathbb{N}$. The cardinalities of $\mathbb{R}$ and $\mathbb{C}$ are uncountable since the elements of these sets cannot be indexed by $\mathbb{N}$ nor any of its subsets.

The Boolean algebra of *discrete* sets (finite or countable) has the operations of unions, intersections, and complements. If $\#(X) = n < \infty$, then $\mathcal{P}(X) = 2^n < \infty$. Thus there are only a finite number of ways to select elements from $X$. Conversely, if $\#(X) = \infty$ then $\#(\mathcal{P}(x))$ is larger than $\#(X)$ since any function $\phi : X \to \mathcal{P}(X)$ cannot be onto i.e., ($\phi(x)$ cannot be a cover for $\mathcal{P}(X)$). Indeed, for any such $\phi$ one can always define a set $S_\phi \subseteq X$ so that $x \in S_\phi \iff x \in \phi(x)$. Then $(\nexists x \in X) \ni (\phi(x) = S_\phi^c = X - S_\phi)$.

**Definition 2.2** A bounded closed interval $I$ from $\mathbb{R}$ is defined by two points $a, b \in \mathbb{R}$ where $a \leq b$, so that $I \equiv [a, b]$, which includes the end points and all points in between. The *measure* (which is different from the cardinality) of $I$ is $m(I) \equiv b - a$ and is the same as the *length* of $I$. The *boundary* of $I$ is the set $\partial I = \{a, b\}$ so that $\#(\partial I) = 2$ but $m(\partial I) = 0$. The *interior* of $I$ is the open interval $I - \partial I = (a, b)$, and $m(I - \partial I) = m(I) = b - a$. A bounded closed rectangle $R$ from $\mathbb{C}$ is defined by two intervals $I, J \subset \mathbb{R}$ where $R = I + iJ$. The *measure* of $R$ is $m(R) = m(I) \cdot m(J)$, which is the same as the *area* of $R$. The *perimeter* $\partial R$ has length $2 \cdot m(I) + 2 \cdot m(J)$.

Let $\mathcal{B}$ denote the collection of open and closed intervals from $\mathbb{R}$, along with their countable unions, intersections, and complements. Then the space $\langle \mathcal{B}, \cup, \cap \rangle$ is called a *$\sigma$-algebra*, which consists of the Borel sets of $\mathbb{R}$.

■ **Theorem 2.2**

Suppose a set $B \subset \mathbb{R}$ is the countable union of countable intersections of open intervals $\{\mathcal{O}_{j,k}\}_{j,k \in \mathbb{N}}$ that are all within the bounded closed interval $[a, b]$;

$$B = \bigcup_{j=1}^{\infty} \bigcap_{k=1}^{\infty} \mathcal{O}_{j,k}, \ \mathcal{O}_{j,k} \subset [a,b] \subset \mathbb{R} \text{ for } -\infty < a < b < \infty. \tag{2.2}$$

Then $B$ can be approximated by a finite union of finite intervals. In particular, there is a countable sequence of disjoint intervals $\{U_\ell\}_{\ell=1}^{\infty}$ so that for $n \in \mathbb{N}$,

$$B_n \equiv \bigcup_{\ell=1}^{n} U_\ell \subset \bigcup_{\ell=1}^{\infty} U_\ell \subseteq B \subset [a,b], \tag{2.3}$$

where the measure of $B_n$ is defined to be $\forall n \in \mathbb{N}$,

$$m(B_n) \equiv \sum_{\ell=1}^{n} m(U_\ell) \leq b - a, \tag{2.4}$$

in terms of the length of $U_\ell$, denoted $m(U_\ell) \geq 0$. Then $m(B_\infty)$ exists as a positive real number, called the measure of $B$ and denoted as $m(B)$.

Measurability extends to any Borel set. Details are given in [56], [60], [62]. Borel sets are needed for most scientific applications.

**Proof** (1) For each $j \in \mathbb{N}$ define

$$L_j \equiv \bigcap_{k=1}^{\infty} \mathcal{O}_{j,k} = \{x \in [a,b] : x \in \mathcal{O}_{j,k}, \forall k \in \mathbb{N}\}.$$

It should be clear that $L_j$ is an interval: *The intersection of two intervals gives another new interval.* The same is not true for unions of intervals if they are disjoint. If uncountable unions are allowed, then sets can be created with a very nonintuitive behavior that cannot be measured.[3] Conversely, the measures $m(L_j)$ are defined as the lower limits of the nonincreasing sequences of nonnegative numbers $\{m(\cap_{k=1}^{n} \mathcal{O}_{j,k})\}_{n=1}^{\infty}$. Thus the limit of this sequence is a nonnegative real number.

(2) An equivalence relation $\sim$ can be defined on $\mathcal{Q} \equiv \{L_j\}_{j \in \mathbb{N}}$, so that

$$L_j \sim L_k \iff [(L_j \cap L_k \neq \emptyset) \vee \exists L_\ell \ni ((L_j \sim L_\ell) \wedge (L_k \sim L_\ell))]. \quad (2.5)$$

(See the exercises in this section.) The consequence is that $\mathcal{Q}$ separates into equivalence classes $\mathcal{U}_\ell$, which are disjoint and measurable. This also defines a partition of $\mathbb{N}$ denoted $\mathcal{N}_\ell$ where $\mathcal{U}_\ell \equiv \cup_{j \in \mathcal{N}_\ell} L_j$. Note that $\#(\mathcal{N}_\ell)$ may be finite or (countably) infinite.

(3) Each set $\mathcal{U}_\ell$ is the limit of nondecreasing intervals $\mathcal{U}_\ell = \cup_{n=1}^{\infty} U_\ell^n$ where

$$U_\ell^n = \cup\{L_j | \forall j \in \mathcal{N}_\ell \wedge j \leq n\} = \{x \in [a,b] : x \in L_j, \forall j \in \mathcal{N}_\ell \cap \mathbb{Z}_n\}.$$

Then $m(U_\ell^n)$ is defined initially by

$$m(U_\ell^1) = \begin{cases} m(L_1) & \text{if } 1 \in \mathcal{N}_\ell, \\ m(U_\ell^1) = m(\emptyset) = 0 & \text{if } 1 \notin \mathcal{N}_\ell. \end{cases}$$

and inductively by

$$m(U_\ell^{n+1}) = \begin{cases} m(L_{n+1}) & \text{if } n+1 \in \mathcal{N}_\ell \wedge U_\ell^n \subseteq L_{n+1}, \\ m(U_\ell^n) + m(L_{n+1} - U_\ell^n) & \text{if } n+1 \in \mathcal{N}_\ell \wedge U_\ell^n \not\subseteq L_{n+1}, \\ m(U_\ell^n) & \text{if } n+1 \notin \mathcal{N}_\ell. \end{cases}$$

---

[3]Such sets are called *unmeasurable* and examples can be found in Royden [62]. It should also be mentioned that the ordering of uncountable operations requires the Axiom of Choice, which is implicitly assumed in this text.

The measures $m(\mathcal{U}_\ell)$ are defined as the upper limits of the nondecreasing sequences of nonnegative numbers $\{m(U_\ell^n)\}_{n=1}^\infty$. The limit of this sequence is a real number no more than $b - a$.

(4) The measure of $B$, defined as $m(B) = \sum_{\ell=1}^\infty m(\mathcal{U}_\ell)$ is no more than $b - a$. $\square$

A function $f : X \subset \mathbb{R} \to \mathbb{R}$ is called *Borel* if and only if the *preimage* of $f$, for each interval $[c, d] \subset \mathbb{R}$,

$$f^{-1}([c, d]) \equiv \{x \in X | f(x) \in [c, d]\},$$

is a Borel set. Typically the domain $\mathcal{D}_f = X$ is also a Borel set.

## ■ Exercises

1. Prove Theorem 2.1, assuming associativity.
2. Define $\mathcal{I}_\mathbb{Q} \equiv [0, 1] \cap \mathbb{Q}$, and suppose that $\{x_n\}_{n=1}^\infty = \mathcal{I}_\mathbb{Q}$ is an enumeration of $\mathcal{I}_\mathbb{Q}$. For each $\epsilon > 0$ define the open intervals

    $$\mathcal{O}_n^\epsilon \equiv (x_n - \epsilon^n, x_n + \epsilon^n).$$

    (a) Show that $\cup_{n=1}^\infty \mathcal{O}_n^\epsilon \supset \mathcal{I}_\mathbb{Q}$.
    (b) Prove that $m\left(\cup_{n=1}^\infty \mathcal{O}_n^\epsilon\right) \leq 2 \cdot \epsilon/(1 - \epsilon)$, for each $\epsilon < 1$.
    (c) Justify the conclusion that $m(\mathcal{I}_\mathbb{Q}) = 0$. (*Hint:* It may help to set $\epsilon = 2^{-k}$ for each $k \in \mathbb{N}_2$.)
    (d) Conclude that $\mathcal{I}_\mathbb{Q} \neq [0, 1]$.

3. A finite collection of open intervals $\{\mathcal{O}_n\}_{n=1}^N$ is called an *open cover* of a Borel set $\mathcal{B}$ if $\cup_{n=1}^N \mathcal{O}_n \supset \mathcal{B}$. For each $N \in \mathbb{N}$ construct a finite open cover $\{\mathcal{O}_n\}_{n=1}^N$ of $[0, 1]$ so that

    $$\left((\forall n \leq N)\left(m(\mathcal{O}_n) < \frac{2}{N}\right)\right) \text{ and } \left(\sum_{n=1}^N m(\mathcal{O}_n) \leq 1 + \frac{2}{N}\right).$$

    (*Hint:* Consider the centers and widths of the open intervals $\mathcal{O}_n$.)

4. Prove that Eq. (2.5) defines an equivalence relation.
5. Let $X = \{1, 2\}$ and verify Theorem 2.1 in this simple case by creating a group table for the $\Delta$ operation, and a table for the $\cup$ operation on $\mathcal{P}(X) - \emptyset$.

## 2.2 A Catalogue of Functions

Walking up a hill, as compared to flat ground, one may experience a sense of physical exertion. Scientists say that this is due to an increase in potential energy that the body is accumulating. Thus it is easily observed that there is a relationship between height and energy, which for small distances, is one of direct proportionality. Let the independent variable $x$ represent the horizontal position of a person. Then, for a variable terrain, let $y$ be their height or vertical position, which is a dependent variable. The region of interest is typically defined to be an interval,

$$\mathcal{D} = [x_-, x_+] \equiv \{x \in \mathbb{R} \mid x_- \leq x \leq x_+\},$$

which will be called the *domain* for the person. As the person walks over the domain, their height $y$ changes in a manner dependent on $x$. In this way, a height function $h : x \to y$ is defined. This is also called a *mapping* and the action of computing or measuring the height is expressed as $y = h(x)$. Note that if the ground is soft, then standing in one spot may result in a slow change in height for a fixed position, so the idealized nature of functional definitions must be kept in mind.

Two hills, $h_1(x)$ and $h_2(x)$, will be considered and defined as *case functions*,

$$h_1(x) \equiv \begin{cases} 0 & \text{for } x < 0 \\ x & \text{for } x \geq 0 \end{cases}, \quad h_2(x) \equiv \begin{cases} 0 & \text{for } x < 0 \\ x^2 & \text{for } x \geq 0 \end{cases}. \tag{2.6}$$

At $x = 0$ the ascent described by $h_2(x)$ is more gradual than with $h_1(x)$, however $h_2(x)$ eventually becomes steeper. Both hill functions are *continuous* because there is no step or cliff. See Fig. 2.1.

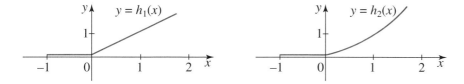

**FIGURE 2.1** ■ The continuous hill functions $h_1(x)$ and $h_2(x)$.

However, *discontinuous* functions, discussed in Section 2.2.2, are certainly an important part of human experience.

Assume that the person is walking on solid ground. The different values of the height is referred to as the *range* of $h$,

$$\mathcal{R} = h\left([x_-, x_+]\right) \equiv \{h(x) \in \mathbb{R} \mid x \in [x_-, x_+]\}.$$

The single-variable concept often used is the *vertical-line test*, which states that over a domain of values where any vertical line touches the graph $y = f(x)$ once is a function on that domain.

**Definition 2.3** The *domain* of a function $f$ is denoted $\mathcal{D}_f$ and its *range* is denoted $\mathcal{R}_f$, also called the *image* of $f$, since $\mathcal{R}_f = f(\mathcal{D}_f)$. Conversely, given the range of a function, the domain is the *preimage* of $f$, since $\mathcal{D}_f = f^{-1}(\mathcal{R}_f)$.

In this text many functions will be presented and used, most of which appear naturally as part of scientific investigation. One family of functions, already discussed in Section 1.2.4, are the $n$th degree polynomials $p(x)$ as defined in Eq. (1.18), which have domain $\mathcal{D}_p = \mathbb{C}$. Such functions will be called *entire* although a precise definition of this term must wait until Chapter 5. The range $\mathcal{R}_p$ of a nonconstant polynomial $p(x)$ is also all of $\mathbb{C}$, since $\forall c \in \mathbb{C}$, the equation $p(z) = c$ is the same as $p(z) - c = 0$, where $p(z) - c$ is another polynomial, and must have at least one zero in $\mathbb{C}$ by the Fundamental Theorem of Algebra.

**Definition 2.4** A function of one variable $f(x)$ is called continuous, if to draw the graph in the $x$–$y$ plane so that $y = f(x)$, it is sufficient to use a single stroke of a pen, without leaving the page, while constantly moving from left to right.

This is a qualitative definition of the concept of continuity, however it should be somewhat intuitive. Polynomial functions are continuous at all points, and are referred to as *entire* functions. Human experience appears continuous for the most part, and so one is typically startled by new events that appear without warning. In other words, discontinuous events are unsettling. However, cliffs, transistors, graduations, and table legs all have discontinuities that have to be dealt with in everyday life. Thus one must be prepared to model continuous and discontinuous functions.

**Definition 2.5** An *operator* on functions with domain $\mathcal{D}$ gives a new function, sometimes with a smaller domain. A *linear* operator $L$ has the property that for functions $f$ and $g$, and constants $a, b$

$$L[a \cdot f + b \cdot g](x) = a \cdot L[f](x) + b \cdot L[g](x). \tag{2.7}$$

A function $h(x)$ with domain $\mathcal{D}$ defines a linear operator $L_h$ by the action of multiplication $L_h[f](x) \equiv h(x) \cdot f(x)$.

In the following list of functions, one also has a list of linear operators, using the preceding definition.

### ■ 2.2.1 Constant Functions

The constant functions $c(x) = a_0 \in \mathbb{R}$ are entire, and the simplest observed in nature. The zero function $0(x) \equiv 0$ in particular appears quite often, as well as the unit function $1(x) \equiv 1$. As operators, zero annihilates every bounded function, and the unit preserves functions,

$$L_0[f](x) = 0 \cdot f(x) = 0(x), \ L_1[f](x) = 1 \cdot f(x) = f(x). \tag{2.8}$$

The domain of a constant function can be any conceivable set, but the range is just the single value $\mathcal{R} = \{a_0\}$. Graphically, these functions are simply horizontal lines.

It is quite natural to think of any constant as being a parameter, rather than a dependent variable or function. Entities like the acceleration due to gravitation, or the color of one's hair may seem to be constant. However, gradual variations, which actually exist, can be hard to detect. Taking an average over time or space gives a good approximation to the true nature of the variable of interest, if little variability is present.

### ■ 2.2.2 Characteristic Functions

Typically one is interested in a domain $\mathcal{D}$ on which many functions will be defined. If $\mathcal{A} \subset \mathcal{D}$ then the characteristic function of $\mathcal{A}$ is defined in terms of cases

$$\chi_{\mathcal{A}} \equiv \begin{cases} 1 & \text{if } x \in \mathcal{A} \\ 0 & \text{if } x \in \mathcal{D}\backslash\mathcal{A} \subseteq \mathcal{A}^c \end{cases} \tag{2.9}$$

Here the complement $\mathcal{A}^c$ may refer to a larger set than the originally intended domain $\mathcal{D}$, and this allows for a natural extension of $\chi_{\mathcal{A}}$ to larger domains $\mathcal{D}' \supset \mathcal{D}$. Even if a characteristic function is extended to $\mathbb{C}$, it is typically not in the class of entire functions, which must at least be continuous. Two exceptions are

$$\chi_{\emptyset}(x) = 0(x), \text{ and } \chi_{\mathcal{D}}(x) = 1(x). \tag{2.10}$$

Characteristic functions have the range $\mathcal{R} = \{0,1\}$ and thus have the *idempotent* property that

$$\chi_{\mathcal{A}}^2(x) = \chi_{\mathcal{A}}(x) \cdot \chi_{\mathcal{A}}(x) = \chi_{\mathcal{A}}(x), \tag{2.11}$$

which holds for the zero function $0(x)$ and the unit function $1(x)$. Idempotency is a property of *projection* operators $P$ where $P[P[f]](x) = P[f](x)$.[4] Thus one observes that

$$L_{\chi_{\mathcal{A}}}\left[L_{\chi_{\mathcal{A}}}[f]\right](x) = \chi_{\mathcal{A}}(x) \cdot L_{\chi_{\mathcal{A}}}[f](x) = \chi_{\mathcal{A}}^2(x) \cdot f(x) = \chi_{\mathcal{A}}(x) \cdot f(x). \tag{2.12}$$

Skillful use of characteristic functions allows for easy handling of case functions, as in Eq. (2.6),

$$h_1(x) = x \cdot \chi_{\mathbb{R}^+}(x), \quad h_2(x) = x^2 \cdot \chi_{\mathbb{R}^+}(x).$$

Very complicated functions, with many discontinuities, can be expressed as sums of characteristic functions multiplied by continuous functions. An important procedure in defining such functions is to decompose a domain $\mathcal{D}$ into a *partition* $\mathcal{P}$, which is a set of sets $\mathcal{P} = \{\mathcal{A}_j\}_{j \in \mathcal{I}}$, so that

$$\mathcal{A}_j \subset \mathcal{D}, \quad \mathcal{A}_j \cap \mathcal{A}_k = \emptyset \text{ for } j \neq k, \quad \bigcup_{j \in \mathcal{I}} \mathcal{A}_j = \mathcal{D},$$

and where $\mathcal{I}$ is an indexing set. An example is the staircase function:

$$Stairs(x) \equiv \lfloor x \rfloor = \sum_{j=-\infty}^{\infty} j \cdot \chi_{[j,j+1)}(x), \tag{2.13}$$

also called the *greatest integer less than or equal to $x$* function (or simply *greatest integer* function), and is also known as the *floor* function. The function

$$Sawtooth(x) \equiv x - \lfloor x \rfloor = \sum_{n=-\infty}^{\infty} (x - n) \cdot \chi_{[n,n+1)}(x), \tag{2.14}$$

is bounded and periodic.

---

[4] The topic of projections is an important concept in this text.

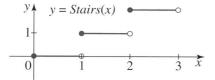 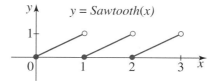

**FIGURE 2.2** ■ The $Stairs(x)$ and $Sawtooth(x)$ functions.

An alternative *stairs* function to the $\lfloor x \rfloor$ is the *ceiling* function $\lceil x \rceil$, or the *least integer, greater-than x* function. The two functions are related by

$$\lceil x \rceil = 1 - \lfloor 1 - x \rfloor. \tag{2.15}$$

### ■ 2.2.3 Linear Functions

The linear function

$$f(x) = a_1 \cdot x + a_0 = m \cdot x + r,$$

varies as $x$ changes, if $a_1 = m \neq 0$. Since constant functions are considered linear, this is a generalization. The graph of a linear function can never be a vertical line $x = b_0 \in \mathbb{R}$. Graphically, linear functions appear to be rotated constant functions, and the angle of rotation $\theta$ is connected with the slope $m$. For example,

$$\theta = 0° \implies m = 0, \; \theta = 45° \implies m = 1, \; \theta = \pm 90° \implies m = \pm\infty,$$
$$\theta = 135° \text{ or } \theta = -45° \implies m = -1. \tag{2.16}$$

A function that is nearly linear can be approximated using the *Mean Value Theorem*, studied in Section 5.13.

A linear function can define a line in the $x$–$y$ plane if the assignment $y = f(x)$ is made. Three forms of a line $\ell$ will be used:

**Slope-Intercept:** $y = m \cdot x + r$ where $m$ = slope, $r$ = $y$-intercept;
**Point-Slope:** $y = m \cdot (x - x_*) + y_*$ where $m$ = slope, $\langle x_*, y_* \rangle$-point on the line; and
**Standard:** $0 = a_2 \cdot y + a_1 \cdot x + a_0$ where $a_2, a_1$ = coefficients, $a_0$ = constant.

Any line $\ell$ is uniquely defined by a point and a direction. In two dimensions the slope $m$ is related to the angle a line makes with the horizon $y = 0$. Given two points on the line $\ell : \langle x_1, y_1 \rangle$ and $\langle x_2, y_2 \rangle$, the slope is determined using the *difference quotient*

$$\text{slope of } \ell \equiv \frac{\Delta y}{\Delta x} = \frac{\text{rise}}{\text{run}} = \frac{y_2 - y_1}{x_2 - x_1}.$$

A line is a continuous function with the property that the slope is the same regardless of which pair of points are chosen to define it, unless the same point is used. However, consider the example of $y = f(x) = 5 \cdot x - 2$, which passes through the point $\langle 1, 3 \rangle$. Then

$$\text{slope} \equiv \frac{\Delta y}{\Delta x} = \frac{y(x) - 3}{x - 1} = \frac{(5 \cdot x - 2) - 3}{x - 1} = \frac{5 \cdot (x - 1)}{x - 1} = 5,$$

independent of $x$. Thus the slope function of $y = 5 \cdot x - 2$ is $y' = 5$.

### ■ 2.2.4 Power Functions

A *power* function will have the form $q(x) = a_* \cdot x^r$, for some $a_*, r \in \mathbb{R}$. For $x > 0$, the definition is clear for any $r \in \mathbb{Z}$, and for $r \in \mathbb{Q}$ the exponent $r$ can be written as $r = n/m$. Then it must be recalled that

$$x^{n/m} = \sqrt[m]{x^n} = \left( \sqrt[m]{x} \right)^n,$$

for it to be clear how to compute this power function. Finally, for an irrational exponent one can choose a rational approximation to $r$ to get an approximate power function. This is the best that computers can do when plotting power functions.

For $x < 0$ there are possibly many acceptable meanings of $q(x)$ whose values are in the complex plane.

### ■ 2.2.5 Nonlinear Polynomials

Any polynomial $y = p(x)$ of degree 2 or greater is called *nonlinear*. About a reference point $(x_*, y_*)$ a polynomial can be expressed as

$$y = p(x) = y_* + m \cdot (x - x_*) + \mathcal{O}\left[(x - x_*)^2\right], \qquad (2.17)$$

where the remainder $\mathcal{O} = p(x) - y_* - m \cdot (x - x_*)$ vanishes quadratically in $(x - x_*)$.

Functions found in nature are expected to have some nonlinear aspects. A deviation from linearity can be due to noise or measurement error, or it can be a property of the system being studied. It is observed that when a rock is dropped off a cliff of height $h_0$, the distance traveled during the first 2 seconds is about 4 times the distance traveled in the first second. The vertical position of the rock $h(t)$ can be fairly accurately modeled by the function

$$h(t) = -g \cdot t^2 + h_0, \ \mathcal{D} = [0, 2], \ \mathcal{R} = [h_0 - 4g, h_0], \tag{2.18}$$

where the time variable $t$ is measured in seconds (s), and the height $h$ is in meters (m). The cliff's height $h_0$ and the acceleration due to gravity is $g = 9.8$ m/s$^2$ are parameters of the system. When convenient, the parameters are set to $h_0 = 0$ m, and sometimes $g \simeq 10$ m/s$^2$ is used for a quick computation. For greater accuracy, one needs to realize that $g$ actually varies by about $\pm 1\%$ over the surface of the Earth. Thus there is a dependence in this experiment on the latitude and longitude. There is also a theoretical dependence of $g$ on altitude.

### ■ 2.2.6 Exponentials

Consider an entire function of the form $f(x) = A_0 \cdot b^x$. Here $b \in \mathbb{R}^+$ is called the *base* and $A_0 \in \mathbb{R}$ the *amplitude* or *initial value* since $f(0) = A_0$. The name, *exponential*, derives from the fact that the independent variable appears in the *exponent*. Although distinct from the power functions, to actually work with exponentials one often approximates $b$ with a rational, and computes $f(x)$ only for $x \in \mathbb{Q}$, using the assumption of continuity to fill in the missing values. One can convert to different bases, and three common choices are $b \in \{2, e, 10\}$. Then

$$b^x = 2^{\alpha_2 \cdot x} = e^{\alpha_e \cdot x} = 10^{\alpha_{10} \cdot x}, \tag{2.19}$$

where the different conversion parameters $\alpha_*$ must satisfy

$$b = 2^{\alpha_2} = e^{\alpha_e} = 10^{\alpha_{10}}. \tag{2.20}$$

Of particular importance is $\alpha_2$. If $b \in (0, 1)$ then $\alpha_2 < 0$ and $\tau = -1/\alpha_2$ is called the *half-life* of the function $f(x)$. Alternatively, if $b > 1$ then $\alpha_2 > 0$ and $\tau = 1/\alpha_2$ is called the *doubling time* of the function $f(x)$.

There are a few basic properties that make this function useful and common. Suppose $b > 1$. Then

$$b^{-\infty} = 0, \ b^0 = 1, \ b^{\infty} = \infty,$$
$$b^x \cdot b^y = b^{x+y}, \ b^x/b^y = b^{x-y}, \ \text{and} \ 1/b^y = b^{-y}.$$

An additional rule, which follows from the preceding, is $(b^x)^n = b^{x \cdot n}$.

**Definition 2.6** For $b > 1$ the mapping $\phi_b : \langle \mathbb{R}, + \rangle \to \langle \mathbb{R}^+, \cdot \rangle$ where $\phi_b(x) \equiv b^x$ defines a group *isomorphism*. In particular $\phi_b$ is a bijection, since $b^x$ is a monotone-increasing function, and it satisfies the *homomorphism* property

$$\phi_b(x+y) = \phi_b(x) \cdot \phi_b(y).$$

This mapping can be extended to an isomorphism on the domain $\mathbb{C}$ depending on the choice of range, also called a *branch*.

Another important property of the exponential function is that

$$A_0 \cdot b^{x+\epsilon}/(A_0 \cdot b^{\epsilon}) = b^x, \tag{2.21}$$

where $\epsilon$ is any number. One way of understanding this expression is to realize that exponential functions are invariant or forgetful of the past values $b^{\epsilon}$.

### ■ 2.2.7 Hyperbolic Trigonometric Functions

The even and odd parts of the exponential function are known as *hyperbolic cosine* and *sine* functions,

$$\cosh(x) \equiv \frac{e^x + e^{-x}}{2}, \ \sinh(x) \equiv \frac{e^x - e^{-x}}{2}, \tag{2.22}$$

where $\cosh(-x) = \cosh(x)$ and $\sinh(-x) = -\sinh(x)$. Special values are $\cosh(0) = 1$ and $\sinh(0) = 0$. For large $x$ both functions behave as $e^x/2$. For $x \ll 0$ behavior is of opposite sign, where $\cosh(x) \simeq e^{-x}/2$ but $\sinh(x) \simeq -e^{-x}/2$. One can easily verify that

$$\cosh^2(x) - \sinh^2(x) = 1. \tag{2.23}$$

### ■ 2.2.8 Trigonometric Functions

In many subareas of engineering, understanding *trigonometry*, the geometry of triangles, is essential. The Pythagorean theorem allows for the accurate construction of a right angle within a triangle. The example of $\triangle(3, 4, 5)$ was discussed in Chapter 1.

Now suppose that a right triangle $\triangle(a, b, c)$ to be studied, is scaled down so that the hypotenuse has unit length $\triangle(C, S, 1)$ as in Fig. 2.3. Then the squares of the other two sides equals 1, i.e., $C^2 + S^2 = 1$. All right triangles are similar to a right triangle with hypotenuse $= 1$. This suggests studying right triangles inscribed in the unit circle $\mathcal{S}^1$, with vertex at the center, and the hypotenuse being the radius.

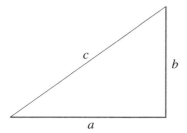
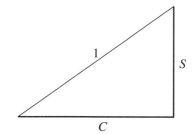

**FIGURE 2.3** ■ Right-angle triangle $\triangle(a,b,c)$ and right-angle triangle with unit hypotenuse $\triangle(C,S,1)$.

The angle of the vertex, denoted as $\theta$, provides a measure for the length along the unit circle, as well as defining the internal angle of the inscribed triangle. In Mesopotamia circa 4000 BC, the circle was divided into 360 equal units, called *degrees*, slightly less than the earth year of 365.242 days. This means that the position of the fixed stars move about 1° per day and allowed the ancients to plan for seeding and harvesting of crops. It is natural to also define the angle $\theta$ as that portion of a circle's perimeter swept out by the triangle. For a circle of radius $r$, with diameter $d = 2r$, the circumference is $2\pi r \equiv \pi d$.

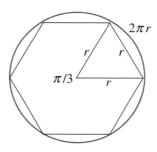

**FIGURE 2.4** ■ Hexagon in a circle.

Thus a full circle is $2\pi$ radii, often just called *radians*. For $\mathcal{S}^1$ a full circle is just $2\pi$. To get a sense of how many radii fit around a circle, inscribe a regular hexagon inside the circle as shown in Fig. 2.4. Then it is clear that there are six equilateral triangles, all with sides $r$. Due to the bending in a circle, $2\pi > 6$ is a lower bound. Now, if the hexagon is increased to just contain the unit circle, then the sides of the equilateral triangle have length $2/\sqrt{3}$, and so $2\pi < 12/\sqrt{3} \simeq 6.93$ is a crude upper bound.[5] Dividing by 2, the following holds:

$$3 < \pi < 22/7 < 6/\sqrt{3}, \qquad (2.24)$$

where $22/7 \simeq 3.143$ is a common approximation, accurate to two decimal places. The arc length of the portion of the circle swept out by angle $\theta$, when expressed

---

[5] At this point it may not be clear why the larger hexagon has a longer perimeter than the circle. This will be discussed in Section 5.6.

in radians, is $r\theta$, and the area is $r^2\theta/2$. The simplicity of these and other formulae compels scientists to use radian measure for angles in trigonometry. However, degrees are still used extensively, because they allow the mind to quickly access the relative sizes of angles. Thus conversion between the two measures is important,

$$x(\text{radians}) = \frac{360 \cdot x}{2\pi}(\text{degrees}) = \left(\frac{180 \cdot x}{\pi}\right)^\circ, \ y(\text{deg}) = y^\circ = \frac{\pi \cdot y}{180}(\text{rad}).$$

However an angle $\theta$ is measured, two natural functions, called the *sine* and the *cosine* of the angle $\theta$, can be defined as

$$\sin(\theta) = S = b/c, \ \cos(\theta) = C = a/c,$$

where $a$, $b$, $c$, $S$, and $C$ are defined in Figure 2.3.

The ranges of the sine and cosine functions are $\mathcal{R} = [-1, 1]$ and the domains are $\mathcal{D} = [0, (\pi/2)\text{rad}] = [0, 90^\circ]$. By the definitions,

$$\sin(0) = 0, \ \sin(\pi/2) = 1, \ \cos(0) = 1, \ \cos(\pi/2) = 0,$$

which should be memorized. By simple geometric considerations of Fig. 2.5,

$$\sin(\pi/4) = \cos(\pi/4) = 1/\sqrt{2}. \tag{2.25}$$

Now, identifying the inscribed triangle with just the point of intersection on the unit circle $\mathcal{S}^1$ allows the extension of the definition of the domains of sine and cosine

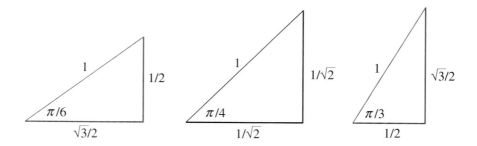

**FIGURE 2.5** ■ Special right-angle triangles: $\triangle(\sqrt{3}/2, 1/2, 1)$, $\triangle(\sqrt{2}/2, \sqrt{2}/2, 1)$, and $\triangle(1/2, \sqrt{3}/2, 1)$.

to all of ℝ. Since tracing the point by $\mathcal{S}^1$ by $2\pi$ radians returns to the same point, sine and cosine are *periodic*, with $period = T = 2\pi$,

$$\sin(\theta + 2\pi) = \sin(\theta), \ \cos(\theta + 2\pi) = \cos(\theta). \tag{2.26}$$

More generally it can be shown that for $\theta, \phi \in \mathbb{R}$

$$\sin(\theta + \phi) = \sin(\theta) \cdot \cos(\phi) + \cos(\theta) \cdot \sin(\phi), \tag{2.27}$$
$$\cos(\theta + \phi) = \cos(\theta) \cdot \cos(\phi) - \sin(\theta) \cdot \sin(\phi), \tag{2.28}$$

which are called *double-angle formulas*. Starting at a value of $\theta$ where sine or cosine is a maximum of 1, the value of $T$ is the minimal amount that $\theta$ has to increase until the maximum of 1 is reached again. The *natural frequency* $= f = T^{-1} = 1/(2\pi)$ is measured by computing the ratio

$$f \equiv \frac{\#\text{maxima from 0 to } \theta}{\text{large value of } \theta} = \frac{1}{T}.$$

A sinusoidal function is defined to be of the form

$$S(t) = A \cdot \sin(\omega t) + B \cdot \cos(\omega t) = C \cdot \sin(\omega t + \phi_0),$$

where the parameters $A, B, C, \phi_0$, and $\omega$ are related by

$$C = \sqrt{A^2 + B^2}, \ \cos(\phi_0) = A/C, \ \sin(\phi_0) = B/C.$$

The quantity $\omega$ is called the *angular frequency* (measured in radians), and is related to the natural frequency $f$ (measured in *cycles*) and period $T$ (measured in reciprocal units of time or distance) by

$$f = \frac{1}{T} = \frac{\omega}{2\pi}, \ \omega = 2\pi \cdot f = \frac{2\pi}{T}. \tag{2.29}$$

The angular frequency is often used in mathematics and theoretical science, but the natural frequency must be used in observational science and engineering. Thus the conversion formula in Eq. (2.29) is important to remember.

## Exercises

1. Use the special triangles to compute the following:

   (a) $\cos(30°), \cos^2(30°)$.
   (b) $\sin(30°), \sin^2(30°)$.
   (c) $\cos^2(30°) + \sin^2(30°)$.
   (d) $\sec(30°) = \dfrac{1}{\cos(30°)}, \sec^2(30°)$.
   (e) $\tan(30°) = \dfrac{\sin(30°)}{\cos(30°)}, \tan^2(30°)$.
   (f) $\sec^2(30°) + \tan^2(30°)$.

2. Draw a large unit circle with abscissa and ordinate. Label all the intersection points $(1, 0), (0, 1), (-1, 0), \ldots$.
   Rotate by $45° = \pi/4$ rad and draw the abscissa and ordinal again. Label all the intersection points $(1, 0), (1/\sqrt{2}, 1\sqrt{2}), (0, 1), \ldots$.
   Finally, identify every $30° = \pi/6$ rad point on the unit circle $(\sqrt{3}/2, 1/2), (1/2, \sqrt{3}/2), (0, 1), \ldots$.

3. Verify Eq. (2.23) by

   (a) computing and simplifying $\cosh(x) + \sinh(x)$,
   (b) computing and simplifying $\cosh(x) - \sinh(x)$,
   (c) computing and simplifying $(\cosh(x) + \sinh(x)) \cdot (\cosh(x) - \sinh(x))$, and
   (d) obtaining an alternate expression using $(a + b) \cdot (a - b) = a^2 - b^2$.

4. Obtain an approximation of $\pi$ using 12 isosceles triangles within a *dodecagon* (12-sided regular polygon), and properties of the special triangles. Where does this approximation fit into Eq. (2.24)?

## 2.3 Euler's Formula

A crowning mathematical achievement of the 18th century was the discovery of *Euler's formula*

$$e^{i\theta} = \cos(\theta) + i \cdot \sin(\theta). \tag{2.30}$$

One consequence is that trigonometric identities follow from the properties of the exponential function. To obtain the first and most fundamental property, take the complex conjugate of Eq. (2.30) to obtain

$$\left(e^{i\theta}\right)^* = e^{-i\theta} = \cos(\theta) - i \cdot \sin(\theta). \tag{2.31}$$

This implies $e^{-i\theta} \cdot e^{i\theta} = e^{-i\theta+i\theta} = e^0 = 1$. Then, clearly $e^{-i\theta} = 1/e^{i\theta}$, so multiplication of Eq. (2.30) and Eq. (2.31) gives $1 = \cos^2(\theta) + \sin^2(\theta)$, which is another form of Pythagorean's theorem. Taking the real and imaginary parts of Eq. (2.31) gives

$$\cos(-\theta) = \cos(\theta), \quad \sin(-\theta) = -\sin(\theta), \tag{2.32}$$

which shows that $\cos(\theta)$ is an even function of $\theta$ whereas $\sin(\theta)$ is an odd function of $\theta$.[6] Adding and subtracting Eq. (2.30) and Eq. (2.31) gives

$$\cos(\theta) = \frac{\exp[i\theta] + \exp[-i\theta]}{2}, \quad \sin(\theta) = \frac{\exp[i\theta] - \exp[-i\theta]}{2 \cdot i}, \tag{2.33}$$

which is the complex-argument version of Eq. (2.22).

### ■ Exercises

1. Find an expression for $\lfloor x \rfloor$ in terms of the function $\lceil \cdot \rceil$. Also, express the difference $\lceil x \rceil - \lfloor x \rfloor$ using a characteristic function.
2. Use Euler's formula to obtain the double-angle formulas in Eq. (2.27) and Eq. (2.28).
3. Prove the identity $\sin(3\,\theta) = 3 \cdot \sin(\theta) - 4 \cdot \sin^3(\theta)$.

### ■ 2.4 Function Rings, Algebras, Ideals, and Fields

Sets of functions can be used to construct extended sets of functions. For example, consider a family of functions $\mathcal{F}$ that are defined on the same domain $\mathcal{D}$ and the

---

[6] An even function is symmetric about the ordinate, whereas an odd function is symmetric about the origin.

same codomain $\mathcal{R}$. If $\mathcal{R}$ is a ring, then the family can be closed under addition and multiplication, defining the new set of functions to be

$$\overline{\mathcal{F}}(\mathcal{D};\mathcal{R}) \equiv \left\{ \sum_{j=1}^{n} \prod_{k=1}^{m} f_{j,k} : \mathcal{D} \to \mathcal{R} \;\middle|\; \forall f_{j,k} \in \mathcal{F},\; \forall n,m \in \mathbb{N} \right\}, \qquad (2.34)$$

where $\overline{\mathcal{F}}$ is the *closure* of $\mathcal{F}$. The space $\langle \overline{\mathcal{F}}(\mathcal{D};\mathcal{R}), +, \cdot \rangle$ is a ring with the multiplicative identity function $1(x) \equiv 1$, and with the additive identity being the zero function $0(x) \equiv 0$, where the additive inverse of $f$ is simply $-f$.

Next, suppose that the codomain $\mathbb{F}$ is a vector space as well as a ring. Then the space can be closed under linear combinations,

$$\overline{\mathcal{F}}(\mathcal{D};\mathbb{F}) \equiv \left\{ \sum_{j=1}^{n} a_j \cdot \prod_{k=1}^{m} f_{j,k} : \mathcal{D} \to \mathbb{F} \;\middle|\; \forall f_{j,k} \in \mathcal{F},\; \forall a_j \in \mathbb{F},\; \forall n,m \in \mathbb{N} \right\}. \qquad (2.35)$$

Such a space is called an *Algebra of Functions* over $\mathbb{F}$ generated by $\mathcal{F}$ [30], and this will be the most common type of space considered in this text. In particular, the set $\overline{\mathcal{F}}(\mathcal{D};\mathbb{F})$ is a vector space. Since the product of functions will often create new, linearly-independent functions, the dimension of $\overline{\mathcal{F}}(\mathcal{D};\mathbb{F})$ will typically be infinite. Furthermore, there are subspaces $\mathcal{I} \subset \overline{\mathcal{F}}(\mathcal{D};\mathbb{F})$, which are invariant under products, and these are called *functional ideals*;

$$\left( (\forall f,g \in \mathcal{I}) \implies (f+g \in \mathcal{I}) \right) \wedge \left( (\forall f \in \mathcal{I}, d \in \overline{\mathcal{F}}) \implies (f \cdot d \in \mathcal{I}) \right). \qquad (2.36)$$

The prime numbers $\mathbb{P}$ generate an infinite number of ideals over the ring of integers $\mathbb{Z}$ in that if $p \in \mathbb{P}$ then $\mathcal{I} \equiv p \cdot \mathbb{Z}$ satisfies the properties in Eq. (2.36). Ideals will be important in our development of calculus. In particular, if $x$ is a variable close to a fixed point $x_0 \in \mathbb{R}$, then the $n^{th}$-*High Order Terms* are expressed

$$HOT_n(x - x_0) = C(x) \cdot (x - x_0)^n = \mathcal{O}\left[(x - x_0)^n\right], \qquad (2.37)$$

where $C(x)$ is a continuous and bounded function at $x_0$, and near $x_0$. The set of $HOT_n$ functions forms an ideal.

Finally, fields of functions can be defined. A good example is the set of all functions $\phi : \mathbb{C} \to \mathbb{C} \cup \{\infty\}$ where $\phi(z) = 0$ and $\phi(z) = 1/0 \equiv \infty$ at only isolated complex numbers, to a finite degree, and has a convergent power series at all other points. Such functions are called *meromorphic* [41], and their reciprocals $1/\phi(z)$ have the same properties. Then the ratio $\phi(z)/\phi(z)$ makes sense on all of $\mathbb{C}$, and equals 1 at all except possibly for an isolated set of $z$. By continuous extension, $(\phi/\phi)(z) \equiv 1$ on $\mathbb{C}$.

## ■ Exercises

1. The set of polynomials $\mathcal{P}$ in Eq. (1.5) has the ideals $\mathcal{I}_n \equiv x^n \cdot \mathcal{P}$. Show that for each $x_0 \in \mathbb{R}$, the sets $\mathcal{I}_n(x_0) \equiv (x - x_0)^n \cdot \mathcal{P}$ are ideals, by briefly explaining why $\mathcal{P}$ is a ring, an algebra, and an ideal over $\mathcal{P}$.

2. Consider the family of decaying exponentials $\mathcal{F} \equiv \{e^{-\lambda t} | \lambda > 0, t \in \mathbb{R}_0^+\}$. Generate the algebra $\overline{\mathcal{F}}$ under $\mathbb{R}$, and show that each $e^{-\lambda_0 t}$ defines an ideal.

## ■ 2.5 Reciprocal Functions

Function rings are very natural spaces. However, function fields are not easy to define because whenever a function $f : \mathcal{D} \to \mathbb{F}$ has a zero in $\mathcal{D}$, its reciprocal cannot have a value in $\mathbb{F}$. Hence the process of dividing by a function has to be performed carefully.

### ■ 2.5.1 Rational Functions

The set of rational functions $\mathcal{R}$ is defined to be all polynomials divided by another polynomial $p(x)/q(x)$. Since polynomials are entire, the only problem with division occurs at the points where the denominator vanishes. In spite of this, the space $\langle \mathcal{R}, +, \cdot \rangle$ forms a field with $0(x)$ and $1(x)$ as the respective identities. There are four basic behaviors that need to be identified:

- The isolated points where division by 0 occurs are called *vertical asymptotes*, but in algebraic computations, they can be ignored.
- When $deg(p) < deg(q)$ then $y = 0$ is called a *horizontal asymptote*.
- Conversely, if $deg(p) > deg(q)$ then the rational function $|p(x)/q(x)|$ grows unboundedly for large $x \in \mathbb{C}$.
- If $deg(p) = deg(q)$ then the horizontal asymptote will be a nonzero constant. In fact, the horizontal asymptote is the same value in all directions. This is a failing of rational functions in modeling different behaviors at $\pm \infty$.

### ■ 2.5.2 Rational Exponentials and Hyperbolic Trigonometrics

A ratio of exponential functions will have two values at the extremes of the domain

$$\frac{A \cdot e^x + B \cdot e^{-x}}{C \cdot e^x + D \cdot e^{-x}} = \frac{A + B \cdot e^{-2 \cdot x}}{C + D \cdot e^{-2 \cdot x}} = \begin{cases} A/C \text{ for } x \to +\infty, \\ B/D \text{ for } x \to -\infty. \end{cases} \quad (2.38)$$

It is sometimes convenient to define the hyperbolic tangent and secant functions,

$$\tanh(x) = \frac{\sinh(x)}{\cosh(x)}, \ \text{sech}(x) = \frac{1}{\cosh(x)}. \tag{2.39}$$

These are bounded functions that become flat as $x \to \pm\infty$. Less commonly used are the hyperbolic cotangent and cosecant functions,

$$\coth(x) = \frac{\cosh(x)}{\sinh(x)}, \ \text{csch}(x) = \frac{1}{\sinh(x)}, \tag{2.40}$$

which have vertical asymptotes at $x = 0$, as well as horizontal asymptotes.

### ■ 2.5.3 The Six Trigonometric Functions

The ratio of the sine and cosine functions gives the tangent function,

$$\tan(\theta) \equiv \frac{\sin(\theta)}{\cos(\theta)} = \frac{b}{a}. \tag{2.41}$$

If a line makes an angle $\theta$ with the horizontal, then the slope of the line is $m = \tan(\theta)$. The reciprocal of the three trigonometric functions sine, cosine, and tangent, are cosecant, secant, and cotangent, where from Fig. 2.3

$$\csc(\theta) \equiv \frac{1}{\sin(\theta)} = \frac{c}{b}, \ \sec(\theta) \equiv \frac{1}{\cos(\theta)} = \frac{c}{a}, \ \cot(\theta) \equiv \frac{\cos(\theta)}{\sin(\theta)} = \frac{a}{b},$$

respectively. Dividing the Pythagorean theorem by $\sin^2(\theta)$ or $\cos^2(\theta)$ gives two more identities

$$\csc^2(\theta) = 1 + \cot^2(\theta), \ \sec^2(\theta) = \tan^2(\theta) + 1.$$

### ■ Exercises

1. Prove that $\tanh^2(x) + \text{sech}^2(x) = 1$ and that $\tan^2(x) + 1 = \sec^2(x)$. What are the corresponding identities for $\coth(x)$, $\text{csch}(x)$, $\cot(x)$, and $\csc(x)$?

## 2.6 A Catalogue of Inverse Functions

The preceding functions discussed will now be considered in terms of their invertibility with respect to composition. In many situations, this is synonymous with the question of solvability for a corresponding equation. The single-variable concept that is important here is the horizontal-line test, which states that over a range of values where any horizontal line touches the graph $y = f(x)$ once, $f$ is invertible on that range.

**Definition 2.7** For $\mathcal{D} \subset \mathbb{R}$ a function $f : \mathcal{D} \to \mathbb{R}$ is called *increasing* on an interval $\mathcal{I}$ if
$$(\forall x, y \in \mathcal{I} \cap \mathcal{D} \ni (x < y)) \implies (f(x) < f(y)).$$
A function $f$ is *decreasing* if $-f$ is increasing.

A function that is increasing or decreasing is invertible since it must always be taking different values. Thus if $f$ is increasing on $\mathcal{I} \cap \mathcal{D}$, then a function $f^{-1} : f(\mathcal{I} \cap \mathcal{D}) \to \mathbb{R}$ can be defined so that $f^{-1}[f(x)] = x$ and where $f[f^{-1}(y)] = y$.

### 2.6.1 Constant Functions

A constant $c(x) = a_0$ function with domain $\mathcal{D}_c$ is not invertible if $\mathcal{D}_c$ contains two or more points. It does not satisfy the horizontal-line test. In the process of solving problems one is lead to two types of equations involving constant functions

$$\text{tautology: } a_0 = a_0, \text{ absurdity: } 0 = 1.$$

A *tautology* is a statement that is always true, and thus gives no new information. An *absurdity* is equivalent to a *contradiction* and can be expressed in many different ways, all of which can be converted to the $0 = 1$ statement. The preimage of $c(x)$ on a set $\mathcal{B} \subset$ codomain, can only be one of two entities,

$$c^{-1}(\mathcal{B}) = \begin{cases} \mathcal{D}_c & \text{if } a_0 \in \mathcal{B} \\ \emptyset & \text{if } a_0 \notin \mathcal{B} \end{cases}.$$

### 2.6.2 Characteristic Functions

A characteristic function $\mathcal{X}_\mathcal{A}(x)$ is not invertible if its domain $\mathcal{D}$ has three or more points. The preimage of 0 is $\mathcal{A}^c$, the complement of $\mathcal{A}$, or simply $\mathcal{X}^{-1}(0) = \mathcal{A}^c$.

The preimage of 1 is $\mathcal{A}$, or simply $\mathcal{X}^{-1}(1) = \mathcal{A}$ and is called the *support* of $\mathcal{X}_\mathcal{A}$. If $\mathcal{X}(x)$ is known to take only values of 0 or 1, then finding its support is important to understanding the true nature of this function.

### ■ 2.6.3 Linear Functions

The linear function $f(x) = a_1 \cdot x + a_0$ has the inverse, for $a_1 \neq 0$,

$$f^{-1}(y) = (1/a_1) \cdot y - a_0/a_1, \tag{2.42}$$

which is also a linear function. Let $\iota(x) \equiv x$ denote the *identity function*, where $x$ is any *entity*. One easily computes that

$$f\left[f^{-1}(y)\right] = \iota(y), \;\; f^{-1}\left[f(x)\right] = \iota(x). \tag{2.43}$$

Due to Eq. (2.42), it should be clear that in the process of finding inverses of linear functions, it is convenient to be working with variables $x$ and $y$ that reside in a field, like $\mathbb{Q}, \mathbb{R}$, or $\mathbb{C}$.

### ■ 2.6.4 Root Functions

A root function $q^{-1}(x)$ is just an inverse of a power function $q(x) = a_* \cdot x^r$. Then it is seen that $q^{-1}(y) = \sqrt[r]{y/a_*}$. Typically the $r$th root can have many values, so $q^{-1}$ is sometimes called a multivalued function. Since one often prefers that a function be single-valued, a branch of the root may be singled out in the definition of $q^{-1}$, depending on the application and requirements due to physical considerations.

### ■ 2.6.5 Nonlinear Polynomials

Inverting a polynomial of degree 2 or greater is not typically easy. To use an iterative procedure, one can write the equation $p(x) = y$, and convert to $p_y(x) \equiv p(x) - y = 0$. Then finding $p^{-1}(y)$ is equivalent to solving for a zero $x_*$ of $p_y(x)$, where $y$ is absorbed into the constant term. In this case one may write $x_* = p^{-1}(y)$, where it must be kept in mind that this may be only representing one branch of the inverse function. The result of an iterative procedure, as described in Chapter 1, will depend on the seed, which if $y$ changes with small enough increments, will typically result in a continuous inverse function.

### ■ 2.6.6 Logarithms

The inverse function of $\phi_b(x) = b^x$ for $x \in \mathbb{R}$ is written $\phi_b^{-1}(y) = \log_b(y)$ for $y \in \mathbb{R}^+$, and is called the *logarithm to the base* $b \in \mathbb{R}^+$. See Fig. 2.6 for $b = e$ or $b = 1/e$.

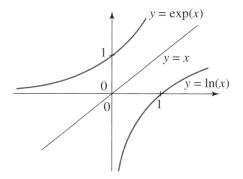 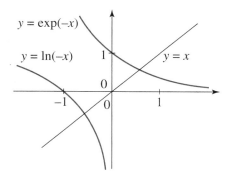

**FIGURE 2.6** ■ Left: $e^x$ and $\ln(x)$ functions; Right: $e^{-x}$ and $\ln(-x)$ functions.

To convert exponential expressions to different bases, as in Eq. (2.19), one needs to solve

$$\alpha_c \equiv 1/\log_b(c) \iff b^x = c^{\alpha_c \cdot x}.$$

Just as all exponential functions can be written to a different base, all logarithms can be expressed to the base of choice.

Two common choices are base 10, for which one often writes $\log = \log_{10}$ and base $e$, where $\ln \equiv \log_e$ is called the *natural logarithm*.[7] Then the property

$$\log_b(x) = \ln(x)/\ln(b),$$

is most valuable for conversion to a different base. As a result of its common use, the number $\ln(2) \simeq 0.693$ is worthy of memorizing. Less common is the number $\ln(10) \simeq 2.303$. If $b \in (0,1)$ then $\tau = -\ln(c)/\ln(b)$ is called the *half-life* of $b^x$. Alternatively, if $b > 1$ then $\tau = \ln(c)/\ln(b)$ is called the *doubling-time* of $b^x$.

There are a few basic properties that make this function useful and common. Suppose $b > 1$. Then

$$\log_b(0) = -\infty, \ \log_b(1) = 0, \ \log_b(\infty) = \infty$$
$$\log_b(x) + \log_b(y) = \log_b(x \cdot y), \ \log_b(x) - \log_b(y) = \log_b(x/y), \ and$$
$$\log_b(1/y) = -\log_b(y).$$

An important rule, which follows from the preceding, is $\log_b(x^n) = n \log_b(x)$.

---

[7] The order of lettering has its origins with the Latin expression *logarithmus naturalis* or the French *logarithmic natural* but might also simply be remembered as *logarithm au natural*.

**Definition 2.8** For $b > 1$ the mapping $\phi_b^{-1} : \langle \mathbb{R}^+, \cdot \rangle \to \langle \mathbb{R}, + \rangle$ where $\phi_b(x) \equiv b^x$ and $\phi_b^{-1}(y) \equiv \log_b(y)$, defines a group *isomorphism*. In particular $\phi_b^{-1}$ is a bijection, since $\log_b(y)$ is a monotone-increasing function, and it satisfies the homomorphism property

$$\phi_b^{-1}(x \cdot y) = \phi_b^{-1}(x) + \phi_b^{-1}(y).$$

This mapping can be extended to an isomorphism defined on certain open subsets of $\mathbb{C}$ that contain $\mathbb{R}^+$ called branches. These domains must all exclude $0 \in \mathbb{C}$ but the ranges are open subsets that do contain 0.

The logarithm functions can be extended to a single-valued function on a domain called the *Riemann surface*, defined as

$$\mathbb{C}_* \equiv \mathbb{R}^+ \cdot e^{i \cdot \mathbb{R}} = \{r \cdot e^{i\theta} \mid r > 0, \theta \in (-\infty, \infty)\}.$$

Then one can define $\ln(r \cdot e^{i\theta}) = \ln(r) + i\theta$ demonstrating that $\ln : \mathbb{C}_* \to \mathbb{C}$ is an isomorphism. However, unless the argument $\theta$ is clearly specified, the logarithm will have a countably infinite number of values. For example,

$$\ln(-1) = i\pi + i2\pi \cdot \mathbb{N} = \{i\pi \cdot (2 \cdot n + 1) \mid n \in \underline{\mathbb{N}}\}. \tag{2.44}$$

### ■ Exercises

1. Note that $\ln(-1)$ is a set, not a number. What is the set $i^i$?
2. Suppose $A, B, C,$ and $D$ are positive in Eq. (2.38). Then rewrite Eq. (2.38) as $E \cosh(x + \alpha)/\cosh(x + \beta)$ by using the expressions $A = e^a$, $B = e^b$, ..., etc.
3. The inverse of the tangent function $\arctan(x) = a(x)$ can be written as $a(x) = i \cdot \ln[b(x)]$ for some complex-valued function $b(x)$. Find this function by using $x = \tan(a(x))$, along with Eq. (2.41) and Eq. (2.33). Use the result to compute the set $\arctan(1)$.

### ■ 2.7 Continuous Functions

To express a function $f : [a, b] \to \mathbb{R}$ one often introduces an independent variable $x \in \mathcal{D} \equiv [a, b]$ and sets the dependent variable $y \equiv f(x)$.

To draw this function a horizontal line segment containing $[a,b]$ is drawn, crossed by a vertical line segment containing $\mathcal{R} \equiv [min(f), max(f)] \subset \mathbb{R}$. A finite sequence of increasing values $\{x_k\}_{k=1}^{N}$ from $[a,b]$, with $x_1 = a$ and $x_N = b$, may be selected, and $\{f(x_k)\}_{k=1}^{N}$ computed, tabulated, and plotted. Then, starting at the point $\langle x_1, f(x_1) \rangle = \langle a, f(a) \rangle$ and ending at $\langle x_N, f(x_N) \rangle = \langle b, f(b) \rangle$ while passing through the points $\langle x_k, f(x_k) \rangle$ one obtains a graphical representation of $f : \mathcal{D} \to \mathcal{R}$, as in Figure 2.7. If this can be done in one smooth stroke, then $f$ is continuous.

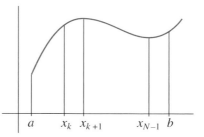

FIGURE 2.7 ■ A continuous function.

**Definition 2.9** The space of continuous functions $f : \mathcal{D} \to \mathcal{R}$ is expressed as $\mathcal{C}^0(\mathcal{D} \to \mathcal{R})$. If the range is unknown, or not specified, then $\mathcal{C}^0(\mathcal{D})$ is used.

A nongraphical method for describing the property of continuity, in order to allow the study of abstract functions, is given as follows:

**Definition 2.10** A function $f : \mathcal{D} \to \mathcal{R}$ is continuous at $x_* \in \mathcal{D}$, if $\forall \epsilon > 0, \exists \delta_\epsilon > 0$ so that

$$(|x - x_*| < \delta_\epsilon \wedge x \in \mathcal{D}) \implies |f(x) - f(x_*)| < \epsilon. \tag{2.45}$$

This is also expressed as

$$\lim_{x \to x_*^{\mathcal{D}}} f(x) = f(x_*), \quad \text{or simply} \quad \lim_{x \to x_*} f(x) = f(x_*),$$

with the understanding that $x \in \mathcal{D}$ is a requirement. If $f$ is continuous for each $x_* \in \mathcal{D}$, then $f \in \mathcal{C}^0(\mathcal{D})$ and one says that $f$ is continuous on $\mathcal{D}$.

The statement in Eq. (2.45) is known as the $\epsilon - \delta$ *criterion for continuity*. It can be tedious to apply to every function, thus a series of properties are easily established.

### ■ Theorem 2.3

1. Linear functions are continuous.
2. Linear combinations of continuous functions are continuous.
3. Products of continuous functions are continuous.
4. Quotients of nonvanishing continuous functions are continuous.
5. Square roots of nonnegative continuous functions are continuous.
6. Compositions of continuous functions are continuous.

The procedure for verifying continuity will be demonstrated in the following example.

### EXAMPLE 2.1

Consider $f(x) = 1/\sqrt{x}$ for $x \geq 1$. Then this function is continuous at $x_* = 4$. Indeed, to find $\delta_\epsilon$, assume that $x \geq 1$ and $|x - 4| < \delta_\epsilon$. Then start with the difference,

$$|f(x) - f(4)| = \left|\frac{1}{\sqrt{x}} - \frac{1}{\sqrt{4}}\right| \tag{2.46}$$

$$= \left|\frac{2 - \sqrt{x}}{2 \cdot \sqrt{x}}\right| \times \left|\frac{2 + \sqrt{x}}{2 + \sqrt{x}}\right| \tag{2.47}$$

$$= \left|\frac{4 - x}{2 \cdot \sqrt{x} \cdot (2 + \sqrt{x})}\right| \leq \frac{1}{6} \cdot |4 - x| < \frac{\delta_\epsilon}{6} \equiv \epsilon, \tag{2.48}$$

where the trick of rationalizing the denominator was used in Eq. (2.47). Then setting $\delta_\epsilon = 6 \cdot \epsilon$ with the caveat that $x \geq 1$, verifies the $\epsilon - \delta$ criterion for continuity of $f(x)$ at $x_* = 4$. To include the domain condition that $x \geq 1$, the alternative definition $\delta_\epsilon \equiv \min\{3, 6 \cdot \epsilon\}$ is often used.

### ■ Exercises

1. Prove that $f(x) = \sqrt{x}$ is continuous at $x = 9$ using the method of Example 2.1.
2. The rational function $r(x) = (x^2 - 4)/(x - 2)$ is not defined at $x = 2$. However, choose a value of $r(2)$ that will make the extended function $r(x)$ continuous at $x = 2$. Use Definition 2.10 to prove continuity at $x = 2$. (*Hint:* Start by considering numbers near $x = 2$.)
3. Consider the function $\text{sgn}(x) = x/|x|$ defined for $x \neq 0$. Explain why no definition for $\text{sgn}(0)$ will make this function continuous, by considering Definition 2.10.

## 2.8 Algebra of Functions

Collections of functions are often considered to be mathematical spaces with the inclusion of operations like addition $f + g$, multiplication $f \cdot g$, and composition $f \circ g$, where,

$$f \circ g(x) = f\left[g(x)\right], \ x \in \mathcal{D}_g \cap g^{-1}(\mathcal{D}_f) \equiv \mathcal{D}_{f \circ g}.$$

Composition is associative, so the expression $f \circ g \circ h(x)$ can have only one meaning. By Theorem 2.3 the space $\mathcal{C}^0(\mathbb{R})$ is an algebra of functions, and for any interval $\mathcal{I}$, so is $\mathcal{C}^0(\mathcal{I} \to \mathcal{I})$.

**Definition 2.11** A function $f : \mathcal{D} \to \mathcal{R}$ is *injective* or *one-to-one* if

$$\forall x_1, x_2 \in \mathcal{D} \implies ((f(x_1) = f(x_2)) \implies (x_1 = x_2)).$$

A function is *surjective* or *onto* if

$$\forall y \in \mathcal{R} \implies ((\exists x \in \mathcal{D}) \ni (f(x) = y)).$$

A function that is one-to-one and onto is called a *bijection*. If $\mathcal{D} = \mathcal{R}$ then a bijection is called an *automorphism* and the set of all such functions is denoted $\mathcal{A}(\mathcal{D})$. If $\#\mathcal{D} \in \mathbb{N}$, then automorphisms are called *permutations*. In this case it is sufficient to consider automorphisms on $\mathbb{Z}_n$ where $\#\mathcal{D} = n$. The set of permutations on $n$ elements is expressed as $S(n) \equiv \mathcal{A}(\mathbb{Z}_n)$.

### ■ Theorem 2.4

For any set $\mathcal{D}$, the space $\langle \mathcal{A}(\mathcal{D}), \circ \rangle$ is a group. The order of the finite permutation group $\langle S(n), \circ \rangle$ is $n!$ as given in Definition 1.9.

**Proof** Every bijection $f : \mathcal{D} \to \mathcal{D}$ has an inverse $f^{-1} : \mathcal{D} \to \mathcal{D}$, which, by the onto property, can be defined $\forall y \in \mathcal{D}$ to be $f^{-1}(y) \equiv x$, where $f(x) = y$, and this implies that

$$y = f(x) = f\left[f^{-1}(y)\right] = f \circ f^{-1}(y). \tag{2.49}$$

There must be only one value of $x \in \mathcal{D}$ where $y = f(x)$, since otherwise it would contradict the one-to-one property. Now that $f^{-1}$ is a well-defined single-valued function, it remains to show that it is also a bijection. The one-to-one property follows from Eq. (2.49) because

$$\forall y_1, y_2 \in \mathcal{D} \implies f^{-1}(y_1) = f^{-1}(y_2) \implies f\left[f^{-1}(y_1)\right] = f\left[f^{-1}(y_2)\right] \implies y_1 = y_2.$$

The onto property follows because $\forall x \in \mathcal{D} \implies f(x) \in \mathcal{D}$, in which case $f^{-1}[f(x)] = x$. One can now conclude that $\forall x, y \in \mathcal{D}$,

$$f^{-1}\left[f(x)\right] = \iota(x), \ f\left[f^{-1}(y)\right] = \iota(y),$$

where $\iota : \mathcal{D} \to \mathcal{D}$ is the identity.

All permutations of $\mathbb{Z}_n$ can be defined by first assigning 0 to some element of $\mathbb{Z}_n$, of which there are $n$ choices. If $n = 1$, then $\iota : \{0\} \to \{0\}$ is the only permutation of $\mathbb{Z}_1$. Now suppose $n \geq 2$. Then 1 has to be assigned to one of the remaining elements of $\mathbb{Z}_n$, of which there are $(n-1)$ choices. Together there are $n \cdot (n-1)$ choices for mapping $\{0, 1\}$ into $\mathbb{Z}_n$. Iterating this counting procedure for the remaining $(n-2)$ numbers verifies the total number of bijections on $\mathbb{Z}_n$ to be $n! = n \cdot (n-1) \cdot (n-2) \ldots (3) \cdot (2) \cdot (1)$. □

## ■ Exercises

**1.** For $A$, $B$, and $k$ positive, find the inverse function of $f(x) = A \cdot e^{-k \cdot x} + B$. Also find the domain of $f^{-1}(x)$. For $A = B = k = 1$ plot both $f(x)$ and $f^{-1}(x)$ along with $y = x$ for $x \in [0, 3]$ as in Fig. 2.6.

## ■ 2.9 Power Series and Products

Many continuous functions, that cannot be expressed as polynomials, or a ratio of polynomials, can be written as an infinite sum or product. Such functions are called *transcendental* and examples are:

$$e^x, \ \ln(x), \ \sin(x), \ \cos(x), \ e^{-x^2}.$$

Since infinite computations cannot be performed in finite time, unless an identity provides a finite expression, one must use truncation and settle for sufficient approximations. These typically involve the use of polynomials, rational functions, characteristic functions, and products of these called *splines*. A common scheme is to identify key points and then interpolate between them using lines or simple curves to construct the approximating function.

The first example is for the exponential function. For any $z = x + i \cdot y \in \mathbb{C}$ define

$$\exp(z) \equiv e^z \simeq 1 + \sum_{j=1}^{M} \frac{z^j}{j!} \simeq \left(1 + \frac{z}{M}\right)^M \quad \text{as } M \to \infty, \tag{2.50}$$

which are worthy of memorization. The exp function is entire since for each $z \in \mathbb{C}$ both approximations converge to the same value in $\mathbb{C}$ as $M \in \mathbb{N}$ gets larger.

The next example is the sine function, which has the two expressions

$$\sin(z) = z + \sum_{j=1}^{\infty} (-1)^j \cdot \frac{z^{2 \cdot j + 1}}{(2 \cdot j + 1)!} = z \cdot \prod_{n=1}^{\infty} \left(1 - \frac{z^2}{\pi^2 \cdot n^2}\right). \tag{2.51}$$

The product expression in Eq. (2.51) is quite simple and reveals the identity that

$$\sin(\pi \cdot n) = 0, \quad \forall n \in \mathbb{Z},$$

and this gives all the zeros of sine. Furthermore, $\sin(z)$ is clearly entire.

The $\Gamma$-function (*Gamma function*) is defined to be

$$\Gamma(z) = \lim_{N \to \infty} \frac{N! \cdot N^z}{z} \prod_{n=1}^{N} \frac{1}{n+z} = \frac{e^{-\gamma \cdot z}}{z} \cdot \prod_{n=1}^{\infty} \frac{e^{z/n}}{1 + z/n}, \tag{2.52}$$

where $\gamma \simeq 0.577$ is the Euler-Mascheroni constant. It is an open mathematical problem as to whether $\gamma$ is irrational, or not. The $\Gamma$-function satisfies the identity $\Gamma(z+1) = z \cdot \Gamma(z)$. However the $\Gamma$-function never vanishes, i.e., has no zeros, since Eq. (2.52) contains no factors in its numerator. This function is not entire, because there are factors in its denominator. The $\Gamma$-function is called *meromorphic* because it fails to be bounded for only isolated points. Indeed, the expressions in Eq. (2.52) are not finite for the negative integers $z = -n$, $\forall n \in \mathbb{N}$. These points are called the *poles* of $\Gamma(z)$. A summation formula is possible for every $z$, except at the poles of $\Gamma(z)$. The product expression demonstrates that

$$\Gamma(z) \cdot \Gamma(1-z) = \frac{\pi}{\sin(\pi \cdot z)}, \tag{2.53}$$

using the identities of Eq. (2.51). Thus the poles of $\Gamma(z)$ and $\Gamma(1-z)$ are connected to the zeros of $\sin(z)$. Note that Eq. (2.52) implies that $\Gamma^2(1/2) = \pi$.

The $\zeta$-function (*Riemann zeta function*) satisfies

$$\zeta(z) = 1 + \sum_{n=2}^{\infty} \frac{1}{n^z} = \prod_{p \in \mathbb{P}} \frac{1}{1 - p^{-z}}, \tag{2.54}$$

and, in these forms, $\zeta(z)$ only converges for $\Re(z) = x > 0$. Indeed, $\zeta(1^+) = +\infty$ because it is the same as the Harmonic series. The product expression follows from the Fundamental Theorem of Arithmetic and the Geometric series formula combined to give Eq. (2.54) called *Euler's identity* [25], not to be confused with Euler's formula in Eq. (2.30). The $\zeta$-function appears in many formulas, like $\zeta(2) = \pi^2/6$. Also, the Euler-Mascheroni constant can be computed using the series

$$\gamma \equiv 1 - \sum_{j=2}^{\infty} \frac{\zeta(j) - 1}{j} \simeq 0.577215665....$$

The *Jacobi theta function* is defined for each parameter $q > 1$ to be

$$\theta(q; z) = \sum_{j=-\infty}^{\infty} \frac{z^j}{q^{j(j-1)/2}} = C_q \cdot \prod_{n=0}^{\infty} \left(1 + \frac{z}{q^n}\right) \cdot \left(1 + \frac{1}{z \cdot q^{n+1}}\right), \tag{2.55}$$

for some $C_q > 0$. Expressions in Eq. (2.55) converge $\forall z \in \mathbb{C}/\{0\}$. The following identities

$$\theta(q; q \cdot z) = q \cdot z \cdot \theta(q; z), \quad \theta(q; 1/(q \cdot z)) = \theta(q; z), \tag{2.56}$$

hold for all $z \neq 0$. The zeros of $\theta(q; z)$ occur for $z \in \{-q^m \mid m \in \mathbb{Z}\}$.

### ■ Exercises

1. Use Eq. (2.52) to show that $\Gamma(0) = \infty$ but that if $z \cdot \Gamma(z)$ is simplified it will approach a finite number as $z \to 0$. Thus, what is $\Gamma(1)$?
2. Use Eq. (2.51) to define the new function $sinc(z) \equiv sin(z)/z$. Use the product formula to draw $sinc(x)$ for $x \in [-2\pi, 2\pi]$.
3. Prove the equality between the sum and product formulas in Eq. (2.54) using a Geometric series for each factor $1/(1 - 1/p^z)$, and the Fundamental Theorem of Arithmetic. Also, briefly explain why $\zeta(z)$ has no zeros, at least according to the definition in Eq. (2.54).
4. Verify Eq. (2.56) using both the sum and product definitions in Eq. (2.55).

## 2.10 Linearly-Independent Spanning Sets of Functions

Various problems in mathematics and science call for a certain type of function with properties such as *linearity, exponentially growing, periodic, continuous, vanishing at $\pm\infty$,* etc. It becomes an important task to identify a sufficiently large family of functions, with the desired characteristics, that will provide a framework to address a problem at hand.

**Definition 2.12** Let $\langle \mathcal{G}(I;\mathbb{R}), + \rangle$ be a group of functions on an interval $I \subseteq \mathbb{R}$. This space extends to the *vector space* of functions

$$\mathcal{V} \equiv \left\{ \sum_{j=1}^{n} c_j \cdot f_j(x) \ \middle| \ c_j \in \mathbb{R}, f_j \in \mathcal{G}, n \in \mathbb{N} \right\}. \tag{2.57}$$

A finite collection of functions $\mathcal{B} \equiv \{e_j(x)\}_{j=1}^{N} \subset \mathcal{V}$ is *linearly independent* if

$$\left( \sum_{j=1}^{n} c_j \cdot e_j(x) = 0 \right) (\forall x \in I) \iff (c_1 = c_2 = \cdots = c_n = 0). \tag{2.58}$$

The elements of $\mathcal{B}$ are called *basis functions* if they form a *spanning set* for $\mathcal{V}$, meaning that

$$(\forall f \in \mathcal{V})\ (\exists \{c_j\}_{j=1}^{N} \subset \mathbb{R}) \quad \ni \quad \left( f(x) = \sum_{j=1}^{N} c_j \cdot e_j(x) \right). \tag{2.59}$$

The *dimension* of $\mathcal{V}$ is $N$, written $dim(\mathcal{V}) = N$.

In the preceding definitions there is a common theme, which is the concept of a *linear combination* of functions. For example, the space of polynomials $\mathcal{P}$ is the set of all linear combinations of integer power functions.

### EXAMPLE 2.2

A basis for all quadratic functions is $\mathcal{B} = \{1, x, x^2\}$. These are linearly independent on any interval $[a, b] \subset \mathbb{R}$ for $a < b$. The spanning set is

$$\mathcal{V} = \{c_1 + c_2 \cdot x + c_3 \cdot x^2 \ | \ c_1, c_2, c_3 \in \mathbb{R}\}. \tag{2.60}$$

The coefficients $\mathcal{C} \equiv \{c_1, c_2, c_3\}$ represent parameters to be determined. Then $dim(\mathcal{V}) = \#\mathcal{B} = 3$.

## Exercises

1. The Legendre basis for cubic polynomials is $\{1, x, (3x^2 - 1)/2, (5x^3 - 3x)/2\}$. Express the polynomial $p(x) = 2x^3 - 4x + 3$ in terms of this basis.

## 2.11 Probability Functions

A *sample space* $\mathcal{S}$ is a set of possible outcomes of a repeatable *experiment*. An *event* $E$ is a subset of the possible outcomes $E \subset \mathcal{S}$. A *probability function* $P$ on $\mathcal{S}$ is a mapping from events $E$ to probabilities $p_E$, expressed as real numbers between 0 and 1, i.e., $P(E) \equiv p_E \in [0, 1]$. A *probability model* for a system requires a sample space and a probability function, and this involves making a choice of parameters. These parameters are either estimated from data collections or computed from theoretical considerations, depending on how a system is to be understood. Regardless of how a probability function is constructed, at its core are the processes of *counting* and *dividing*,

$$P(E) = \frac{\#\{\phi \in E\}}{\#\{\psi \in \mathcal{S}\}},$$

where it is assumed that the outcomes $\phi, \psi, \ldots$ are all equally likely, and that $\#E \leq \#\mathcal{S} < \infty$. If $\#E$ is very large, infinite, or indeterminant[8] then the probability $P(E)$ can be estimated from a finite data sample as part of an experiment.

An example of a well-understood system, which will be used here, is the process of flipping $n$ fair coins simultaneously. The probability of each coin landing heads is $p_H = 0.5$ and of the coin landing tails is $p_T = 0.5$. If one of the coins is unfair, then there is a sufficient number of trials that will show that the model cannot explain the results of experiments. The decision to adopt a model for a system requires testing the system until a sufficient *confidence level* is reached. The amount of experimentation required to achieve a desired level of confidence will be one issue addressed in this text.

If $\mathcal{S}$ is finite or countably infinite, then every event $E \subset \mathcal{S}$ can be assigned a probability. This is done under the following conditions:

**Definition 2.13** The *axioms* or *rules of probability* on a sample space $\mathcal{S}$, for arbitrary events $E, E', E'', \ldots, E_i, E_j, \cdots \subset \mathcal{S}$, and indexing set $\mathcal{N} \subset \mathbb{N}$ are

- **Probability range:** $P(E) \in [0, 1]$, or simply $0 \leq P(E) \leq 1$;

---

[8] For example, the human world population is *large*, the number of directions one can point at is *infinite*, and the number of lightbulbs that a factory will eventually produce is *indeterminant*.

## 2.11 Probability Functions

- **Extreme cases:** $P(\emptyset) = 0, P(\mathcal{S}) = 1$;
- **Disjoint unions:** $(\forall \{E_j\}_{j \in \mathcal{N}} \ni E_j \cap E_k = \emptyset) \implies P(\bigcup_{j \in \mathcal{N}} E_j) = \sum_{j \in \mathcal{N}} P(E_j)$;
- **Complement rules:** $1 = P(\mathcal{S}) = P(E \cup E^c) = P(E) + P(E^c)$,
  $P(E^c) = 1 - P(E)$;
- **Containment rule:** $E \subset E' \implies P(E) \leq P(E')$, or more specifically
  $P(E') = P(E \cup (E' \backslash E)) = P(E) + P(E' \backslash E)$;
- **Addition rule:** $P(E' \cup E'') + P(E' \cap E'') = P(E') + P(E'')$,
  $P(E' \cup E'') = P(E') + P(E'' \backslash E') = P(E') + P(E'') - P(E' \cap E'')$;
- **Product rule:** If statements $\phi$ and $\psi$ are independent,
  then $P(\phi \wedge \psi) = P(\phi) \cdot P(\psi)$.

Given the partial knowledge that $E'$ has occurred, where $P(E') \neq 0$, the conditional probability that $E$ also occurs is defined to be

$$P(E|E') \equiv P(E \cap E')/P(E').$$

These definitions and observations will be familiarized with Example 2.3;

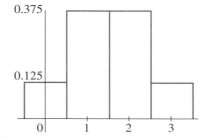

**FIGURE 2.8** ■ Probability distribution of tails $T$ for three fair coins.

### EXAMPLE 2.3

Consider the experiment of flipping three coins simultaneously, and recording the number of heads $H$, and tails $T$ that appear. Keeping track of which coin produced which result gives $2 \cdot 2 \cdot 2 = 2^3 = 8$ possible outcomes. However, ignoring which coin produced which result, gives the sample space

$$\mathcal{S} \equiv \{3H0T, 2H1T, 1H2T, 0H3T\}.$$

The probability distribution of tails $T$ in Figure 2.8 assumes that the three coins are fair, giving

$$P(3H0T) = 1/8, \ P(2H1T) = 3/8, \ P(1H2T) = 3/8, \ P(0H3T) = 1/8.$$

Consider the events $E \equiv \{\# \text{ H is even}\}$, $E' \equiv \{\# \text{ T is not } 0\}$. Then $P(E) = 1/2$, $P(E') = 7/8$. Furthermore, $P(E|E') = (3/8 + 1/8)/(7/8) = 4/7$.

In the situation where $\mathcal{S}$ is uncountable, like $\mathcal{S} = \mathbb{R}$, a $\sigma$-algebra of events must be used to define the action of $P$ on $\mathcal{S}$, in order to avoid contradictions.

### ■ 2.11.1 Combinations and Permutations

Consider the set of $n$ elements $\mathbb{Z}_n \equiv \{0, 1, 2, \ldots, (n-1)\}$. The number of ways to create an ordered list of $m$ elements from $\mathbb{Z}_n$ is

$$\#\{\langle n_1, n_2, n_3, \ldots, n_m\rangle \mid n_j \in \mathbb{Z}_n\} = n \cdot n \cdot n \cdots n = n^m.$$

Thus the function $Pow : \mathbb{N} \times \mathbb{N} \to \mathbb{N}$ defined by $Pow(n, m) \equiv n^m$ is important for counting the number of ways that objects can be arranged. For example, the number of possible four-letter words, in the English language, is $26^4 = 456{,}976$ which is about the number of all words that are defined in English. Of course not all words are of length 4, and not all strings of 4 letters have a meaning in English.

The number of ways of creating four-letter words with all different letters is $Perm(26, 4) = 26 \cdot 25 \cdot 24 \cdot 23 = 358{,}800$. The function of permutations $Perm : \mathbb{N} \times \mathbb{N} \to \mathbb{N}$ is defined for $n \geq m \geq 1$ as

$$Perm(n, m) \equiv {}_nP_m = n!/(n-m)! = n \cdot (n-1) \cdots (n-m+1).$$

The notation ${}_nP_m$ is common in probability theory, but will not be used here. The first choice is the first letter in the word, the second choice is the second letter, etc.

The total number of four-letter collections with all different letters, where the order of the letters is not important, is $Comb(26, 4) \equiv Perm(26, 4)/24 = 14{,}950$. The quantity $24 = 4!$ is the number of ways that four numbers can be ordered; 4 for the first, 3 for the second, 2 for the third, and the last number has only 1 place to go. This is called the factorial function,

$$Fact : \mathbb{N} \to \mathbb{N}, \quad Fact(n) \equiv n \cdot (n-1) \cdots 3 \cdot 2 \cdot 1.$$

The function of combinations $Comb : \mathbb{N} \times \mathbb{N} \to \mathbb{N}$ is defined for $n \geq m \geq 1$ as

$$Comb(n, m) = \frac{n!}{m!(n-m)!} = \frac{n \cdot (n-1) \cdots (n-m+1)}{m \cdot (m-1) \cdots (1)} = \binom{n}{m},$$

and this is the same as the binomial coefficient as defined in Eq. (1.43). It may not be obvious, but this quantity is always a positive integer. A suggestion of this fact comes from observing that the numerator and denominator are products of the same number of consecutive integers. This quantity is the number of ways one can choose $m$ elements from $n$, or simply,

$$Comb(n, m) = \binom{n}{m} = \binom{n}{n-m} = \text{from } n \text{ choose } m.$$

Here the order of choosing is not important.

Next, suppose that four letters are to be chosen in alphabetical order. Then consider $26 + 4 - 1 = 29$ positions for $25 - x$'s and $4 - O$'s. The $O$ represents a letter, and the $x$ represents a separator for the different letters, which are in alphabetical order. Then, for example

$$xxxxOOxxxxxxxOxxxxxxxOxxxxxxx \equiv eels$$

The number of ways of doing this are $Comb(29, 4) = 23,751$.

Finally, if there are $k$ objects with occurrences $\{n_1, n_2, \ldots, n_k\}$ then the number of ways that these can be arranged is

$$Cmulti(n_1, n_2, \ldots, n_k) = \binom{n}{n_1, n_2, \ldots, n_k} = \frac{(n_1 + n_2 + \cdots + n_k)!}{n_1! \cdot n_2! \cdot \cdots \cdot n_k!}. \quad (2.61)$$

In the case of two objects $Cmulti(n_1, n_2) = Comb(n_1+n_2, n_1) = Comb(n_1+n_2, n_2)$.

Once counting and collecting have been mastered, one can start asking questions;

$$\phi = \text{What is the probability of randomly choosing four-letters,}$$
$$\text{and have them be in alphabetical order?} \quad (2.62)$$

The answer is $P(\phi) = Comb(29, 4)/26^4 \simeq 0.052 \simeq 5\%$ or about 1 in 20.

### ■ Exercises

1. Ten data points are collected from a never-ending random process that can only have six different values. If the values are collected one at a time and placed in a line, how many different patterns can be created? If the order is now ignored, how many different ways can the 10 data points distribute into the 6 different values? Decide which is larger, and prove the claim.

2. Five green, four blue, and two yellow beads are mixed and placed in a row, from left to right, along a string to form a necklace. How many different patterns can be created? (*Hint:* Use Eq. (2.61).)

### 2.11.2 Discrete Distributions

A *discrete probability distribution* is a function $P : \mathcal{D} \to [0,1]$ where $\mathcal{D}$ is a finite or a countably infinite set, so that

$$\sum_{j \in \mathcal{D}} P(j) = 1, P(j) \geq 0.$$

This function extends to $P : \mathcal{P}(\mathcal{D}) \to [0,1]$, where $\mathcal{P}(\mathcal{D})$ is the power set of $\mathcal{D}$. Thus for any $\mathcal{A} \subset \mathcal{D}$, or $\mathcal{A} \in \mathcal{P}(\mathcal{D})$, then

$$P(\mathcal{A}) \equiv \sum_{j \in \mathcal{A}} P(j) \leq 1.$$

The $n$th moment of a discrete distribution is

$$M_n \equiv \sum_{j \in \mathcal{D}} j^n \cdot P(j). \tag{2.63}$$

The *population mean* is denoted $\mu$ and the *population variance* is $\sigma^2$, where $\sigma$ is the *population standard deviation*. Then

$$1 = M_0, \ \mu = M_1, \ \sigma^2 = M_2 - M_1^2. \tag{2.64}$$

### 2.11.3 Discrete Uniform Distributions

The simplest distribution on the finite discrete set $\mathbb{Z}_n = \{0, 1, \ldots, (n-1)\}$ is called *uniform* and will be denoted $U(n)$. It has

$$U(n)[j] \equiv \begin{cases} 0 & \text{for } j < 0 \\ 1/n & \text{for } j \in \mathbb{Z}_n \\ 0 & \text{for } j \geq n \end{cases}, \ P(j \leq j_*) = \begin{cases} 0 & \text{for } j_* < 0 \\ (j_* + 1)/n & \text{for } j_* \in \mathbb{Z}_n \\ 1 & \text{for } j_* \geq (n-1) \end{cases}$$

It is easy to see that for a uniform distribution on $\mathbb{Z}_n$,

$$M_0 = 1, \ M_1 = (n-1)/2, \ M_2 = (n-1) \cdot (2 \cdot n - 1)/6, \tag{2.65}$$

so that the parameters of this model are

$$\mu = (n-1)/2, \ \sigma^2 = (n^2 - 1)/12. \tag{2.66}$$

Given no *a priori* information, one would assume a uniform distribution for a set of outcomes. Probability questions in this model are not very difficult to answer. However, a uniform distribution cannot be defined on infinite domains like $\mathbb{N}$. In particular, its support must be finite or have finite measure.

### ■ Exercises

1. For the uniform distribution $U(n)$, the probabilities on points are $P(j) = 1/n$ for each $j \in \mathbb{Z}_n$.

   (a) Use induction to prove $\sum_{j=0}^{n-1} j = (n-1) \cdot n/2$.
   (b) Use induction to prove $\sum_{j=0}^{n-1} j^2 = (n-1) \cdot n \cdot (2n-1)/6$.
   (c) Next, verify statements in Eq. (2.65) using definitions in Eq. (2.63).
   (d) Now verify statements in Eq. (2.66) using definitions in Eq. (2.64).

### ■ 2.11.4 Bernoulli and Binomial Distributions

When performing an experiment with two possible outcomes, the outcomes are often denoted as $\{failure, success\}$. If one is attempting to diagnose a disease, the terms used might be $\{negative, positive\}$. If one is flipping a coin, then the outcomes are $\{tails, heads\}$. To quantify the two possible outcomes, it is convenient to choose $\mathcal{S} \equiv \mathbb{Z}_2 = \{0, 1\}$ as the sample space.

Associated with the experiment, there is a probability of success $p_s$ and a probability of failure $q_f$, so that for each trial $p_s + q_f = 1$. This information is contained in a single probability function $B(1, p_s) : \mathcal{S} \to \{q_f, p_s\}$ called the *Bernoulli* distribution, so that with $0 \equiv$ failure, and $1 \equiv$ success,

$$B(1, p_s)[0] = q_f, \quad B(1, p_s)[1] = p_s.$$

This is also expressed as $B(1, p_s)[j] = p_s^j \cdot q_f^{1-j}$ for $j \in \mathbb{Z}_2$. Unless costs or safety issues are prohibitive, an experiment is typically performed more than once, in which case the *Binomial* distribution $B(n, p_s) : \mathbb{Z}_{n+1} \to [0, 1]$ is appropriate where

$$B(n, p_s)[j] = \binom{n}{j} \cdot p_s^j \cdot q_f^{n-j}. \tag{2.67}$$

From the binomial expansion Eq. (1.42), it is clear that

$$\sum_{j=0}^{n} B(n, p_s)[j] = (p_s + q_f)^n = 1^n = 1, \qquad (2.68)$$

as required of any probability distribution. Hence $B(n, p_s)[j]$ gives the probability of $j$ successes, after $n$ trials, where order is not important, and where the probability of success for each trial is the fixed value of $p_s$.

### ■ 2.11.5 Poisson Distribution

Let $\mu > 0$ be a single parameter. Then the Poisson distribution, written $Pois(\mu) : \mathbb{N}_0 \to [0, 1]$, is defined so that

$$Pois(\mu)[k] = \frac{\mu^k \cdot e^{-\mu}}{k!}, \quad P(k \leq k_*) = e^{-\mu} \cdot \sum_{j=0}^{k_*} \frac{\mu^j}{j!},$$

for $k, k_* \in \mathbb{N}_0$, and where $0! \equiv 1$. An application is found in the approximation

$$B(n, p)[k] \simeq Pois(\mu)[k] \qquad \text{where} \qquad \mu = n \cdot p,$$

which is often used in the parameter ranges

$$p \leq 0.05, \quad n \geq 20, \quad \text{so that} \quad \mu = n \cdot p \leq 10.$$

There is a slight problem with this approximation, in that if the variable $k > n$, then the probabilities should vanish. For the Poisson distribution the value of $P(k > n)$ is usually small enough to ignore.

---

### ■ Exercises

1. Consider flipping three unfair coins with $P(H) = 0.4$ and $P(T) = 0.6$.

   (a) Compute the probabilities $P(3H0T)$, $P(2H1T)$, $P(1H2T)$, and $P(0H3T)$ as in Example 2.3.
   (b) Plot the coin distribution for tails $T$, as in Figure 2.8.
   (c) Compute $Pois(n \cdot p)[k]$ for $k \in \{0, 1, 2, 3\}$ where $p = P(T) = 0.6$.
   (d) Use the four values $\{Pois(n \cdot p)[k]\}_{k=0}^{3}$ to compute $P(k \geq 4)$.

### 2.11.6 Continuous Distributions

A *continuous probability distribution* is a function $P : B \to [0, 1]$ where $B$ is a Borel set. Associated with a continuous distribution is a *cumulative distribution function* or simply a *cdf* $F : \mathbb{R} \to [0, 1]$ where

$$F(r) \equiv P\left((-\infty, r]\right), \ \forall r \in \mathbb{R}.$$

Given a cdf, probabilities can be computed using the definition

$$P\left((a, b]\right) \equiv F(b) - F(a).$$

One then extends the definition to any Borel set using Theorem 2.2 and the containments in Eq. (2.3) to write

$$\sum_{j=1}^{n} P(U_j) = P(B_n) \leq P(B).$$

### 2.11.7 Continuous Uniform Distributions

For a continuous variable $x$, the uniform distribution on an interval $\mathcal{I} = [a, b]$, is denoted as $U([a, b]) = U(\mathcal{I})$ and is defined for $a < b$ as

$$pdf = \frac{1}{b-a} \cdot \chi_{\mathcal{I}}(x), \ cdf = \frac{x-a}{b-a} \cdot \chi_{\mathcal{I}}(x) + \chi_{(b,\infty)}(x). \tag{2.69}$$

Two real parameters, $a$ and $b$, must be specified, and the set $\mathcal{I}$ is the support of $U(\mathcal{I})$. In any subinterval of $\mathcal{I}$, denoted $\mathcal{I}' \equiv [a', b'] \subset \mathcal{I}$, the probability of obtaining a value in $\mathcal{I}'$ is found by taking the relative sizes of the intervals,

$$P(x \in \mathcal{I}') = \frac{b' - a'}{b - a} = \text{probability that } x \text{ is in } \mathcal{I}'. \tag{2.70}$$

If $\mathcal{I} \cap \mathcal{I}' = \emptyset$, then $P(x \in \mathcal{I}') = 0$. Using integral calculus, it is easy to compute

$$\mu = (b+a)/2, \ \sigma = (b-a)/\sqrt{12}. \tag{2.71}$$

There is an obvious extension of uniform distributions to any Borel set.

### 2.11.8 Arbitrary Continuous Distributions

Any continuous random variable $x$, with cumulative distribution function $F(x)$, can be modeled using a random variable $y$ with uniform distribution $U((0, 1))$ using the transformation $x = F^{-1}(y)$. The inverse can only be defined on subsets of the open interval $(0, 1)$ where $F^{-1}$ is a function, which is where $F$ is monotone increasing. This holds since, for $r \in \mathbb{R}$, the cdf of $x$ is

$$P(x \leq r) = P(y \leq F(r)) = F(r),$$

which holds because the cdf of $y$ is the identity function on $[0, 1]$. For example, every random variable $x$ with uniform distribution on an interval $\mathcal{I} = [a, b]$ can be transformed into the random variable

$$y = (x - a)/(b - a) \iff x = (b - a) \cdot y + a,$$

where $y$ has a uniform distribution on $U([0, 1])$. For more general continuous distributions, nonlinear functions and their inverses must be used.

### ■ Exercises

1. Consider the uniform distribution $U([0.4, \ 3.2])$.

   (a) Plot the distribution $U([0.4, 3.2])(x)$ for $x \in [0, 4]$ using the *pdf* in Eq. (2.69).
   (b) Compute the $\mu$ and $\sigma$ using Eq. (2.71).
   (c) Draw and label the lines $\mu$, $\mu - \sigma$, and $\mu + \sigma$ on the distribution plot.
   (d) Compute $P(x \leq 1.5)$ using Eq. (2.70) and shade the region on the plot.

### ■ 2.11.9 Exponential Distributions

Let $\mu > 0$ be a single parameter. Then the exponential distribution is defined as

$$\begin{aligned} pdf &= (1/\mu) \cdot e^{-t/\mu} \cdot \mathcal{X}_{\mathbb{R}_0^+}(t), \\ P(t \leq t_*) &= (1 - e^{-t_*/\mu}) \cdot \mathcal{X}_{\mathbb{R}_0^+}(t_*). \end{aligned} \quad (2.72)$$

Since this distribution appears in the study of waiting times, negative values of $t$ are often meaningless, so the distribution is set to 0. Note that $\mu$ is the mean, and it can be shown that $\sigma = \mu$. The parameter $\mu$ is estimated from experiments by counting the number of events in a fixed period of time and determining the frequency

$$f \equiv \frac{\#\text{events}}{\text{long time period}} = \frac{1}{\mu}. \quad (2.73)$$

The longer the time period, the better one expects the estimate of the average *waiting time for a single event* to be, which is the quantity $\mu$.

### ■ 2.11.10 Gaussian (Normal) Distribution

Consider a population with variable $x$ that has mean $\mu$ and variance $\sigma^2$, with a symmetric bell-shape distribution. Such distributions are normally seen in applications, and the most common one is denoted as $N(\mu, \sigma)$, where

$$\text{pdf} \equiv N(\mu, \sigma)[x] = \frac{e^{-(x-\mu)^2/(2\sigma^2)}}{\sigma \cdot \sqrt{2\pi}}, \qquad (2.74)$$
$$\text{cdf} \equiv \Phi_{\mu,\sigma}(x) = \text{area under pdf from } -\infty \text{ to } x.$$

The special case where $\mu = 0$ and $\sigma = 1$ gives the *standard normal* distribution $N(0,1)[x]$. Otherwise $N(\mu, \sigma)[x]$ is a Gaussian or normal distribution.

### Student's *t*-Distribution

Consider a sample of size $n$ taken from a population, and the average of the variable $x$ is found to be $\bar{x}$. Suppose the sample variance is also computed to be

$$s_x^2 = \frac{1}{n-1} \sum_{j=1}^{n} (x_j - \bar{x})^2, \text{ where } \bar{x} = \frac{1}{n} \sum_{j=1}^{n} x_j.$$

Then the Student's *t*-distribution is

$$\begin{aligned}\text{pdf} \equiv T(\bar{x}, s_x, n-1)[t] &= \frac{\Gamma(n/2) \cdot \left[1 + (t-\bar{x})^2 / \left(s_x^2 \cdot (n-1)\right)\right]^{-n/2}}{s_x \cdot \sqrt{\pi \cdot (n-1)} \cdot \Gamma((n-1)/2)} \\ &\simeq \frac{e^{-(t-\bar{x})^2/(2 \cdot s_x^2)}}{s_x \cdot \sqrt{2\pi}} \quad \text{as } n \to \infty, \end{aligned} \qquad (2.75)$$

where $\Gamma$ is defined in Eq. (2.52) and the approximation to the Gaussian distribution holds due to the basic observation in Eq. (2.50).

The computations in this textbook will only use the Gaussian approximation in Eq. (2.75). However, practicing statisticians must use the *t*-distribution when called for in an application. An example is if $n < 30$.

### ■ Exercises

1. A sample of waiting times for customers at a store is taken and found to be $\{3 \text{ min}, 2 \text{ min}, 7 \text{ min}, 5 \text{ min}\}$. Find the average waiting time, based on this sample, using Eq. (2.73). Use Eq. (2.72) to determine the probability that a person will wait twice the average time.
2. Compute and simplify $N(\mu, \sigma)[\mu]$, $N(\mu, \sigma)[\mu - \sigma]$, and $N(\mu, \sigma)[\mu + \sigma]$. Also, compute and simplify the ratio $N(\mu, \sigma)[\mu + \sigma]/N(\mu, \sigma)[\mu]$.
3. Verify the approximation in Eq. (2.75) in the special case that $\bar{x} = 0$, $s_x = 1$. In particular, suppose that $(n-1) = 2 \cdot M$ and use formulas in Eq. (2.52) and Eq. (2.50) for finite $M$. Isolate and simplify $\Gamma^2(M+1/2)/(\Gamma(M) \cdot \Gamma(M+1))$ into a form suggesting that it approaches 1 as $M \to \infty$. Assume this fact to obtain the Gaussian distribution on the right-hand side of Eq. (2.75).

### ■ 2.11.11 Bayes' Formula and Inference

As an introduction to *inference*, consider an experiment with two sample spaces $\mathcal{S}_F$ and $\mathcal{S}_R$ that have associated probability distributions. Choose a partition from each sample space where the possible factors are $\{F^j\}_{j \in \mathcal{A}_F}$ and the responses are $\{R^k\}_{k \in \mathcal{A}_R}$, respectively. After many trials the data is organized so that all conditional probabilities are known for $\mathcal{S}_F$ given $\mathcal{S}_R$,

$$P(R^k|F^j) \in [0,1] \ \forall F^j, \ j \in \mathcal{A}_F \ \text{ and } \ \forall R^k, \ k \in \mathcal{A}_R. \qquad (2.76)$$

Now suppose that a specific $R^K \subset \mathcal{S}_R$ is observed. A typical question in science is

*What is the probability that $F^J \subset \mathcal{S}_F$ has occurred?*

Bayes' formula says that the answer is

$$P(F^J|R^K) = \frac{P(F^J \cap R^K)}{P(R^K)} = \frac{P(R^K|F^J) \cdot P(F^J)}{\sum_{j \in \mathcal{A}_F} P(R^K|F^j) \cdot P(F^j)}. \qquad (2.77)$$

The proof of this formula will not be shown here. However, it follows by application of basic set-theoretic principles.

> **EXAMPLE 2.4**
>
> Consider three coins; a nickel, a dime, and a quarter. An experiment consists of flipping all three coins $F$, and observing the result $R$.
>
> Let $\mathcal{S}_F$ be the total number of heads showing $= \{3H0T, 2H1T, 1H2T, 0H3T\}$.
> Let $\mathcal{S}_R$ be the total value showing for coins that landed tails $= \{0, 5, 10, 15, 25, 30, 35, 40\}$;
>
> The two spaces will be partitioned into only two parts:
>
> Let $F^1$ be the event of an even number of heads $= \{2H1T, 0H3T\}$;
> Let $F^2 = (F^1)^c = \{3H0T, 1H2T\}$.
> Let $R^1$ be the event of showing a value greater than $25 = \{30, 35, 40\}$;
> Let $R^2 = (R^1)^c = \{0, 5, 10, 15, 25\}$.
>
> Now, suppose experiments have revealed the conditional probabilities
>
> $$P(R^1|F^1) = 1/4, \; P(R^2|F^1) = 3/4, \; P(R^1|F^2) = 1/2, \; P(R^2|F^2) = 1/2.$$
>
> Then Bayes' formula allows the computation
>
> $$P(F^2|R^1) = \frac{P(R^1|F^2) \cdot P(F^2)}{P(R^1|F^1) \cdot P(F^1) + P(R^1|F^2) \cdot P(F^2)} \quad (2.78)$$
> $$= \frac{(1/2) \cdot (1/2)}{(1/4) \cdot (1/2) + (1/2) \cdot (1/2)} = \frac{(1/4)}{(3/8)} = \frac{2}{3} \simeq .667.$$
>
> The direct calculation gives $P(F^2 \cap R^1)/P(R^1) = (1/4)/(3/8) = 2/3$. In conclusion, if the monetary total is greater than 25, then there is a two-thirds chance that the number of heads is odd. However, there remains a one-third possibility that the number of heads is even, i.e., $P(F^1|R^1) = 1/3$.

The lesson here is that partial knowledge about factors input to a system can sometimes be obtained indirectly using the responses of the system. It is required that there be some dependence between the factors and responses, but such dependence may be weak. However, building upon partial information obtained through data collection, one can piece together a useful picture of the system.

## ■ Exercises

1. Repeat Example 2.4, but for unfair coins, where $P(H|nickel) = 0.4$, $P(H|dime) = 0.3$, and $P(H|quarter) = 0.2$. In particular, compare the Bayes's

result to the direct computation. Is this distribution better at inferring an odd number of heads? (*Hint:* Compute the probabilities $P(TTT), P(THT), \ldots$ where $THT$ means the nickel is tails, the dime is heads, and the quarter is tails, so $P(THT) = P(30)$. Then use Eq. (2.77) to obtain conditional probabilities.)

## ■ 2.12 Relations

A relation between two variables $x$ and $y$ is often expressed as an equation $f = 0$ where $f(x, y)$ is a function of the two variables. For example, the unit circle is expressed as the set of points $(x, y)$ that solve

$$0 = f(x,y) \equiv x^2 + y^2 - 1. \tag{2.79}$$

If this describes the path of a peddle on a bicycle, where the origin stays at the crank, then to move the bicycle to the *right* requires pushing the peddle to the *right* when $y > 0$. To continue this motion of the bicycle, the peddle must be moved to the *left* for $y < 0$. Such a loop is called a *hysteresis*. In this example, if $x = 0$ then the direction that the peddle moves is not known unless $y$ is also specified.

A relation may be observed through the collection and comparison of data pairs $\{(x_j, y_j)\}$. Suppose that $x_j$ takes values mainly in $\{-1, 0, 1\}$, and

$$x_i \simeq 1 \implies y_i \simeq 0, \ x_j \simeq 0 \implies \begin{cases} P(y_j \simeq 1) = 0.5 \\ P(y_j \simeq -1) = 0.5 \end{cases}, \ x_k \simeq -1 \implies y_k \simeq 0.$$

Then one may suspect that the variables are related according to Eq. (2.79).

Near a fixed point $(x_*, y_*)$ where $f(x_*, y_*) = 0$, there may be a function $y = y(x)$ that fits the relation $f(x, y) = 0$, where $y(x_*) = y_*$. For example, the relation

$$0 = g(x,y) \equiv x^4 - x^2 + y^2, \tag{2.80}$$

passes through the point $(0, 0)$. However, if $x_* \neq 0$ and $y_* \neq 0$ but $g(x_*, y_*) = 0$, then the four points $(\pm x_*, \pm y_*)$ also solve Eq. (2.80). See Fig. 2.9. Indeed, Eq. (2.80) can be solved to obtain two functions

$$y_+(x) = x \cdot \sqrt{1 - x^2}, \ y_-(x) = -x \cdot \sqrt{1 - x^2}.$$

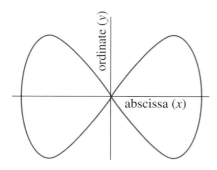

**FIGURE 2.9** ■ A Lissajous relation.

Hence, if $(x_*, y_*) \simeq (0,0)$ then $y_+ \simeq x$ or $y_- \simeq -x$ and without more information, it is impossible to predict which path the system actually follows.

## ■ Exercises

1. Relations defined by $(x,y) = (\cos(t), A \cdot \sin(\nu \cdot t + \delta))$, for an independent variable $t \in \mathbb{R}$ and fixed parameters $\{A, \nu, \delta\}$ are known as Jules Lissajous figures. An example is given in Figure 2.9.

   (a) Show that $(x,y) = (\cos(t), \sin(t))$ solves Eq. (2.79).
   (b) By modifying Eq. (2.79) find a relation involving just $x$ and $y$ that $\forall t \in \mathbb{R}$ is solved by $(x,y) = (\cos(t), 2 \cdot \sin(t))$. (*Hint:* Look at $x^2$ and $y^2$.)
   (c) Show that $(x,y) = (\cos(t), \sin(t) \cdot \cos(t))$ solves Eq. (2.80). Also, use Eq. (2.27) to find the $A$, $\nu$, and $\delta$ that shows this is a Lissajous figure.

2. Use the relation Eq. (2.23) to find a relation that $(x,y) = (\cosh(t), \sinh(t))$ satisfies $\forall t \in \mathbb{R}$. Also, find a point $(x_*, y_*)$ that solves this relation where $(\pm x_*, \pm y_*)$ gives four different solutions.

# 3 Statistics

Inference, based on a statistical approach, incorporates uncertainty as a natural part of critical questioning. By first displaying data properly, one can begin looking for patterns and developing opinions. Then the process of asking questions, which is the consequence of newly formed ideas, proceeds with the design of experiments that will check for the validity of statements. Probability theory is used to assign likelihood measurements, based on data collection, for models of natural phenomena.

## Introduction

Coping with large amounts of data is part of what humans are able to do naturally. The eye receives millions of bits of information per second; however, this is reduced to around 10 bits of cognition per second once accepted into the conscious brain.

In order to properly collect data, an experiment must be performed in a controlled environment so that only the independent variables are changed, while the dependent variables are measured. In a perfectly determined system, two measurements under the same conditions will lead to the same answer. This assumes *fatalism* of the system. However, there are often *hidden*, *lurking*, or *extraneous* variables that cannot be measured, even under the best conditions. This leads to a distribution of measurements, obtained under the same conditions. Thus the objective of experiments in general is not to find the exact relationship between variables, but rather the *multivariate distributions*.

## ■ 3.1 The Problem with Precision

All of science is founded upon measurement, where uncertainty is a vital aspect. If a person measures the width of their left hand to be 43 mm, then this really means, based on the *rules of rounding*, that the true value is between 42.5 mm and 43.49 mm. Another way to express this is to say that the true width of the hand is in the set $\mathcal{I} = [42.5 \text{ mm}, 43.5 \text{ mm})$, with 100% confidence, based on our measurement. Someone else may not have a strong belief in the first measurement and so they will perform it themselves. Suppose they obtain a value of 44 mm, giving an interval of $\mathcal{I}' = [43.5 \text{ mm}, 44.5 \text{ mm})$. The problem is that $\mathcal{I} \cap \mathcal{I}' = \emptyset$. Now the question becomes, "Who is correct?"

A better approach to measurement is to back away from any sense of certainty. We all make unintentional mistakes. Accounting for the blurriness of our eyes, and the unsteadiness of our hands, 100% confidence is unreasonable. Suppose instead that one is 60% confident that $\mathcal{I}$ contains the true value of the hand width, and the other person is 70% confident that $\mathcal{I}'$ contains the true value. A third person, not making a measurement, may want to assess the situation and produce an interval of 90% confidence. How can this be done with the two measurements taken by different people? This is not an easy question to answer. The third party has to assign a *belief* or *trust* value to the two people that already made their measurements. Furthermore, how the uncertainty is distributed in the measurement is very important, and often requires an assumption on the part of the investigator. This chapter will address this issue specifically.[1]

## 3.2 Random Variables, Models, and Parameters

In statistics, both dependent $X$ and independent $Y$ variables are thought of as having associated uncertainties. Actual readings are denoted $x$ and $y$. In this case, the term *random variable* is used, and with both $X$ and $Y$ there will be corresponding *probability density functions* (abbreviated as pdf or PDF) denoted $P_X, P_Y$. These entities are also written as $f_X(x), f_Y(y)$ and sometimes simply called *distributions*. There are cases where a PDF is obtained theoretically, but often in practice it is assumed to take a form that is suggested by theory or experiment. In this setting, variables do not have a strict relationship between each other. Rather, data will suggest a relationship $Y = \mathcal{M}(X)$, and one is motivated to compute a confidence level for the model. The simplest, nontrivial example is a linear relationship between the random variables,

$$Y = a \cdot X + b = \mathcal{M}(X). \tag{3.1}$$

Then $a$ and $b$ are parameters of the model $\mathcal{M}$, and these now have uncertainties (distributions) associated with their values. Thus the linear model in Eq. (3.1) is

---

[1] An excellent discussion about this issue, with regard to government weights and measures, can be found in Moore and MaCabe [49].

not a precise mathematical equation. Instead, if $x$ were known, then the model would give a distribution of $Y$ that is specific to this value of $x$, written

$$f_{Y|x}[y] = P(y|x) = \text{probability of } (Y = y) \text{ given } (X = x). \tag{3.2}$$

The consequence is a parametric family of distributions, where $x$ is the parameter.

In statistics, a relationship like Eq. (3.1) expresses an association between $x$ and $y$, and not necessarily a causation. For example, if $x$ represents the diastolic blood pressure, and $y$ is the systolic blood pressure, then typically the two variables are either both high, or both low. However both readings may have an underlying causation like the extent of heart disease.

### ■ 3.2.1 Terminology

The objects of study in a statistical analysis are *individuals* of a *population*. An individual's property can be *qualitative* (also called *categorical*), or *quantitative*. A *parameter* of a population can be a *count*, like the total number of mature oak trees in a forest, or a *measurement*, like the average heights of these trees in the forest. To obtain parameters, every tree has to be identified as part of a *census*. However, this may be time consuming, expensive, and even intractable. To deal with these problems, a *survey* can be performed whereby a *sample* of the population is taken and parameters of the *subpopulation* are computed. With respect to the full population, however, these parameters are called *statistics*. The set of numbers from a census or survey is called *data*. Categorical data is also called *discrete*. Alternatively, measurements of quantities result in data with a *continuous* range of values. An *observational study* is the act of collecting data by counting or measuring a property of individuals in a (sub)population. However, an *experimental study* involves applying a *treatment* to individuals and recording the *effect*, if any.

### ■ 3.2.2 States of a System

Variables are observed during a counting or measuring process, at different points in time. If times between observations are short, then the data represents a *state of the system*. Two situations are of interest over long periods of time:

**(1)** Are the parameters of a system changing?
**(2)** If so, what is the nature of this change?

To monitor the states of a system, different forms of display can be used, associated with the different variable types:

**Nominal:** No ordering. Use a bar graph or pie chart, (e.g., race, gender);
**Ordinal:** Weak ordering. Use a bar graph, dot plot, (e.g., grades, happiness);
**Interval:** Measurable differences. Histogram, (e.g., lengths, temperatures); and
**Ratio:** Size with zero point. Scatter plot, (e.g., height vs. weight, mass vs. volume).

Quantitative variables, like height, weight, or hand widths, have continuous numerical values. These aspects of a population can be measured and ordered sequentially, but often do not fit into strict states. Categorical variables, like hair color, gender, and ethnicity are qualitative and discrete, so they function better as state variables. Over time, some of the state variables can change, either by personal choice, a natural process, or by an external input. A common application of this awareness is in the field of medical diagnosis:

In this situation, the states of a system are often distinct. However, diagnoses may be difficult since a transition to illness may involve several changes in one's quantitative variables. Someone becoming diabetic, for example, may notice fluctuations in their blood glucose levels that at first appear to be part of normal daily variations. The question becomes, "How can one detect a true change of state?"

### ■ Exercises

1. Other than in a medical situation, give a system that has discrete states, whereby continuous numerical measurements can be used to identify the state.

### ■ 3.3 Descriptive Statistics

Counts and measurements have inherent uncertainty and variability.

**Definition 3.1** A *random variable* is an entity that may take different values each time it is observed. All else being equal and unchanging, a random variable has a fixed range of values and fixed probability of each value being observed. The information of the likely results from multiple observations is contained in an associated distribution function. A random variable is called *quantitative* if it takes numerical values in a field $\langle \mathbb{F}, +, \cdot \rangle$, *ordinal* if the different values can be ordered, and *categorical* if values are used simply for labeling.

Students in a class are asked to measure their left hands, in millimeters, across their four fingers, excluding the thumb. There is a range of values reported by the class members. The list of numerical values is referred to as *quantitative data*. Suppose that each data point is listed with the class member's gender. Then this is referred to as *qualitative* (or *categorical*) data. The following was collected by the students in a class on a college campus:[2]

$$57(m) \quad 44(m) \quad 43(f) \quad 35(f) \quad 30(f)$$
$$54(m) \quad 29(f) \quad 33(f) \quad 63(m) \quad 52(f)$$
(3.3)

It is now a challenge to organize and display the data in a manner that can be understood and then presented to others.

### ■ 3.3.1 Stem-Leaf Plots

To create an effective *stem-leaf* plot of quantitative data, units are chosen so that there are two significant figures that vary over the data set, where the first significant figure is the *stem* and the second is the *leaf*. In the example of the student hand widths, a measurement of 43 mm corresponds to a stem of 4 and a leaf of 3. In Eq. (3.3), there are four stems $\{6, 5, 4, 3, 2\}$ and the leaves range from 0 to 9. These are displayed in Table 3.1 by first ordering the data set from smallest to largest, and also separating the female $(F)$ data from the male $(M)$ data.

**TABLE 3.1** ■ Hand widths for a class of statistics students.

| Leaves | | | Stem |
|---|---|---|---|
|   |   | 3 | 6 |
| 7 | 4 | 2 | 5 |
|   | 4 | 3 | 4 |
| 5 | 3 | 0 | 3 |
|   |   | 9 | 2 |

| Females | | | Stem | Males | |
|---|---|---|---|---|---|
|   |   |   | 6 | 3 |   |
|   |   | 2 | 5 | 4 | 7 |
|   |   | 3 | 4 | 4 |   |
| 5 | 3 | 0 | 3 |   |   |
|   |   | 9 | 2 |   |   |

---

[2]This is based on an actual class study. However, data has been adjusted for demonstrational purposes, and is not intended to be representative of true subpopulations.

For quantitative information expressed as values in $\mathbb{F} = \mathbb{R}$, the *mean* of a data set $\{x_j\}_{j=1}^n$ is the *average* of the numerical values, expressed as integers, fractions, or decimals,

$$x_{mean} = \bar{x} \equiv \frac{x_1 + x_2 + \cdots + x_n}{n} = \frac{1}{n} \cdot \sum_{j=1}^{n} x_j. \tag{3.4}$$

If the data set is a sample from a larger population, then a measure of the width, or spread of the data, is the *sample variance*

$$var(x) = s_x^2 \equiv \frac{(x_1 - \bar{x})^2 + \cdots + (x_n - \bar{x})^2}{n-1} = \frac{1}{n-1} \cdot \sum_{j=1}^{n} (x_j - \bar{x})^2, \tag{3.5}$$

which is slightly more than the average of the *squared deviations* $\{(x_j - \bar{x})^2\}$. The value $(n-1)$ is the number of *degrees of freedom* for the statistic $var(x)$, where the average in Eq. (3.4) takes 1 degree of freedom from the $n$ data points. This corrects for the uncertainty associated with using a sample of a population, rather than a complete census. The *sample standard deviation* is $s_x \equiv \sqrt{var(x)}$. The numbers $\bar{x}$ and $s_x$ are *statistics* of the population but parameters of the sample.

The *median*, defined to be the midpoint (or halfway point) of a *sequentially-ordered* data set, is a measure of the center of the distribution of a random variable, which is different from $\bar{x}$. These are computed using the formulae

$$x_{med} \equiv \begin{cases} x_{(n+1)/2} & \text{for } n \text{ odd} \\ (x_{n/2} + x_{1+n/2})/2 & \text{for } n \text{ even} \end{cases} \tag{3.6}$$

For the data in Eq. (3.3),

$$\text{mean} = \frac{440}{10} = 44.0, \quad \text{median} = \frac{43 + 44}{2} = 43.5. \tag{3.7}$$

One observes from these statistics that mean > median and in such cases the data is said to be *skewed* or *stretched* toward the larger values. The skewing in Eq. (3.7) is not significant (relative error $\simeq 1\%$), but in cases of significant skewing, the median is a better measure of center than the mean.

In Table 3.1, the distribution of the leaves for the entire sample is *bimodal*, whereas, once separated as in the second table, each distribution appears symmetric and the centers are discernable.

For categorical information, one is typically interested in proportions $p_i \in [0,1]$ of the population that are in each category. In the preceding example, the proportion of students who are female, that have hand widths greater than 4 cm, is

$$p = P(F > 4 \text{ cm}) = 2/6 = 1/3 \simeq 0.33 = 33\%.$$

Note that this is subtly different from the proportion of students who are female *and* have hand widths greater than 4 cm,

$$p = P((gender = f) \wedge (H > 4 \text{ cm})) = 2/10 = 1/5 \simeq 0.20 = 20\%.$$

### ■ 3.3.2 Dot and Bar Plots

One can take the *interval* data in Table 3.1 and rotate them so that the larger hand widths are to the right. If each data point is stacked over its stem value, then one obtains a *dot plot* as shown in Figure 3.1.

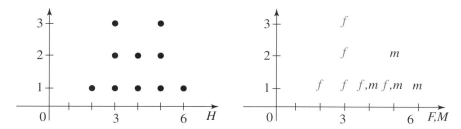

**FIGURE 3.1** ■ Left: Dot plot distribution for hand widths ($H$). Right: Dot plot with female ($f$) and male ($m$) students identified.

The heights within a stem category can also be displayed as a rectangle or *bar*. In this approach the height corresponds to the number of observations within a stem category, and so the area in the rectangle corresponds to the number of observations.

### ■ 3.3.3 Five-Number Summary and Box Plots

In the field of statistics, plotting data $\{x_j\}_{j=1}^n$ will often suggest a distribution. Since data sets are discrete, visualization must also be discrete. One approach, used for sets of size $n \geq 10$, is to arrange the data into increasing order $\langle x_j \rangle_{j=1}^n$ using re-indexing. Five statistics will now be defined. The minimum is defined to be $x_{min} \equiv x_1$ and the maximum $x_{max} \equiv x_n$. The median $x_{med}$ is the *middle* point of the ordered sequence if $n$ is odd, and the average of the middle two points if $n$ is even. Finally, the *quarter points* or *quartiles* are defined to be

$$x_{1q} \equiv x_{\lfloor (n+1)/4 \rfloor}, \; x_{3q} \equiv x_{n-\lfloor (n+1)/4 \rfloor}. \tag{3.8}$$

The *five-number summary* is the ordered sequence of the following five numbers,

$$\langle x_{min}, x_{1q}, x_{med}, x_{3q}, x_{max} \rangle \;=\; \langle min, Q1, M, Q3, max \rangle^3 \;=\; \langle 29, 33, 44.0, 54, 66 \rangle.$$

This sequence provides a quick visual assessment of the data's distribution. To graphically display the five-number summary, first use the three middle numbers $\langle Q1, M, Q3 \rangle$ to draw a box, with a central partition along $M$. In this way, around 50% of the data is enclosed between $Q1$ and $Q3$. Outside the box, compute the difference $IQR \equiv Q3 - Q1$, which is called the *interquartile range*. For the $H$ data $IQR = 54 - 33 = 21$. It has been suggested that all data beyond $1.5 \cdot IQR$ units of the two sides of the box should be called *outliers* and be labeled individually, possibly for further consideration [31]. The remaining points are within a nonoutlier range, which for the $H$ data is

$$[Q1 - 1.5 \cdot IQR, Q3 + 1.5 \cdot IQR] \;=\; [33 - 1.5 \cdot 21, 54 + 1.5 \cdot 21] \;=\; [1.5, 85.5]. \tag{3.9}$$

All of the $H$ data is within this range. Thus the data outside the $Q1 - Q3$ box are displayed using a *whisker* drawn from the minimum 29 and maximum 66 of the *nonoutliers* to the box.

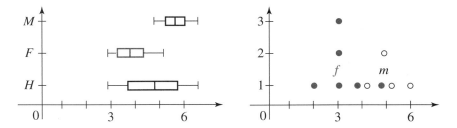

**FIGURE 3.2** ■ Left: Box plot distributions in cm for class's hand widths ($H$), for just females ($F$), and just males ($M$). Right: Dot plots for female and male students.

A box plot will indicate spread and skewness in data, as seen in the plots of $H$, $F$, and $M$ (see Figure 3.2). However a box plot will hide the presence of multiple

---

[3] This notation is more commonly used in descriptive statistics for the five-number summary.

modes, as are clearly present in Figure 3.1 and Table 3.1. By analyzing the within-gender categories one sees that the variations of $M$ and $F$ are much smaller than for the entire class ($H$) as a whole. A further observation is that there is only a limited overlap of the box plots for $M$ and $F$. In fact the overlap only occurs for the whiskers. A natural question that arises from the box plots is, "Can gender be estimated by measuring a student's hand width?"

### ■ Exercises

1. Four new individuals, whose genders are unknown, attend the class. Their hand widths are 55 mm, 47 mm, 6 mm, and 62 mm. Combine this into the $H$ data in Eq. (3.3), obtain the new five-number summary, the IQR, and the nonoutlier range. Draw a box plot just for $H$ in cm, as in Figure 3.2.

2. Six different individuals, whose genders are unknown, attend the original class. Five of their hand widths are all 46 mm and one is 6 mm. Combine this into the $H$ data in Eq. (3.3), obtain the five-number summary, the IQR, and the nonoutlier range. Identify the outlier. Draw a box plot just for $H$, as in Figure 3.2, indicating the outlier with a symbol "•". Give a possible reason for the outlier.

3. Measure the lengths of your fingers, to the nearest millimeter, and make a box plot of the data. Be sure to label the locations of the values for the five-number summary.

## ■ 3.4 Distributions

When categorical data is collected, it can be sorted into various separate bins, and the number of individuals in each category totaled. For example, if one uses hair color as an identifier, then a bar graph is one way to display the number of occurrences for each color type. Alternatively, if a quantity is measured from a population, one can artificially separate the data into numerical categories by using intervals. In this way one obtains a frequency diagram or histogram, with no separation between the bins.

### ■ 3.4.1 Frequency Diagrams and Histograms

For a detailed description of a distribution of an ordered data set $\langle x_j \rangle_{j=1}^n$ with range $\mathcal{R} = [x_1, x_n]$, consider a partition of $\mathcal{R}$ into $m \ll n$ parts. If the intervals are of equal length $(x_n - x_1)/m$, then a fixed unit can be chosen and bars of height, corresponding to the counted number of data points in the respective intervals, can be used to indicate the frequency of observations. The actual range used often stretches beyond $\mathcal{R}$ so the partition has integer or finite decimal values. This is a *frequency diagram*, which is a distribution that can be visualized in better detail than is possible with the box plot.

To allow for greater flexibility two issues are considered:

**(1)** Allow partitions of the abscissa (horizontal axis) to have unequal widths; and
**(2)** Index the ordinate (vertical axis) so that the total area is 1.

When looking at a data set $\{x_j\}_{j=1}^n$ corresponding to a quantitative random variable of a population, a special part is played by the *Gaussian*, or *normal*, distribution, written $N(\mu, \sigma)$. The distribution curve has a bell shape, which is symmetric about the *mean*, or *average*, of the population $\mu$, defined to be

$$\mu \equiv \frac{1}{n} \cdot \sum_{j=1}^n x_j. \tag{3.10}$$

Furthermore, about two-thirds of the area under the curve is within one standard deviation $\sigma$ of the mean $\mu$. The population variance is defined to be

$$\sigma^2 \equiv \frac{1}{n} \cdot \sum_{j=1}^n (x_j - \mu)^2 = \frac{\text{SSD}}{n}, \tag{3.11}$$

so that $\sigma$ is the population standard deviation. The variance $\sigma^2$ has additive properties that $\sigma$ does not have. The entities $(x_j - \mu)$ are called *deviations* from the mean, and $(x_j - \mu)^2$ are *squared* deviations, which are never negative. SSD refers to the *sum of the squared deviations*. The numbers $\mu$ and $\sigma$ are *parameters* of the population.

Specifically, the notation $N(\mu, \sigma)$ represents the distribution of a random variable $X$, which has a *distribution function*

$$N(\mu, \sigma)[x] = \frac{1}{\sqrt{2\pi} \cdot \sigma} \cdot \exp\left[\frac{-(x-\mu)^2}{2 \cdot \sigma^2}\right]. \tag{3.12}$$

In the study of variability, one can approximate $\mu$ and $\sigma$, under regularity assumptions of the true distribution of $X$, using $\bar{x}$ and $s_x$ computed from a large sample of the population.

**Remark 3.1** The notation $N(\mu, \sigma)$ may be confusing because it suggests that $N$ is a function of $\mu$ and $\sigma$. However, these are parameters of the statistical model. The expression $N(\mu, \sigma)[x]$ will be used to indicate the dependence of the distribution on the variable $x$. The distinction between parameters and variables will be emphasized throughout this chapter.

---

**Definition 3.2** A *probability density function* (PDF) is a nonnegative real-valued function on $\mathbb{R}$ whose total area is 1. It is sometimes just called a *probability distribution*. The corresponding *cumulative distribution function* (CDF) is the area under the PDF up to the value of interest. Thus a CDF is nonnegative, nondecreasing, with values in $[0, 1]$.

---

### EXAMPLE 3.1

Returning to the hand width example, one can compute the mean and standard deviations for the class's population to be

$$\mu_H = 44.0 \text{ mm}, \quad \sigma_H = 10.9 \text{ mm}. \tag{3.13}$$

The corresponding normal-distribution model $N(44.0, 10.9)$ can be compared with the actual distributions in Figure 3.1. The shape in Figure 3.3 is similar to Figure 3.2. However, as with a box plot, a normal distribution cannot display the bimodality of the hand-width data.

---

By accumulating the area under this distribution, from the left to the right, one obtains the CDF. One should become familiar with the normal distribution, and practice drawing both the PDF as in Figure 3.3 and the CDF as in Figure 3.4. Be sure to indicate the parameters $\mu$ and $\sigma$.

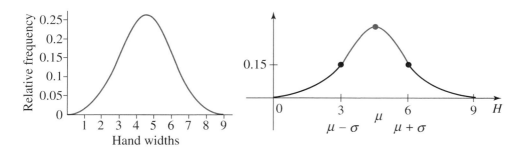

**FIGURE 3.3** ■ Probability distribution (PDF) using $\mu$ and $\sigma$ for hand widths ($H$).

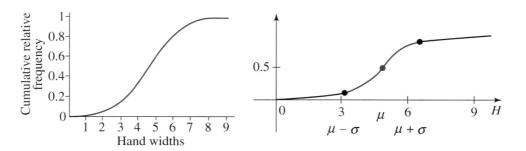

**FIGURE 3.4** ■ Cumulative distribution function (CDF) using $\mu$ and $\sigma$ for $H$.

If a random variable $G$ can only take a finite set of values $\{g_i\}_{i=1}^n$, which are found to distribute according to the proportions $\{p_i\}_{i=1}^n$, then

$$\mu = \langle G \rangle = \sum_{i=1}^n g_i \cdot p_i, \tag{3.14}$$

$$\sigma^2 = \langle (G-\mu)^2 \rangle = \sum_{i=1}^n (g_i - \mu)^2 \cdot p_i.$$

These definitions are often useful to think of when considering distributions because the calculations are finite. The proportions $p_i$ are easily estimated from data sets. Indeed, one counts $c_i$ the number of occurrences of category $i$, and divides by the sample size $n$, to obtain $p_i \simeq c_i/n$. For example, a random sample from a normal distribution of a variable $X$ will have the expected value

$$E(X) = \langle X \rangle = \mu. \tag{3.15}$$

Then a linear function of $X$ is a new random variable $Y = aX + b$ as in Eq. (3.1) that, by Eq. (3.14), satisfies

$$E(Y) = \langle a \cdot X + b \rangle = a \cdot \langle X \rangle + b = a \cdot \mu + b. \tag{3.16}$$

The variance can also be written as an expectation, so that the linearity property in Eq. (3.16) gives

$$\begin{aligned} Var(X) &\equiv E\left((X - E(X))^2\right) = \langle (X - \langle X \rangle)^2 \rangle \\ &= \langle X^2 - 2X\langle X \rangle + \langle X \rangle^2 \rangle = \langle X^2 \rangle - 2\langle X \rangle\langle X \rangle + \langle X \rangle^2 \\ &= \langle X^2 \rangle - \langle X \rangle^2 = E\left(X^2\right) - E^2(X) = \sigma^2. \end{aligned} \tag{3.17}$$

It easily follows that $Var(a \cdot X + b) = a^2 \cdot Var(X)$.

## Exercises

1. By computing, as in Eq. (3.17), prove that $Var(a \cdot X + b) = a^2 \cdot Var(X)$.
2. Consider a random variable $C$ that takes the values $c_1 = 1$, $c_2 = 3$, $c_3 = 4$, and $c_4 = 7$ with associated probabilities $p_1 = 0.1$, $p_2 = 0.3$, and $p_3 = 0.4$.

   (a) Deduce the probability of $C$ being 7, i.e., $p_4 \equiv P(C = 7)$.
   (b) What is the probability of obtaining a 4 or less, i.e., $P(C \leq 4)$?
   (c) What is the probability that $C$ will be odd?
   (d) Use the formulae from Eq. (3.14) to compute $\mu$ and $\sigma$.

3. Measure the lengths of your fingers, and compute the mean $\mu$ and standard deviation $\sigma$. Draw a dot plot of the lengths, using a partition that reveals a nonuniform distribution (use bins around size $\sigma$). Also, roughly draw the best-fit normal distribution for this data. Do finger lengths appear normally distributed?

## 3.5 Introduction to Inference

Suppose a new student is admitted to the class, and their hand width measured. It is found to be 55 mm. What can be inferred from this information? This is facilitated by *Bayes formula* presented in Eq. (2.77).

### 3.5.1 Bayes Rule for Uniformly Distributed States

Inference involves estimating the answer to questions like, "What is the probability that the student is a male?" To begin the analysis it will be assumed that the student is equally likely to be a male or a female. This is an assumption that the states of gender are uniformly distributed in the population. Assuming that the uncertainty in the hand widths is $\pm 0.5$ mm, no measurement of either females or males has a value in the range $[54.5, 55.5]$. This is a problem of insufficient data. To proceed with an analysis however, it will be assumed that the distributions of $F$ and $M$ are actually normal, and that the classroom sample was not completely representative. The respective sample means and sample standard deviations are computed to be, using Eq. (3.4) and Eq. (3.5),

$$x_F = 37.0 \text{ mm}, \; s_F = 8.9 \text{ mm}, \; x_M = 54.5 \text{ mm}, \; s_M = 7.9 \text{ mm}. \tag{3.18}$$

Now, the probabilities that the person is a female or male, given that their hand width is in the set $\mathcal{I} \equiv [54.5, 55.5)$ can be approximated as

$$P(x \in I|F) = N(37.0, 8.9)[55] \simeq \frac{1}{\sqrt{2\pi} \cdot (8.9)} \cdot \exp\left[\frac{-(55-37.0)^2}{2 \cdot (8.9)^2}\right] = 0.0058,$$

$$P(x \in I|M) = N(54.5, 7.9)[55] \simeq \frac{1}{\sqrt{2\pi} \cdot (7.9)} \cdot \exp\left[\frac{-(55-54.5)^2}{2 \cdot (7.9)^2}\right] = 0.0504.$$

At this point Bayes rule can be applied to conclude that

$$\begin{aligned} P(M|x \in I) &= \frac{P(x \in I|M)}{P(x \in I|F) + P(x \in I|M)} = \frac{0.0504}{0.0058 + 0.0504} \\ &= 0.0504/0.0562 = 0.897 \simeq 90\%. \end{aligned} \quad (3.19)$$

This is strong support for the belief that the new student is a male. However, one must be concerned about the small sample size. The purpose of this example is to demonstrate the mental process that a data sample (regardless of how small) induces, and how it is that opinions are made. An honest statistical study requires repeated trials. Confidence then increases as other studies obtain similar results.

### ■ 3.5.2 Bayes Rule for Non-Uniform *A Priori* Distributions

Suppose that the campus, where the class is held, has admitted 55% females and 45% males. Then the answer to the question, "What is the probability that the student is a male?" changes due to this *a priori* information. Bayes formula then states

$$\begin{aligned} P(M|x \in I) &= \frac{P(x \in I|M) \cdot P(M)}{P(x \in I|F) \cdot P(F) + P(x \in I|M) \cdot P(M)} \\ &= \frac{0.0504 \cdot 0.45}{0.0058 \cdot 0.55 + 0.0504 \cdot 0.45} \\ &= 0.0227/0.0272 = 0.834 \simeq 83\%. \end{aligned} \quad (3.20)$$

This adjustment still gives strong support for the belief that the student is a male. However, now the *a priori* information has been incorporated and should result in more consistent results when used in comparisons with other studies across campus.

## Exercises

1. An individual, whose gender is unknown, but whose hand width is 26 mm, attends the original class.

   (a) Using the statistics in Eq. (3.18) and Bayes rule for a uniform distribution of genders, compute the probability that this student is a female.
   (b) For the *a priori* distribution of 55% females and 45% males, compute the probability that this student is a female as in Eq. (3.20).
   (c) Is the new student's hand width an outlier with respect to the entire class's distribution (use the result in Example 3.13 and the $3 \cdot \sigma$ criterion)?

## 3.6 A Catalogue of Distributions

Due to the calculations just performed, it should be clear how properly modeled distributions can be used to answer questions about a system. The following are a collection of distributions that are commonly observed in histograms of quantitative data. Using techniques of descriptive statistics to understand data sets will often suggest some PDF. The selection of a PDF from the catalogue below allows an easy computation of probabilities, using the corresponding CDF. The process can be demonstrated via

$$\text{Histogram} \xrightarrow{Induction} \text{PDF} \xrightarrow{Computation} \text{CDF} \xrightarrow{Deduction} \text{Probabilities.}$$

The CDF is always monotone increasing on the support of its PDF. Thus the CDF can be inverted, which is necessary in order to make a statistical inference.

### 3.6.1 Gaussian (Normal) Distribution

Consider a population with variable $X$ that has mean $\mu$ and variance $\sigma^2$, with a symmetric bell-shaped distribution. Such distributions are normally seen in applications, and the most common one is $N(\mu, \sigma)$, as defined in Eq. (3.12) and Eq. (2.74)

$$\text{pdf} = \frac{e^{-(x-\mu)^2/(2\sigma^2)}}{\sqrt{2\pi} \cdot \sigma}, \quad \text{cdf} \equiv \Phi_{\mu,\sigma}[x] = \text{ area from } -\infty \text{ to } x. \tag{3.21}$$

The basic shape of $\Phi_{\mu,\sigma}[x]$ is given in Fig. 3.4. In general, suppose that an observation $x_*$ is made of an $N(\mu,\sigma)$ process $X$. Then the *Z-score* is defined to be the statistic

$$z_* \equiv (x_* - \mu)/\sigma, \qquad (3.22)$$

which is a linear transformation of $x_*$. Questions about $X$ are now answered using $N(0,1)$, the standard normal distribution. For very negative values of $z \ll 0$, the corresponding CDF is closely approximated by

$$\Phi_{0,1}[z] \simeq \sqrt{\frac{2}{\pi}} \cdot \frac{\exp(-z^2/2)}{|z| + \sqrt{z^2 + \pi}} \quad \text{for } z \leq 0, \qquad (3.23)$$

and $\Phi_{0,1}[z] = 1 - \Phi_{0,1}[-z]$ for $z > 0$. Due to this symmetry it is sufficient to formulate probability questions for normally-distributed random variables in the tail regions ($z \ll 0$ or $z \gg 0$). It is also very useful to have a formula for the inverse $\Phi_{0,1}^{-1}[y]$, which can be estimated using the linear approximation, for $y \in (0.2, 0.8)$,

$$z \simeq \sqrt{2\pi} \cdot (y - 0.5), \qquad (3.24)$$

and $y \in (0, 0.2)$, corresponding to the tail region, by the iteration scheme,

$$\begin{aligned} z_0 &= 1, \\ z_{n+1} &= \sqrt{-2 \cdot \ln(y) - \ln(\pi/2) - 2 \cdot \ln(z_n + \sqrt{(z_n^2 + \pi)})}. \end{aligned} \qquad (3.25)$$

Symmetry implies that values for $y \in (0.8, 1)$ can now be estimated as well. Convergence of Eq. (3.25) is rather slow, so a couple of iterations is typically needed.

**Remark 3.2** The $68 - 95 - 99.7$ rule for normally distributed data says that approximately

- 68% of the data will be one standard deviation $\sigma$ within the mean $\mu$,
- 95% of the data will be $2\sigma$ within $\mu$, and
- 99.7% of the data are expected to be within the interval $[\mu - 3\sigma, \mu + 3\sigma]$.

In particular, the standard normal distribution of Eq. (3.23) gives

$$1 - 2 \cdot \Phi_{0,1}[1] = 0.681, \ 1 - 2 \cdot \Phi_{0,1}[2] = 0.954, \ 1 - 2 \cdot \Phi_{0,1}[3] = 0.997.$$

Data beyond three standard deviations from the mean will be called outliers.

**EXAMPLE 3.2**

A company makes lightbulbs to the specification that their lifetimes are $\mu = 2200$ hrs with a standard deviation of $\sigma = 100$ hrs. What is the probability that a customer will have the lightbulb they purchased last more than $t_* = 2450$ hrs?
**Answer:** One easily computes the Z-score to be $z_* = 2.50$. Then from Eq. (3.23)

$$P(T > t_*) = P(Z > z_*) \simeq 0.0063 \simeq 0.6\%.$$

Customers begin to complain if the lightbulbs last less than $t_* = 1900$ hrs. If 2 million lightbulbs are sold each year, about how many complaints are expected?
**Answer:** In this case the Z-score is $z_* = -3.00$. Now from Eq. (3.25)

$$P(T < t_*) = P(Z < z_*) \simeq 0.0014 \simeq 0.1\%.$$

The number of complaints expected should be around $2 \times 10^6 \cdot 0.0014 = 2800$.

### ■ 3.6.2 Uniform Distribution

The simplest distribution $U(\mathcal{I})$ is defined for $\mathcal{I} = [a, b]$, $a < b$ from Eq. (2.69)

$$\text{pdf} = U([a,b])[x] \equiv \frac{1}{b-a} \cdot \chi_{[a,b]}(x), \ \text{cdf} = \frac{x-a}{b-a} \cdot \chi_{[a,b]}(x) + \chi_{(b,\infty)}(x). \quad (3.26)$$

Using a continuous version of (3.14) it is rather easy to show that

$$\mu = \langle X \rangle = (b+a)/2, \ \sigma^2 = Var(X) = (b-a)^2/12. \quad (3.27)$$

### 3.6.3 Exponential Distribution

Let $\mu > 0$ be a single parameter and $T$ a random variable. Then the single-event Poisson, or exponential distribution is defined as $Exp(\mu)$, where Eq. (2.72) gives

$$\begin{aligned} \text{pdf} &= Exp(\mu)[t] \equiv (1/\mu) \cdot e^{-t/\mu} \cdot \mathcal{X}_{\mathbb{R}_0^+}(t), \\ \text{cdf} &= (1 - e^{-t/\mu}) \cdot \mathcal{X}_{\mathbb{R}_0^+}(t). \end{aligned} \qquad (3.28)$$

See Fig. 3.5. The support of $Exp(\mu)$ is $\mathbb{R}_0^+ = [0, \infty)$.

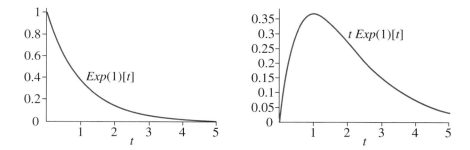

**FIGURE 3.5** ■ Exponential distribution and its first moment function.

It can be shown that the mean is $\langle T \rangle = \mu$ and the variance is $\sigma^2 = \text{Var}(T) = \mu^2$.

#### EXAMPLE 3.3

Suppose that the average waiting time at a bank is 15 min. Assuming an exponential distribution, the probability of waiting no more than 3 min, on any particular visit, is

$$P(t \leq 3 \text{ min}) = (1 - e^{-3\text{min}/15\text{min}}) \cdot \mathcal{X}_{\mathbb{R}_0^+}(3) = 1 - e^{-1/5} \simeq 0.181,$$

or around an 18% chance. The probability of waiting over a half an hour is

$$P(t > 30 \text{ min}) = 1 - P(t \leq 30 \text{ min}) = e^{-30\text{min}/15\text{min}} \cdot \mathcal{X}_{\mathbb{R}_0^+}(30) = e^{-2} \simeq 0.1353,$$

or about a 14% chance.

## 3.6.4 Bernoulli and Binomial Distributions

Consider performing an experiment with two possible outcomes, or states

$$\text{success: } S \equiv 1, \text{ failure: } F \equiv 0. \tag{3.29}$$

In a Bernoulli experiment, there is a probability of success $p_s$ and a probability of failure $q_f$, so that for each trial $p_s + q_f = 1$. For example, if a fair coin is flipped, then the probability of obtaining heads $p_H = 0.5 = 50\%$. The Bernoulli distribution corresponds to one trial where, by definition, $\mu = \langle S \rangle = p_s$, which is the expected probability of success. The variance is easily computed since there are only two possible states. A linear combination of squared deviations, where the coefficients are the proportions of occurrences, gives

$$\begin{aligned}\sigma^2 = \text{Var}(S) &= q_f \cdot (0-\mu)^2 + p_s \cdot (1-\mu)^2 = q_f \cdot (-p_s)^2 + p_s \cdot (q_f)^2 \\ &= (p_s + q_f) \cdot (q_f \cdot p_s) = p_s \cdot q_f.\end{aligned} \tag{3.30}$$

Thus the standard deviation is $\sigma = \sqrt{p_s \cdot q_f}$.

Next, suppose that a Bernoulli experiment is performed $n$ times. An example is the case of flipping a coin $n$ times. If order is not important, then the distribution function is given by Eq. (2.67). Then, for $k \in \mathbb{Z}_{n+1} = \{0, 1, 2, \ldots, n\}$,

$$\text{pdf} = \binom{n}{k} \cdot p_s^k \cdot q_f^{n-k}, \quad \text{cdf} = \sum_{j=0}^{k} \binom{n}{j} \cdot p_s^j \cdot q_f^{n-j}. \tag{3.31}$$

The expectation value is found by modifying the CDF,

$$\langle k \rangle \equiv \mu = \sum_{j=0}^{n} \binom{n}{j} \cdot j \cdot p_s^j \cdot q_f^{n-j} = n \cdot p_s, \tag{3.32}$$

with variance $Var(k) = \sigma^2 = n \cdot p_s \cdot q_f$.

## 3.6.5 Poisson Distribution

Let $\mu > 0$ be a single parameter and $K$ a discrete random variable. Then the *multi-event* Poisson distribution, written $Pois(\mu)$, is defined

$$\text{pdf} = Pois(\mu)[k] \equiv e^{-\mu} \cdot \frac{\mu^k}{k!}, \quad \text{cdf} = e^{-\mu} \cdot \sum_{j=0}^{k} \frac{\mu^j}{j!}, \tag{3.33}$$

for $k \in \mathbb{N}_0$, where $0! \equiv 1$. The definition for $k < 0$ is $Pois(\mu)[k] = 0$. It can be shown that the mean is $\langle K \rangle = \mu$ and the variance is $\sigma^2 = \text{Var}(K) = \mu$.

## ■ Exercises

1. (Gaussian) From the statistics in Eq. (3.18) use the CDF inversion method in Eq. (3.25), and the transformation Eq. (3.22), to compute the range of male hand widths in the largest 10% of the population.

2. (Gaussian) A random process $X$ has values $\langle 2, 4, 5, 7, 11 \rangle$ with associated probabilities $\langle 0.1, 0.3, 0.1, 0.3, 0.2 \rangle$.

   (a) Use the formula in Eq. (3.14) to compute $\mu$ and $\sigma$.
   (b) Assuming a normal distribution, use the $Z$-score in Eq. (3.22) and the approximate inverse Eq. (3.25) to compute the probability of $X$ being 9 or more.
   (c) Based on the actual distribution, what is $P(X \geq 9)$ actually?

3. (Uniform) Estimate the variance of $U([a, b])$ by dividing $\mathcal{I} = [a, b]$ into three equal parts and using Eq. (3.14). Simplify and compute the relative percent error compared to the true value of $\sigma^2 = (b - a)^2 / 12$.

4. (Uniform) Roughly draw $U([0.2, 3.4])$ and the normal distribution in comparison, as in Fig. 3.3, by computing the parameters $\mu$ and $\sigma$ in Eq. (3.27).

5. (Exponential) The average waiting time at a bank is found to be 23 mins. It is 3:49 pm as you enter the establishment with a check. Bank transactions at or after 4:00 pm are processed the next day. What is the probability that the check will be processed today?

6. (Bernoulli) Suppose that there are three possible results from an experiment $S = 1$, $N = 0$, and $F = -1$ (for *success*, *neither*, and *failure*). The probabilities that these states will occur is found to be $p_S = 0.3$, $p_N = 0.5$, and $p_F = 0.2$. Find the mean $\mu$ and standard deviation $\sigma$ for this distribution. Also compute the second moment $M_2 \equiv \sum_{j=1}^{3} g_j^2 \cdot p_j$ and verify that the variance satisfies $\sigma^2 = M_2 - \mu^2$.

7. (Binomial) Four fair coins are flipped. Draw the dot plot by computing $B(4, 0.5)[k]$ for $k \in \{0\} \cup \mathbb{Z}_5$. Also, roughly draw the $N(n \cdot p, \sqrt{n \cdot p \cdot q})$ distribution in comparison, as demonstrated in Fig. 3.3.

8. (Poisson) Suppose there are on average three people in line at the bank. What is the chance that five or more people will be in line? Use a Poisson distribution.

## 3.7 Parameter Estimation

To model a system, one must identify variables and parameters. By observing a phenomenon, a relationship between variables may be postulated. In this way, a model is created. If the model is sufficiently flexible, it will have adjustable parameters that must be estimated. Thus the model will inherit uncertainty due to the estimation process, so one can have only *less* than 100% confidence in any mathematical model that attempts to describe a phenomenon in observational science.

In this section, only probability models will be considered. In particular, let $x$ be a random variable where the PDF is unknown, but will be estimated based on data. Then an assumption must be made about the functional form of the distribution.

### 3.7.1 Exponential Distribution

Suppose that, by inspection, it appears that a random variable $T$ has an exponential distribution $Exp(\mu)[t]$. In this case $\mu > 0$ is a single parameter for the model. If $n$ samples are collected $\{t_i\}_{i=1}^n$, then the likelihood function

$$\begin{aligned} L_{Exp}(t_1, t_2, \ldots t_n; \mu) &\equiv \prod_{i=1}^n Exp(\mu)[t_i] \\ &= \mu^{-n} \cdot \exp\left(-(t_1 + t_2 + \cdots + t_n)/\mu\right), \end{aligned} \qquad (3.34)$$

has a maximum when

$$\mu = (t_1 + t_2 + \cdots + t_n)/n \equiv \bar{t}, \qquad (3.35)$$

which is just the sample average. Typical computations of a best-fit estimation are formulated as an optimization problem in calculus [31]. Although the function $L_{Exp}$ will not be discussed in detail here, it has many applications in parameter estimation [3].

### 3.7.2 Normal Distribution

Suppose that the histogram of a data set $\{x_j\}_{j=1}^n$ appears to be symmetric, unimodal, and bell-shaped. Then a normal distribution $N(\mu, \sigma)$ may be postulated, giving two parameters $\mu$ and $\sigma$ to be estimated. Here the likelihood function

$$L_N(x_1, x_2, \ldots x_n; \mu, \sigma) \equiv \prod_{j=1}^{n} N(\mu, \sigma)[x_j] \qquad (3.36)$$

$$= \frac{Exp\left([(x_1 - \mu)^2 + (x_2 - \mu)^2 + \cdots + (x_n - \mu)^2]/(2 \cdot \sigma^2)\right)}{\sigma^n \cdot (2\pi)^{n/2}},$$

is used to obtain the corresponding probability model. In this case it can be shown that two equations

$$\mu = \sum_{j=1}^{n} x_j \cdot \frac{1}{n} = \overline{x}, \quad \sigma^2 = \sum_{j=1}^{n}(x_j - \overline{x})^2 \cdot \frac{1}{n} = \frac{n-1}{n} \cdot s_x^2, \qquad (3.37)$$

hold, for $s_x$ defined in Eq. (3.5). The reason for the difference between the sample variance $s_x^2$ and the population variance $\sigma^2$ can now be addressed. Indeed, leaving the details for the exercises, it can be shown that

$$\langle s_x^2 \rangle \equiv \langle \sum_{i=1}^{n} \frac{(x_i - \overline{x})^2}{n-1} \rangle = \frac{n}{n-1} \cdot \left[\sum_{i=1}^{n} \langle x_i^2 \rangle \cdot \frac{1}{n} - \overline{x}^2\right] = \frac{n}{n-1} \cdot \sigma^2. \qquad (3.38)$$

An alternative analysis is provided by the *method of modes* [31].

### ■ 3.7.3 Tests for Probability Models

Once an appropriate probability model has been chosen, and the parameters estimated based on preliminary data, it can be tested for accuracy as part of further experiments. To do this, a statistic must be computed from the data, which can be used as a measure of appropriateness. One suggestion involves rearranging the data $\{x_j\}$ so that it is indexed in increasing order $\langle x_{\sigma(j)} \rangle$. Here $\sigma : \mathbb{Z}_n \to \mathbb{Z}_n$ is a *permutation*, which ensures that

$$j \leq j' \implies x_{\sigma(j)} \leq x_{\sigma(j')}.$$

Since this can always be done, the notation $\langle x_j \rangle_{j=1}^{n}$ will suggest a set that is arranged in increasing order. The empirical CDF, constructed from the $n$ data points, is

$$F(x_1, x_2, \ldots x_n)[x] \equiv \begin{cases} 0, x < x_1 \\ k/n, x_k \leq x < x_{k+1} \\ 1, x \geq x_n \end{cases} \quad \text{for } k \in \mathbb{Z}_n. \qquad (3.39)$$

So that there is no ambiguity, if $x_{k-1} = x_k < x_{k+1}$, then the intention of this definition is to imply that $F(x_1, x_2, \ldots, x_n)[x_{k-1}] = F(x_1, x_2, \ldots, x_n)[x_k] = k/n$. The following result, not proven here, is known as the *Glivenko-Cantelli theorem*, and gives a sense as to why the simple inspection of data can lead to a good choice for a population's distribution.

■ **Theorem 3.1**

Suppose that $F(x)$ is a continuous CDF for a random variable $X$. Choose $n$ data points from this distribution and express them as an ordered data set $\langle x_j \rangle_{j=1}^n$. Let $F(x_1, \ldots, x_n)[x]$ be constructed from this data set, as defined in Eq. (3.39). Then $|F(x) - F(x_1, \ldots, x_n)[x]| \to 0$ for each $x$ as $n \to \infty$ (almost surely).[4]

Consequently, for large sample sizes, a poorly chosen distribution will reveal itself. Suppose that the domain $\mathcal{D}$ of $F(x)$, a continuous CDF, is partitioned into $n$ regions, each with proportion $\{p_k\}_{k=1}^m$. Now make $n$ observations, where typically $n > m$, and record the numbers counted in each partition $\{C_k\}_{k=1}^m$. Note that $\sum_{k=1}^m C_k = n$. Then construct the statistic

$$Q(m-1) = \sum_{k=1}^m \frac{(C_k - n \cdot p_k)^2}{n \cdot p_k}. \tag{3.40}$$

This has an approximate $\chi^2(m-1)$ distribution [31], defined for $t \geq 0$ as

$$\text{pdf} = \chi^2(m-1)[t] = \frac{1}{2 \cdot \Gamma((m-1)/2)} \cdot \left(\frac{t}{2}\right)^{(m-3)/2} \cdot e^{-t/2}. \tag{3.41}$$

Note that, for example, $\chi^2(2)[t] = Exp[1/2](t)$. Also, from the definition in Eq. (2.52), $\Gamma(1/2) = \sqrt{\pi}$, $\Gamma(1) = 1$, and $\Gamma(x+1) = x \cdot \Gamma(x)$.

The CDF of the $\chi^2$ distribution is very important, however not easy to compute exactly. For a large number of partitions, say $m > 20$, $\chi^2(m-1) \simeq N(m-1, \sqrt{2(m-1)})$. However, for small samples this distribution is not symmetric,

---

[4]The likelihood of a condition holding, over samples of size $n$, is the probability of it holding for all configurations. Thus, a condition holds almost surely if the proportion of configurations where it does not hold, vanishes as $n$ increases.

as seen in Fig. 3.6, so tables should be consulted.[5] If a distribution $\{p_k\}$ is hypothesized, then it is rejected, at the 5% level, if

$$[Q(m-1) - (m-1)]/\sqrt{2(m-1)} \geq z_{0.05} \simeq 1.65.$$

This is an approximate condition that is considered acceptable if $min\{n \cdot p_k\} \geq 5$.

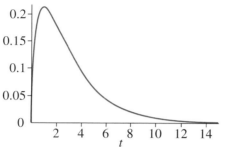 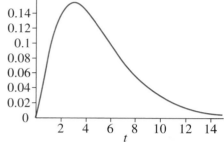

**FIGURE 3.6** ■ Left: The $\chi^2(1)[t]$ distribution; Right: The $\chi^2(3)[t]$ distribution.

### EXAMPLE 3.4

Recall the distributions of hand widths displayed in Figure 3.2, and the statistics in Eq. (3.18) for data in Eq. (3.3). In this example there are two approximately normal distributions. To show this, one can use Eq. (3.40) for the male population. The four partitions will be chosen based on the standard deviation, according to a normal distribution $N(54.5, 7.9)$:

$$P(x < 46.5) = 16\%, \ P(46.5 \leq x < 54.5) = 34\%, \ P(54.5 \leq x < 62.4) = 34\%, \ P(62.4 \leq x) = 16\%,$$

so, with only four data points, the statistic

$$Q_M(4-1) = \frac{(1 - 4 \cdot 0.16)^2}{4 \cdot 0.16} + \frac{(1 - 4 \cdot 0.34)^2}{4 \cdot 0.34} + \frac{(1 - 4 \cdot 0.34)^2}{4 \cdot 0.34} + \frac{(1 - 4 \cdot 0.16)^2}{4 \cdot 0.16},$$

is $Q_M(3) \simeq 0.6$ for males. However $\chi^2(3) \simeq N(3, \sqrt{6})$ so to be 95% confident that the male distribution is not normal, one needs $Q(3) \geq 1.65 \cdot \sqrt{(6)} + 3 = 7.0$, which clearly does not hold. Note that $min\{n \cdot p_k\} = 0.64$, which is quite less than 5. Thus this conclusion is questionable.

---

[5] The normal approximation will be used here only for demonstration purposes.

Similarly, for the female hand widths, using $N(54.5, 8.9)$,

$$Q_F(4-1) = \frac{(0 - 6 \cdot 0.16)^2}{6 \cdot 0.16} + \frac{(4 - 6 \cdot 0.34)^2}{6 \cdot 0.34} + \frac{(1 - 6 \cdot 0.34)^2}{6 \cdot 0.34} + \frac{(1 - 6 \cdot 0.16)^2}{6 \cdot 0.16},$$

resulting in $Q_F(3) \simeq 3.4$, which again is less than $\chi^2(3) \simeq 7.0$. Thus the data in Eq. (3.3) does not reject the normal distribution for the female hand widths. Here $min\{n \cdot p_k\} = 0.96 < 5$, calling into question any conclusions.

## ■ Exercises

1. For the hand-width data in Eq. (3.3) test for normality by partitioning the normal distribution into four parts as in Example 3.4, using the parameters computed in Example 3.1. Is there reason to reject the suggestion that the class's hand widths are normally distributed?
2. Suppose that the data $\mathcal{D} = \{0.3, 1.2, 0.7, 0.9\}$ is believed to come from a uniform distribution. Construct the likelihood function $L_U(\mathcal{D}; a, b)$ and find the parameters $a < b$ that maximizes $L_U$.
3. Suppose 10 visits to a bank result in waiting under 1 min three times, between 1 and 2 min four times, between 2 and 3 min twice, about 5 min once, and 15 min once. Compute the most likely waiting time using Eq. (3.35) and Eq. (3.14). What is the difference between these formulas?
4. Complete the details in Eq. (3.38) using the definitions in Eqs. (3.15) and (3.17).

## ■ 3.8 Statistical Inference

The *rules of statistical inference* are formulated in order to measure levels of confidence in statements. The approach is to phrase questions in a manner that will allow their truth-values to be estimated from probabilistic calculations. Then one must collect data and compute statistics. If the essential information involves counting the number of individuals in two different categories, then the proportion in one category $\hat{p}_i = \langle p_i \rangle$ is the statistic to be computed that estimates the true proportion of category-$i$ in a population. If the measurements are quantitative, then the average $\overline{x} = \langle x \rangle$ gives a statistic that estimates the expected value of $x$ for future measurements.

### 3.8.1 Confidence Intervals

To estimate the information in a data set, one can look for a peak value, or *mode*, of a distribution. This statistic is an example of a *point estimate*. The objective is to identify a reasonable center for a population. Given a random variable, its expected value provides a reasonable start at estimating a center for its distribution. The example of hand widths shows that the center is not sufficient to give a realistic picture of the class's distribution. However, if a distribution is nearly normal, then expected values $\langle p \rangle \in (0,1)$ or $\langle X \rangle \in \mathbb{R}$ can reasonably be used as a measure of center, and the variances $Var(p) > 0$ or $Var(X) > 0$ can be used to measure variability. Choosing a level of confidence $C \in (0,1) = (0\%, 100\%)$ necessitates an expanded region of uncertainty, or *margin of error*, which is

$$\delta_C(p) = z_C^* \cdot \sqrt{\frac{Var(p)}{n}} \quad \text{or} \quad \delta_C(X) = z_C^* \cdot \sqrt{\frac{Var(X)}{n}},$$

where $z_C^*$ must be computed from Eq. (3.24) or Eq. (3.25)[6] using $y = (1-C)/2$. The most commonly used values are,

$$z_{80}^* = 1.28, \ z_{90}^* = 1.65, \ z_{95}^* = 1.96, \ z_{98}^* = 2.33, \ z_{99}^* = 2.57. \tag{3.42}$$

The corresponding confidence intervals for proportions $p$ or measurements $X$ are

$$\mathcal{I}_C(p) = [\langle p \rangle - \delta_C(p), \langle p \rangle + \delta_C(p)] \quad \text{or} \quad \mathcal{I}_C(X) = [\langle X \rangle - \delta_C(X), \langle X \rangle + \delta_C(X)].$$

These are known as *interval estimates* for a population, and are interpreted under various assumptions [49], for example:

- The random variable is (nearly) normally distributed within the population;
- The population size is large so that the distributions are (nearly) continuous.

In the practice of statistical methods, data is first collected using an unbiased broadly-reaching method, referred to as a *simple random sample* (SRS). Then the confidence intervals are computed from the data.

#### Proportions

A sample proportion $\hat{p}$, and its conjugate $\hat{q} = 1 - \hat{p}$, are used to compute the standard error $SE_{\hat{p}} \equiv \sqrt{\hat{p} \cdot \hat{q}/n}$, where $n$ is the sample size. Then the $C$-level

---

[6]Tables are also available in most books on statistics [31], [49], [75].

confidence interval for a data set is

$$\mathcal{I}_C(p) \simeq \left[\hat{p} - z_C^* \cdot \sqrt{\hat{p} \cdot \hat{q}/n},\ \hat{p} + z_C^* \cdot \sqrt{\hat{p} \cdot \hat{q}/n}\right]. \tag{3.43}$$

A large collection of studies will result in a distribution of sample proportions according to $N(\hat{p}, SE_{\hat{p}})$, by the Law of Large Numbers.

Choosing a margin of error $\delta_C$ for a fixed $C$ necessitates a sample size of approximately $n \simeq \hat{p} \cdot \hat{q} \cdot (z_C^*/\delta_C)^2$. A value for $\hat{p}$ needs to be chosen *a priori*, but in many circumstances a reasonable value exists. For example, in an election, typically the leading candidate will not have much more than 50% support, so $\hat{p} \simeq 0.5$ is a reasonable choice to estimate a sample size for polling the electorate.

## Means

When a quantitative random variable $X$ is sampled, the sample mean $\bar{x}$, from Eq. (3.4), and standard deviation $s_x$ given in Eq. (3.5) are commonly computed statistics. The standard error, for a data set of size $n$, is defined to be $SE_x \equiv \sigma/\sqrt{n} \simeq s_x/\sqrt{n}$, where $\sigma$ is the population standard deviation. Supposing that $\sigma$ is unknown, as is often the case, the $C$-level confidence interval is approximately,

$$\mathcal{I}_C(x) \simeq \left[\bar{x} - z_C^* \cdot s_x/\sqrt{n},\ \bar{x} + z_C^* \cdot s_x\sqrt{n}\right]. \tag{3.44}$$

A large collection of studies will result in a distribution of sample means according to $N(\mu, SE_x)$, by the Central Limit theorem.

For a margin of error $\delta_C$, a sample size of approximately $n \simeq (z_C^* \cdot s_x/\delta_C)^2$ is required. A preliminary study may be needed to choose an *a priori* value for $s_x$. Also, this formula requires adjustments for small sample sizes.[7]

### ■ 3.8.2 Hypotheses Testing

This book began with a discussion of how the critical question,

*"Is the diagonal of a unit square a fraction?"*

has a definite answer, as a result of a logical contradiction. The conclusion is a certainty, in the negative.

In the observational sciences one may suppose that a long-held belief is actually false. Such beliefs may cluster around schools of thought like those that have occurred with theories for the extinction of the dinosaurs, or the true age and size

---

[7] In particular, the student's $t$ distribution should be used. However, for this introductory presentation, the normal distribution is sufficiently accurate.

of the universe. In these situations, the support for one theory may be increased by establishing contradictions in competing theories. This motivates one group to gather evidence for a contradiction. Using probability, an estimate of how significant a contradiction appears to be, can be computed as a *statistical hypothesis test*. The more that questions are formulated and tested statistically, the more that the truth-value of a *scientific hypothesis* can finally be realized.

## Proportions

To measure a distribution of qualities of individuals, one needs to divide a population into at least two categories. Suppose, for example, that an SRS of 250 people results in 143 supporters of one candidate in an upcoming election. What is the likelihood that this is a rare event, and that the candidate would not win an election if held on the day of the data gathering? There are two possibilities, expressed as

$$H_0 : p = p_0 = 0.5, \quad H_1 : p > p_0 = 0.5, \tag{3.45}$$

where statement $H_0$ is called the *null hypothesis* and statement $H_1$ is called the *alternate hypothesis*. The statistical inference process consists of assuming $H_0$, computing the probability of $\hat{p} \geq 144/250 = 0.572$ occurring, and comparing to a specified *significance level* $\alpha \in (0, 0.5) = (0\%, 50\%)$. The level of significance is related to the chance of making an error in judgment based on the data.

**Type I Error:** Rejecting $H_0$ when it is in fact true. $P(\text{type I error}) = \alpha$.

The value $(1-\alpha)$ is known as the *specificity* meaning that a smaller $\alpha$ will reduce the probability of making a decision to reject $H_0$. However, this will result in rejecting $H_1$ more often, even when it is in fact true.

**Type II Error:** Do not reject $H_0$ when $H_1$ is true. $P(\text{type II error}) = \beta$.

The value $(1 - \beta)$ is known as the *sensitivity* or *power* of the test. A smaller value of $\beta$ will reduce the probability of rejecting $H_1$ when it is true.

To begin an analysis of proportions, calculate the standardized *test statistic*

$$t_* \equiv \frac{\hat{p} - p_0}{SE_{p_0}} = \frac{\hat{p} - p_0}{\sqrt{p_0 \cdot q_0/n}} = \frac{(\hat{p} - p_0) \cdot \sqrt{n}}{\sqrt{p_0 - p_0^2}}.$$

In the polling example $t_* = (0.572 - 0.5) \cdot \sqrt{250}/\sqrt{0.5 - 0.25} \simeq 2.28$. Assuming that $t_*$ is nearly $N(0, 1)$, the likelihood of getting at least this value is

$$P(t \geq t_*) \simeq P(z \geq 2.28) \ = \ \Phi_{0,1}[-2.28] \simeq 0.0115 \simeq 1\%,$$

using the formula in Eq. (3.23). It is said that the data gives a $P$-value of 0.0115. Here the $P$-value is quite small. Thus $H_0$ is rejected at the $\alpha = 5\%$ level, meaning that this size poll would get this high a support 1 out of 87 times, assuming that the actual support is only 50%.[8] This is strong evidence that the assumption $H_0$ is false. However, it is not a proof, and in fact one must be concerned with how the poll was taken. For instance, if data was gathered outside of the candidate's campaign office, or due to a response enticement, then bias may have distorted the study. Note that the size of the electorate does not enter this calculation. However, the larger the population, the more difficult it may be to obtain an SRS. In these cases a *stratification of the population* is needed to control sampling costs while avoiding convenience bias.[9]

The null hypothesis in Eq. (3.45) only tests $H_0$. However, one may wish to ask, "What is the probability of the poll detecting 60% support for the candidate?" The alternative hypothesis test can now be written as

$$H_0 : p \leq p_0 = 0.5, \quad H_1 : p = p_1 = 0.6,$$

where $p_1$ is suggested by the data value of $\hat{p} = 0.572 \simeq 0.6$. Now the probability of making the wrong conclusion, if $p_1 = 0.6$ is correct, is

$$\beta \equiv P(p \leq p_0) = P(z < -3.16)) = \Phi_{0,1}[-3.16] \simeq 0.0008,$$

which means that a sample size of $n = 250$ is sufficient to detect such a large support. This is no surprise because the standard error is $SE_{0.50} = \sqrt{250}/\sqrt{0.5 \cdot 0.5} = 0.0361 \simeq 3\%$. So the power of detecting $p_1 = 55\%$ is still quite high. However, for $p_1 = 51\%$, detecting a majority support will occur with frequency

$$(1 - \beta) = P(p > p_0) = P(z > -0.316) = 1 - \Phi_{0,1}[-0.316] \simeq 1 - 0.359 = 0.641,$$

which is a rather low power. In many cases $(1 - \beta) \geq 0.80$ is desirable [49]. The goal, in general, is to make both $\alpha$ and $\beta$ as small as possible for the two different states of the system $p_0$ and $p_1$.

## Means

In measuring a quantitative variable, variation can be due to instrumentation inaccuracy, reading error, or natural environmental effects. Some questions, like *"What is the length of one's finger?"* depend on specific definitions that may have a variety

---

[8] Here $1/87 \simeq 0.0115$. The commonly used $P$-value of 0.05 corresponds to 1 out of 20.

[9] Some discussion on strata within a population can be found in [49] and [16].

of interpretations by different people. Microscopically, the end of one's finger is not really definable. This is a common issue in weights and measures [16]. For example, suppose a campus-wide study is performed to measure hand widths and the mean is found to be $\mu = 47.2$ mm. How does this compare to the sample mean of $\bar{x} = 44.0$ mm? In particular, is the class exceptionally different from the rest of the campus? The question can be formulated as

$$H_0 : \mu = \bar{x} = 44.0 \text{ mm}, \ H_1 : \mu \neq \bar{x} = 44.0 \text{ mm},$$

which is called a *two-sided* or *two-tailed* hypothesis test. If $H_0$ is true, this really means that the difference $\mu - \bar{x} = 47.2 - 44.0 = 3.2$ is not significant at a chosen level $\alpha$. In this case a comparison with $z_{1-\alpha}$ must be made so that

$$P(|z| > |z_{1-\alpha}|) = P(z > z_{1-\alpha}) + P(z < -z_{1-\alpha}) = \alpha.$$

Thus one needs to compute the probability of $\mu - \bar{x} \geq 3.2$. This requires a measure of the variability of data, which was estimated by the sample standard deviation to be $s_x = 11.54$, from data in Eq. (3.3).[10] If the population standard deviation $\sigma$ is available and the population distribution is nearly normal, then the Z-score definition in Eq. (3.22) suggests a natural, standardized test statistic. Otherwise, the approximation

$$t_* = \frac{\bar{x} - \mu}{SE_x} = \frac{\bar{x} - \mu}{\sigma/\sqrt{n}} = \frac{(\bar{x} - \mu)\sqrt{n}}{\sigma} \simeq \frac{(\bar{x} - \mu)\sqrt{n}}{s_x},$$

can be used.[11] The data for this example gives $t_* = (47.2 - 44.0) \cdot \sqrt{10}/11.54 \simeq 0.88$. Thus, from formula (3.23),

$$P(|t| \geq |t_*|) \simeq P(|z| \geq 0.88) = 2 \cdot \Phi_{0,1}[-0.88] \simeq 2 \cdot (0.1895) \simeq 38\%,$$

which is not small. Thus, $H_0$ cannot be rejected,[12] because the $P$-value 0.379 suggests that the sample obtained for the class will occur about 2 out of 5 times. Hence the hand widths of the class are typical for the campus. However, a deeper investigation would involve comparing the actual distributions of the class versus the campus population. In this study only the means were compared, but this is how one typically starts with a statistical analysis.

---

[10] Compare this value with the population standard deviation of $\sigma_H$ computed in Eq. (3.13).

[11] If $s_x$ is used instead of $\sigma$, then the student's $t$ distribution, with $(n-1)$ degrees of freedom, is considered to be the accurate value for probability in real applications, unless the sample size is large, $n \geq 30$.

[12] One might also say that the data failed to reject the null, as suggested in [75].

### 3.8.3 Differences of States

To characterize states, parameters like means, proportions, and variances must be estimated from data. If the parameters change, a continued monitoring of the system through data collection and analysis should be expected to reveal the transition from one state to another. One method for keeping track of transitions is to compute a *differences statistic*.

#### Proportions

For two different states of proportions $p_0$ and $p_1$ the variance is defined to be

$$Var(p_0, p_1) = \frac{Var(p_0)}{n_0} + \frac{Var(p_1)}{n_1} = \frac{p_0 \cdot q_0}{n_0} + \frac{p_1 \cdot q_1}{n_1}, \quad (3.46)$$

and the standard error is $SE(p_0, p_1) = \sqrt{Var(p_0, p_1)}$. Then the differences test statistic is

$$t_* = \frac{p_0 - p_1}{SE(p_0, p_1)}, \text{ or in general: } t_*^{\Delta p} = \frac{(p_0 - p_1) - \Delta p}{SE(p_0, p_1)}, \quad (3.47)$$

where $\Delta p$ is a suggested difference between the proportions. For example, the gender difference in the class was found to be $p_F - p_M = 0.2$. However the true campus value was $\Delta p = 0.55 - 0.45 = 0.10$. Thus to inquire whether there is a gender preference for the class, compute

$$t_*^{\Delta p} = \frac{(p_F - p_M) - \Delta p}{\sqrt{p_F \cdot q_F/n_F + p_M \cdot q_M/n_M}} = \frac{(0.2) - 0.1}{\sqrt{0.6 \cdot 0.4/6 + 0.4 \cdot 0.6/4}} = \frac{0.1}{\sqrt{0.1}} = 0.316.$$

Under various assumptions $t_*^{\Delta p}$ is approximately $N(0, 1)$, which in this case implies a *P*-value of, using the approximation in Eq. (3.24),

$$P(|t| > |t_*^{\Delta p}|) = 2 \cdot P(t > t_*^{\Delta p}) = 2 \cdot \Phi_{0,1}[-0.316] = 0.748 \simeq 75\%.$$

This expresses a lack of evidence for a significant difference in gender preference.

#### Means

Suppose two different states of a system are characterized by means $x_0$ and $x_1$. The *difference variance* is defined to be

$$Var(x_0, x_1) = \frac{Var(x_0)}{n_0} + \frac{Var(x_1)}{n_1} = \frac{s_0^2}{n_0} + \frac{s_1^2}{n_1}, \quad (3.48)$$

giving the standard error $SE(x_0, x_1) = \sqrt{Var(x_0, x_1)}$. Then the differences test statistic for means is

$$z_* = \frac{x_0 - x_1}{SE(x_0, x_1)}, \text{ or in general: } x_*^{\Delta x} = \frac{(x_0 - x_1) - \Delta x}{SE(x_0, x_1)}, \tag{3.49}$$

where $\Delta x$ is a proposed difference between the means. In the hand-width example, to suggest that female hand widths are not significantly less than those of male students can be tested using the statistics computed in Eq. (3.18),

$$z_* = \frac{x_F - x_M}{SE(x_F, x_M)} = \frac{37.0 - 54.5}{\sqrt{8.9^2/6 + 7.9^2/4}} = \frac{-17.5}{5.37} = -3.26,$$

which, assuming an $N(0, 1)$ distribution for $z_*$ gives a $P$-value of

$$P(z < z_*) = P(z < -3.26) = \Phi_{0,1}[-3.26] = 0.0006.$$

This is significant evidence for a difference in hand widths due to gender.

### ■ Exercises

1. Compute the 95% confidence interval for the proportion of males in the class, based on the data in Eq. (3.3) using the appropriate value in Eq. (3.42). Does this indicate strong evidence that the entire campus has more females than males?

2. Compute the 90% confidence interval for the class's hand-width data in Eq. (3.3) using the appropriate value in Eq. (3.42). Does it contain the campus value of $\mu = 47.2$? What is the power of detecting $\mu$ from the class's data?

3. A claim is made that females prefer the class, as compared to males. Set up a hypothesis test for this claim given that the campus proportion of females is $p = 0.55$. Compute the $P$-value, and compare it to a 5% significance level.

4. A claim is made that female hand widths from the class are representative of the campus values for both genders. Test this hypothesis using the statistics in Eq. (3.18) and the campus mean of $\mu = 47.2$. Be sure to compute the $P$-value, and compare to a 10% significance level.

### ■ 3.9 Linear Regression

It is now possible to make a connection between input, output, and a model that endeavors to explain the process. The procedure consists of a sequence of straightforward computations involving the model and some data. The interpretation of a model, however, is part of the art of science, and is always open for debate.

### ■ 3.9.1 Example of the Deaf Frog

A scientist observes that frogs with three legs do not jump as far, when a loud noise is made, as frogs with four legs. The scientist suspects that this is because the ears of frogs are in their legs. The goal is to test the belief that this is true. The design of the experiment will consist of:

**(1)** Comparing the *distances traveled* to the *number of ears*; and
**(2)** Normalizing the data using the distance achieved with four ears.

Suppose $n$ frogs are chosen for this experiment. A loud noise is applied five times near frog-$j$, and the distances that it jumps are measured and averaged, giving the number $D_{j,4}$.

Now one of the legs is muffled using sound-proof tape. The leg chosen for each frog is decided by a random process. The same loud noise is applied five times giving the average distance traveled $D_{j,3}$. Similarly $D_{j,2}$ and $D_{j,1}$ are found. When all four legs are muffled, the frog does not jump, thus $D_{j,0} = 0$, because in this case, the frog is deaf! This is an example of *ratio data*

$$\{\langle D_{j,0} = 0, D_{j,1}, D_{j,2}, D_{j,3}, D_{j,4} \rangle\}_{j=1}^n, \tag{3.50}$$

which suggests new variables $d_{j,e} \equiv D_{j,e}/D_{j,4}$ that normalize the data sequence,

$$\{\langle 0, D_{j,1}D_{j,4}, D_{j,2}/D_{j,4}, D_{j,3}/D_{j,4}, 1 \rangle\}_{j=1}^n = \{\langle 0, d_{j,1}, d_{j,2}, d_{j,3}, 1 \rangle\}_{j=1}^n.$$

The mathematical modeling process can now be stated as:

**Factor:** $e$ = number of ears;
**Response:** $d$ = distance traveled (normalized); and
**Model:** $\hat{d} = \mathcal{M}(e)$.

The simplest model corresponding to a linear response for ratio data is $\mathcal{M}(e) = m \cdot e$ where $m$ is the *parameter of the model*. Here the model is forced to go through the origin $\mathcal{M}(0) = 0$, and the data must be used to determine the $C$-confidence interval $I_C = (m - \delta_C, m + \delta_C)$ where $\delta_C > 0$ is the margin of error.

### ■ 3.9.2 Linear Models

Suppose a list of $m \in \mathbb{N}$ (dependent) variables $\mathcal{R} \equiv \langle y^k \rangle_{k=1}^m$ are believed to behave mainly in response to $n \in \mathbb{N}$ (independent) variables $\mathcal{F} \equiv \langle x^j \rangle_{j=1}^n$. (The superscript indexes $j$ and $k$ are different from the subscript data-point indices.) The set $\mathcal{R}$ consists of the *responses* and $\mathcal{F}$ are the *factors* of the system. A *linear model* proposes the existence of slopes $a^k{}_j$ and intercepts $b^k$

$$\mathcal{S} \equiv \left\{ \langle a^k{}_j \rangle_{j=1,k=1}^{n,m},\ \langle b^k \rangle_{k=1}^{m} \right\},$$

so that the linear model between the $\mathcal{F}$ and $\mathcal{R}$ variables becomes

$$y^k = \sum_{j=1}^{n} a^k{}_j \cdot x^j + b^k.$$

The model parameters $\mathcal{S}$ are elements of a *parameter space*, also called the *model space*, and it is $(n \cdot m + m)$ *dimensional*. Typical questions in science are:

**(1)** How should the parameters (coefficients, constants) be determined?
**(2)** What is the range of applicability of the model?
**(3)** Within its range of application, how confident can one be with its predictions?

### ■ 3.9.3 Scatter Plots

The simplest nontrivial example is that of one factor $x^1 = x$ and one response $y^1 = y$. These are assumed to be random variables, but with a possible association. Given a data set $\{\langle x_i, y_i \rangle\}_{i=1}^{n}$ one can generate a scatter plot. Suppose three frogs are used. The data gathered is shown in Table 3.2.

| Relative Distances | Number of Ears | | | | |
|---|---|---|---|---|---|
| | 0 | 1 | 2 | 3 | 4 |
| $d_{1,k}$ | 0 | 0.20 | 0.41 | 0.69 | 1 |
| $d_{2,k}$ | 0 | 0.31 | 0.57 | 0.77 | 1 |
| $d_{3,k}$ | 0 | 0.23 | 0.53 | 0.72 | 1 |

**TABLE 3.2** ■ Normalized deaf frog data.

Here the number of ears $k$ is the factor, and the relative distances jumped $d_{j,k}$ are the responses. These data are now easily plotted as shown in Figure 3.7.

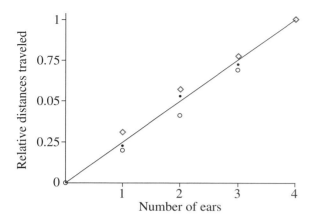

**FIGURE 3.7** ■ Scatter plot for relative distances traveled vs. number of ears.

### ■ 3.9.4 Control Charts

Connecting two associated variables may be a third variable $t$ so that

$$Data \equiv \{\langle x_j, y_j \rangle\}_{j=1}^n = \{\langle x(t_j), y(t_j) \rangle \mid \{t_j\}_{j=1}^n\}. \quad (3.51)$$

In this case $t$ is called a *lurking variable*. For example, a high temperature and body aches are often associated with each other, but the lurking variable may be the presence of a flu. In this case, $x$ and $y$ are called *symptoms* whereas $t$ is the *illness*. However, one can have body aches without a fever, and vice versa, without the presence of the flu. Thus $x$ and $y$ are not necessarily correlated. Still, one may be able to show that a high body temperature can cause aching joints and muscles.

Lurking variables are sometimes identifiable by plotting a data set as a time sequence, known as a *control chart*. For example, consider the counts for the number of *weather-extreme events* between 1900–1995, as published in [37],[13]

$$\begin{array}{ccccccccc} 20 & 22 & 17 & 18 & 23 & 20 & 20 & 18 & 19 \\ 23 & 18 & 19 & 16 & 19 & 22 & 21 & 22 & 22 \end{array} \quad (3.52)$$

If a frequency plot is used for the data then it appears to be bimodal, which may be due to a natural climatic cycle, like that which occurs between El Niño and La

---

[13]Five-year averages are used. Subtle issues on how a *count* is made will be ignored. Also, the data is presented for purposes of discussion.

Niña. One may suspect that there are two equilibria that climate switches between. However, a time series chart shows that there have been more events recently. This suggests a switching that may be occurring from one climactic state to another.[14]

One question that may arise is whether there is indeed an upward trend in the number of extreme weather events per year. This would suggest that there is a change over time in the parameters governing the system. Thus, if one supposes that the data in Figure 3.8, for years before 1980, imply that the system is in *statistical control*, then for the years starting in 1980, the system is behaving *out of control*. However, a measure needs to be created to assess the seriousness of this new behavior.

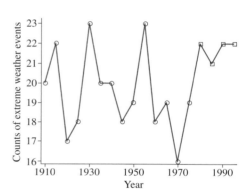

| frequency | | | | counts |
|---|---|---|---|---|
| | | o | o | 23 |
| x | x | x | o | 22 |
| | | | x | 21 |
| o | o | o | | 20 |
| o | o | o | | 19 |
| o | o | o | | 18 |
| | | | o | 17 |
| | | | o | 16 |

**FIGURE 3.8** ■ Left: Frequency plot of recent counts "x" and more past counts "o" of weather extremes. Right: Time series for the same data.

### ■ 3.9.5 Correlation Coefficient

In many systems there are too many hidden factors for a scientist to consider when trying to be deterministic about the actual relationship between variables. Thus, one starts by looking for associations between a few variables rather than correlations between all variables. However, the goal remains to establish a cause-and-effect relationship. With this in mind, the *correlation coefficient* $r$ between an independent variable $x$ and a *presumed* dependent variable $y$ is defined to be

$$r \equiv \frac{c_{x,y}}{s_x \cdot s_y}, r \in [-1, 1], \tag{3.53}$$

---

[14]More examples applied to global warming can be found in [75].

where the *covariance* $c_{x,y} \equiv Cov(Data)$ of the sequence in Eq. (3.51) is defined as

$$Cov(Data) \equiv \frac{1}{n-1} \cdot \sum_{j=1}^{n}(x_j - \overline{x}) \cdot (y_j - \overline{y}), |Cov(Data)| \leq s_x \cdot s_y. \qquad (3.54)$$

The linear relationship, obtained using Eq. (3.53) and Eq. (3.54),

$$y = a \cdot x + b, \ a \equiv r \cdot \frac{s_y}{s_x} = \frac{c_{x,y}}{s_x^2}, \ b \equiv \overline{y} - a \cdot \overline{x}, \qquad (3.55)$$

is called the *least-squares* fit of the data to a line. The quantities $a$ and $b$ are estimated parameters of the model. Note that when $r = 1$ or $-1$, the usual equation of a line is obtained, but this can only occur if the data points lie exactly on a line. Here $r$ is a quantitative measure of the linear relationship between $x$ and $y$. Furthermore, it can be shown that [31]

$$r^2 = \text{proportion of variability in } y \text{ explained by variations in } x.$$

The statistic $r^2$ can thus be thought of as the confidence level that one can have in the linear model in Eq. (3.55).

### ■ 3.9.6 Significant Association

Suppose that a model has been suggested, and parameters for a *best-fit model* obtained. How does one measure the confidence in the model? As discussed, computing $r^2$ is one method. If this value is small, a level of significance may need to be assessed in order to justify further pursuing an association.

Suppose that an examination of data indicates some association. Then the data's correlation coefficient $r \neq 0$ can be used to construct the *confidence statistic*

$$Z_{conf}[r] = \frac{\ln(1+r) - \ln(1-r)}{2/\sqrt{n-3}}, \qquad (3.56)$$

where $n \in \mathbb{N}$ is the sample size.[15] A hypothesis test that addresses the question of whether a significant correlation exists, is expressed as

$$H_0 : Z_{conf} = 0, \ H_1 : Z_{conf} \neq 0,$$

---

[15]The statistic $Z_C$ in Eq. (3.56) is presented on page 544 of [31] with a more detailed discussion. It is related to Fisher's z-transform [55].

which is now a *two-sided* test. To be 95% confident that an association between $x$ and $y$ has been found, it is sufficient that $|Z_{conf}| > Z_{0.95} \simeq 1.96$. Here $Z_{conf}$ is designed so that it is approximately $N(1,0)$, so for different levels of confidence, one must use a different Z-score. Different areas of science, medicine, and engineering have different standards for levels of confidence that typically range from 90% to 99%. A professional must assess the risk or cost of a model's failure in deciding on a confidence level, keeping in mind that 100% is not possible, and that there are increasing time and resource costs associated with trying to be too confident.

### ■ Exercises

1. Consider a system $y = f(x)$ where the $x$-inputs $\langle 1, 2, 3, 4 \rangle$ result in the corresponding $y$-outputs $\langle 1, 2, 4, 5 \rangle$.

   (a) Find the best-fit *linear* model $y = a \cdot x + b$ using linear regression, by first computing the correlation coefficient $r$.
   (b) Compute the $Z_{conf}$ statistic for the linear model.
   (c) Find the best-fit *exponential* model $y = A \cdot \exp(k \cdot x)$. Be sure to create a new variable $w \equiv \ln(y) = k \cdot x + \ln(A)$, and new output sequence. Then use linear regression to find an association between the variables $w$ and $x$.
   (d) Compute the $Z_{conf}$ statistic for the exponential model, and compare it to the linear model. Which model is more reasonable to use?

2. It is proposed that the length of a pendulum $L$ is directly related to the period of oscillations $T$. Data is collected in Table 3.3, where counts correspond to the number of completed cycles of the pendulum, and $Hz \equiv cycles/seconds$.

   (a) Find the average length $\overline{L}$ and period $\overline{T}$.
   (b) Make a table of the deviations $\{L_i - \overline{L}\}$ and $\{T_i - \overline{T}\}$, the squared deviations $\{(L_i - \overline{L})^2\}$ and $\{(T_i - \overline{T})^2\}$, and the products $\{(L_i - \overline{L}) \cdot (T_i - \overline{T})\}$.
   (c) Compute the sum-of-the-squared deviations $SSD_L = \sum(L_i - \overline{L})^2$, $SSD_T = \sum(T_i - \overline{T})^2$, and the sum-of-the-product deviations $SPD_{L,T} = \sum(L_i - \overline{L}) \cdot (T_i - \overline{T})$.

| Variables | Data | | | | |
|---|---|---|---|---|---|
| $L$ (cm) | 30 | 50 | 80 | 100 | 110 |
| $counts/15$ sec | 13 | 10 | 9 | 8 | 7 |
| $T$ (sec) | 1.15 | 1.50 | 1.67 | 1.88 | 2.14 |

**TABLE 3.3** ■ Length of pendulum vs. period of oscillations

(d) For $n = 5$ compute the sample variances $s_L^2 = SSD_L/(n-1)$ and $s_T^2 = SSD_T/(n-1)$, and the covariance $C_{L,T} = SPD_{L,T}/(n-1)$.
(e) Compute the sample standard deviations $s_L$ and $s_T$.
(f) Make a table of the standardized deviations $\{(L_i - \bar{L})/s_L\}$ and $\{(T_i - \bar{T})/s_T\}$.
(g) Compute the correlation coefficient, for $n = 5$,

$$r = \frac{SPD_{L,T}}{\sqrt{SSD_L \cdot SSD_T}} = \frac{1}{n-1} \sum_{i=1}^{n} \frac{(L_i - \bar{L})}{s_L} \cdot \frac{(T_i - \bar{T})}{s_T},$$

and $r^2$. What percentage of the variability is explained by the model?
(h) Assume the relationship between the variables

$$s_L \cdot (\hat{T} - \bar{T}) = r \cdot s_T \cdot (L - \bar{L}),$$

and use algebraic rearrangement to find the linear regression model $\hat{T} = a \cdot L + b$.
(i) Compute the confidence statistic $Z_{conf}$ in the model from Eq. (3.56). Does the data in Table 3.3 suggest a correlation between the length of a pendulum $L$ and the period $T$?

3. Consider the data in Eq. (3.52) displayed in Figure 3.8.

   (a) Compute the average frequency of weather extremes for the most recent 4 data values $\bar{f}_r$ and for the 14 past data values $\bar{f}_p$.
   (b) Compute the variance for the most recent 4 data values $s_r^2$, and the past 14 data values $s_p^2$, along with the combined variance $v_{p,r} = s_p^2/14 + s_r^2/4$.
   (c) Is there a significant difference between the recent 4 data values, and the past 14 values at the $\alpha = 5\%$ level?
   (d) Suppose that the recent data represents a new state. By making appropriate assumptions, compute the power $(1-\beta)$ of detecting a transition away from the past average frequency of weather extremes to the new state.

4. Recent data on the ocean's heat content, presented in [36], gives the sequence $\{0, -1, -3, -1, 4, 2, 3, 4, 7, 8, 12\}$ (in $10^{22} J$), at 5-year intervals from 1955–2005.

   (a) Plot the time series and dot plot, as in Figure 3.8 on page 124.
   (b) Compute the sample mean and variance for the past years 1955–1985.
   (c) Compute just the mean for the recent years 1990–2005.
   (d) Compute the difference variance and test for the significance of the difference between the temperature means at the $\alpha = 1\%$ level.

## 3.10 Main Theorems

A possible misinterpretation of the field of statistics is that it is a nonrigorous area of study. In fact, the field of statistics rests on the firm foundation of probability theory. A series of results are presented here with the intention of making it clear as to what extent one can know anything from the observation and analysis of collected data. For any observational statement $\phi$, the notation $P(\phi)$ will refer to the *probability* that $\phi$ is true. Then

$$P : \{\text{statements of a system }\} \to [0,1], \; P(\text{tautology}) = 1, P(\text{contradiction}) = 0.$$

**Definition 3.3** A sequence of statements $\langle \phi_n \rangle$ holds true, *asymptotically almost certainly* (a.a.c.), or *asymptotically almost surely* (a.a.s.), if

$$\forall \epsilon \in (0,1) \implies \exists N_\epsilon \in \mathbb{N} \ni (\forall n \geq N_\epsilon \implies P(\phi_n) > 1 - \epsilon).$$

One says that the sequence $\{\phi_n\}$ converges in probability to a *truth-value* of true.

### 3.10.1 Law of Large Numbers and Central Limit Theorems

Variations in counts or measurements of a population are potentially due to a multitude of reasons. So much so, that a type of regularization occurs for large surveys or sample sizes. Consequently, if many samples of size $n$ are taken from the same population (with replacement), then the distribution of either proportions or averages will be (nearly) normally distributed. The statements in Table 3.4 and Table 3.5 make these observations precise. An outline of the proofs, showing the connection to results from probability theory, is given in the next section. For fixed $\epsilon > 0$ the *Law of Large Numbers* (LLN) says that,

$$\phi_n^p \equiv (|\hat{p} - p| \leq \epsilon) \qquad \text{and} \qquad \phi_n^x \equiv (|\overline{x} - \mu| \leq \epsilon), \qquad (3.57)$$

converge in probability to be true. This follows rigorously by defining,

$$N_\epsilon^p \equiv p \cdot q / (\epsilon^2 \cdot (1 - \epsilon)) \qquad \text{and} \qquad N_\epsilon^x \equiv \sigma^2 / (\epsilon^2 \cdot (1 - \epsilon))$$

respectively. More specific results, involving the normal distribution, are called *Central Limit theorems* (CLT). Here, two special cases are stated that have been used in this chapter. A detailed proof of the CLT using calculus can be found in Ross [61].

| Law of Large Numbers | Law of Large Numbers |
|---|---|
| for *proportions* | for *averages* |
| As a population is surveyed repeatedly, the relative frequency of success $\hat{p}$ approaches the true proportion $p$ so that for fixed $\epsilon > 0$, and survey total $n$, sufficiently large, $P(|\hat{p} - p| \geq \epsilon) \leq 1 - p \cdot q/(\epsilon^2 \cdot n),$ where $p \cdot q = p \cdot (1-p) = Var(p)$. | The larger a sample taken from a population, the more likely the sample mean $\overline{x}$ is close to the true mean $\mu$ so that for fixed $\epsilon > 0$, and sample size $n$, sufficiently large, $P(|\overline{x} - \mu| \geq \epsilon) \leq 1 - \sigma^2/(\epsilon^2 \cdot n),$ where $\sigma^2 = Var(x)$. |

**TABLE 3.4** ■ LLN for proportions and means.

| Central Limit Theorem | Central Limit Theorem |
|---|---|
| for *proportions* | for *averages* |
| $\hat{p}$ is nearly $N(p, \sqrt{p \cdot q/n})$. | $\overline{x}$ is nearly $N(\overline{x}, \sigma/\sqrt{n})$. |
| For fixed $\epsilon > 0$ and $n$ sufficiently large, $P(|\hat{p} - p| \geq \epsilon) \leq 2 \cdot \Phi_{0,1}\left[-\epsilon \cdot \sqrt{n/(p \cdot q)}\right].$ | For fixed $\epsilon > 0$ and $n$ sufficiently large, $P(|\overline{x} - \mu| \geq \epsilon) \leq 2 \cdot \Phi_{0,1}\left[-\epsilon \cdot \sqrt{n}/\sigma\right].$ |

**TABLE 3.5** ■ CLT for proportions and averages.

### ■ 3.10.2 Theoretical Details

In the following, $X$ will be a random variable with a distribution that allows the computation of the mean and variance using expectations. In particular, the expectation values $\langle X \rangle$, $\langle X^2 \rangle$ are required to be finite real numbers. It may help to think of a random variable with a discrete distribution, which justifies the use of the formulas in Eq. (3.14).

■ **Lemma 3.1**
**Markov Inequality:** Suppose a nonnegative random variable $X$ has mean, or average $\langle X \rangle = \mu \geq 0$. Then $\forall r > 0$,

$$P(X \geq r) \leq \min\{1, \mu/r\}. \tag{3.58}$$

This result, which can be improved, holds because the mean is strongly affected by outliers. Thus, if the mean is small, and the random variable is nonnegative, then there is little chance of obtaining a very large value for $X$.

**Proof** The trick is to define a random discrete variable $Y_r$ in terms of $X$, as

$$(X/r \geq 1) \implies (Y_r = 1), \ (X/r < 1) \implies (Y_r = 0),$$

which can also be written as $Y_r(X) \equiv \chi_{[r,\infty)}(X)$. By assumption $X/r < 0$ is impossible. Thus the inequality $Y_r \leq X/r$ always holds. Taking the expectation of both sides of $r \cdot Y_r \leq X$ gives

$$r \cdot P(X \geq r) = \langle r \cdot Y_r \rangle \leq \langle X \rangle = \mu. \tag{3.59}$$

Also, $P(X \geq r) \leq 1$ for all probabilities. Thus Eq. (3.58) holds. □

■ **Lemma 3.2**
**Chebyshev's Inequality:** Suppose a real-valued random variable $X$ has mean $\mu$ and variance $\sigma^2 \in \mathbb{R}_0^+$. Then $\forall r > 0$,

$$P(|X - \mu| \geq r) \leq \min\left\{1, \sigma^2/r^2\right\}. \tag{3.60}$$

This slightly more informative result is still rather weak and can be improved if more details of the distribution of the random variable are known [61].

**Proof** Here, the trick is to define the random variable

$$W_r \equiv \chi_{[r,\infty)}(|X - \mu|) = \chi_{[r^2,\infty)}\left((X - \mu)^2\right),$$

which is clearly nonnegative. The Markov inequality in Eq. (3.58) implies that

$$P(|X - \mu| \geq r) = P((X-\mu)^2 \geq r^2) \leq \min\{1, \sigma^2/r^2\}, \quad (3.61)$$

where, by definition, $\langle (X-\mu)^2 \rangle = Var(X) = \sigma^2$. □

**Remark 3.3** By taking the positive square root, $(W_r \geq r^2) \iff (|X-\mu| \geq r)$. Furthermore, since $r$ was arbitrary, one can make the replacement $r = k \cdot \sigma$. This gives the more common form of the Chebyshev's inequality

$$P(|X - \mu| \geq k \cdot \sigma) \leq 1/k^2. \quad (3.62)$$

The interpretation of Eq. (3.62) is as follows:

> The *probability* of observing the value of a random variable being $k$ *standard deviations* from the *mean* is less than $1/k^2$.

An observation that is three standard deviations from the mean is considered to be an outlier. Chebyshev's inequality says that, typically, no more than one outlier is expected out of every nine observations. Of course one can never rule out a possible *clustering* of outlier events. However, these events are even more rare. For example, two outliers in a row would not be expected to occur more often than 1 out of 81 pairwise successive trials.

■ **Theorem 3.2**

**Weak Law of Large Numbers:** Suppose a random variable $X$ has mean $\mu$ and variance $\sigma^2$. For each $n$ suppose a sequence of observations $\{x_j\}_{j=1}^n$ are made of $X$ with averages $\bar{x}$. For each $\epsilon \in (0,1)$, define the corresponding sequence of statements

$$\phi_n(\epsilon) \equiv (|\bar{x} - \mu| \leq \epsilon).$$

Then the sequence $\langle \phi_n(\epsilon) \rangle$ holds true a.a.c., or simply $\lim_{n \to \infty} P(\phi_n(\epsilon)) = 1$.

**Proof** The trick is to define a random variable $U_n$, for each $n \in \mathbb{N}$, in terms of $n$-copies of $X$, as

$$U_n = (X_1 + X_2 + \cdots + X_n)/n = \frac{1}{n}\sum_{i=1}^n X_i.$$

Then from Eq. (3.16) $\langle U_n \rangle = \mu$. Furthermore, $\langle X_i \cdot X_j \rangle = \langle x_i \rangle \cdot \langle x_j \rangle = \mu^2$ for $i \neq j$ since the $i$th sample is *independent* of the $j$th sample. Thus

$$Var(n \cdot U_n) = \sum_{i=1}^{n} Var(X_i) = n \cdot \sigma^2 \implies Var(U_n) = \frac{\sigma^2}{n}, \quad (3.63)$$

using Eq. (3.17). Now, by Eq. (3.60) and Eq. (3.63),

$$P(|U_n - \mu| \geq \epsilon) = P\left((U_n - \mu)^2 \geq \epsilon^2\right) \leq \min\left\{1, \sigma^2/\left(n \cdot \epsilon^2\right)\right\}.$$

For fixed $\epsilon > 0$, the right-hand side vanishes as $n \to \infty$. □

To apply this result to a categorical property of a population, a random variable $X$ has to be defined. Suppose $x = 1$ if an individual selected has the property, and $x = 0$ otherwise. Then the distribution of $X$ is the same as the proportional distribution in the population. This implies the LLN as stated in Eq. (3.57).

For quantitative random variables $X$ that are real-valued, Theorem 3.2 gives a weak form of the LLN. Note that there is no specific reference to the distribution of $X$. Thus, these results have broad application. Stronger statements require methods of calculus and specifics of the distribution of the random variable $X$.

### ■ Exercises

1. Prove Markov's inequality in Eq. (3.58) in the special case of a finite, discrete distribution using the definition of expected value given in Eq. (3.14).

2. Prove, for a finite discrete distribution Eq. (3.14), that $Var(X) = \langle X^2 \rangle - \langle X \rangle^2$. Also explain why $\langle X^2 \rangle \geq \langle X \rangle^2$ for all these distributions.

3. Prove, for a finite discrete distribution Eq. (3.14), that $Var(X_1 + X_2) = 2 \cdot Var(X)$, where $X_1$ and $X_2$ are two different copies of the same random variable $X$. This requires independence of the two samples, i.e., $\langle X_1 \cdot X_2 \rangle = \langle X \rangle^2$. Also, using only expectations, prove the general case in Eq. (3.63), by induction.

4. Six ping-pong balls are marked with numbers $x \in \{7, 8, 8, 9, 10, 12\}$. A class of 10 students is asked to sample from the bag twice. The first time they are to blindly choose 2 and compute their average. Replacing the balls, mixing, then blindly choosing 4 balls, the average is computed again. The results are recorded in Table 3.6.

   (a) Compute the population mean $\mu$ and variance $\sigma^2$.
   (b) Compute the mean of the sample means $\overline{\overline{x_2}}$ for $n = 2$, and $\overline{\overline{x_4}}$ for $n = 4$.

| Stats ($n$) | | | | Sample | | Means | | | |
|---|---|---|---|---|---|---|---|---|---|
| $\bar{x}$ (2) | 11 | 9 | 9.5 | 11 | 8.5 | 10 | 10.5 | 8 | 8 | 10 |
| $\bar{x}$ (4) | 9.75 | 9 | 8 | 9 | 9.25 | 8.5 | 9.75 | 8.75 | 8.75 | 9.75 |

**TABLE 3.6** ■ Sampling of ping-pong balls with Top: $n = 2$ and Bottom: $n = 4$.

(c) Compute the sample variances $s_2^2$ for $n = 2$, and $s_4^2$ for $n = 4$.
(d) Compute the ratios $\sigma^2/(n \cdot s_n^2)$ for $n = 2$ and $n = 4$.
(e) Plot the $\bar{x}_2$ and $\bar{x}_4$ data on a side-by-side stem-leaf plot, with stems $\{7, 8, 9, 10, 11\}$ and using leaves $\{0, 3, 5, 7\}$.
(f) Roughly draw an $N(\bar{\bar{x}}_2, \sigma/\sqrt{2})$ over the stem-leaf plot for $n = 2$, and draw an $N(\bar{\bar{x}}_2, \sigma/2)$ over the stem-leaf plot for $n = 4$.
(g) Is this consistent with the CLT for averages?

## 3.11 Bivariate Analysis

Consider a system with two random variables $X$ and $Y$ that are to be measured and compared. To see if there is a linear association between these variables, a scatter plot of a data set $\{\langle x_j, y_j \rangle\}_{j=1}^n$ can be examined. To test whether a significant association exists, the scatter plot can be compared with the bivariate function

$$B(x, y) = B_0 \cdot e^{-S(x,y)} = \frac{B_0}{e^{S(x,y)}}, \quad (3.64)$$

where the normalization constant $B_0$ ensures that the volume under the distribution is 1. The exponent $S(x, y) \geq 0$ increases quadratically as $x, y \to \pm\infty$ to ensure that it behaves similarly to a normal distribution in each variable. The shape of this distribution depends on the following parameters of the random variables $X$ and $Y$:

$$\mu_X \equiv \langle X \rangle, \qquad \mu_Y \equiv \langle Y \rangle, \quad (3.65)$$
$$\sigma_X^2 \equiv \langle X^2 \rangle - \langle X \rangle^2, \qquad \sigma_Y^2 \equiv \langle Y^2 \rangle - \langle Y \rangle^2, \quad (3.66)$$
$$\rho_{X,Y} \equiv \frac{Cov(X, Y)}{\sigma(X) \cdot \sigma(Y)} = \frac{\langle X \cdot Y \rangle}{\sigma_X \cdot \sigma_Y}. \quad (3.67)$$

The distributions of $X$ and $Y$ must be used to compute the $\mu$, $\sigma$, and $\rho$. Note that it is not difficult to show that $\rho(X, Y) \in [-1, 1]$, which is referred to as the *correlation coefficient*. The normalization is found to be

$$B_0 = \left(2\pi \cdot \sigma_X \cdot \sigma_Y \cdot \sqrt{1 - \rho_{X,Y}^2}\right)^{-1}. \tag{3.68}$$

The exponent of the bivariate distribution is best expressed in terms of the transformed variables

$$Z_X \equiv \frac{X - \mu_X}{\sigma_X}, \ Z_Y \equiv \frac{Y - \mu_Y}{\sigma_Y}, \tag{3.69}$$

in which case

$$S(X, Y) = \frac{Z_X^2 - 2 \cdot \rho_{X,Y} \cdot Z_X \cdot Z_Y + Z_Y^2}{2 \cdot (1 - \rho_{X,Y}^2)}. \tag{3.70}$$

If $X$ and $Y$ have the distribution given by Eq. (3.64), Eq. (3.68), and Eq. (3.70), then either variable can be considered independent, suggesting the other is dependent. Suppose $Y$ is considered dependent on $X$. Then for each fixed $x$, the random variable $Y$ has a normal distribution with expectation

$$E(Y|x) = \rho_{X,Y} \cdot \frac{\sigma_Y}{\sigma_X} \cdot (x - \mu_X) + \mu_Y = \alpha \cdot x + \beta, \tag{3.71}$$

which is a linear function of $x$. Thus only $\alpha$ and $\beta$ need to be estimated from the data to obtain a model for the dependence of $y$ on $x$. Using the data to obtain an estimate of $\rho_{X,Y}$ gives a level of confidence for the model.

### ■ 3.11.1 Confidence in a Linear Model

From the bivariate analysis, a correlation coefficient $\rho \in [-1, 1]$ can be computed with sufficient information. If $\rho \neq 0$ is suspected, then the *confidence statistic* in Eq. (3.56) must be modified to

$$Z_{conf}(\rho)[r] = \frac{\ln[(1+r)/(1-r)] - \ln[(1+\rho)/(1-\rho)]}{2/\sqrt{n-3}}, \tag{3.72}$$

for sample size $n \in \mathbb{N}$.[16] Here $r$ is the statistic computed from data, and $\rho$ is the model parameter. For different levels of confidence $C \cdot 100\%$, a Z-score of $z_C = \Phi_{0,1}^{-1}[1 - C/2]$ can be estimated from Eq. (3.24) or Eq. (3.25). For example

$$z_{0.95} = \Phi_{0,1}^{-1}[1 - 0.95/2] = \Phi_{0,1}^{-1}[0.025] \simeq 1.96,$$

---

[16]The statistic $Z_{conf}(\rho)$ in Eq. (3.72) is given on page 546 of [31].

so the bound $|Z_{conf}(\rho)[r]| \geq 1.96$ would have to be solved. The result is a 95% confidence interval $\mathcal{I}_{0.95} = [\rho_-, \rho_+]$. If $0 \in \mathcal{I}_{0.95}$ then the data, which resulted in a value of $r$, for a sample of size $n$, implies no significant association between $X$ and $Y$.

### ■ Exercises

1. In Eq. (3.72) solve for $\rho$ in terms of $Z$, $r$, and $n$. Simplify to obtain the expression

$$\rho = \frac{r - \tanh(Z/\sqrt{n-3})}{1 - r \cdot \tanh(Z/\sqrt{n-3})}. \tag{3.73}$$

2. The data in Table 3.3 indicates a nonlinear relationship between period $T$ and pendulum length $L$. Define the *squared period* $S \equiv T^2$.

   (a) Compute the list of data $S = T^2$ from Table 3.3.
   (b) Find the mean, variance, and standard deviation for $S$.
   (c) Compute the covariance $Cov(L, S)$.
   (d) Find the correlation coefficient $r$ for the ratio model $L = a \cdot S = a \cdot T^2$.
   (e) Compute the 95% confidence interval for $\rho$ using Eq. (3.73).
   (f) Find $a = r \cdot s_L/s_S$, which is an estimate of $\alpha$ in Eq. (3.71), and convert to $m/s^2$. Also, use the confidence interval for $\rho$ to obtain a 95% confidence interval for $a$. Does it contain the accepted value of $g_{true} = 9.8 m/s^2$?
   (g) Suppose that the acceleration-due-to-gravity parameter $g$ is the constant of proportionality between $L$ and $\omega^{-2}$ where the angular frequency $\omega$ is given by the relation in Eq. (2.29). Transform the 95% confidence interval for $a$ to a 95% confidence interval for $g$. Does this interval contain the value of $g_{true} = 9.8 m/s^2$ now?

# II Multidimensionality

In Part II, the issues of working with many variables is considered. Linear systems of equations are well understood using a matrix formulation. Collections of functions, like polynomials, form vector spaces using linear combinations. Calculus is used to make sense of functions and nonlinear equations. Once combined into the single subject of vector calculus, many objects in three-dimensional space, along with their motion, can be modeled and understood.

# 4 Linear Algebra and Matrices

Organizing data is essential for a clear analysis and proper interpretation of observations. Lists become vectors and operations become matrix multiplication. Linear operations allow one to model interactions between many variables.

## Introduction

Matrix algebra was developed in the early 1800s as a way of organizing data and solving systems of linear equations involving many unknowns. In the mid-1900s the theory was used to solve complex problems in science by translating solution techniques into computer programs. This is now standard practice in science.

**Definition 4.1** An *algorithm* is a sequence of steps that take a system from an initial state (the input) to a final state (the output). The steps must follow formal rules of logic that are consistent. An algorithm or calculation is *effective* if it can proceed in a purely mechanical way.

Being able to express a problem and its solution as a system of linear equations allows one to write an algorithm to solve the problem. Thus, throughout this section, one may try to consider how a computation could be automated through a program.

## ■ 4.1 Lists of Data

Scientists who collect data for analysis will often express the information as an ordered list. This data can be written as a row of numbers,

$$\underline{x} = (x_1,\ x_2,\ \ldots,\ x_n) = (x_1\ x_2\ \ldots\ x_n) \in \mathbb{R}_n, \tag{4.1}$$

where each data point $x_i$ is a real number. The extension of $\mathbb{R}$ to multivariables involves ordered copies of the reals,

$$\mathbb{R} \times \mathbb{R} \times \ldots \mathbb{R} = \mathbb{R}_n.$$

The quantity $\underline{x}$ is called a *row vector*. The collection of all $n$-arrays of ordered numbers $\mathbb{R}_n$ is called the *row space*. The column spaces used here are:

$$\begin{pmatrix} x_1 \\ x_2 \end{pmatrix} \in \mathbb{R}^2, \quad \begin{pmatrix} x_1 \\ x_2 \\ x_3 \end{pmatrix} \in \mathbb{R}^3, \quad \ldots \quad \vec{x} = \begin{pmatrix} x_1 \\ x_2 \\ \cdot \\ x_n \end{pmatrix} \in \mathbb{R}^n,$$

and the connection with the corresponding row spaces is made using the transpose "$T$" operation,

$$(x_1 \ x_2 \ \ldots \ x_n)^T = \begin{pmatrix} x_1 \\ x_2 \\ \cdot \\ x_n \end{pmatrix} \in \mathbb{R}^n, \quad \begin{pmatrix} y_1 \\ y_2 \\ \cdot \\ y_n \end{pmatrix}^T = (y_1 \ y_2 \ \ldots \ y_n) \in \mathbb{R}_n.$$

For display purposes the row notation will often be employed in the text, but the $T$ symbol will be used to indicate if a column vector is intended.

The spaces $\mathbb{R}_n$ and $\mathbb{R}^n$ can contain the same information, but they are distinguished due to their dual nature. This is expressed by

$$\mathbb{R}_n^T = \mathbb{R}^n, \quad (\mathbb{R}^n)^T = \mathbb{R}_n.$$

Note that $(\mathbb{R}_n^T)^T = \mathbb{R}_n$ so $T^2 = identity$ on both $\mathbb{R}_n$ and $\mathbb{R}^n$.

## ■ 4.2 Vector Spaces

The set $\mathbb{R}^n$ can be given a group structure using *vector addition*, defined as

$$\vec{x} + \vec{y} = \begin{pmatrix} x_1 \\ x_2 \\ \cdot \\ x_n \end{pmatrix} + \begin{pmatrix} y_1 \\ y_2 \\ \cdot \\ y_n \end{pmatrix} = \begin{pmatrix} x_1 + y_1 \\ x_2 + y_2 \\ \cdot \\ x_n + y_n \end{pmatrix} \in \mathbb{R}^n. \tag{4.2}$$

Clearly, the column space $\langle \mathbb{R}^n, + \rangle$ is closed, has the zero vector $\vec{0}$ as the identity, and has $-\vec{x}$ as the inverse of $\vec{x}$;

$$\vec{0} \equiv \begin{pmatrix} 0 \\ 0 \\ \cdot \\ 0 \end{pmatrix}, \quad -\vec{x} \equiv \begin{pmatrix} -x_1 \\ -x_2 \\ \cdot \\ -x_n \end{pmatrix} \implies \vec{x} + (-\vec{x}) = \begin{pmatrix} x_1 - x_1 \\ x_2 - x_2 \\ \cdot \\ x_n - x_n \end{pmatrix} = \vec{0}. \quad (4.3)$$

The last expression is simply written as $\vec{x} - \vec{x} = \vec{0}$. Associativity of vector addition inherits its associativity from addition in $\mathbb{R}$. Since "+" is commutative, the group is abelian. The row space $\langle \mathbb{R}_n, + \rangle$ is similarly shown to be an abelian group.

---

**Definition 4.2** An abelian group $\langle \mathcal{V}, +_\mathcal{V} \rangle$ is a *vector space* over a field $\langle \mathbb{F}, +_\mathbb{F}, \times_\mathbb{F} \rangle$ if a map exists @ : $\mathbb{F} \times \mathcal{V} \to \mathcal{V}$ that is compatible with both the abelian group and field operations. Compatibility is defined, for any $a, b \in \mathbb{F}$ and $\vec{x}, \vec{y} \in \mathcal{V}$ by the properties:

(1) $\mathbb{F}$ linear: $@\langle a +_\mathbb{F} b, \vec{x} \rangle = @\langle a, \vec{x} \rangle +_\mathcal{V} @\langle b, \vec{x} \rangle$,
(2) $\mathcal{V}$ linear: $@\langle a, \vec{x} +_\mathcal{V} \vec{y} \rangle = @\langle a, \vec{x} \rangle +_\mathcal{V} @\langle a, \vec{y} \rangle$,
(3) $\mathbb{F}$ associative: $@\langle a \times_\mathbb{F} b, \vec{x} \rangle = a \times_\mathbb{F} \langle b, \vec{x} \rangle$, and
(4) $\mathbb{F}$ multiplicative identity: $@\langle 1, \vec{x} \rangle = \vec{x}$.

Two consequences are the following extra properties:

(5) $\mathbb{F}$ zero element: $@\langle 0, \vec{x} \rangle = \vec{0}$, and
(6) $\mathcal{V}$ zero element: $@\langle a, \vec{0} \rangle = \vec{0}$.

The row $\mathbb{R}_n$ and column $\mathbb{R}^n$ spaces are vector spaces over $\langle \mathbb{R}, +, \cdot \rangle$ where the symbol "$\cdot$" is used for $\times_\mathbb{F}, \times_\mathcal{V}$, and @, without inconsistency.

---

A vector space is also called a *linear space* or a *linear manifold* because it is closed under linear combinations over its field. In particular, $\forall a, b \in \mathbb{R}$ and $\forall \vec{x}, \vec{y} \in \mathbb{R}^n$, then $a \cdot \vec{x} + b \cdot \vec{y} \in \mathbb{R}^n$, which is clear from the explicit computation,

$$a \cdot \begin{pmatrix} x_1 \\ x_2 \\ \cdot \\ x_n \end{pmatrix} + b \cdot \begin{pmatrix} y_1 \\ y_2 \\ \cdot \\ y_n \end{pmatrix} = \begin{pmatrix} a \cdot x_1 \\ a \cdot x_2 \\ \cdot \\ a \cdot x_n \end{pmatrix} + \begin{pmatrix} b \cdot y_1 \\ b \cdot y_2 \\ \cdot \\ b \cdot y_n \end{pmatrix} = \begin{pmatrix} a \cdot x_1 + b \cdot y_1 \\ a \cdot x_2 + b \cdot y_2 \\ \cdot \\ a \cdot x_n + b \cdot y_n \end{pmatrix}. \quad (4.4)$$

This procedure for combining an array of numbers is natural and appears in many computations. Fortunately, computers can perform these calculations very quickly. Thus if a problem can be expressed in a vector space, it can often be solved numerically with great speed and accuracy.

### ■ 4.2.1 The Dot Product

The *dot product* (also called the *inner product*) is a contraction of a row vector with a column vector resulting in a scalar (i.e., a number). This is defined only between two vectors of the same size,

$$\begin{aligned}\vec{x} \cdot \vec{y} &= \vec{x}^T \vec{y} = \underline{x}\, \vec{y} = (x_1\ x_2\ \ldots\ x_n) \cdot \begin{pmatrix} y_1 \\ y_2 \\ \cdot \\ y_n \end{pmatrix} \\ &= x_1 y_1 + x_2 y_2 + \cdots + x_n y_n = \sum_{i=1}^{n} x_i \cdot y_i. \end{aligned} \quad (4.5)$$

In this way the dot product appears as a natural extension of the usual product in $\mathbb{R}$ and is a reason that the "·" notation was used earlier. Other common expressions for the dot product are

$$\vec{x} \cdot \vec{y} = \langle \vec{x}, \vec{y} \rangle = \langle \vec{x} | \vec{y} \rangle. \quad (4.6)$$

The dot product has various properties:

(**1**) symmetric: $\vec{x} \cdot \vec{y} = \vec{y} \cdot \vec{x}$,
(**2**) bilinearity: $(a \cdot \vec{x} + b \cdot \vec{y}) \cdot \vec{z} = a \cdot \vec{x} \cdot \vec{z} + b \cdot \vec{y} \cdot \vec{z}$,
$\qquad \vec{x} \cdot (a \cdot \vec{y} + b \cdot \vec{z}) = a \cdot \vec{x} \cdot \vec{y} + b \cdot \vec{x} \cdot \vec{z}$,
(**3**) positivity: $\vec{x} \cdot \vec{x} \geq 0$, and
(**4**) normalizability: $\vec{x} \cdot \vec{x} = 0 \iff \vec{x} = \vec{0}$.

---

**Definition 4.3** The *norm, magnitude,* or *Euclidean length* of a vector $\vec{x} \in \mathbb{R}^n$ is

$$\|\vec{x}\| \equiv \sqrt{\vec{x} \cdot \vec{x}} = \sqrt{\langle \vec{x} | \vec{x} \rangle} = \left( \sum_{j=1}^{n} x_j^2 \right)^{1/2}. \quad (4.7)$$

A vector $\vec{x}$ is called *normalizable* if $\vec{x} \neq \vec{0}$, in which case the *direction* of $\vec{x}$ is defined to be

$$\hat{x} \equiv \frac{1}{\|\vec{x}\|}\vec{x} = (x_1/\|\vec{x}\|, x_2/\|\vec{x}\|, \cdots, x_n/\|\vec{x}\|)^T.$$

The direction $\hat{x}$ always has the property that $\|\hat{x}\| = 1$. Any vector $\vec{x}$ where $\vec{x} \cdot \vec{x} = \|\vec{x}\|^2 = 1$ is called a *unit vector*, and will be written as $\hat{x}$. In this case, $\hat{x}$ is said to have been *normalized*.

---

Two natural properties of the norm of a vector, which follow from Eq. (4.6) and Eq. (4.7), are the *constant rule* and *triangle inequality*:

$$\|a \cdot \vec{x}\| \leq |a| \cdot \|\vec{x}\|, \quad \|\vec{x} + \vec{y}\| \leq \|\vec{x}\| + \|\vec{y}\|, \tag{4.8}$$

for $a \in \mathbb{R}$ and $\forall \vec{x}, \vec{y} \in \mathbb{R}^n$. Any nonzero vector $\vec{x}$ can be expressed as $\vec{x} = \|\vec{x}\| \cdot \hat{x}$, i.e., as a product of its magnitude and direction. The dot product of two unit vectors is restricted to the interval $[-1, 1]$. In particular, using the dot-product properties

$$0 \leq \|\hat{x} \pm \hat{y}\|^2 = \langle \hat{x} \pm \hat{y} | \hat{x} \pm \hat{y} \rangle = \|\hat{x}\|^2 \pm 2 \cdot \langle \hat{x} | \hat{y} \rangle + \|\hat{y}\|^2 = 2 \pm 2 \cdot \langle \hat{x} | \hat{y} \rangle.$$

Thus, rearranging and combining the inequality gives

$$-1 \leq \langle \hat{x} | \hat{y} \rangle \leq 1 \iff |\langle \vec{x} | \vec{y} \rangle| \leq \|\vec{x}\| \|\vec{y}\|, \tag{4.9}$$

which is called the *Cauchy-Schwarz* inequality.

Significant geometrical quantities are obtained using the dot product. For example, the angle $\theta$ between two vectors $\vec{x}$ and $\vec{y}$ is obtained from the *cosine law*

$$\cos(\theta) = \frac{\vec{x} \cdot \vec{y}}{\sqrt{(\vec{x} \cdot \vec{x})(\vec{y} \cdot \vec{y})}} = \frac{\langle \vec{x} | \vec{y} \rangle}{\|\vec{x}\| \|\vec{y}\|} = \langle \hat{x} | \hat{y} \rangle = \hat{x} \cdot \hat{y}. \tag{4.10}$$

To obtain the angle from this expression requires inverting the cosine function, and applying the $arccos \equiv \cos^{-1}$ function to both sides, giving $\theta = arccos(\hat{x} \cdot \hat{y})$ whose values are restricted to the range $[0, \pi] = [0°, 180°]$.

### EXAMPLE 4.1

**The angle between vectors:** Consider the two vectors $\vec{u} = (1, 3, -2)^T$ and $\vec{w} = (-1, 0, 2)^T$ from $\mathbb{R}^3$. Then

$$\|\vec{u}\| = \sqrt{14}, \quad \|\vec{w}\| = \sqrt{5}, \quad \langle \vec{u}, \vec{w} \rangle = (1)(-1) + (3)(0) + (-2)(2) = -1 + 0 - 4 = -5, \qquad (4.11)$$

and the corresponding unit vectors are

$$\hat{u} = \left(1/\sqrt{14}, 3/\sqrt{14}, -2/\sqrt{14}\right)^T, \quad \hat{v} = \left(-1/\sqrt{5}, 0, 2/\sqrt{5}\right)^T. \qquad (4.12)$$

From the cosine law it follows that,

$$\cos(\theta) = \hat{u} \cdot \hat{v} = -5/\sqrt{70}, \qquad (4.13)$$

which gives $\theta \simeq 120^0$ as the angle between $\vec{u}$ and $\vec{w}$.

## ■ Exercises

1. Consider the two vectors $\vec{u} = (1, -2)^T$ and $\vec{v} = (3, 1)^T$.

   (a) Compute and simplify $3 \cdot \vec{u} - 2 \cdot \vec{v}$.
   (b) Compute and simplify $\vec{u} \cdot \vec{v}$.
   (c) Compute and simplify $\|\vec{u}\|$ and $\|\vec{v}\|$.
   (d) Compute the angle between $\vec{u}$ and $\vec{v}$.

2. Consider the two vectors $\vec{a} = (-1, 1, -2)^T$ and $\vec{b} = (0, 3, 1)^T$.

   (a) Compute and simplify $2 \cdot \vec{a} + \vec{b}$.
   (b) Compute and simplify $\vec{a} \cdot \vec{b}$.
   (c) Compute and simplify $\|\vec{a}\|$ and $\|\vec{b}\|$.
   (d) Compute the angle between $\vec{a}$ and $\vec{b}$.

3. Obtain the constant rule in Eq. (4.8) using the definition in Eq. (4.7).

4. Obtain the squared form of the triangle inequality in Eq. (4.8) using the Cauchy-Schwarz inequality in Eq. (4.9) and properties of the dot product and the norm.

## 4.3 Bases, Dimension, Linear Independence, and Vector Sums

A one-dimensional vector space $\mathcal{V}$ over a field $\mathbb{F}$ is generated by a single, nonzero vector $\vec{v} \in \mathcal{V}$ so that

$$\mathcal{V} = span[\vec{v}] \equiv \{c \cdot \vec{v} : c \in \mathbb{F}\}. \tag{4.14}$$

This is written as $dim(\mathcal{V}) = 1$.

Two vectors $\vec{x}$ and $\vec{y}$ from a vector space $\mathcal{U}$ are called *linearly independent* if

$$a \cdot \vec{x} + b \cdot \vec{y} = \vec{0} \iff a = b = 0.$$

In this case the *span* of the set $\{\vec{x}, \vec{y}\} \subset \mathcal{U}$ is

$$span\left[\{\vec{x}, \vec{y}\}\right] \equiv \{a \cdot \vec{x} + b \cdot \vec{y} : a, b \in \mathbb{F}\}, \tag{4.15}$$

and this defines a vector space over $\mathbb{F}$. Since $span\left[\{\vec{x}, \vec{y}\}\right] \subset \mathcal{U}$, the space $\langle span\left[\{\vec{x}, \vec{y}\}\right], + \rangle$ is called a *two-dimensional subspace* of $\mathcal{U}$. In particular, $dim(span\left[\{\vec{x}, \vec{y}\}\right]) = 2$.

Suppose a vector space $\mathcal{W}$ over a field $\mathbb{F}$ has three linearly independent vectors, i.e., $\exists \{\vec{x}, \vec{y}, \vec{z}\} \subset \mathcal{W}$ so that

$$a \cdot \vec{x} + b \cdot \vec{y} + c \cdot \vec{z} = \vec{0} \iff a = b = c = 0.$$

Then $span\left[\{\vec{x}, \vec{y}, \vec{z}\}\right]$ is a three-dimensional subspace of $\mathcal{W}$ and $dim(\mathcal{W}) \geq 3$.

**Definition 4.4** A vector space $\mathcal{V}$ over a field $\mathbb{F}$ is *n-dimensional* if a sequence of $n$ vectors $\{\vec{v}_1, \vec{v}_2, \ldots, \vec{v}_n\}$ from $\mathcal{V}$ can be found that are *linearly independent*

$$\sum_{j=1}^{n} c_j \cdot \vec{v}_j = \vec{0} \iff c_1 = c_2 = \cdots = c_n = 0,$$

and the sequence $\{\vec{v}_1, \vec{v}_2, \ldots, \vec{v}_n\}$ *spans* $\mathcal{V}$

$$span\left[\{\vec{v}_1, \vec{v}_2, \ldots, \vec{v}_n\}\right] \equiv \left\{\sum_{j=1}^{n} c_j \cdot \vec{v}_j : c_j \in \mathbb{R}, \forall j \in \mathbb{Z}_n\right\} = \mathcal{V}.$$

This information about $\mathcal{V}$ is written as $dim(\mathcal{V}) = n$. The sequence $\{\vec{v}_1, \vec{v}_2, \ldots, \vec{v}_n\}$ is called a *basis* for $\mathcal{V}$.

The span of a sequence of vectors is the set of all linear combinations of the sequence. This gives the closure property of the span, but also ensures that the identity vector $\vec{0}$ and the inverses are included, so that the span is an abelian group. The remaining vector space properties follow from the requirement that the sequence comes from a known vector space. In particular, a *subspace* $\mathcal{S}$ of a vector space $\mathcal{V}$ inherits properties from $\mathcal{V}$. These facts are summed up in,

■ **Theorem 4.1**

A subset $\mathcal{S}$ of a vector space $\mathcal{V}$ over a field $\mathbb{F}$ is itself a vector space (called a subspace) if it is closed under linear combinations. That is,

$$(\forall \vec{x}, \vec{y} \in \mathcal{S}) \text{ and } (\forall a, b \in \mathbb{F}) \implies a \cdot \vec{x} + b \cdot \vec{y} \in \mathcal{S}. \tag{4.16}$$

Often a set of elements arise naturally in an application. In order to perform mathematical operations, one endeavors to put these elements into the context of a vector space.

**Definition 4.5** The *vector space sum* of two vector spaces $\mathcal{U}$ and $\mathcal{W}$ over a field $\mathbb{F}$, is a new vector space over $\mathbb{F}$ defined as,

$$\mathcal{U} \oplus \mathcal{W} = \{a \cdot \vec{u} +_\mathbb{F} b \cdot \vec{w} \mid \forall \vec{u} \in \mathcal{U}, \forall \vec{w} \in \mathcal{W}, \forall a, b \in \mathbb{F}\},$$

which is similar to the span in Eq. (4.15). Thus

$$\mathcal{U} \cap \mathcal{W} = \{\vec{0}\} \implies dim(\mathcal{U} \oplus \mathcal{W}) = dim(\mathcal{U}) + dim(\mathcal{W}).$$

Otherwise, the new vector space will have a lower dimension than the combination of its summands. If $\mathcal{V}$ has an inner product structure, and $\mathcal{U} \subset \mathcal{V}$ is a subspace, then the *orthogonal complement* is the subspace of $\mathcal{V}$ defined so that

$$\mathcal{U}^\perp \equiv \{\vec{v} \in \mathcal{V} \mid \forall \vec{u} \in \mathcal{U} \implies \vec{v} \cdot \vec{u} = 0\}.$$

In this case $dim(\mathcal{V}) = dim(\mathcal{U}) + dim(\mathcal{U}^\perp)$ and $\mathcal{V} = \mathcal{U} \oplus \mathcal{U}^\perp$.

Often a vector space is decomposed into a vector sum of many vector subspaces, where one or more of the spaces has a special property.

## 4.3 Bases, Dimension, Linear Independence, and Vector Sums

**EXAMPLE 4.2**

The standard basis for $\mathbb{R}^N$ over $\mathbb{R}$ are the vectors

$$\hat{e}_1 \equiv \begin{pmatrix} 1 \\ 0 \\ \cdot \\ 0 \end{pmatrix}, \; \hat{e}_2 \equiv \begin{pmatrix} 0 \\ 1 \\ \cdot \\ 0 \end{pmatrix}, \; \ldots, \hat{e}_N \equiv \begin{pmatrix} 0 \\ 0 \\ \cdot \\ 1 \end{pmatrix},$$

and they generate the corresponding subspaces $\mathcal{E}_j \equiv \mathbb{R} \cdot \hat{e}_j$, so that

$$\mathbb{R}^N = \mathcal{E}_1 \oplus \mathcal{E}_2 \oplus \cdots \oplus \mathcal{E}_N, \; \mathcal{E}_i \cap \mathcal{E}_j = \vec{0} \text{ for } i \neq j.$$

Thus $\mathbb{R}^N$ is an orthogonal direct sum. Similarly, $\mathbb{R}_N$ has the standard basis $\{\underline{\delta}^i\}_{i=1}^N$ where $\underline{\delta}^i \equiv \hat{e}_i^T$, which is called the dual basis to $\{\hat{e}_j\}_{j=1}^N$. In particular, from the inner product structure on $\mathbb{R}^n$,

$$\underline{\delta}^i \hat{e}_j = \hat{e}_i^T \hat{e}_j = \delta_j^i \equiv \begin{cases} 0, \text{if } i \neq j \\ 1, \text{if } i = j \end{cases}$$

where $\delta_j^i \equiv \delta_{ij}$ is called the *Kronecker delta function*.

Conversely, a new vector space can be created by adjoining vector spaces.

**Definition 4.6** The *vector space product*, or *tensor product*, of two vector spaces $\mathcal{U}$ and $\mathcal{W}$ over a field $\mathbb{F}$, is a new vector space over $\mathbb{F}$ defined as,

$$\mathcal{U} \otimes \mathcal{W} \equiv \{\vec{u} \otimes \vec{w} \mid \vec{u} \in \mathcal{U} \text{ and } \vec{w} \in \mathcal{W}\}.$$

The new entities $\vec{u} \otimes \vec{w}$ span a new vector space where $dim(\mathcal{U} \otimes \mathcal{W}) = dim(\mathcal{U}) \cdot dim(\mathcal{W})$ by extrapolating the natural properties [55]:

**Bilinearity:** $(a \cdot \vec{u} + b \cdot \vec{v}) \otimes \vec{w} = a \cdot \vec{u} \otimes \vec{w} + b \cdot \vec{v} \otimes \vec{w};$
$\vec{u} \otimes (c \cdot \vec{v} + d \cdot \vec{w}) = c \cdot \vec{u} \otimes \vec{v} + d \cdot \vec{u} \otimes \vec{w}.$

If three (or more) vector spaces are joined using a tensor product, then the additional property:

**Associativity:** $(\vec{u} \otimes \vec{v}) \otimes \vec{w} = \vec{u} \otimes (\vec{v} \otimes \vec{w}),$

is imposed. If $\mathcal{U}$ and $\mathcal{W}$ have inner product structures, then a new inner product is inherited from $\mathcal{U}$ and $\mathcal{W}$ so that $\forall \vec{u}_1 \otimes \vec{w}_1, \vec{u}_2 \otimes \vec{w}_2 \in \mathcal{U} \otimes \mathcal{W}$,

$$\langle \vec{u}_1 \otimes \vec{w}_1, \vec{u}_2 \otimes \vec{w}_2 \rangle_{\mathcal{U} \otimes \mathcal{W}} \equiv \langle \vec{u}_1, \vec{u}_2 \rangle_{\mathcal{U}} \cdot \langle \vec{w}_1, \vec{w}_2 \rangle_{\mathcal{W}}.$$

Suppose $\mathcal{U}$ and $\mathcal{W}$ have dual spaces $\mathcal{U}^T$ and $\mathcal{W}^T$, respectfully. Then $\mathcal{U}^T \otimes \mathcal{W}^T$ is a dual space to $\mathcal{U} \otimes \mathcal{W}$ by defining the operation of

**Contraction:** $(\underline{x} \otimes \underline{y})[\vec{u} \otimes \vec{w}] = (\underline{x} \cdot \vec{u}) \cdot (\underline{y} \cdot \vec{w})$.

Thus there is a natural way to create new vector spaces from known spaces.

### ■ Exercises

1. Consider the two vectors $\vec{u} = (1\ -2)^T$ and $\vec{w} = (3\ 1)^T$.
   (a) Compute the contraction $\underline{\delta}^1 \otimes \underline{\delta}^2 [\vec{u} \otimes \vec{w}]$ (see Example 4.2).
   (b) Compute the contraction $(\underline{\delta}^1 \otimes \underline{\delta}^1 + \underline{\delta}^2 \otimes \underline{\delta}^2)[\vec{u} \otimes \vec{w}]$.
   (c) Compute the contraction $(\underline{\delta}^1 \otimes \underline{\delta}^2 - \underline{\delta}^2 \otimes \underline{\delta}^1)[\vec{u} \otimes \vec{w}]$.

2. Consider the two vectors $\vec{a} = (-1\ 1\ -2)^T$ and $\vec{b} = (0\ 3\ 1)^T$.
   (a) Compute the contraction $(\underline{\delta}^1 \otimes \underline{\delta}^1 + \underline{\delta}^2 \otimes \underline{\delta}^2 + \underline{\delta}^3 \otimes \underline{\delta}^3)[\vec{a} \otimes \vec{b}]$.
   (b) Compute the contraction $(\underline{\delta}^1 \otimes \underline{\delta}^2 - \underline{\delta}^2 \otimes \underline{\delta}^1)[\vec{a} \otimes \vec{b}]$.
   (c) Compute the contraction $(\underline{\delta}^2 \otimes \underline{\delta}^3 - \underline{\delta}^3 \otimes \underline{\delta}^2)[\vec{a} \otimes \vec{b}]$.
   (d) Compute the contraction $(\underline{\delta}^3 \otimes \underline{\delta}^1 - \underline{\delta}^1 \otimes \underline{\delta}^3)[\vec{a} \otimes \vec{b}]$.

### ■ 4.4 Projections

There are special transformations on vector spaces.

**Definition 4.7** Every nonzero vector $\vec{u}$ in $\mathbb{R}^n$ generates a *projection* that is a function or transformation $Proj_{\vec{u}} : \mathbb{R}^n \to \mathbb{R}^n$. This projection is defined as

$$Proj_{\vec{u}} = \|\vec{u}\|^{-2}\ \vec{u}\ \vec{u}^T = \hat{u}\ \hat{u}^T = \hat{u} \otimes \hat{u}^T, \quad (4.17)$$

where the tensor product is sometimes simply written as $\hat{u} \otimes \hat{u}$. For any $\vec{w} \in \mathbb{R}^n$, the projection operator acts as

$$Proj_{\vec{u}}(\vec{w}) = \frac{\vec{u} \cdot \vec{w}}{\|\vec{u}\|^2} \cdot \vec{u} = (\hat{u} \cdot \vec{w}) \cdot \hat{u}. \quad (4.18)$$

In general, a projection $P$ on a vector space $\mathcal{V}$ is any function $P : \mathcal{V} \to \mathcal{V}$ where

(1) **$P$ is linear:** $P(a\vec{v}_1 + b\vec{v}_2) = aP(\vec{v}_1) + bP(\vec{v}_2)$;
(2) **$P$ is idempotent:** $P^2 = P$.

If $\mathcal{V}$ has an inner product structure, then $P$ is called an *orthogonal projection* if

**(3) $P$ is symmetric:** $\langle \vec{v}_1 | P(\vec{v}_2) \rangle = \langle P(\vec{v}_1) | \vec{v}_2 \rangle$.

When $\mathcal{V} = \mathbb{R}^n$, then $P$ can be expressed as an $n \times n$ matrix, in which case the property of orthogonality can be expressed as $P = P^T$, or simply that $P$ is symmetric, or *self-adjoint*. The *rank* of a projection is $rank(P) = dim(Range(P))$.

---

Due to Eq. (4.18) the projections on $\mathbb{R}^n$ inherit properties from the inner product. For example, the projection onto a vector $\vec{u}$ is a rank 1 linear operator on $\mathbb{R}^n$,

$$Proj_{\vec{u}}(a \cdot \vec{x} + b \cdot \vec{y}) \;=\; a \cdot Proj_{\vec{u}}(\vec{x}) \;+\; b \cdot Proj_{\vec{u}}(\vec{y}) \;=\; (a\,\hat{u} \cdot \vec{x} + b\,\hat{u} \cdot \vec{y})\,\hat{u}. \tag{4.19}$$

Furthermore, the projection onto $\vec{u}$ is *idempotent* in that

$$\begin{aligned}(Proj_{\vec{u}})^2(\vec{w}) \;&=\; Proj_{\vec{u}}[Proj_{\vec{u}}(\vec{w})] \;=\; Proj_{\vec{u}}(\hat{u} \cdot \vec{w}\,\hat{u}) \;=\; (\hat{u} \cdot \vec{w}) Proj_{\vec{u}}(\hat{u}) \\ &=\; (\hat{u} \cdot \vec{w})(\hat{u} \cdot \hat{u})\,\hat{u} \;=\; (\hat{u} \cdot \vec{w})\hat{u} \;=\; Proj_{\vec{u}}(\vec{w})\,. \end{aligned} \tag{4.20}$$

It is left as an exercise to show that $Proj_{\vec{u}}$ is symmetric.

In general, consider a one-dimensional subspace $\ell \subset \mathbb{R}^n$ with generator $\vec{u} \neq \vec{0}$ so that $\ell \equiv span[\vec{u}]$. Then for a vector $\vec{v} \notin \ell$, one finds the *closest* vector $\vec{v}_\ell \in \ell$ to $\vec{v}$ by using a projection onto $\ell$

$$\vec{v}_\ell \;\equiv\; Proj_\ell(\vec{v}) \;=\; Proj_{\vec{u}}(\vec{v}) \;=\; \frac{\langle \vec{u}|\vec{v}\rangle}{\langle \vec{u}|\vec{u}\rangle}\,\vec{u} \;=\; \langle \hat{u}|\vec{v}\rangle\,\hat{u} \;\in \ell \subset \mathbb{R}^n. \tag{4.21}$$

To actually show that this produces the vector in $\ell$, which is the closest to $\vec{v}$, one must minimize $\|t \cdot \vec{u} - \vec{v}\|$ with respect to $t \in \mathbb{R}$. This is easily achieved using calculus, although a geometric argument is given in the exercises.

Next, consider a two-dimensional subspace $\mathcal{P} \subset \mathbb{R}^n$ with generators $\vec{u}_1 \neq \vec{0}$ and $\vec{u}_2 \notin span[\vec{u}_1]$. In this situation one writes $\mathcal{P} \equiv span[\vec{u}_1, \vec{u}_2]$. Then any vector $\vec{v} \in \mathbb{R}^n$ generates a vector $\vec{v}_\mathcal{P}$, by

$$\vec{v}_\mathcal{P} \;\equiv\; \langle \hat{u}_1|\vec{v}\rangle\,\hat{u}_1 \;+\; \langle \hat{u}_2|\vec{v}\rangle\,\hat{u}_2 \in \mathcal{P}. \tag{4.22}$$

Although $\vec{v}_\mathcal{P}$ is not necessarily the closest vector in $\mathcal{P}$ to $\vec{v}$, it does generate the line $\ell_{\vec{v}} \equiv span[\vec{v}_\mathcal{P}]$, which is closest to $\vec{v}$ in the plane $\mathcal{P}$. Thus the vector

$$Proj_{\ell_{\vec{v}}}(\vec{v}) \;\equiv\; Proj_{\vec{v}_\mathcal{P}}(\vec{v}) \;=\; \frac{\langle \vec{v}_\mathcal{P}|\vec{v}\rangle}{\langle \vec{v}_\mathcal{P}|\vec{v}_\mathcal{P}\rangle}\,\vec{v}_\mathcal{P} \;\in \ell_{\vec{v}} \subset \mathcal{P} \subset \mathbb{R}^n, \tag{4.23}$$

is the closest vector in $\ell_{\vec{v}}$ to $\vec{v}$. If $\mathcal{P}$ is generated by *orthogonal* vectors $\vec{u}_1 \perp \vec{u}_2$, then $\vec{v}_{\mathcal{P}}$ is the closest vector in $\mathcal{P}$ to $\vec{v}$. This motivates the process of taking $m \leq n$ linearly independent vectors $\{\vec{u}_1, \vec{u}_2, \ldots, \vec{u}_m\}$ and creating $m$ *orthonormal* vectors $\{\hat{w}_1, \hat{w}_2, \ldots, \hat{w}_m\}$ according to the *Gram-Schmidt* iteration process

$$\hat{w}_1 \equiv \frac{\vec{u}_1}{\|\vec{u}_1\|}, \quad \vec{w}_k \equiv \vec{u}_k - \sum_{j=1}^{k-1} Proj_{\hat{w}_j}(\vec{u}_k), \quad \hat{w}_k \equiv \frac{\vec{w}_k}{\|\vec{w}_k\|}. \tag{4.24}$$

■ **Exercises**

1. Consider the function $p(t) \equiv \|t \cdot \vec{u} - \vec{v}\|^2$, for fixed vectors $\vec{u}, \vec{v} \in \mathbb{R}^n$. Expand the norm into a quadratic polynomial in $t \in \mathbb{R}$, using the square of Eq. (4.7), and find the minimum value of $p(t)$ using the vertex value $t_{vertex}$. How is $t_{vertex}$ related to a projection between $\vec{u}, \vec{v}$? Explain why one can conclude that $|\langle \vec{u}|\vec{v}\rangle| \leq \|\vec{u}\| \cdot \|\vec{v}\|$.

2. Consider arbitrary vectors $\vec{u}, \vec{v}, \vec{w} \in \mathbb{R}^n$. Show, using Eq. (4.21), that

$$\langle \vec{u}|Proj_{\vec{v}}(\vec{w})\rangle = \langle Proj_{\vec{v}}(\vec{u})|\vec{w}\rangle.$$

3. Consider the two vectors $\vec{u} = (1 \ -2)^T$ and $\vec{v} = (3 \ 1)^T$.
   (a) Compute and simplify $Proj_{\vec{u}}(\vec{v})$ and $Proj_{\vec{v}}(\vec{u})$.
   (b) Compute $Proj_{\hat{e}_1}(\vec{u})$ and $Proj_{\hat{e}_2}(\vec{u})$.

4. Consider the vectors $\vec{u}_1 = (-1 \ 1 \ -2)^T$, $\vec{u}_2 = (0 \ 3 \ 1)^T$, and $\vec{u}_3 = (2 \ 0 \ 0)^T$. Use the Gram-Schmidt iteration process in Eq. (4.24) to find an orthonormal basis $\{\hat{w}_1, \hat{w}_2, \hat{w}_3\}$ for $\mathbb{R}^3$, up to three decimal places for each component.

■ **4.5 Vectors in Two- and Three-Dimensional Space**

Consider two points $P_1, P_2$ in $\mathbb{R}^n$ for $n = 2$ or $3$. Then the displacement from $P_1$, the tail, to $P_2$, the head, is the difference

$$\overrightarrow{P_1P_2} \equiv P_2 - P_1 \iff P_2 = P_1 + \overrightarrow{P_1P_2}. \tag{4.25}$$

Although this may suggest that $P_1$, $P_2$, and $\overrightarrow{P_1P_2}$ are the same entities, it is better to think of $\overrightarrow{P_1P_2}$ as representing a new object, called a *displacement vector*. Generically, this entity may be written as

$$\overrightarrow{\Delta P} \equiv \overrightarrow{P_1P_2} = P_2 - P_1,$$

since it corresponds to the difference between the points. In particular, if $P_1$ is a point in a solid object, and the object is moved rigidly so that $P_1$ goes to $P_2$, then all points in the object are *displaced* by the same vector $\overrightarrow{P_1P_2}$. Any other point $P_1'$ in the object gets moved to $P_2' = P_1' + \overrightarrow{P_1P_2}$ or simply $P_2' = P_1' + \overrightarrow{\Delta P}$.

Next, suppose that a solid object gets moved by one vector $\vec{u}$ and then by another $\vec{w}$. The net result will be the same as movement by a single *resultant* vector $\vec{u} + \vec{w}$. For example, using any two vectors $\vec{u} = (u_1, u_2, u_3)^T$ and $\vec{w} = (w_1, w_2, w_3)^T$ from $\mathbb{R}^3$, and numbers $a$, $b$ from $\mathbb{R}$, a *linear combination* is defined to be

$$a \cdot \vec{u} + b \cdot \vec{w} = a \begin{pmatrix} u_1 \\ u_2 \\ u_3 \end{pmatrix} + b \begin{pmatrix} w_1 \\ w_2 \\ w_3 \end{pmatrix} = \begin{pmatrix} a \cdot u_1 + b \cdot w_1 \\ a \cdot u_2 + b \cdot w_2 \\ a \cdot u_3 + b \cdot w_3 \end{pmatrix} \in \mathbb{R}^3, \quad (4.26)$$

for any $a, b \in \mathbb{R}$. This gives a vector that has characteristics of both $\vec{u}$ and $\vec{w}$. Graphically, this can be seen by plotting $\vec{u}$ and $\vec{w}$, then rescaling to obtain $a \cdot \vec{u}$ and $b \cdot \vec{w}$, and then repositioning either $b \cdot \vec{w}$ or $a \cdot \vec{u}$, to create the resultant vector

$$\text{Resultant} = a \cdot \vec{u} + b \cdot \vec{w} = b \cdot \vec{w} + a \cdot \vec{u}, \quad (4.27)$$

as indicated in Figure 4.1. Since $\mathbb{R}^3$ is closed under linear combinations, the resultant must be a new vector in $\mathbb{R}^3$. The zero vector $\vec{0} = (0, 0, 0)^T$ corresponds to the origin in this visualization, and does not move a point $P_1 + \vec{0} = P_1$, or change a vector $\vec{u} + \vec{0} = \vec{0} + \vec{u} = \vec{u}$.

Every vector $\vec{u} \in \mathbb{R}^3$ can be written in terms of the *standard basis* $\{\hat{i}, \hat{j}, \hat{k}\}$ as,

$$\vec{x} = \begin{pmatrix} x_1 \\ x_2 \\ x_3 \end{pmatrix} = x_1 \cdot \hat{i} + x_2 \cdot \hat{j} + x_3 \cdot \hat{k}, \; \hat{i} \equiv \begin{pmatrix} 1 \\ 0 \\ 0 \end{pmatrix}, \; \hat{j} \equiv \begin{pmatrix} 0 \\ 1 \\ 0 \end{pmatrix}, \; \hat{k} \equiv \begin{pmatrix} 0 \\ 0 \\ 1 \end{pmatrix}. \quad (4.28)$$

From Example 4.2, these definitions correspond to $\hat{i} = \hat{e}_1$, $\hat{j} = \hat{e}_2$, and $\hat{k} = \hat{e}_3$. Geometrically, multiplication by a scalar changes the length of a vector, and if the scalar is negative, then multiplication reverses its orientation. The addition of two nonzero vectors gives a resultant vector that retains some aspects of both input vectors. In particular, two vectors, and their sum, must remain in the same plane.

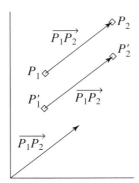 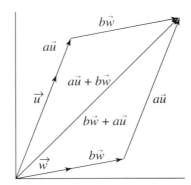

**FIGURE 4.1** ■ Left: Vector $\overrightarrow{P_1P_2}$ as the difference between the points $P_1$, $P_2$, or the points $P_1'$, $P_2'$. Right: Vectors $\vec{u}$ and $\vec{w}$, $a\vec{u}$ and $b\vec{w}$, and their resultant $a\vec{u} + b\vec{w}$.

In two or more dimensions $n \geq 2$ a single nonzero vector $\vec{u} \in \mathbb{R}^n$ generates a line by constant multiplications

$$\ell \equiv \{c \cdot \vec{u} \mid \forall c \in \mathbb{R}\}, \tag{4.29}$$

which is a subspace of $\mathbb{R}^n$. Geometrically, if $\ell$ is now thought of as a collection of points, then the definition in Eq. (4.29) is a line parameterized by $c \in \mathbb{R}$, although $c$ can also be thought of as an independent variable. Since $\ell$ is also closed under scalar multiplication, it will be referred to as a *subvector space*, a *linear subspace*, or just a *subspace*. In three or more dimensions $n \geq 3$, two nonzero vectors $\vec{u}, \vec{w} \in \mathbb{R}^n$ generate a collection of points by linear combinations, which is also a subspace,

$$\mathcal{P} \equiv \{c \cdot \vec{u} + d \cdot \vec{w} \mid \forall c, d \in \mathbb{R}\}, \tag{4.30}$$

which is parameterized by $c, d \in \mathbb{R}$. The parameters $c$ and $d$ can also be thought of as independent variables if and only if $\vec{u}$ and $\vec{w}$ point in different directions. This means that the angle between the vectors cannot be 0 or $180^0 = \pi \; radians$. In the case that

$$|\langle \vec{u}|\vec{w}\rangle| = \|\vec{u}\|\|\vec{w}\| \iff \vec{u} \parallel \vec{w},$$

one says that $\vec{u}$ and $\vec{w}$ are *parallel* or *collinear*. Conversely, when

$$\langle \vec{u}|\vec{w}\rangle = 0 \iff \vec{u} \perp \vec{w},$$

one says that $\vec{u}$ and $\vec{w}$ are *perpendicular* or *orthogonal* to each other.

## 4.5.1 Cross Product in $\mathbb{R}^2$ and $\mathbb{R}^3$

In $\mathbb{R}^3$ the *cross product* (also called the outer product) can be defined as an operation $\times : \mathbb{R}^3 \times \mathbb{R}^3 \to \mathbb{R}^3$ where

$$\times \langle \vec{x}, \vec{y} \rangle = \vec{x} \times \vec{y} = \begin{pmatrix} x_1 \\ x_2 \\ x_3 \end{pmatrix} \times \begin{pmatrix} y_1 \\ y_2 \\ y_3 \end{pmatrix} \equiv \begin{pmatrix} x_2 \cdot y_3 - x_3 \cdot y_2 \\ x_3 \cdot y_1 - x_1 \cdot y_3 \\ x_1 \cdot y_2 - x_2 \cdot y_1 \end{pmatrix} \in \mathbb{R}^3. \quad (4.31)$$

The main useful geometric property of the cross product is that

$$(\vec{x} \times \vec{y}) \cdot \vec{x} = (\vec{x} \times \vec{y}) \cdot \vec{y} = 0,$$

which implies that $\vec{x} \times \vec{y} \perp \vec{x}$ and $\vec{x} \times \vec{y} \perp \vec{y}$. In terms of the standard basis,

$$\hat{i} \times \hat{j} = \hat{k}, \ \hat{j} \times \hat{k} = \hat{i}, \ \hat{k} \times \hat{i} = \hat{j}.$$

After careful consideration of the definition in Eq. (4.31) it should be clear that it is antisymmetric $\vec{x} \times \vec{y} = -\vec{y} \times \vec{x}$. Thus

$$\hat{j} \times \hat{i} = -\hat{k}, \ \hat{k} \times \hat{j} = -\hat{i}, \ \hat{i} \times \hat{k} = -\hat{j}.$$

The cross product is *bilinear*, like the dot product, so that

$$(a\vec{x} + b\vec{y}) \times \vec{z} = a\vec{x} \times \vec{z} + b\vec{y} \times \vec{z}, \ \vec{x} \times (a\vec{y} + b\vec{z}) = a\vec{x} \times \vec{y} + b\vec{x} \times \vec{z}. \quad (4.32)$$

The consequence is that

$$\hat{i} \times \hat{i} = \vec{0}, \ \hat{j} \times \hat{j} = \vec{0}, \ \hat{k} \times \hat{k} = \vec{0},$$

indicating that $\vec{x} \parallel \vec{y} \iff \vec{x} \times \vec{y} = \vec{0}$.

In $\mathbb{R}^2$ only $\hat{i}$ and $\hat{j}$ are needed to describe the geometry,

$$\hat{i} \equiv \begin{pmatrix} 1 \\ 0 \end{pmatrix}, \hat{j} \equiv \begin{pmatrix} 0 \\ 1 \end{pmatrix} \implies \hat{i} \cdot \hat{i} = \hat{j} \cdot \hat{j} = 1, \ \hat{i} \cdot \hat{j} = \hat{j} \cdot \hat{i} = 0. \quad (4.33)$$

However the cross product is simply a scalar so that

$$\hat{i} \times \hat{j} = 1, \ \hat{j} \times \hat{i} = -1, \ \hat{i} \times \hat{i} = \hat{j} \times \hat{j} = 0.$$

If $\vec{x}, \vec{y} \in \mathbb{R}^2$ then $\vec{x} \times \vec{y}$ can be thought of as pointing out of the plane, if a direction needs to be associated with the product. The magnitude is related to the angle between the vectors according to the *sine law*:

$$\sin(\theta) = \frac{|\vec{x} \times \vec{y}|}{\sqrt{(\vec{x} \cdot \vec{x})(\vec{y} \cdot \vec{y})}} = |\hat{x} \times \hat{y}|, \tag{4.34}$$

or inverting the sine function, and applying the $arcsin \equiv \sin^{-1}$ function to both sides, gives $\theta = arcsin(|\hat{x} \times \hat{y}|)$ whose values are restricted to the range $[0, \pi/2] = [0°, 90°]$. The formula in Eq. (4.34) also works in $\mathbb{R}^3$:

$$\sin(\theta) = \frac{\|\vec{x} \times \vec{y}\|}{\|\vec{x}\| \cdot \|\vec{y}\|} = \|\hat{x} \times \hat{y}\|. \tag{4.35}$$

In higher dimensions the angle $\theta$ is best obtained from the cosine law in Eq. (4.10).

### ■ Exercises

1. Consider the two vectors $\vec{u} = (1 \ -2)^T$ and $\vec{v} = (3 \ 1)^T$.
   
   (a) Express $\vec{u}$ and $\vec{v}$ as linear combinations of $\{\hat{i}, \hat{j}\}$.
   (b) Use this expression to compute and simplify $\vec{u} \times \vec{v}$.
   (c) Compute the angle between $\vec{u}$ and $\vec{v}$ using the sine law in Eq. (4.34).

2. Consider the two vectors $\vec{a} = (-1 \ 1 \ -2)^T$ and $\vec{b} = (0 \ 3 \ 1)^T$.
   
   (a) Express $\vec{a}$ and $\vec{b}$ as linear combinations of $\{\hat{i}, \hat{j}, \hat{k}\}$.
   (b) Use this expression to compute and simplify $\vec{a} \times \vec{b}$.
   (c) Compute the angle between $\vec{a}$ and $\vec{b}$ using the sine law in Eq. (4.35).

### ■ 4.6 Linear Systems and Matrix Equations

As introduction to this applied area, consider the population of any living species. These can be broken up into three basic categories,

$$Y = \#of\ young, \quad A = \#of\ adults, \quad G = \#of\ elderly.$$

Then the total population is a sum of the three categories $P = Y + A + G$. The change in each category from season to season depends on characteristics of the

## 4.6 Linear Systems and Matrix Equations

species. The new subpopulation levels of the young is expected to be of the form

$$Y_{new} = m_{11} Y_{past} + m_{12} A_{past}.$$

The parameters $m_{11}$ and $m_{12}$ have the following meanings:

$m_{11} \equiv$ proportion of young surviving to the next season, and
$m_{12} \equiv$ proportion of adult population that sired young.

Each parameter is intended to capture one source of changes to the subpopulation of young individuals. Similarly, in a new season, the number of adults will be

$$A_{new} = m_{21} Y_{past} + m_{22} A_{past},$$

where the parameters have the meanings:

$m_{21} \equiv$ proportion of young surviving to adulthood, and
$m_{22} \equiv$ proportion of adults surviving to the next season.

Finally, the elderly population becomes

$$G_{new} = m_{32} A_{past} + m_{33} G_{past},$$

with the interpretations:

$m_{23} \equiv$ proportion of adults that become infertile, and
$m_{33} \equiv$ survivability of elderly to the next season.

To compactly represent this mathematical model, define the population state vectors

$$\vec{P}_{new} = \begin{pmatrix} Y_{new} \\ A_{new} \\ G_{new} \end{pmatrix}, \quad \vec{P}_{past} = \begin{pmatrix} Y_{past} \\ A_{past} \\ G_{past} \end{pmatrix}, \tag{4.36}$$

and the species *parameter* vectors

$$\underline{m}^1 = (m_{11} \; m_{12} \; 0), \quad \underline{m}^2 = (m_{21} \; m_{22} \; 0), \quad \underline{m}^3 = (0 \; m_{32} \; m_{33}). \tag{4.37}$$

The connection between the previous generations from season to season can then be expressed as

$$\vec{P}_{new} = M\vec{P}_{past} = \begin{pmatrix} \mathbf{m}^1 \vec{P}_{past} \\ \mathbf{m}^2 \vec{P}_{past} \\ \mathbf{m}^3 \vec{P}_{past} \end{pmatrix}.$$

The species parameter matrix or stepping matrix $M$ is a $3 \times 3$ array

$$M = \begin{pmatrix} m_{11} & m_{12} & 0 \\ m_{21} & m_{22} & 0 \\ 0 & m_{11} & m_{12} \end{pmatrix} = \begin{pmatrix} \mathbf{m}^1 \\ \mathbf{m}^2 \\ \mathbf{m}^3 \end{pmatrix}. \tag{4.38}$$

This matrix contains information that must be measured and can be known only to a limited degree of certainty. Still, predictions can be made about the future population levels if it is assumed that components of $M$ remain constant from season to season. In two seasons the new population levels will be $\vec{P}_2 = M^2 \vec{P}_{past}$, and in three seasons $\vec{P}_3 = M^3 \vec{P}_{past}$, etc.

### EXAMPLE 4.3

Suppose a species is observed to have an initial state vector $\vec{P}_0 = (100\ 50\ 20)^T$ with stepping matrix

$$M = \begin{pmatrix} 0.4 & 1.5 & 0 \\ 0.2 & 0.8 & 0 \\ 0 & 0.1 & 0.8 \end{pmatrix}, \quad M^2 = \begin{pmatrix} 0.46 & 1.8 & 0 \\ 0.24 & 0.94 & 0 \\ 0.02 & 0.16 & 0.64 \end{pmatrix}. \tag{4.39}$$

The derivation of $M^2$ will be shown in the next section. Instead, consider the corresponding population predictions

$$\vec{P}_1 = M\vec{P}_0 = \begin{pmatrix} 115 \\ 60 \\ 21 \end{pmatrix}, \quad \vec{P}_2 = M\vec{P}_1 = M^2\vec{P}_0 = \begin{pmatrix} 136 \\ 71 \\ 22.8 \end{pmatrix}. \tag{4.40}$$

In this way future population values can be predicted. However, errors and uncertainties will certainly affect the validity of the model for higher iterations. (Continued in Example 4.4.)

## 4.6.1 Matrix Multiplication

The size or dimension of a matrix $M$ is expressed as $n \times m$ where $n$ is the number of rows and $m$ is the number of columns. The space of such matrices with real-valued components is denoted $\mathcal{M}_{n,m}(\mathbb{R})$. It is instructive to express $M$ in terms of $n$-row vectors and $m$-column vectors, as

$$M = \begin{pmatrix} m_{11} & m_{12} & \cdots & m_{1m} \\ m_{21} & m_{22} & \cdots & m_{2m} \\ . & . & \cdots & . \\ m_{n1} & m_{n2} & \cdots & m_{nm} \end{pmatrix} = \begin{pmatrix} \underline{r}^1 \\ \underline{r}^2 \\ . \\ \underline{r}^n \end{pmatrix} = \begin{pmatrix} \vec{c}_1 & \vec{c}_2 & \cdots & \vec{c}_m \end{pmatrix}. \quad (4.41)$$

This is also expressed as $M = (m_{ij})$. This allows a way of defining matrix multiplication as an extension of the dot product.

### EXAMPLE 4.4

(Continued from Example 4.3.) The square of $M$ from the previous example is computed as

$$M^2 = M \cdot M = \begin{pmatrix} \boxed{0.4} & \boxed{1.5} & \boxed{0} \\ 0.2 & 0.8 & 0 \\ 0 & 0.1 & 0.8 \end{pmatrix} \cdot \begin{pmatrix} \boxed{0.4} & 1.5 & 0 \\ \boxed{0.2} & 0.8 & 0 \\ \boxed{0} & 0.1 & 0.8 \end{pmatrix} = \begin{pmatrix} 0.4^2 + (1.5)(0.2) + 0 \cdot 0 & . & . \\ . & . & . \\ . & . & . \end{pmatrix}.$$

This allows the derivation of $M^2$. Then matrix multiplication of $M$ and $M^2$ with $\vec{P}_0$ gives the result stated in Eq. (4.40).

Indeed, if $A$ is a matrix with dimension $m \times k$, then the product is defined

$$M \cdot A = \begin{pmatrix} \underline{r}^1 \\ \underline{r}^2 \\ . \\ \underline{r}^n \end{pmatrix} \cdot \begin{pmatrix} \vec{a}_1 & \vec{a}_2 & \cdots & \vec{a}_k \end{pmatrix} = \begin{pmatrix} \underline{r}^1 \vec{a}_1 & \underline{r}^1 \vec{a}_2 & \cdots & \underline{r}^1 \vec{a}_k \\ \underline{r}^2 \vec{a}_1 & \underline{r}^2 \vec{a}_2 & \cdots & \underline{r}^2 \vec{a}_k \\ . & . & \cdots & . \\ \underline{r}^n \vec{a}_1 & \underline{r}^n \vec{a}_2 & \cdots & \underline{r}^n \vec{a}_k \end{pmatrix}, \quad (4.42)$$

which is a new matrix with dimension $n \times k$. The components are dot products. For brevity one can express the product as $M \cdot A = (\underline{r}^i \vec{a}_j)$. However, this can only make sense if the number of components in each $\underline{r}^i$ and $\vec{a}_j$ agree. In this example,

$$dim(\underline{r}^i) = 1 \times m, \ dim(\vec{a}_j) = m \times 1, \ dim(\underline{r}^i \vec{a}_j) = 1 \times 1 \in \mathbb{R},$$

and for the matrices,

$$dim(M) = n \times m, \ dim(A) = m \times k, \ dim(M \cdot A) = n \times \boxed{m \times m} \times k = n \times k.$$

For properly formulated applications these equations must hold true.

### ■ Exercises

1. Consider the two vectors $\vec{u} = (1\ -2)^T$, $\vec{v} = (3\ 1)^T$, and matrix $A = \begin{pmatrix} 2 & 3 \\ -1 & 4 \end{pmatrix}$.

    (a) Compute $A \cdot \vec{u}$ and $A \cdot \vec{v}$.
    (b) Compute $\vec{u}^T \cdot A$ and $\vec{v}^T \cdot A$.
    (c) Compute $\vec{u}^T \cdot A \cdot \vec{v}$ and $\vec{v}^T \cdot A \cdot \vec{u}$.

2. Consider the vectors $\vec{a} = (-1\ \ 1\ \ -2)^T$, $\vec{b} = (0\ 3\ 1)^T$, and $M$ defined in Eq. (4.39).

    (a) Compute $M \cdot \vec{a}$ and $M \cdot \vec{b}$.
    (b) Compute $\vec{a}^T \cdot M$ and $\vec{b}^T \cdot M$.
    (c) Compute $\vec{a}^T \cdot M \cdot \vec{b}$ and $\vec{b}^T \cdot M \cdot \vec{a}$.

### ■ 4.7 Vector, Matrix, and Linear Algebra

Matrix multiplication and the dot product are the more difficult barriers of this subject, so they are dealt with early in this chapter. However, vectors and matrices have basic algebraic properties that are extensions of the reals $\mathbb{R}$. For example, just as two vectors add component-wise, so can two matrices of the same size. Thus the set of all $n \times m$ matrices with real components $\mathcal{M}_{n,m}(\mathbb{R})$, with the operation of component-wise addition, forms an abelian group, with the *zero matrix* as the identity. However, $\langle \mathcal{M}_{n,m}, + \rangle$ is not a cyclic group and cannot have a finite number of generators. Still, as a vector space, invariant under scalar multiplication, $\mathcal{M}_{n,m}$ can be generated by a linear combination of $n \cdot m$ basis matrices. Thus $dim\,(\mathcal{M}_{n,m}) = n \cdot m$. Furthermore, the set of square matrices $\mathcal{M}_{n,n}$ forms a ring under the operations of addition and multiplication.

Multivariable linear equations can have multiple solutions or no solutions. Overdetermined equations can have a closest fitting solution. When the number of variables exceeds three, one cannot directly visualize the solutions. However, all operations proceed as practiced in two and three dimensions.

### 4.7.1 Matrix Equations

Let $A$ be an $n \times n$ square *parameter* matrix and $\vec{b}$ an $n \times 1$ *output* vector. Suppose $\exists A^{-1}$, an $n \times n$ matrix called the *inverse* of $A$, where $A^{-1} \cdot A = I$ is the $n \times n$ (multiplicative) identity matrix

$$I \equiv diag(1, 1, \ldots, 1) = \begin{pmatrix} 1 & 0 & \ldots & 0 \\ 0 & 1 & \ldots & 0 \\ \vdots & \vdots & \ddots & 0 \\ 0 & \ldots & 0 & 1 \end{pmatrix} = (\delta^i_j). \tag{4.43}$$

Then there is an $n \times 1$ vector of variables $\vec{x}$ that solves the matrix equation

$$A \cdot \vec{x} = \vec{b} \quad \Longleftrightarrow \quad \vec{x} = A^{-1} \cdot \vec{b}. \tag{4.44}$$

One way that $A^{-1}$ may be obtained is by using a Geometric series,

$$A^{-1} = \lambda \cdot (I - I + \lambda \cdot A)^{-1} = \lambda \cdot I + \lambda \cdot \sum_{k=1}^{\infty} (I - \lambda \cdot A)^k, \tag{4.45}$$

assuming that it converges for some $\lambda \neq 0$. Truncation gives an approximate inverse.

### Theorem 4.2

Given a system of $n$ linear equations in $m$ unknown variables, then the system can be written as a matrix equation $A \cdot \vec{x} = \vec{b}$, where $A$ is an $n \times m$ matrix, $\vec{x} \in \mathbb{R}^m$ and $\vec{b} \in \mathbb{R}^n$. There are several cases:

**$A$ is square** where $n = m$, number of rows (equations) equal the number of columns (variables):
$rank(A) = n = m \implies \exists!$ solution $\vec{x} = A^{-1} \cdot \vec{b}$, or if $\vec{b} \in Range(A)$ and
$rank(A) = k < n = m \implies \exists$ a vector space of solutions with dimension $n - k$;

**$A$ is not square** where $n < m$, number of rows (equations) is less than the number of columns (variables) and $\vec{b} \in Range(A)$:
$rank(A) = n \implies \exists$ a vector space of solutions with dimension $m - n$, or
$rank(A) = k < n \implies \exists$ a vector space of solutions with dimension $m - k$;

**$A$ is not square** where $n > m$, number of rows (equations) is greater than the number of columns (variables):
$rank(A) = m \implies \exists!$ solution $\vec{x} = (A^T A)^{-1} A^T \vec{b} \iff \vec{b} \in Range(A)$, or
$rank(A) = k < m \implies \exists$ solution $\iff \vec{b} \in Range(A)$.

The proof of these statements becomes clear after considering how solutions to a matrix equation are found, systematically. As a special case, consider the general $2 \times 2$ matrix $A = \begin{pmatrix} a & b \\ c & d \end{pmatrix}$. There are two important characteristics of $A$, the trace and determinant of $A$, defined as

$$tr(A) \equiv a + d, \ det(A) \equiv a \cdot d - b \cdot c. \tag{4.46}$$

If $det(A) \neq 0$, then the characteristic $\kappa = 1/det(A)$ can be defined, in which case

$$A^{-1} = \kappa \cdot \begin{pmatrix} d & -b \\ -c & a \end{pmatrix}. \tag{4.47}$$

For $n \times n$ matrices with $n \geq 3$, inversion formulae exist but are progressively more complicated as $n$ increases.

### ■ Exercises

1. Show that for any $2 \times 2$ matrix $A$, one has $2 \cdot det(A) = tr(A)^2 - tr(A^2)$.
2. Consider the matrix $A = \begin{pmatrix} 2 & 3 \\ -1 & 4 \end{pmatrix}$.

   (a) Compute $det(A)$ and find $A^{-1}$ using the formula in Eq. (4.47). Also express the components up to two decimal places.
   (b) Let $\lambda = 1/\sqrt{det(A)}$ and compute an approximation of $A^{-1}$ using the first two terms in Eq. (4.45) up to two decimal places for each component. In particular, compute and simplify $B \equiv \lambda \cdot I + \lambda \cdot (I - \lambda \cdot A)$.
   (c) For $\vec{u} = (1 - 2)^T$ solve $A \cdot \vec{x} = \vec{u}$ for $\vec{x}$ up to two decimal places.
   (d) Compute the relative error $\|\vec{x} - B \cdot \vec{u}\|/\|\vec{x}\|$.

### ■ 4.7.2 Gaussian Elimination

There is a systematic way of studying linear equations that allows one to easily solve for the variables in terms of the parameters using an algorithm called Gaussian Elimination. To understand the steps of this method, consider the following *elementary operations* of $N$ equations listed as $\{①,②,\ldots,Ⓝ\}$:

$$\begin{aligned} \rho \text{ multiples of a row:} &\quad ⓙ \to \boxed{\rho} \cdot ⓙ \quad (\text{for } \rho \neq 0), \\ \text{switch order of two rows:} &\quad ⓘ \leftrightarrow ⓙ, \\ \text{add one row to another row:} &\quad ⓘ \to ⓘ + ⓙ. \end{aligned} \qquad (4.48)$$

These three elementary operations can never change the solutions of a system of linear equations. In fact, if performed properly, they will lead to a solution, a set of solutions, or a proof that no solution exists. Consider a standard $3 \times 3$ case, with parameter matrix $A$, *input* vector $\vec{x}$, and *output* vector $\vec{b}$:

$$\begin{aligned} a_{11}x_1 + a_{12}x_2 + a_{13}x_3 &= b_1 \\ a_{21}x_1 + a_{22}x_2 + a_{23}x_3 &= b_2 \quad \Longleftrightarrow \quad A \cdot \vec{x} = \vec{b} \\ a_{31}x_1 + a_{32}x_2 + a_{33}x_3 &= b_3. \end{aligned} \qquad (4.49)$$

This system can be rewritten as an *augmented matrix*, where the symbols $+$ and $-$ and the components of the input $x_k$ are excluded to give

$$\begin{aligned} a_{11}x_1 + a_{12}x_2 + a_{13}x_3 &= b_1 \\ a_{21}x_1 + a_{22}x_2 + a_{23}x_3 &= b_2 \quad \Longleftrightarrow \quad \begin{pmatrix} a_{11} & a_{12} & a_{13} & | & b_1 \\ a_{21} & a_{22} & a_{23} & | & b_2 \\ a_{31} & a_{32} & a_{33} & | & b_3 \end{pmatrix}, \\ a_{31}x_1 + a_{32}x_2 + a_{33}x_3 &= b_3 \end{aligned} \qquad (4.50)$$

where only the parameters $a_{ij}$ and the components of the output vector $\vec{b}$ appear. Now, the elementary operations in Eq. (4.48) can be used to change the augmented matrix into the identity, while changing the output vector $\vec{b}$. Thus, the goal of Gaussian Elimination (GE) is to have a final form

$$(\,I \mid \vec{c}\,) = \begin{pmatrix} 1 & 0 & 0 & | & c_1 \\ 0 & 1 & 0 & | & c_2 \\ 0 & 0 & 1 & | & c_3 \end{pmatrix} \Longleftrightarrow \begin{pmatrix} x_1 \\ x_2 \\ x_3 \end{pmatrix} = \begin{pmatrix} c_1 \\ c_2 \\ c_3 \end{pmatrix} \Longleftrightarrow \vec{x} = \vec{c},$$

where $\vec{c}$ is the solution to the matrix equation $A \cdot \vec{x} = \vec{b}$, and $I$ is the $3 \times 3$ identity.

## EXAMPLE 4.5

From Example 4.3 suppose that an output vector $\vec{P}_1 = (115\ 60\ 21)^T$ was the result of the matrix equation $M \cdot \vec{P}_0 = \vec{P}_1$. To solve for the input vector $\vec{P}_0$, the objective is to perform a series of elementary operations from Eq. (4.48) to obtain the transformation

$$\text{Start: } \left( M \mid \vec{P}_1 \right) \xrightarrow{\text{elementary operations}} \left( I \mid \vec{P}_0 \right) \text{ :Goal.} \tag{4.51}$$

This can be demonstrated as follows:

$$\begin{pmatrix} 0.4 & 1.5 & 0 & | & 115 \\ 0.2 & 0.8 & 0 & | & 60 \\ 0 & 0.1 & 0.8 & | & 21 \end{pmatrix} \xrightarrow{①↔②} \begin{pmatrix} 0.2 & 0.8 & 0 & | & 60 \\ 0.4 & 1.5 & 0 & | & 115 \\ 0 & 0.1 & 0.8 & | & 21 \end{pmatrix} \tag{4.52}$$

$$\xrightarrow[-2.5\times②]{5\times①} \begin{pmatrix} 1 & 4 & 0 & | & 300 \\ -1 & -3.75 & 0 & | & -287.5 \\ 0 & 0.1 & 0.8 & | & 21 \end{pmatrix} \xrightarrow{②+①} \begin{pmatrix} 1 & 4 & 0 & | & 300 \\ 0 & 0.25 & 0 & | & 12.5 \\ 0 & 0.1 & 0.8 & | & 21 \end{pmatrix} \tag{4.53}$$

$$\xrightarrow[-10\times③]{4\times②} \begin{pmatrix} 1 & 4 & 0 & | & 300 \\ 0 & 1 & 0 & | & 50 \\ 0 & -1 & -8 & | & -210 \end{pmatrix} \xrightarrow[①-4\times②]{③+②} \begin{pmatrix} 1 & 0 & 0 & | & 100 \\ 0 & 1 & 0 & | & 50 \\ 0 & 0 & -8 & | & -160 \end{pmatrix} \tag{4.54}$$

To complete the application of GE, apply the elementary operation $(-1/8) \times ③$, which gives the goal $(I|\vec{P}_0)$ with the solution, or input vector $\vec{P}_0 = (100\ 50\ 20)^T$.

---

**Definition 4.8** The *rank* of an $n \times m$ matrix $A$, denoted $rank(A)$, is the number of nonzero rows obtained by performing GE to the extent possible. These nonzero row vectors are a basis of the *row space*, denoted $Row(A)$, whose dimension is the same as $rank(A)$.

---

This definition gives a method for finding a basis of any finite dimensional vector space, by re-expressing elements in terms of row vectors, and then performing GE toward the goal.

---

**Definition 4.9** The *column space* of an $n \times m$ matrix $A$, denoted $Col(A)$, is the *transpose* of elements in $Row(A^T)$. This can be written as

$$Col(A) = (Row(A^T))^T \equiv \{\overline{a}^T : \overline{a} \in Row(A^T)\}.$$

It can be shown that $dim(Col(A)) = dim(Row(A)) = rank(A)$. Considering $A : \mathbb{R}^m \to \mathbb{R}^n$ as a function, $Range(A) = Col(A)$.

---

This last statement follows by considering $M$ in Eq. (4.41) acting on a vector $\vec{x}$,

$$M \cdot \vec{x} = \begin{pmatrix} m_{11}x_1 + & \ldots & +m_{1m}x_m \\ \cdot & \ldots & \cdot \\ m_{n1}x_1 + & \ldots & +m_{nm}x_m \end{pmatrix} \qquad (4.55)$$

$$= \begin{pmatrix} m_{11} \\ \cdot \\ m_{n1} \end{pmatrix} \cdot x_1 + \cdots + \begin{pmatrix} m_{1m} \\ \cdot \\ m_{nm} \end{pmatrix} \cdot x_m$$

$$= x_1 \cdot \vec{c}_1 + \cdots + x_m \cdot \vec{c}_m.$$

Thus, if $M$ is $n \times m$ and $\vec{x} \in \mathbb{R}^m$; then $M \cdot \vec{x} \in Col(M) \subseteq \mathbb{R}^n$. Note that $M \cdot M^T$ is an $n \times n$ square matrix and $M^T \cdot M$ is an $m \times m$ square matrix. However,

$$rank(M) = rank(M \cdot M^T) = rank(M^T \cdot M) = dim(Col(M)) = dim(Row(M)).$$

---

**Definition 4.10** The *null space* of an $n \times m$ matrix $M$ is defined to be

$$Null(M) \equiv \{\vec{x} \in \mathbb{R}^m : M \cdot \vec{x} = \vec{0}\}.$$

This is also called the *kernel* of $M$ and written $ker(M)$. The dimension of the null space, $dim(Null(M))$ is called the *nullity* of $M$.

---

The following fact, called the *Rank-Nullity theorem* [23],

$$rank(M) + nullity(M) = m,$$

implies that the *domain space of* $M$, written as $Dom(M) = \mathbb{R}^m$, can be decomposed into two parts, $Null(M)$ and its complement in $Dom(M)$. Although $Null(M)$ is a subspace of $Dom(M)$, its complement is not. However, taking a $rank(M)$ number of linearly independent vectors $\{\vec{v}_j\}_{j=1}^{rank(M)}$ from the complement

$$Null(M)^c \equiv Dom(M) - Null(M) = \{\vec{v} \in Dom(M) : M \cdot \vec{v} \neq \vec{0}\},$$

one can write the decomposition of $Dom(M) = \mathbb{R}^m$ as a vector sum

$$Dom(M) = Null(M) \oplus span\left[\{\vec{v}_j\}_{j=1}^{rank(M)}\right],$$

where now $span\left[\{\vec{v}_j\}_{j=1}^{rank(M)}\right]$ is a subspace of $Dom(M)$.

## Exercises

1. Consider the matrix $A = \begin{pmatrix} 2 & 3 \\ -1 & 4 \end{pmatrix}$ and the column vector $\vec{u} = \begin{pmatrix} 1 \\ -2 \end{pmatrix}$.

    (a) Write the matrix equation $A \cdot \vec{x} = \vec{u}$ as an augmented matrix, as outlined in Eq. (4.49) and Eq. (4.50).
    (b) Solve for $\vec{x}$ using Gaussian Elimination.

2. Solve the system $\begin{pmatrix} 1 & 3 & 2 \\ -1 & 4 & 0 \\ 2 & 1 & 2 \end{pmatrix} \cdot \begin{pmatrix} x \\ y \\ z \end{pmatrix} = \begin{pmatrix} 1 \\ 1 \\ -2 \end{pmatrix}$ by writing as an augmented matrix and then applying GE.

3. Use GE to find the rank of $B = \begin{pmatrix} 1 & 3 & 2 \\ -1 & 4 & 0 \\ 2 & -1 & 2 \end{pmatrix}$. Also, pick a basis for $Col(B)$ and express this subspace as a linear combination of the basis.

## 4.8 Matrix Inverse and the Gauss-Jordan Method

For a square $n \times n$ matrix $A$, an inverse matrix $A^{-1}$ can be defined *iff* $rank(A) = n$. In this case, $A$ has full rank, and is said to be *invertible*, so that

$$A \cdot A^{-1} = A^{-1} \cdot A = I.$$

When an inverse exists, then one can solve *matrix equations* of the form

$$A \cdot \vec{x} = \vec{b} \implies \text{solution} = \vec{x} = A^{-1} \cdot \vec{b}.$$

Thus, an inverse allows one to solve the same matrix equation, for many different output vectors $\vec{b}$, without having to perform GE each time. The procedure for finding an inverse matrix is called the *Gauss-Jordan* (GJ) method.

## EXAMPLE 4.6

Recall Example 4.3. To solve for the inverse $M$, a series of elementary operations from Eq. (4.48) are performed so that

$$\text{Start:} \ ( M \mid I ) \xrightarrow{\text{elementary operations}} ( I \mid M^{-1} ) \ \text{:Goal} \tag{4.56}$$

This can be demonstrated as follows:

$$\begin{pmatrix} 0.4 & 1.5 & 0 & | & 1 & 0 & 0 \\ 0.2 & 0.8 & 0 & | & 0 & 1 & 0 \\ 0 & 0.1 & 0.8 & | & 0 & 0 & 1 \end{pmatrix} \xrightarrow{① \leftrightarrow ②} \begin{pmatrix} 0.2 & 0.8 & 0 & | & 0 & 1 & 0 \\ 0.4 & 1.5 & 0 & | & 1 & 0 & 0 \\ 0 & 0.1 & 0.8 & | & 0 & 0 & 1 \end{pmatrix}. \tag{4.57}$$

$$\xrightarrow[5 \times ①]{② - 2 \times ①} \begin{pmatrix} 1 & 4 & 0 & | & 0 & 5 & 0 \\ 0 & -0.1 & 0 & | & 1 & -2 & 0 \\ 0 & 0.1 & 0.8 & | & 0 & 0 & 1 \end{pmatrix}. \tag{4.58}$$

$$\xrightarrow[① + 40 \times ②]{③ + ②} \begin{pmatrix} 1 & 0 & 0 & | & 40 & -75 & 0 \\ 0 & -0.1 & 0 & | & 1 & -2 & 0 \\ 0 & 0 & 0.8 & | & 1 & -2 & 1 \end{pmatrix}. \tag{4.59}$$

To complete the application of the GJ method, apply the elementary operations $(-10) \times ②$ and $(5/4) \times ③$, which gives the goal $(I|M^{-1})$ with

$$M^{-1} = \begin{pmatrix} 40 & -75 & 0 \\ -10 & 20 & 0 \\ 5/4 & -5/2 & 5/4 \end{pmatrix}, \tag{4.60}$$

which, one can verify, satisfies $M^{-1}M = I$.

## ■ Exercises

1. Find the inverse of $A = \begin{pmatrix} 2 & 3 \\ -1 & 4 \end{pmatrix}$ using the GJ method.

2. Find the inverse of $C = \begin{pmatrix} 1 & 3 & 2 \\ -1 & 4 & 0 \\ 2 & 1 & 2 \end{pmatrix}$ using the GJ method.

## 4.9 Valuations of a Matrix, Determinant, and Trace

There are several ways of measuring the size of a matrix. Each way, called a *matrix valuation*, gives some valuable information about the matrix. An example is the *maximum norm* and the *maximum row-sum norm* of the matrix $M \in \mathcal{M}_{n \times m}$:

$$|M|_\infty \equiv max\,\{|m_{ij}| : i \in \mathbb{Z}_n, j \in \mathbb{Z}_m\}, \quad \|M\|_\infty \equiv max\,\left\{\sum_{j=1}^m |m_{ij}| : i \in \mathbb{Z}_n\right\}.$$

The same can be defined for a vector $\vec{v} \in \mathbb{R}^m$. It can be seen, with little effort, that

$$|M \cdot \vec{v}|_\infty \leq m \cdot |M|_\infty \cdot |\vec{v}|_\infty \quad \text{and} \quad \|M \cdot \vec{v}\|_\infty \leq \|M\|_\infty \cdot \|\vec{v}\|_\infty.$$

A more commonly used matrix norm is, as defined in [67],

$$\|M\| \equiv \max\{\|M \cdot \vec{x}\| : (\vec{x} \in \mathbb{R}^m) \wedge (\|\vec{x}\| = 1)\}, \qquad (4.61)$$

which satisfies the *constant rule*, and the *triangle* and the *product inequalities*,

$$\|a \cdot M\| = |a| \cdot \|M\|, \quad \|M + N\| \leq \|M\| + \|N\|, \quad \|M \cdot K\| \leq \|M\| \cdot \|K\|. \qquad (4.62)$$

In particular, for a square matrix $A$ and positive integer $r$, $\|A^r\| \leq \|A\|^r$. This allows any polynomial expression $p(A)$ to be bounded as a linear operator;

$$\left[p(x) = \sum_{k=0}^r a_r \cdot x^k \iff p_{abs}(x) \equiv \sum_{k=0}^r |a_r| \cdot |x|^k\right] \implies [\|p(A)\| \leq p_{abs}(\|A\|)].$$

Another measure for invertible matrices $A \in \mathcal{M}_{n \times n}$ is the *condition number*

$$cond(A) \equiv \|A\|_\infty \cdot \|A^{-1}\|_\infty \geq 1. \qquad (4.63)$$

This allows one to estimate the uncertainty $\Delta \vec{x}$, of the input $\vec{x} \in \mathbb{R}^n$, for a matrix equation $A \cdot \vec{x} = \vec{b}$, based on the measured uncertainty $\Delta \vec{b}$ of the output $\vec{b} \in \mathbb{R}^n$. The formula is expressed as an inequality between relative uncertainties,

$$\|\Delta \vec{x}\|_\infty / \|\vec{x}\|_\infty \leq cond(A) \cdot \left(\|\Delta \vec{b}\|_\infty / \|\vec{b}\|_\infty\right). \qquad (4.64)$$

**Definition 4.11** The *trace* of a square $n \times n$ matrix $A$, written $tr(A) \in \mathbb{R}$, is the sum of its diagonal elements.

It should be clear that $\forall A, B \in \mathcal{M}_{n \times n}$, with $n \in \mathbb{N}$,

$$tr(A + B) = tr(A) + tr(B).$$

**Definition 4.12** Let $U$ be a square $n \times n$ matrix, which is *upper triangular* (having only zeros below the diagonal). The *determinant*, written $det(U) \in \mathbb{R}$, is the product of its diagonal elements.

The product of two upper-triangular matrices $U_1 \cdot U_2$ is also upper triangular, so that the diagonal elements simply multiply. Thus

$$det(U_1 \cdot U_2) = det(U_1) \cdot det(U_2).$$

If a matrix $L$ is lower triangular (having only zeros above the diagonal) then $L^T$ is upper triangular. Thus it is reasonable to conclude $det(L) \equiv det(L^T)$.

The determinant of a diagonal matrix $D$ (both upper and lower triangular) is the product of its diagonal elements. A matrix $A$ is called *diagonalizable* if $\exists S \in \mathcal{M}_{n \times n}$ which is invertible, where $S \cdot A \cdot S^{-1} = D$ is a diagonal matrix. Then $det(A) = det(D)$. In general the following can be shown [23], [53]:

### ■ Theorem 4.3

For $A, B \in \mathcal{M}_{n \times n}$, with $n \in \mathbb{N}$,

$$det(A \cdot B) = det(A) \cdot det(B).$$

The matrix $A$ is invertible *iff* $det(A) \neq 0$, in which case $det(A^{-1}) = 1/det(A)$.

**Remark 4.1** The notation $det(A) = |A|$ is also used for the determinant of square matrices $A$. However, it appears to imply that the determinant is a positive real number, where in fact $|A| \in \mathbb{C}$ in general. If the components of $A$ are real, then so is $|A|$.

A matrix equation $A \cdot \vec{x} = \vec{b}$ of $n$ equations with $n$ unknowns, has a unique solution *iff* $det(A) \neq 0$. This allows for a quick test of solvability, even before trying to find a solution. In general, the procedure for computing $det(A)$ for large matrices $A$ uses an iterative scheme. In the $2 \times 2$ case

$$det\begin{pmatrix} a & b \\ c & d \end{pmatrix} = a \cdot d - c \cdot b \neq 0 \implies \begin{pmatrix} a & b \\ c & d \end{pmatrix}^{-1} = \frac{1}{a \cdot d - c \cdot b} \cdot \begin{pmatrix} d & -b \\ -c & a \end{pmatrix}.$$

For larger matrices, first express in terms of column vectors of length $n - 1$,

$$A = \begin{pmatrix} a_{11} & a_{12} & \ldots & a_{1n} \\ \vec{c}_1 & \vec{c}_2 & \ldots & \vec{c}_n \end{pmatrix} \quad \vec{c}_{n+1} \quad \vec{c}_{n+2} \quad \ldots \quad \vec{c}_{2 \cdot n - 1} \quad \text{where } \vec{c}_{n+j} = \vec{c}_j.$$

Here the column vectors have been rewritten to the right of the matrix, and represent an $(n-1) \times (n-1)$ array. Then the formula in general is

$$det(A) \equiv \sum_{j=1}^{n} a_{1j} \cdot det\begin{pmatrix} \vec{c}_{j+1} & \vec{c}_{j+2} & \ldots & \vec{c}_{j+n-1} \end{pmatrix}. \tag{4.65}$$

### EXAMPLE 4.7

Consider Example 4.3 again and compute,

$$det\begin{pmatrix} 0.4 & 1.5 & 0 \\ 0.2 & 0.8 & 0 \\ 0 & 0.1 & 0.8 \end{pmatrix} \quad \begin{matrix} 0.2 & 0.8 \\ 0 & 0.1 \end{matrix} \tag{4.66}$$

$$= (0.4) \cdot \begin{vmatrix} 0.8 & 0 \\ 0.1 & 0.8 \end{vmatrix} + (1.5) \cdot \begin{vmatrix} 0 & 0.2 \\ 0.8 & 0 \end{vmatrix} + (0) \cdot \begin{vmatrix} 0.2 & 0.8 \\ 0 & 0.1 \end{vmatrix}$$

$$= (0.4) \cdot (0.64 - 0) + (1.5) \cdot (0 - 0.16) + (0) \cdot (0.02 - 0)$$

$$= 0.256 - 0.24 + 0 = 0.016.$$

Since $det(A) = 0.016 \neq 0$, the matrix $M$ is invertible. However, finding this inverse requires an application of, for example, the GJ method.

# Exercises

1. Consider the matrix $A = \begin{pmatrix} 2 & 3 \\ -1 & 4 \end{pmatrix}$.

   (a) Compute $tr(A)$ and $det(A)$.
   (b) Compute $|A|_\infty$ and $\|A\|_\infty$.
   (c) Compute $|A^{-1}|_\infty$, $\|A^{-1}\|_\infty$, and $cond(A)$.

2. Let $A = \begin{pmatrix} 2 & 3 \\ -1 & 4 \end{pmatrix}$ be a parameter matrix, and $\vec{b}_1 = \begin{pmatrix} 11 \\ -23 \end{pmatrix}$, $\vec{b}_2 = \begin{pmatrix} 9 \\ -25 \end{pmatrix}$ the extreme values of the output vector for the equation $A \cdot \vec{x} = \vec{b}$.

   (a) Compute $\Delta \vec{b} = \vec{b}_2 - \vec{b}_1$ and $\vec{b} = (\vec{b}_1 + \vec{b}_2)/2$.
   (b) Compute $\|\Delta \vec{b}\|_\infty$ and $\|\vec{b}\|_\infty$.
   (c) Compute $cond(A)$.
   (d) Compute the relative output uncertainty $\|\Delta \vec{b}\|_\infty / \|\vec{b}\|_\infty$.
   (e) Using the inequality in Eq. (4.64) estimate the relative input uncertainty.

3. Let $C = \begin{pmatrix} 1 & 3 & 2 \\ -1 & 4 & 0 \\ 2 & 1 & 2 \end{pmatrix}$ be a parameter matrix, and $\vec{d}_1 = \begin{pmatrix} 6 \\ -9 \\ 1 \end{pmatrix}$, $\vec{d}_2 = \begin{pmatrix} 8 \\ -7 \\ -1 \end{pmatrix}$ the extreme values of the output vector for the matrix equation $C \cdot \vec{y} = \vec{d}$.

   (a) Compute $\|C\|_\infty$, $\|C^{-1}\|_\infty$, and $cond(C)$.
   (b) Compute $det(C)$ using the method of Eq. (4.65) and Eq. (4.66), as in Example 4.7.
   (c) Compute $\Delta \vec{d} = \vec{d}_2 - \vec{d}_1$ and $\vec{d} = (\vec{d}_1 + \vec{d}_2)/2$.
   (d) Compute $\|\Delta \vec{d}\|_\infty$ and $\|\vec{d}\|_\infty$, and compute the relative output uncertainty $\|\Delta \vec{d}\|_\infty / \|\vec{d}\|_\infty$.
   (e) Using the inequality in Eq. (4.64) estimate the relative input uncertainty $\|\Delta \vec{y}\|_\infty / \|\vec{y}\|_\infty$.

4. Prove all three statements in Eq. (4.62) by carefully using Eq. (4.61).

## 4.10 Elementary Matrices

The elementary operations can be encoded as *multiplicatively* invertible matrices, referred to as *elementary matrices* [53]. These are simple $2 \times 2$ modifications of the

$n \times n$ identity matrix $I$. Examples for $n = 3$ are

$$E_{\boxed{\rho} \cdot ②} \equiv \begin{pmatrix} 1 & 0 & 0 \\ 0 & \rho & 0 \\ 0 & 0 & 1 \end{pmatrix}, \; E_{①\leftrightarrow②} \equiv \begin{pmatrix} 0 & 1 & 0 \\ 1 & 0 & 0 \\ 0 & 0 & 1 \end{pmatrix}, \; E_{③\leftrightarrow③+①} \equiv \begin{pmatrix} 1 & 0 & 0 \\ 0 & 1 & 0 \\ 1 & 0 & 1 \end{pmatrix},$$

corresponding to the elementary operation given in Eq. (4.48). These help in computing the determinant because

$$det\left(E_{\boxed{\rho} \cdot ②}\right) = \rho, \; det\left(E_{①\leftrightarrow②}\right) = -1, \; det\left(E_{③\leftrightarrow③+①}\right) = 1.$$

These matrices have obvious inverses.

### EXAMPLE 4.8

The elementary operations required to decompose $M$ as elementary matrices are found in Eq. (4.57).

$$M = \begin{pmatrix} 0.4 & 1.5 & 0 \\ 0.2 & 0.8 & 0 \\ 0 & 0.1 & 0.8 \end{pmatrix} \rightarrow \left\{ \begin{array}{c} ① \leftrightarrow ② \\ ② - 2 \times ① \\ 5 \times ① \\ ③ + ② \\ ① + 40 \times ② \end{array} \right\} \rightarrow \begin{pmatrix} 1 & 0 & 0 \\ 0 & -0.1 & 0 \\ 0 & 0 & 0.8 \end{pmatrix} = D_M$$

$$S \equiv E_5 \cdot E_4 \cdot E_3 \cdot E_2 \cdot E_1 = \begin{pmatrix} 1 & 40 & 0 \\ 0 & 1 & 0 \\ 0 & 0 & 1 \end{pmatrix} \cdot \begin{pmatrix} 1 & 0 & 0 \\ 0 & 1 & 0 \\ 0 & 1 & 1 \end{pmatrix} \cdot \begin{pmatrix} 5 & 0 & 0 \\ 0 & 1 & 0 \\ 0 & 0 & 1 \end{pmatrix} \cdot \begin{pmatrix} 1 & 0 & 0 \\ -2 & 1 & 0 \\ 0 & 0 & 1 \end{pmatrix} \cdot \begin{pmatrix} 0 & 1 & 0 \\ 1 & 0 & 0 \\ 0 & 0 & 1 \end{pmatrix}$$

$$S^{-1} \equiv E_1^{-1} \cdot E_2^{-1} \cdot E_3^{-1} \cdot E_4^{-1} \cdot E_5^{-1} = \begin{pmatrix} 0 & 1 & 0 \\ 1 & 0 & 0 \\ 0 & 0 & 1 \end{pmatrix} \cdot \begin{pmatrix} 1 & 0 & 0 \\ 2 & 1 & 0 \\ 0 & 0 & 1 \end{pmatrix} \cdot \begin{pmatrix} 1/5 & 0 & 0 \\ 0 & 1 & 0 \\ 0 & 0 & 1 \end{pmatrix} \cdot \begin{pmatrix} 1 & 0 & 0 \\ 0 & 1 & 0 \\ 0 & -1 & 1 \end{pmatrix} \cdot \begin{pmatrix} 1 & -40 & 0 \\ 0 & 1 & 0 \\ 0 & 0 & 1 \end{pmatrix}$$

Then it can be shown that $D_M = S \cdot M$ and $M = S^{-1} \cdot D_M$. The inverse of each elementary matrix is rather easy to obtain. Thus, for example, the inverse of $E_5$ can be found as

$$E_5^{-1} = \begin{pmatrix} 1 & 40 & 0 \\ 0 & 1 & 0 \\ 0 & 0 & 1 \end{pmatrix}^{-1} = \begin{pmatrix} 1 & -40 & 0 \\ 0 & 1 & 0 \\ 0 & 0 & 1 \end{pmatrix}, \quad (4.67)$$

where just the 2 × 2 portion of the matrix needs to be inverted. This allows for inversion to be reduced to a product of inverted elementary matrices, in reverse order. Similarly, the determinant is simply

$$det(E_1) = \begin{vmatrix} 0 & 1 & 0 \\ 1 & 0 & 0 \\ 0 & 0 & 1 \end{vmatrix} = -1, \; det(E_3) = \begin{vmatrix} 1 & 0 & 0 \\ 0 & 5 & 0 \\ 0 & 0 & 1 \end{vmatrix} = 5, \; det(E_5) = \begin{vmatrix} 1 & 40 & 0 \\ 0 & 1 & 0 \\ 0 & 0 & 1 \end{vmatrix} = 1.$$

Thus $det(M) = det(S^{-1}) \cdot det(D_M) = (-1) \cdot (1) \cdot (1/5) \cdot (1) \cdot (1) \cdot (-0.08) = 0.016$.

## ■ Exercises

1. Express $A = \begin{pmatrix} 2 & 3 \\ -1 & 4 \end{pmatrix}$ as a product of elementary matrices using the GJ method as in Example 4.8. Also, find the determinant using this decomposition.

2. Express $C = \begin{pmatrix} 1 & 3 & 2 \\ -1 & 4 & 0 \\ 2 & 1 & 2 \end{pmatrix}$ as a product of elementary matrices using the GJ method as in Example 4.8, and find the determinant using this decomposition.

## ■ 4.11 Eigenvalues and Eigenvectors

A matrix $A \in \mathcal{M}_{n \times n}$ acts on an $n$-component column vector, in a linear manner, to give a new $n$-component column vector. This action can appear complicated, but actually is uniquely determined by the way $A$ operates on any basis of $\mathbb{R}^n$.

**Remark 4.2** The components of a square matrix $A$ are often real-valued parameters of a system. There are several characteristics of a matrix that summarize its components, and the spectrum, which are the *eigenvalues* of $A$, is an example. The eigenvalues can be complex, even though the components of $A$ are real.

The simplest family of square matrices are *diagonal*,

$$\mathcal{D}_{n \times n} = \left\{ diag(\lambda_1, \lambda_2, \ldots \lambda_n) \equiv \begin{pmatrix} \lambda_1 & 0 & \cdots & 0 \\ 0 & \lambda_2 & \cdots & 0 \\ 0 & 0 & \cdots & 0 \\ 0 & 0 & \cdots & \lambda_n \end{pmatrix} : \lambda_j \in \mathbb{C}, j \in \mathbb{Z}_n \right\},$$

which is closed under addition and matrix multiplication. The algebra of diagonal matrices is very similar to that of $\langle \mathbb{C}, +, \cdot \rangle$. Most matrices can be transformed into a diagonal form using GE. However, when a matrix cannot be diagonalized, it is called *defective*, and is transformable to an upper or lower triangular matrix. The Jordan canonical form is an example [23], [53].

Geometrically, a diagonal matrix transforms a standard-basis vector into a stretched or contracted version of itself, while leaving the direction invariant,

$$diag(\lambda_1, \lambda_2, \ldots, \lambda_n) \cdot \hat{e}_j = \lambda_j \cdot \hat{e}_j. \tag{4.68}$$

Thus a unit $n$-cube, with *edges* defined by the orthonormal basis $\{\hat{e}_j\}_{j=1}^n$, is transformed into an $n$-dimensional rectangle, with edges defined by the basis $\{\lambda_j \cdot \hat{e}_j\}_{j=1}^n$, whose $n$-volume is defined to be

$$Vol\left(\{\lambda_j \cdot \hat{e}_j\}_{j=1}^n\right) = \prod_{j=1}^n \lambda_j = det\left(diag(\lambda_1, \lambda_2, \ldots, \lambda_n)\right). \tag{4.69}$$

Now, for an arbitrary vector $\vec{v} \in \mathbb{R}^n$, the action of a diagonal matrix is simple and will typically change the direction of $\vec{v}$, so that

$$\begin{pmatrix} \lambda_1 & 0 & \cdots & 0 \\ 0 & \lambda_2 & \cdots & 0 \\ 0 & 0 & \cdots & 0 \\ 0 & 0 & \cdots & \lambda_n \end{pmatrix} \cdot \begin{pmatrix} v_1 \\ v_2 \\ \cdot \\ v_n \end{pmatrix} = \begin{pmatrix} \lambda_1 \cdot v_1 \\ \lambda_2 \cdot v_2 \\ \cdot \\ \lambda_n \cdot v_n \end{pmatrix}. \tag{4.70}$$

To proceed with the same analysis for more general matrices, one can try to solve an equation similar to Eq. (4.68).

---

**Definition 4.13** A complex number $\lambda \in \mathbb{C}$ is an eigenvalue for a square matrix $A \in \mathcal{M}_{n \times n}(\mathbb{C})$ if $\exists \vec{v} \in \mathbb{C}^n \setminus \{\vec{0}\}$ so that

$$A \cdot \vec{v} = \lambda \cdot \vec{v} \iff (A - \lambda \cdot I) \cdot \vec{v} = \vec{0}. \tag{4.71}$$

In this case $\vec{v}$ is called an eigenvector of $A$. Every constant multiple of an eigenvector is another eigenvector. The *eigenspace* of $A$ associated with eigenvalue $\lambda$ is defined

$$\mathbb{E}_\lambda^A \equiv \{\vec{v} \in \mathbb{R}^n | A \cdot \vec{v} = \lambda \cdot \vec{v}\}. \tag{4.72}$$

The quantity $dim(\mathbb{E}_\lambda^A)$ is called the *geometric multiplicity* of $\lambda$. The sum of the geometric multiplicities of an $n \times n$ matrix is at most $n$. However, if it is less than $n$, then $A$ is called *defective*. The set of eigenvalues is denoted $\sigma(A)$ and is called the spectrum of $A$.

---

The eigenspaces of $A$ are *invariant* under the action of $A$, which if $\lambda \neq 0$ is expressed simply as $A \cdot \mathbb{E}_\lambda^A = \mathbb{E}_\lambda^A$, even though individual vectors are often changed in magnitude by $A$. An exception occurs if $\lambda = 1$, in which case $A \cdot \vec{v} = 1 \cdot \vec{v} = \vec{v}$.

If $dim(\mathbb{E}_\lambda^A) = p > 1$, then $p$ linearly independent vectors in the subspace $\mathbb{E}_\lambda^A$ can be found $\{\vec{v}_k\}_{k=1}^p$, which solve $A \cdot \vec{v}_k = \lambda \cdot \vec{v}_k$. Using the Gram-Schmidt process, an orthonormal basis $\{\hat{u}_k\}_{k=1}^p$ for $\mathbb{E}_\lambda^A$ can be constructed. The projection onto $\mathbb{E}_\lambda^A$ has rank $p$ and is defined to be

$$Proj_\lambda^A \equiv Proj_{\mathbb{E}_\lambda^A} = \sum_{k=1}^p Proj_{\hat{u}_k} = \sum_{k=1}^p \hat{u}_k \otimes \hat{u}_k^T.$$

For any vector $\vec{v} \in \mathbb{R}^n$, the projected vector $Proj_\lambda^A(\vec{v}) \in \mathbb{E}_\lambda^A$ is an eigenvector of $A$ with eigenvalue $\lambda$.

---

**Definition 4.14** A complex number $\lambda \in \mathbb{C}$ is an eigenvalue for a square matrix $A$ if and only if it solves the $n$th degree polynomial equation $p_A(\lambda) = 0$, where the characteristic polynomial of $A$ is defined to be

$$p_A(\lambda) = det(A - \lambda \cdot I). \tag{4.73}$$

Since the eigenvalues $\{\lambda_j\}_{j=1}^n$ of $A$ are the zeros of $p_A(\lambda)$, the Fundamental Theorem of Algebra implies that

$$p_A(\lambda) = \prod_{j=1}^n (\lambda_j - \lambda).$$

The *algebraic multiplicity* of each eigenvalue $\lambda_k$ is the number of times $\lambda_k$ appears in the set $\{\lambda_j\}_{j=1}^n$. The sum of algebraic multiplicities is always $n$. The spectrum of $A$ is $\sigma(A) = \{\lambda \in \mathbb{C} | p_A(\lambda) = 0\}$.

---

### EXAMPLE 4.9

The characteristic polynomial of $M$ is denoted $p_M(\lambda)$. The eigenvalues are found by solving

$$0 = p_M(\lambda) \equiv det(M - \lambda \cdot I) = det \begin{pmatrix} 0.4 - \lambda & 1.5 & 0 \\ 0.2 & 0.8 - \lambda & 0 \\ 0 & 0.1 & 0.8 - \lambda \end{pmatrix}$$

$$= (0.4 - \lambda) \cdot \begin{vmatrix} 0.8 - \lambda & 0 \\ 0.1 & 0.8 - \lambda \end{vmatrix} + (1.5) \cdot \begin{vmatrix} 0 & 0.2 \\ 0.8 - \lambda & 0 \end{vmatrix} + (0) \cdot \begin{vmatrix} 0.2 & 0.8 - \lambda \\ 0 & 0.1 \end{vmatrix}$$

$$= (0.4 - \lambda) \cdot (0.8 - \lambda)^2 + (1.5) \cdot (-(0.2) \cdot (0.8 - \lambda)) + 0$$

$$= -(\lambda - 0.8) \cdot (\lambda^2 - 1.2\lambda + 0.02). \tag{4.74}$$

Since a linear factor was found for $p_M(\lambda)$, there is no need to use the method for solving a cubic equation as in Eq. (1.9). Solving the quadratic gives the three positive, real eigenvalues,

$$\lambda_1 = 0.8, \quad \lambda_2 = 0.6 + \sqrt{0.34} \simeq 1.18, \quad \lambda_3 = 0.6 - \sqrt{0.34} \simeq 0.0169.$$

The eigenspace for $\lambda_1$ is found by using the general form of an eigenvector $\vec{v} = (a\ b\ c)^T$ and solving

$$M \cdot \vec{v} = 0.8 \vec{v} \iff \begin{pmatrix} 0.4 & 1.5 & 0 \\ 0.2 & 0.8 & 0 \\ 0 & 0.1 & 0.8 \end{pmatrix} \begin{pmatrix} a \\ b \\ c \end{pmatrix} = (0.8) \cdot \begin{pmatrix} a \\ b \\ c \end{pmatrix}$$

$$\iff \begin{pmatrix} 0.4a + 1.5b \\ 0.2a + 0.8b \\ 0.1b + 0.8c \end{pmatrix} = \begin{pmatrix} 0.8a \\ 0.8b \\ 0.8c \end{pmatrix} \iff \begin{cases} 1.5b = 0.4a \\ 0.2a = 0.0 \\ 0.8b = 0.0 \end{cases} \iff \begin{cases} a = 0.0 \\ b = 0.0 \end{cases}$$

This system of equations must always be underdetermined to give an eigenspace. In this analysis, $c$ is unrestricted. Thus

$$\mathbb{E}_{0.8}^M = \left\{ \begin{pmatrix} 0 \\ 0 \\ c \end{pmatrix} : c \in \mathbb{R} \right\} \implies Proj_{0.8}^M = \hat{v} \otimes \hat{v}^T = \begin{pmatrix} 0 & 0 & 0 \\ 0 & 0 & 0 \\ 0 & 0 & 1 \end{pmatrix}.$$

The eigenspaces and projections for $\lambda_2$ and $\lambda_3$ are more complicated.

## Exercises

1. Suppose $A$ is an $n \times n$ matrix, and $S$ is an $n \times n$ invertible matrix.

   (a) Show that if $\vec{v}$ is an eigenvector of $A$, then $S\vec{v}$ is an eigenvector of $S \cdot A \cdot S^{-1}$.
   (b) Why is $\sigma(A) = \sigma(S \cdot A \cdot S^{-1})$?
   (c) Show that a characteristic polynomial of a $2 \times 2$ matrix is quadratic and monic.
   (d) Explain why the negative of the characteristic polynomial of a $3 \times 3$ matrix is monic, using a generalization of the computation in Eq. (4.74).
   (e) For an invertible matrix $S$, use Theorem 4.3, definition in Eq. (4.73), and identity $I = S \cdot S^{-1}$ to show that $p_A(\lambda) = p_{S \cdot A \cdot S^{-1}}(\lambda)$ for any $n \times n$ matrix $A$.
   (f) For all $n \times n$ matrices, what is the value of $p_A(0)$?

2. Consider the matrix $B = \begin{pmatrix} 2 & 3 \\ -1 & 4 \end{pmatrix}$.

   (a) Find the characteristic polynomial for $B$ and the eigenvalues $\sigma(B)$.
   (b) Find the eigenspaces for the corresponding eigenvalues of $B$.
   (c) Is $B$ defective?

3. Consider $C = \begin{pmatrix} 1 & 3 & 2 \\ -1 & 4 & 0 \\ -2 & 1 & -2 \end{pmatrix}$.

   (a) Find and simplify the characteristic polynomial for $C$.
   (b) For $C$, find the eigenvalues $\sigma(C)$. (Note: One eigenvalue should be easy to find, while the other two are complex.)

4. Find the eigenspaces $\mathbb{E}^M_{\lambda_2}$ and $\mathbb{E}^M_{\lambda_3}$ in Example (4.9). (*Hint:* Start with eigenvector $\vec{v} = (a\ b\ c)^T$ and then solve for components $a$ and $c$ in terms of $b$.)

## 4.12 Sum and Product Matrix Representations

Just as there are different representations for functions, multilinear transformations can be expressed in various ways, depending on the application.

### 4.12.1 Spectral and $QDQ^T$ Representations of a Symmetric Matrix

The eigenvalues of a real-valued symmetric matrix $A = A^T$ are sometimes referred to as the *spectral values* of $A$ in which case the set of all eigenvalues of $A$ is written

$\sigma(A)$. In general, if $A$ has complex components it is called *Hermitian* if $A^* = A^T$. This is also expressed as $A^\dagger \equiv (A^*)^T = A$, since $(A^*)^* = A$ and $(A^T)^T = A$.

■ **Theorem 4.4**

A symmetric $n \times n$ matrix $A$ with real components, has only real eigenvalues, expressed as $\sigma(A) \subset \mathbb{R}$. Eigenvectors, corresponding to different eigenvalues, are orthogonal. If the components of $A$ are complex, then these statements hold if $A^\dagger = A$ with a modification of the inner product.

**Proof** Suppose that $\vec{v} \in \mathbb{R}^n$ is an eigenvector of $A$ with eigenvalue $\lambda \in \mathbb{C}$. Then

$$0 \leq \|A\vec{v}\|^2 = \langle A\vec{v} | A\vec{v} \rangle = \langle \lambda\vec{v} | \lambda\vec{v} \rangle = \lambda^2 \cdot \|\vec{v}\|^2, \tag{4.75}$$

which implies that $\lambda \in \mathbb{R}$. However, if $\vec{v} \in \mathbb{C}^n$ is an eigenvector of $A$, then this argument fails. Define $\vec{v}^*$ to be the vector where each component is the complex conjugate of the components of $\vec{v}$. Then the vector norm and inner product can be extended[1]

$$\|\vec{v}\|^2 \equiv \vec{v}^* \cdot \vec{v} = \sum_{j=1}^{n} |v_j|^2, \ \langle \vec{v}, \vec{w} \rangle \equiv \vec{v}^* \cdot \vec{w} = \sum_{j=1}^{n} v_j^* w_j. \tag{4.76}$$

Then $\langle \vec{v}, \vec{v} \rangle$ is a nonnegative real number $\forall \vec{v} \in \mathbb{C}^n$. Furthermore, if $A$ is Hermitian,

$$\begin{aligned}\langle \vec{v}, A\vec{v} \rangle^* &= (\vec{v}^\dagger A \vec{v})^* = (\vec{v}^\dagger)^*(A\vec{v})^* = \vec{v}^T A^* \vec{v}^* = \vec{v}^T A^T \vec{v}^* \\ &= (A\vec{v})^T \vec{v}^* = \left((\vec{v}^*)^T A \vec{v}\right)^T = \vec{v}^\dagger A \vec{v} = \langle \vec{v}, A\vec{v} \rangle,\end{aligned} \tag{4.77}$$

which ensures that $\langle \vec{v}, A\vec{v} \rangle \in \mathbb{R}$. Here the trivial fact that $z^T = z$, $\forall z \in \mathbb{C}$, is used.

The eigenspaces of a symmetric matrix are necessarily orthogonal. Indeed, suppose $\vec{v}_1 \in \mathbb{E}_{\lambda_1}^A$ and $\vec{v}_2 \in \mathbb{E}_{\lambda_2}^A$ where $\lambda_1 \neq \lambda_2$. Then

$$\lambda_2 \cdot \vec{v}_2 \cdot \vec{v}_1 = (A \cdot \vec{v}_2) \cdot \vec{v}_1 = \vec{v}_2 \cdot (A \cdot \vec{v}_1) = \lambda_1 \cdot \vec{v}_2 \cdot \vec{v}_1, \tag{4.78}$$

which implies $0 = (\lambda_1 - \lambda_2) \cdot \vec{v}_2 \cdot \vec{v}_1$. This can only hold if $\vec{v}_2 \cdot \vec{v}_1 = 0$. The argument can be extended to Hermitian matrices as well. □

---

[1] The inner product notation will now be changed to $\langle \cdot, \cdot \rangle$, which allows complex vectors.

## 4.12 Sum and Product Matrix Representations

It can be shown that a Hermitian $n \times n$ matrix $A$ is never defective. This allows the construction of a matrix $Q$ that is

**Orthogonal:** $Q^\dagger = Q^{-1}$, and
**Diagonalizing:** $Q^\dagger \cdot A \cdot Q = diag(\lambda_1, \lambda_2, \ldots, \lambda_n) \equiv D$.

Indeed, choosing a unit eigenvector from each eigenspace, $\hat{v}_j \in \mathbb{E}^A_{\lambda_j}$, the matrix $A$ can be expressed using a *spectral representation*

$$A = \sum_{j=1}^{n} \lambda_j \cdot \hat{v}_j \otimes \hat{v}_j^\dagger = \sum_{\lambda \in \sigma(A)} \lambda \cdot Proj^A_\lambda = \sum_{\lambda \in \sigma(A)} \lambda \cdot Proj_{\mathbb{E}^A_\lambda} \quad (4.79)$$

$$= (\hat{v}_1 \ \hat{v}_2 \ \cdots \ \hat{v}_n) \cdot diag(\lambda_1, \lambda_2, \ldots, \lambda_n) \cdot \begin{pmatrix} \hat{v}_1^\dagger \\ \hat{v}_2^\dagger \\ \cdot \\ \hat{v}_n^\dagger \end{pmatrix} = Q \cdot D \cdot Q^\dagger. \quad (4.80)$$

For diagonal matrices $D = diag(\lambda_1, \lambda_2, \ldots, \lambda_n)$, or the identity $I = diag(1, 1, \ldots, 1)$

$$D = \sum_{j=1}^{n} \lambda_j \cdot \hat{e}_j \otimes \underline{\delta}^j, \ I = \sum_{j=1}^{n} \hat{e}_j \otimes \underline{\delta}^j. \quad (4.81)$$

The identity matrix $I$ is symmetric, Hermetian, and orthogonal. Diagonal matrices $D$ are symmetric, but are Hermetian only if $\lambda_j \in \mathbb{R}$, and orthogonal only if $\lambda_j \in \mathcal{S}^1$, $\forall j \in \mathbb{Z}_n$.

If $dim(\mathbb{E}^A_{\lambda_k}) > 1$ for some $\lambda_k \in \sigma(A)$, then an orthonormal basis for $\mathbb{E}^A_{\lambda_k}$ must be chosen so that the formula in Eq. (4.79) holds. This can be done using the Gram-Schmidt orthonormalization process.

■ **Theorem 4.5**

**The Spectral Theorem**
Let $f \in \mathcal{C}^0(\mathbb{R})$ and $A$ be a symmetric $n \times n$ square matrix. Then, $\forall t \in \mathbb{R}$

$$f(t \cdot A) \equiv \sum_{\lambda \in \sigma(A)} f(t \cdot \lambda) \cdot Proj_{\mathbb{E}^A_\lambda} = Q \cdot f(t \cdot D) \cdot Q^\dagger, \quad (4.82)$$

is also a square matrix. In particular, $f(D) = diag(f(\lambda_1), f(\lambda_2), \ldots, f(\lambda_n))$. For any vectors $\vec{v}, \vec{w} \in \mathbb{R}^n$, the function $g(t) \equiv \vec{v} \cdot f(t \cdot A) \cdot \vec{w}$ is continuous in $t \in \mathbb{R}$.

### 4.12.2 Perm × LDU Decomposition

Not all matrices can be diagonalized like a symmetric matrix.[2] However, using GE, a representation can be obtained that allows for quick application of the matrix as a linear operator. This is useful in programming linear transformations.

Suppose $A$ is an $n \times m$ matrix. Then there is an $n \times n$ permutation matrix $Perm$, an $n \times n$ lower triangular matrix $L$, a diagonal $n \times m$ matrix $D$, and an $m \times m$ upper triangular matrix $U$ so that

$$A = Perm \cdot L \cdot D \cdot U.$$

To obtain this form, GE needs to be applied using elementary matrices until the reduced row-echelon form of $A$ is derived. Then $Perm^{-1}$ is the product of just the elementary permutations. Thus $Perm^{-1} \cdot A$ has an $LDU$ decomposition.

In the situation that $A$ is an $n \times n$ square matrix, note that a square permutation matrix only has $\pm 1$ as eigenvalues. Thus the eigenvalues of $Perm^{-1} \cdot A$ will only differ from those of $A$ by sign. Furthermore, since elementary permutations are idempotent, they can be pulled out of the elementary-matrix decomposition of $A$. This is achieved through a process of *conjugation* by permutations on elementary matrices. For example, since matrix multiplication is noncommutative

$$\begin{pmatrix} 0 & 1 \\ 1 & 0 \end{pmatrix} \begin{pmatrix} 1 & k \\ 0 & 1 \end{pmatrix} \begin{pmatrix} 0 & 1 \\ 1 & 0 \end{pmatrix} = \begin{pmatrix} 0 & 1 \\ 1 & k \end{pmatrix} \begin{pmatrix} 0 & 1 \\ 1 & 0 \end{pmatrix} = \begin{pmatrix} 1 & 0 \\ k & 1 \end{pmatrix} \quad (4.83)$$

$$\begin{pmatrix} 0 & 1 \\ 1 & 0 \end{pmatrix} \begin{pmatrix} k & 0 \\ 0 & 1 \end{pmatrix} \begin{pmatrix} 0 & 1 \\ 1 & 0 \end{pmatrix} = \begin{pmatrix} 0 & 1 \\ k & 0 \end{pmatrix} \begin{pmatrix} 0 & 1 \\ 1 & 0 \end{pmatrix} = \begin{pmatrix} 1 & 0 \\ 0 & k \end{pmatrix}.$$

Thus, although pulling a permutation through a *nonpermutation* elementary matrix changes it, the type of elementary matrix is not changed. In this way, all permutations in $A$ can be pulled to the left. The remaining matrices, which can be analyzed using GE, are either upper triangular or lower triangular (or both, i.e., diagonal). Now, the lower triangular matrices can be pulled to the left, and the upper triangular matrices to the right, because

$$\begin{pmatrix} 1 & 0 \\ 0 & \lambda \end{pmatrix} \begin{pmatrix} 1 & \mu \\ 0 & 1 \end{pmatrix} \begin{pmatrix} 1 & 0 \\ 0 & 1/\lambda \end{pmatrix} = \begin{pmatrix} 1 & \mu \\ 0 & \lambda \end{pmatrix} \begin{pmatrix} 1 & 0 \\ 0 & 1/\lambda \end{pmatrix} = \begin{pmatrix} 1 & \mu/\lambda \\ 0 & 1 \end{pmatrix} \quad (4.84)$$

$$\begin{pmatrix} 1 & 0 \\ 0 & \lambda \end{pmatrix} \begin{pmatrix} 1 & 0 \\ \mu & 1 \end{pmatrix} \begin{pmatrix} 1 & 0 \\ 0 & 1/\lambda \end{pmatrix} = \begin{pmatrix} 1 & 0 \\ \lambda \cdot \mu & \lambda \end{pmatrix} \begin{pmatrix} 1 & 0 \\ 0 & 1/\lambda \end{pmatrix} = \begin{pmatrix} 1 & 0 \\ \lambda \cdot \mu & 1 \end{pmatrix}.$$

---

[2] One can come close with the Jordan Canonical Form discussed in specialized texts [23].

## 4.12 Sum and Product Matrix Representations

In this way, the diagonal elementary matrices can be pulled toward the middle of the lower and upper triangular matrices. There is never a need to commute an upper and lower triangular matrix if $Perm^{-1}A$ can be transformed into a diagonal matrix using only GE, without permutations. These identities hold for higher dimensional matrices as well [53].

### Exercises

1. Use Eq. (4.82) to resolve the following:

   (a) For a diagonal matrix $D$, show that $det(exp(D)) = exp(tr(D))$.
   (b) For symmetric $n \times n$ matrices $A$, show that $p_A(A) = 0_{n \times n}$.
   (c) If $A$ is a symmetric $n \times n$ invertible matrix, find an expression for $A^{-1}$ using the Spectral theorem.

2. Find the spectral representation of $A = \begin{pmatrix} 2 & -2 \\ -2 & 4 \end{pmatrix}$ as defined in Eq. (4.79).

3. Find the spectral representation of $B = \begin{pmatrix} 0 & 1 & -1 \\ 1 & 1 & 0 \\ -1 & 0 & 1 \end{pmatrix}$ as defined in Eq. (4.79) and the $Q \cdot D \cdot Q^T$ representation as in Eq. (4.80).

4. For $C = \begin{pmatrix} -2 & 6 & 2 \\ 6 & 7 & -1 \\ 2 & -1 & -1 \end{pmatrix}$ first show that $det(C) = 0$. Then find the spectral representation of $C$ and the $Q \cdot D \cdot Q^T$ representation.

5. A matrix $K$ is called *skew* Hermitian if $K^\dagger = -K$. How can a skew Hermitian matrix be converted to obtain a Hermitian matrix? Give a proof that the construction works, using properties of complex conjugation and the transpose.

6. Recall that an $n \times n$ matrix $H$ with complex components is called Hermitian if $H^* = H^T$.

   (a) Show that $H = \begin{pmatrix} 1 & -i \\ i & 1 \end{pmatrix}$ is a Hermitian matrix.
   (b) Find the determinant and trace of $H$.
   (c) Find the eigenvalues of $H$.
   (d) Find an eigenvector for each eigenvalue of $H$, and show that they are orthogonal *if* the inner product in Eq. (4.76) is used.
   (e) Find the spectral representation and the $Q \cdot D \cdot Q^T$ representation of $H$.

7. Compute the conjugate of $E = \begin{pmatrix} 1 & 0 & 0 \\ 0 & 1 & k \\ 0 & 0 & 1 \end{pmatrix}$ with $P = \begin{pmatrix} 0 & 1 & 0 \\ 1 & 0 & 0 \\ 0 & 0 & 1 \end{pmatrix}$ i.e., $P \cdot E \cdot P$, and state what property of $E$ is preserved.

8. Recall the elementary matrix decomposition of $M$, given in Example 4.8. Use the elementary matrices and commutation relations in Eq. (4.83) and Eq. (4.84) to write $M = Perm \cdot L \cdot D \cdot U$. Be sure to identify all four matrices clearly. Also, use the expression to compute the determinant of $M$.

## 4.13 Linear Regression

Given a data set consisting of $n$ pairs of dependent $y$ and independent $x$ variables, $\{\langle x_j, y_j \rangle\}_{j=1}^n$ a scatter plot will have a center determined by the sample means $(\overline{x}, \overline{y})$. The *goodness-of-fit* for a model $f : \mathbb{R} \to \mathbb{R}$ is initially obtained by inspecting the triple $\mathcal{T} \equiv \{\langle x_j, y_j, f(x_j) \rangle\}_{j=1}^n$. Quantitative measures of a model's fitness often involves computing and comparing the *Sum of the Squared-Deviations of the Data* statistic,

$$\text{SSD}(\mathcal{T}) \equiv \sum_{j=1}^n (y_j - \overline{y})^2, \text{ defining } \overline{y} \equiv \frac{1}{n} \sum_{j=1}^n y_j, \quad (4.85)$$

the *Sum of the Squared-Deviations of the Model* statistic,

$$\text{SSM}(\mathcal{T}) \equiv \sum_{j=1}^n \left( f(x_j) - \overline{f} \right)^2, \text{ requiring } \overline{y} \simeq \overline{f} \equiv \frac{1}{n} \sum_{j=1}^n f(x_j), \quad (4.86)$$

and the *Sum of the Squared Errors* statistic,

$$\text{SSE}(\mathcal{T}) \equiv \sum_{j=1}^n (y_j - f(x_j))^2, \text{ where } 0 \equiv \sum_{j=1}^n (y_j - f(x_j)). \quad (4.87)$$

A goal of modeling is to choose a model function $f(x)$ that will minimize $\text{SSE}(\mathcal{T})$. The class of functions considered here are linear in their defining parameters $a_j$,

$$f(x) \equiv a_1 \cdot e_1(x) + a_2 \cdot e_2(x) + \cdots + a_m \cdot e_m(x), \quad (4.88)$$

where the basis functions $\{e_j(x)\}_{j=1}^m$ are linearly independent on the domain of interest $I \subset \mathbb{R}$ for the model $f(x)$. It is important that the basis elements be sufficiently independent for the system under consideration [55]. The objective is to obtain the best values of $a_j$ based on the data for the model $y = f(x)$ being proposed. From the data in $\mathcal{T}$ and the model in Eq. (4.88), the linear system of equations

$$\begin{aligned} y_1 &= a_1 \cdot e_1(x_1) + a_2 \cdot e_2(x_1) + \cdots + a_m \cdot e_m(x_1), \\ &\phantom{=} \cdots \vdots \cdots \\ y_n &= a_1 \cdot e_1(x_n) + a_2 \cdot e_2(x_n) + \cdots + a_m \cdot e_m(x_n), \end{aligned} \quad (4.89)$$

follows. If a lot of data is available and a simple model with few parameters is considered, then the condition $n \gg m$ should be expected. This implies that the system in Eq. (4.89) is *over-determined*, and has no solution in general. This is typically the case in scientific applications. System (4.89) is handled by first writing it as a matrix equation

$$\vec{y} = B \cdot \vec{a} + \vec{\mathcal{E}}, \quad (4.90)$$

where the components of $B = (b_{ij})$ are $b_{ij} \equiv e_i(x_j)$, and the components of the *error vector* $\vec{\mathcal{E}}$, although unknown, are responsible for the deviations of the data from the model as observed in $\mathcal{T}$. Multiplying the matrix equation in Eq. (4.90) on the left by $B^T$, gives

$$B^T \cdot \vec{y} = B^T \cdot B \cdot \vec{a} + B^T \cdot \vec{\mathcal{E}}. \quad (4.91)$$

■ **Theorem 4.6**

Suppose a set of basis functions $\{e_i(x)\}_{i=1}^m$ are linearly independent on an interval $I = [a, b]$ for some $a < b$. Then there are sequences $\{x_j\}_{j=1}^n$ for $n \geq m$ so that the $m \times m$ matrix $B^T \cdot B$ is invertible, where $B = (e_i(x_j))$.

**Proof** Choose points $\{x_j\}_{j=1}^n$ so that the vectors $\{\vec{b}_i\}_{i=1}^m$ are linearly independent, where $\vec{b}_i \equiv (e_i(x_1) \ldots e_i(x_n))^T$. It is certainly possible to choose $m$ of these points from $I$, on which the basis functions are linearly independent. This ensures that $rank(B) = m$ where $B^T = (\vec{b}_1 \ldots \vec{b}_m)$. Thus $rank(B^T \cdot B) = m$, and since $B^T \cdot B$ is an $m \times m$ matrix, it is invertible. □

The parameters of the model function in Eq. (4.88) have an estimated solution

$$\vec{a} = \left(B^T \cdot B\right)^{-1} \cdot B^T \cdot \vec{y} - \left(B^T \cdot B\right)^{-1} \cdot B^T \cdot \vec{\mathcal{E}}. \tag{4.92}$$

The main condition for a matrix version of linear regression is that

$$\left(B^T \cdot B\right)^{-1} \cdot B^T \cdot \vec{\mathcal{E}} = \vec{0}. \tag{4.93}$$

This gives a linear combination of the basis functions, called the *regression function*

$$\hat{y}(x) = f(x) \equiv \vec{a} \cdot \vec{e}(x) = \sum_{i=1}^{m} a_i \cdot e_i(x), \tag{4.94}$$

which is derivable from the *mathematical model* and the *experimental data*. A measure of the usefulness of this model is given by the $r$-squared statistic

$$r^2 \equiv 1 - \frac{\text{SSE}}{\text{SSD}} = 1 - \frac{\sum_{j=1}^{n}(y_j - f(x_j))^2}{\sum_{j=1}^{n}(y_j - \overline{y})^2}, \tag{4.95}$$

which takes values in $[0, 1]$. If $\overline{f} = \overline{y}$ and the identity

$$\text{SSD} = \text{SSM} + \text{SSE}, \quad \text{where} \quad \text{SSD} \geq 0, \; \text{SSM} \geq 0, \; \text{SSE} \geq 0, \tag{4.96}$$

holds, then one can use the alternative definition

$$r^2 = \frac{\text{SSM}}{\text{SSD}} = \frac{\sum_{j=1}^{n}(f(x_j) - \overline{y})^2}{\sum_{j=1}^{n}(y_j - \overline{y})^2} = \frac{\sum(\text{model deviations})^2}{\sum(\text{data deviations})^2}, \tag{4.97}$$

which is an estimate of the proportion of variation in the data that is explained by the model. In the case of linear regression with $m = 2$, $e_1(x) = 1$, $e_2(x) = x$, the correlation-coefficient statistic $r$ can also be used to compute $r^2$.

### Confidence Regions for Parameters

To compute a level of desired confidence for the model in Eq. (4.94), an associated *confidence region* around the parameters $\vec{a}$ can be obtained. This issue is one focus of the field of study called *metrology*, or *measurement theory*.

---

**Definition 4.15** The components of a *vector of random variables* $\vec{\zeta} = (\zeta_1 \ldots \zeta_m)^T$ are said to be *independent and identically distributed* (IID or i.i.d.) if each component has the same distribution function, and the components are pairwise independent. If the distributions are normal $N(\mu, \sigma)$, then $\vec{\zeta}$ is called a normal IID (NIID) vector of random variables. If the distributions are standard normal $N(0, 1)$, then $\vec{\zeta}$ is called a standard normal IID (SNIID) vector of random variables.

Suppose that the components of the vector of parameters $\vec{a}$ are NIID. Then it can be transformed into an SNIID vector $\vec{\zeta}$ if the means $\{\mu_i\}_{i=1}^m$ and the variances $\{\sigma_i^2\}_{i=1}^m$ are known. The standardized components are naturally,

$$\zeta_i \equiv (a_i - \mu_i)/\sigma_i, \ \forall i \in \mathbb{Z}_m. \tag{4.98}$$

Consider a desired confidence level $(1 - \alpha) \times 100\%$, where $\alpha \in (0, 1)$ is the corresponding significance level. Suppose $B$ is a unitless $n \times m$ matrix, with $n \geq m$, that connects $n$-dimensional data information to $m$-dimensional modeling choices. Theoretically, there is a $C = (1 - \alpha)$-*confidence-level value* $v_C^*$ so that

$$C - level \text{ confidence region} \equiv \{\vec{\zeta} \in \mathbb{R}^m : \vec{\zeta}^T B^T B \vec{\zeta} \leq m \cdot v_C^*\}, \tag{4.99}$$

is a confidence region for the model parameters $\vec{a} = (a_1, a_2, \ldots, a_m)^T$. Note that there are alternative expressions for the vector-matrix product in Eq. (4.99),

$$\vec{\zeta}^T B^T B \vec{\zeta} = \langle \vec{\zeta}, B^T B \vec{\zeta} \rangle = (B\vec{\zeta})^T B \vec{\zeta} = \langle B\vec{\zeta}, B\vec{\zeta} \rangle = \|B\vec{\zeta}\|^2. \tag{4.100}$$

Using the inverses of the transforms in Eq. (4.98),

$$a_i = \mu_i + \zeta_i \cdot \sigma_i, \ \forall i \in \mathbb{Z}_m, \tag{4.101}$$

the confidence region in Eq. (4.99) represents an $m$-dimensional ellipsoid with the center at $\vec{a}_* = (\mu_1, \mu_2, \ldots \mu_m)^T$. Further discussion can be found in [55].

An experiment designed to test the validity of a model will produce an estimate of the center for a set of parameters $\vec{a}_*$. To apply the theory of confidence regions to the results of such an experiment, compute the *Mean of the Squared Errors* MSE = SSE/$(n - m)$ where $n$ is the **n**umber of data points, and $m$ is the number of parameters in the **m**odel. The confidence-level value $v_C^*$ is defined as the solution to $\mathcal{F}(m, n)[v_C^*] = 1 - C$, where the function $\mathcal{F}$ is defined in the next section. If the model results in a less than perfect fit, then MSE $> 0$. Thus, based on a single study, the actual best-fit parameter set $\vec{a} \in \mathbb{R}^m$ will satisfy the condition

$$m \cdot \text{MSE} \cdot v_C^* \geq \langle (\vec{a} - \vec{a}_*), B^T B \cdot (\vec{a} - \vec{a}_*) \rangle, \tag{4.102}$$

with confidence $C = (1 - \alpha) \times 100\%$, where $\alpha$ is the significance level.

## Comparing Two Models

Suppose $f_1(x)$ is a model with $m_1$ parameters, and $f_2(x)$ is a model with $m_2$ parameters where $m_2 > m_1$ and $f_2(x)$ is a modification of $f_1(x)$. One says that $f_1$ is *nested* in the more sophisticated model $f_2$. The objective is to measure the significance of the improvement in modeling using $f_2$ rather than $f_1$. This is done by first noting that the *degrees of freedom* (*df*) of the two models, for data sets of size $n$, where $n > m_2 > m_1$, are

$$df_1 \equiv n - m_1, \; df_2 \equiv n - m_2, \tag{4.103}$$

so that $df_1 > df_2 \geq 1$. Then the statistic

$$v_* = \frac{df_2}{df_1} \times \frac{\text{SSM}_1}{\text{SSM}_2} = \frac{n - m_2}{n - m_1} \times \frac{\sum_{j=1}^n (y_j - f_1(x_j))^2}{\sum_{j=1}^n (y_j - f_2(x_j))^2}, \tag{4.104}$$

is expected to satisfy $v_* \geq 1$ since $f_2$ must fit the data better than $f_1$. To test the significance of the improvement compute the *P*-value $\mathcal{F}(df_1, df_2)[v_*]$. The function $\mathcal{F}$ is defined next.

### ■ 4.13.1 Discussion of the *F*-Distribution

Suppose two experiments are performed to collect data on a single variable $X$, either through counts or measurements. Then a ratio of the sum-of-squares statistic can be constructed with the objective of measuring a difference in the variations between the two data sets.[3] The *F*-distribution, named after the statistician Ronald A. Fisher, arises from the ratio of two $\chi^2$-distributions. In this case, one constructs the statistic

$$w_* = \frac{\sum_{j=1}^m (x_j(1) - \overline{x}(1))^2}{m - 1} \cdot \frac{n - 1}{\sum_{k=1}^n (x_k(2) - \overline{x}(2))^2} \tag{4.105}$$

where $df_1 = m - 1$ and $df_2 = n - 1$ are the degrees of freedom of the respective experiments. A requirement here is that $w_* \geq 1$; only significance tests will be discussed. For $a = df_1$ and $b = df_2$ the *F*-distribution is defined as:

$$\text{pdf} = F(a,b)[w] \equiv \frac{\Gamma((a+b)/2)}{\Gamma(a/2) \cdot \Gamma(b/2)} \cdot \sqrt{a^a \cdot b^b} \cdot \frac{w^{(a-2)/2}}{(a \cdot w + b)^{(a+b)/2}},$$

$\text{cdf} = 1 - \mathcal{F}(a,b)[w]$ where $\mathcal{F}(a,b)[w]$ = area of *pdf* from $w$ to $\infty$.

---

[3]The general theory of Analysis of Variances (ANOVA) can be found in [31].

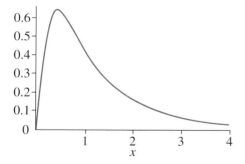
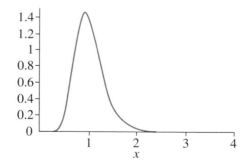

**FIGURE 4.2** ■ Left: The $F(5,5)[w]$ distribution; Right: $F(50,50)[w]$.

The definition in Eq. (2.52) can be used to compute the $\Gamma$-coefficients. The $F$-distribution is unimodal and skewed, as seen in Figure 4.2. For large $a, b \in \mathbb{N}$ the PDF function $F(a, b)[w]$ can be approximated by a normal distribution $N(\mu_\mathcal{F}, \sigma_\mathcal{F})[w]$, using

$$\mu_\mathcal{F} = \langle F(a,b) \rangle = \frac{b}{b-2}, \quad (4.106)$$

$$\sigma_\mathcal{F}^2 = Var\,(F(a,b)) = \frac{2 \cdot (a+b-2)}{a \cdot (b-4)} \cdot \left(\frac{b}{b-2}\right)^2.$$

Then for large samples with degrees of freedom $a = df_1 > 20$, $b = df_2 > 20$, and for significance levels $\alpha \in (0.01,\ 0.10)$

$$\mathcal{F}(df_1, df_2)[w_\alpha] \simeq \Phi_{\mu_\mathcal{F}, \sigma_\mathcal{F}}[-w_\alpha] = \Phi_{0,1}[-(w_\alpha - \mu_\mathcal{F})/\sigma_\mathcal{F}] = \alpha. \quad (4.107)$$

Suppose the experiment gives $s_1 > s_2$. Then the suggestion that the sample standard deviation $s_1$ is equal to $s_2$ is rejected, at the 5% significance level, if

$$\frac{w_* - \mu_\mathcal{F}}{\sigma_\mathcal{F}} \geq z_{0.05} \simeq 1.65, \text{where } w_* = s_1^2/s_2^2 \quad (4.108)$$

For small sample sizes $\sigma_\mathcal{F}$ is too large to use Eq. (4.107), however the following

$$\mathcal{F}(a,b)[w_\alpha] \simeq \frac{\Gamma((a+b)/2)}{\Gamma(a/2) \cdot \Gamma(b/2)} \cdot \sqrt{a^a \cdot b^b} \cdot \frac{2}{b} \cdot \frac{w_\alpha^{a/2}}{(a \cdot w_\alpha + b^2/(b+2))^{(a+b)/2}}, \quad (4.109)$$

gives a crude approximation for $w_\alpha \geq 1$. To solve for $w_\alpha$ given $\alpha \in (0,1)$, one can write the equation $\mathcal{F}(a,b)[w_\alpha] = \alpha$, then multiply by the denominator and take both sides to the power $2/(a+b)$. Writing an algorithm for $w$ gives a convergent sequence $\{w_n\}$, with $w_0 = 1$ as seed.[4]

### EXAMPLE 4.10

The sample standard deviations of hand widths displayed in Figure 3.2, were computed in Eq. (3.18) and the $F$-statistic is

$$w_* = s_M^2/s_F^2 = 7.9^2/8.9^2 = 0.788.$$

The degrees of freedom for the data are $df_M = 4 - 1 = 3$ and $df_F = 6 - 1 = 5$. Thus, to obtain an approximate $P$-value for the equality of $s_M$ to $s_F$, first compute

$$\mu_\mathcal{F} = \frac{5}{5-2} = 1.67, \; \sigma_\mathcal{F}^2 = \frac{2 \cdot (3+5-2)}{3 \cdot (5-4)} \cdot \left(\frac{5}{5-2}\right)^2 = 11.11, \; \sigma_\mathcal{F} = 3.33.$$

Thus, one determines from the approximation in Eq. (4.107) that

$$P(w \geq 0.788) \simeq P\left(z \geq \frac{w_* - \mu_\mathcal{F}}{\sigma_\mathcal{F}}\right) = P(z_* \geq -0.265) = 1 - \Phi_{0,1}[-0.265] \simeq 1 - 0.374 = 0.626,$$

or a 63% chance of randomly obtaining a statistic of $w_* \simeq 0.788$. Thus the variations of the male and of the female hand widths, about their respective means, is not significantly different.

Conversely, consider the reciprocal $F$-statistic

$$w_* = s_F^2/s_M^2 = 8.9^2/7.9^2 = 1.27.$$

---

[4]Since multiple solutions are possible, programs or tables should be used to compute the $F$-distribution and its associated probability statements, in general. In particular, confidence intervals for the ratios of squared errors must take into account the skewness of the $F$-distribution about the approximate center $w \simeq 1$.

Since the degrees of freedom are $df_F = 5$ and $df_M = 3$, one determines $\mu_{\mathcal{F}} = 3/(3-2) = 3$. However $\sigma_{\mathcal{F}}$ is undefined. To obtain an approximate $P$-value use Eq. (4.107), and compute

$$P(w \geq 1.27) = \mathcal{F}(5,3)[1.27] \simeq \frac{\Gamma(4)}{\Gamma(5/2) \cdot \Gamma(2/2)} \cdot \sqrt{5^5 \cdot 3^3} \cdot \frac{2}{3} \cdot \frac{1.27^{5/2}}{(5 \cdot 1.27 + 3^2/5)^4} = 0.406,$$

or a 41% chance of randomly obtaining a statistic of $w_* \geq 1.27$. Again one concludes that the variations of the male and of the female hand widths is not significantly different.

## ■ Exercises

1. Show that SSD = SSM + SSE if $\overline{f} = \overline{y}$ and Eq. (4.93) holds. Consider starting with SSD in Eq. (4.85). Then add and subtract $f(x_j)$.

2. A voltage $V$ (in *volts*) is applied to a resistor and the resulting current $I$ (in *amps*) is measured giving three data points:

$$\{(V_i, I_i)\}_{i=1}^2 = \{(2.0, 0.3), (3.0, 0.4), (4.0, 0.5)\}.$$

   (a) For the $\ell$inear model $\hat{I}_\ell = m \cdot V + k$, obtain an equation for each data point, and express the triple as a $3 \times 2$ matrix equation.
   (b) Use matrix multiplication to solve for the parameter vector $(m\ k)^T$.
   (c) Give the predicted model $\hat{I}_\ell(V)$ and find the correlation coefficient $r$ and the $r^2$-statistic.
   (d) Next, consider a *r*atio model that assumes the relation $\hat{I}_r = C \cdot V$ between the variables. Using the data and the ratio model obtain a $3 \times 1$ matrix equation.
   (e) Define the $3 \times 1$ matrix $B$ as in Eq. (4.90) and use vector multiplication to solve for the parameter $C$, then give the predicted model $\hat{I}_r(V)$.
   (f) Alternatively, use the ratio model $\hat{I}_r = C \cdot V$ to construct the SSE($C$) function. Expand and simplify into a quadratic polynomial in $C$.
   (g) The vertex of the SSE($C$) function occurs at the minimum. Find the value of the *conductance* $C$ that minimizes the SSE polynomial.
   (h) Compute SSD ($C$), SSM ($C$), and SSE ($C$). Does Eq. (4.96) hold?
   (i) Compute $r^2$ for the ratio model.
   (j) Use Eq. (4.102) and Eq. (4.109), where $B$ is the $3 \times 1$ matrix from (e), to obtain a 95% confidence region (interval) for $C$. This will require using $\alpha = 0.05$ and writing the solution to Eq. (4.109) as an algorithm. Compute several iterations to estimate $w_{0.05}$, and then employ Eq. (4.102).

(k) Compute $v_*$ in Eq. (4.104) and decide from the result which model fits best.

(l) Draw both models on the same graph. Use a basic physical principle to explain why the ratio model can be used, but the linear model must never be used.

3. Consider a system with $x$-inputs $\langle 1, 2, 3, 4 \rangle$ and $y$-outputs $\langle 1, 2, 4, 5 \rangle$.

   (a) Consider the *linear* model $f_2(x) = \alpha \cdot x + \beta$ and write an equation for each data point. Also express the quadruple as a $4 \times 2$ matrix equation.
   (b) Use matrix multiplication to solve for the parameter vector $(\alpha \ \beta)^T$.
   (c) Compute the $SSE_3$ for the quadratic model.
   (d) Find the equation that defines the 85% confidence region for $(\alpha \ \beta)^T$ using Eq. (4.102) and the results from (a), (b), and (c). An iterative solution to Eq. (4.109) needs to be written and run about three times. Use the result to give an approximate confidence rectangle, using the two cases where $\alpha$ and $\beta$ equal their values from (b).
   (e) Next consider the quadratic model $f_3(x) = a \cdot x^2 + b \cdot x + c$ and obtain an equation for each data point. Express the system as a $4 \times 3$ matrix equation.
   (f) Use Gaussian Elimination to solve for the parameter vector $(a \ b \ c)^T$, and obtain the model $f_3(x)$.
   (g) Next consider the cubic model $g_3(x) = d \cdot x^3 + e \cdot x + f$ and modify the procedure used for the quadratic model to solve for the parameter vector $(d \ e \ f)^T$. Keep a minimum of five decimal places in each matrix component.
   (h) Use $g_3(x)$ to compute $SSE_3$ for the cubic model, keeping three decimal places.
   (i) Compute $v_*$ using Eq. (4.104). Explain why approximation in Eq. (4.107) cannot be used.
   (j) Reciprocate the $v_*$ statistic in (i) and use Eq. (4.109) to obtain the $P$-value. Give a qualitative assessment as to how one should proceed to obtain a model that better fits the data?

4. Sixty ping-pong balls, marked with numbers, are randomly selected 20 at a time (20), and 40 at a time (40). The average of the sample averages, in the 20-sample case, is $\bar{x}(20) = 9.6$ with sample variance $s_{20}^2 = 0.684$. The averages of the sample averages, in the 4-sample case, is $\bar{x}(40) = 9.1$ with sample variance $s_{40}^2 = 0.233$.

   (a) Compute the ratio of sample variances $w_* \equiv s_{20}^2 / s_{40}^2$.
   (b) What are the degrees of freedom, $df_{20}$ and $df_{40}$, for samples (20) and (40).

(c) Estimate the probability $P(w \geq w_*)$ using the approximation for the cumulative distribution function $\mathcal{F}(\mathit{df}_1, \mathit{df}_2)[w]$ in Eq. (4.107).
(d) Is there reason to believe that the (20) samples were selected much differently than the (40) samples?
(e) Suppose samples of size two give a variance $s_2^2 = 1.30$, and samples of size four give $s_4^2 = 0.303$. Use Eq. (4.109) to look for a significant difference between the variances.

# 5 Calculus

## Introduction

A main purpose of calculus is to approximate curves by lines called tangent lines. If you graph a function such as a polynomial, say $f(x)$, and you zoom in repeatedly on a point $(p, f(p))$ on your graph, you will see the graph appear to straighten out on a small scale, appearing linear.

Thus, your graph should be well approximated by a line, called the tangent line near $x = p$. The study of such tangent lines and their slopes is called differential calculus. Differential calculus gives us the power to approximate complicated curves by their much simpler tangent lines, a process called linear approximation.

We can interpret slopes of lines as rates of change, and thus the slope of a tangent line will measure the rate of change of our function on a small scale. For instance, let's drive along a straight road at 40 miles per hour. Then, by the distance formula, distance = velocity × time, $D(t) = 40$ mph × $t$ hr. $D(t)$ is a linear function, and we can consider the slope of this line to be $m = 40$ mph. So our slope of the line is our rate of change of distance with respect to time, or more simply, our velocity. For a curve well-approximated by a tangent line, the slope of a tangent line is considered to be an instantaneous rate of change at the point $(p, f(p))$. Thus differential calculus will let us study rates of change.

Another main purpose of calculus is to find areas under curves. This part of calculus is called integral calculus. It turns out that computation of areas is intimately related to the process of finding tangent lines, a surprising fact that is a cornerstone of calculus called the *Fundamental Theorem of Calculus*. It basically says that if you know how to compute slopes of tangent lines well enough to go forward and backward, then you know how to find areas under curves. It turns out that a host of applications are equivalent to finding areas under curves. These include finding probability in statistics, finding averages, finding volumes, finding balance points called centers of mass, describing planetary motion, and many other computations that do not appear to be area computations at first glance.

We undertake our exploration of calculus while keeping our eyes peeled for applications in our core studies, the sciences, and engineering.

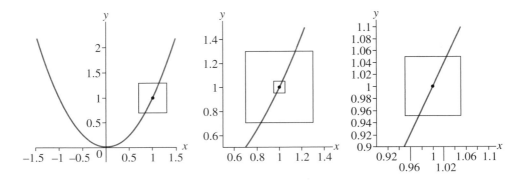

**FIGURE 5.1** ■ A smooth function viewed at different scales near the same point.

## 5.1 The Advancing Operation

Given a function $f : \mathbb{R} \to \mathbb{R}$ and small number $\epsilon > 0$, a *translation* by $\epsilon$ of $f(x)$ [2], or simply the $\epsilon$-*advance* of $f(x)$, is the operation

$$A_\epsilon[f](x) = f(x + \epsilon).$$

For example, $A_1[x^2](x) = (x+1)^2 = x^2 + 2x + 1$. Also, $A_2[e^x](x) = e^{x+2} = e^2 e^x$. If $\epsilon < 0$ then $A_\epsilon$ is a *delay* operation, so $A_{-3}[2 \cdot x + 4](x) = 2 \cdot (x - 3) + 4 = 2x - 2$.

## Exercises

1. Compute and simplify the following:
   (a) $A_\epsilon[x^2](x) - A_{-\epsilon}[x^2](x)$
   (b) $A_\epsilon[e^x](x) \cdot A_{-\epsilon}[e^x](x)$
   (c) $A_x[4x^3 + 3x^2 - 5x + 3](x)$

## 5.2 Continuity

A function from a connected set $\mathcal{S} \subset \mathbb{R}$ to $\mathbb{R}$ is continuous if it can be drawn without removing the writing instrument from the display surface. If the starting point is $(x_0, y_0)$, where $y_0 = f(x_0)$, then advancing to the final point $(x_1, y_1)$ involves advancing along a curve

$$\mathcal{C}_f \equiv \{ (A_\epsilon[\iota](x_0), \ A_\epsilon[f](x)) \ = \ (x_0 + \epsilon, \ f(x_0 + \epsilon)) \ : \ 0 \leq \epsilon \leq x_1 - x_0 \}.$$

The interval $\mathcal{S} = [x_0, x_1]$ is connected. All functions $f : \mathcal{S} \to \mathbb{R}$, where $\mathcal{C}_f$ is a connected curve on the $x$-$y$ plane, forms the *space of continuous functions* on $\mathcal{S}$ and is denoted $\mathcal{C}^0(\mathcal{S})$. All polynomials, the sine and cosine functions, and the exponential functions are continuous on $\mathbb{R}$ and so fall into the class $\mathcal{C}^0(\mathbb{R})$.

If $f(x)$ is continuous on the interval $[a, b]$, and if the height $h$ lies between the heights $f(a)$ and $f(b)$, then when we trace out the graph of $f(x)$ without lifting our writing instrument from the page we must pass through height $h$ at some point on our graph. We conclude that if $h$ is intermediate to $f(a)$ and $f(b)$ there must be a point $c$ with $a < c < b$ and $f(c) = h$. This is the idea behind the following Intermediate Value theorem from advanced calculus [59].

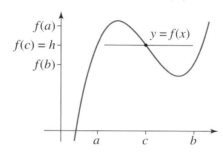

**FIGURE 5.2** ■ A smooth function $f(x)$ with intermediate value $f(c) = h$.

### ■ Theorem 5.1
### Intermediate Value Theorem

Let $f(x)$ be continuous on the interval $[a, b]$, and let $h$ be any number between $f(a)$ and $f(b)$, then there is a point $c$ with $a < c < b$ and $f(c) = h$.

---

An equivalent way to check that a function is continuous is to check that for each $p$ in $[x_0, x_1]$ we have that $f(x)$ tends to $f(p)$ as $x$ tends to $p$, that is, there are no gaps in the graph of $f$ at $p$. The notation for this is

$$\lim_{x \to p} f(x) = f(p). \tag{5.1}$$

This is read as: the limit of $f(x)$ as $x$ approaches $p$ is $f(p)$. In order for this to occur, $f(p)$ must be defined, and all values $f(x)$ must approach $f(p)$ as $x$ approaches $p$. If Eq. (5.1) holds, we say $f$ is continuous at $p$. If $f$ is continuous at each $p$ in $[x_0, x_1]$ we say $f$ is continuous on the interval $[x_0, x_1]$. For instance, since $f(x) = x^2$ tends to $p^2 = f(p)$ as $x$ tends to $p$, then we say that the limit of $x^2$ as $x$ approaches $p$ is $p^2$, and we write $\lim_{x \to p} x^2 = p^2$. Since this holds for all $p$, $f(x) = x^2$ is continuous on $\mathbb{R}$.

## ■ Exercises

1. For $f(x) = x^2$ on the interval $[1, 2]$ and any $h$ between $f(1) = 1^2 = 1$ and $f(2) = 2^2 = 4$, compute algebraically the point $c$ guaranteed by the Intermediate Value theorem.
2. Evaluate $\lim_{x \to p} x^3$ intuitively, and verify that $f(x) = x^3$ is continuous at any point $p$.

## ■ 5.3 Slope Function, Linear Approximation, and the Derivative

If we take a small fraction $a$ with $|a| < 1$, then it is not hard to see that higher powers $a^n$ of $a$ tend toward 0 as $n$ grows large. For instance, for $1 > a > 0$, when multiplying both sides of the inequality $1 > a$ by $a$, we obtain the inequality $a > a^2$. Again, multiplying the resulting inequality $a > a^2$ by $a$, we obtain $a^2 > a^3$. This pattern repeats indefinitely and we obtain

$$1 > a > a^2 > a^3 > ... > a^n > ... \tag{5.2}$$

For instance if $a = .1$ then Eq. (5.2) becomes

$$1 > .1 > (.1)^2 = .01 > (.1)^3 = .001 > ... > .000000001 > ... \tag{5.3}$$

We note the pattern that higher powers $a^n$ for $n \geq 2$ are much smaller than $a$, for small $a$, as is illustrated in Eq. (5.3). That is $a^2 = (.1)^2 = .01$, $a^3 = (.1)^3 = .001$, etc., are all much smaller than $a = .1$ (and also much smaller than $a^0 = (.1)^0 = 1$). This leads to the following mantra of approximating:

High powers $a^n$ for $n > 1$ of a small number $a$ are negligible and can be ignored in an approximation of an expression involving $a$.

So to approximate any expression involving $a$'s we keep constants and expressions with a single power of $a$ and drop all higher powers. For instance, if we are approximating the expression

$$f(a) = 4 + 3a + 5a^2 - 2a^3 + a^4$$

for small values of $a$, we will simply drop the higher-order terms (involving powers of $a$ of 2 or more here) to obtain the approximation

$$f(a) \approx 4 + 3a.$$

Here the expression $4 + 3a$ is called the linear part of $f(a)$ or the *linearization* of $f(a)$. We refer to the higher-order terms as $HOT(a)$ (in this case $HOT(a) = 5a^2 - 2a^3 + a^4$), and write

$$f(a) = 4 + 3a + HOT(a).$$

The process of dropping the higher-order terms, $HOT(a)$, to obtain the linearization is called modding out by the higher-order terms, or more technically, modding out by the ideal of higher-order terms in $a$. We say $f(a) = 4 + 3a$ modulo higher-order terms. We also say we are modding out the higher-order terms in $a$ to obtain the linearization $4 + 3a$. We further consider any expression $E(a)$ to be a higher-order term in $a$ if $E(0) = 0$ and $E(a)/a$ tends to 0 as $a$ tends to 0. Thus $HOT(a)/a$ tends to 0 as $a$ tends to 0. In our example, $HOT(a)/a = (5a^2 - 2a^3 + a^4)/a = 5a - 2a^2 + a^3 = a(5 - 2a + a^2)$, which tends to 0 as $a$ tends to 0. What this indicates is that $E(a)$ is much smaller than $a$ for $a$ near 0. In limit notation we have that $\lim_{x \to a} E(a)/a = 0$, and also $\lim_{x \to a} HOT(a)/a = 0$.

It is an amazing fact that dropping the higher-order terms, as in our mantra, can be put to work as a main tool in the finding of tangent lines. To see this, we first remind you of the process of expanding a function about a point $p$. As an illustration, to expand $x^2$ about the point $p$, we use the binomial expansion to write $x^2 = (p + [x - p])^2 = p^2 + 2p[x - p] + [x - p]^2$. To find the tangent line of $f(x) = x^2$ at the point $p$, we simply expand about $p$ as previously shown, and mod out the higher-order terms in $[x - p]$, in this case dropping the $[x - p]^2$ term, to obtain the linearization $p^2 + 2p[x - p]$. We have just computed our first tangent line $y = p^2 + 2p[x - p]$ by taking the linearization! This is the best line for approximating $f(x)$ near $p$. We call this the tangent line of the function $f(x) = x^2$ at the point $p$. The slope of the tangent line, in our case $2p$, the coefficient of $[x - p]$, is called the *derivative* of $f(x) = x^2$ at $p$ and it is denoted by $f'(p) = 2p$.

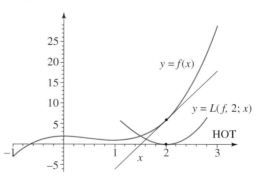

**FIGURE 5.3** ■ A smooth function with its linearization at a point, and the $HOT$.

## EXAMPLE 5.1

Expand $f(x) = x^2$ about the point $p = 1$, and find the tangent line at $p = 1$. First we have

$$x^2 = (1 + [x-1])^2 = 1^2 + 2[x-1] + [x-1]^2$$

and the linearization is $1 + 2[x-1]$, where we have dropped the term $HOT([x-1]) = [x-1]^2$. The equation of the tangent line is

$$y = 1 + 2\,[x-1],$$

while the derivative at $p = 1$ is $f'(1) = 2$, the coefficient of $[x-1]$ in the linearization, which is the slope of our line. This is in agreement with our general formula stating that for the function $f(x) = x^2$ the derivative at $p$ is $f'(p) = 2p$, where in our case we have let $p = 1$ be the point of expansion. Notice also that $HOT([x-1])/[x-1] = [x-1]^2/[x-1] = [x-1]$ tends to 0 as $x$ tends to 1, confirming that the higher degree term $[x-1]^2$ is indeed the higher-order term.

We further pursue this process for higher-degree monomials.

## EXAMPLE 5.2

Expand $f(x) = x^3$ about the point $p$; find the linearization of $f(x)$ at $p$; identify the higher-order terms; give the equation of the tangent line to $f(x)$ at $p$; find the derivative $f'(p)$ at $p$.

**Answer:** Using the binomial expansion, we obtain the expansion of $f(x)$ about the point $p$:

$$\begin{aligned}
f(x) &= x^3 = (p + [x-p])^3 = \sum_{k=0}^{3} \binom{3}{k} p^{3-k}[x-p]^k \\
&= \binom{3}{0} p^3[x-p]^0 + \binom{3}{1} p^2[x-p]^1 + \binom{3}{2} p^1[x-p]^2 + \binom{3}{3} p^0[x-p]^3 \\
&= \frac{3!}{3!0!} p^3[x-p]^0 + \frac{3!}{2!1!} p^2[x-p]^1 + \frac{3!}{1!2!} p^1[x-p]^2 + \frac{3!}{0!3!} p^0[x-p]^3 \\
&= 1p^3 + 3p^2[x-p]^1 + 3p[x-p]^2 + 1[x-p]^3.
\end{aligned}$$

From this we conclude:

**The linearization is:** $p^3 + 3p^2[x-p]$.
**The higher-order terms are:** $HOT([x-p]) = 3p[x-p]^2 + [x-p]^3$.

The equation of the tangent line is:

$$y = p^3 + 3\,p^2\,[x-p]. \tag{5.4}$$

## 5.3 Slope Function, Linear Approximation, and the Derivative

And the derivative is the slope of the tangent line, or equivalently the coefficient of the first-order term $[x-p]$, is:

$$f'(p) = 3p^2.$$

Finally, note that our tangent line in Eq. (5.4) can be expressed in terms of $f(p) = p^3$ and $f'(p) = 3p^2$ as

$$y = p^3 + 3p^2[x-p] = f(p) + f'(p)[x-p],$$

where the last expression will generalize to all functions $f(x)$ having a derivative. Since $f'(p) = (p^3)' = 3p^2$ for all choices of $p$ we write

$$(x^3)' = 3x^2. \tag{5.5}$$

We are going to be building a catalogue of derivatives for all the basic functions. In this spirit, and based on the preceding examples, we begin with the derivative of the monomial $x^n$, for $n$ a positive integer, to obtain the so-called Power Rule for positive integer exponents.

■ **Theorem 5.2**

For the monomial $f(x) = x^n$ with $n$ a positive integer, the derivative at the point $p$ is $f'(p) = (p^n)' = np^{n-1}$. Since this holds for all choices of $p$ we conclude:

$$(x^n)' = n\, x^{n-1}, \tag{5.6}$$

for all positive integers $n$.

**Proof** Expanding $x^n$ about the point $p$, the binomial theorem gives

$$\begin{aligned} x^n &= (p + [x-p])^n \\ &= p^n + \binom{n}{1} p^{n-1}[x-p]^1 + \sum_{k=2}^{n} \binom{n}{k} p^{n-k}[x-p]^k \\ &= p^n + n\, p^{n-1}[x-p] + HOT([x-p]). \end{aligned}$$

The derivative is now seen to be $np^{n-1}$, the coefficient of $[x-p]$. And $f'(p) = (p^n)' = np^{n-1}$. Since $(p^n)' = np^{n-1}$ is true for all $p$, we conclude $(x^n)' = nx^{n-1}$ for all $x$ and for all positive integers $n$. □

After our preceding examples, we are now in a position to make some fundamental definitions.

---

**Definition 5.1** If a function $f(x)$ expanded about $p$ can be written as a linear function plus higher-order terms

$$f(x) = c_0 + c_1[x-p] + HOT([x-p]), \tag{5.7}$$

where $HOT$ satisfies $HOT(0) = 0$ and $HOT([x-p])/[x-p]$ tends to 0 as $x$ tends to $p$, then we say $f(x)$ is *differentiable* at $p$. We define the *derivative* of $f(x)$ at $p$ to be

$$f'(p) = c_1, \tag{5.8}$$

where $c_1$ is the coefficient of the linear term $[x-p]$. Observing that $f(p) = c_0$, we have that $f(x)$ is differentiable at $p$ with derivative $f'(p)$ precisely when $f(x)$ can be written as

$$f(x) = f(p) + f'(p)[x-p] + HOT([x-p]), \tag{5.9}$$

where $HOT$ satisfies $HOT(0) = 0$ and $HOT([x-p])/[x-p]$ tends to 0 as $x$ tends to $p$.

The process of finding the derivative $f'(p)$ is called *differentiation*. In doing so, we say we are differentiating $f$ at $p$.

---

An alternative approach to defining a derivative is to express our preceding properties in limit notation. In this setting $f$ is differentiable at $p$ is equivalent to being able to find a linear function $c_0 + c_1[x-p]$ so that

$$\lim_{x \to p} \frac{f(x) - (c_0 + c_1[x-p])}{[x-p]} = 0, \tag{5.10}$$

which follow from the fact that $(f(x)-(c_0+c_1[x-p]))/([x-p]) = HOT(x-p)/[x-p]$, which tends to 0 as $x$ tends to $p$. This alternate definition of Eq. (5.10) is the one that extends to functions of several variables.

A differentiable function, by definition, is one that can be well approximated by the linear function $f(p)+f'(p)[x-p]$, since the higher-order terms $HOT([x-p])$ are small when compared with $[x-p]$ for small values of $[x-p]$, equivalently for $x$ near

## 5.3 Slope Function, Linear Approximation, and the Derivative

$p$. This line of approximation has several names including: the *linear approximation*, the *linearization* of $f(x)$ at $p$, and the *tangent line* to $f(x)$ at $p$.

---

**Definition 5.2** For any function $f(x)$ differentiable at $p$, the linear approximation, or the linearization, of $f(x)$ at $p$ is

$$L(f; p; x) \equiv f(p) + f'(p) \cdot [x - p].$$

The equation of the tangent line to $f(x)$ at $p$ is

$$y = f(p) + f'(p) \cdot [x - p].$$

---

To see that the linearization is unique, we first make some observations. Let $I_p$ be the ideal of higher-order terms at $p$, that is $I_p$ consists of those functions $HOT([x-p])$ with $HOT(0) = 0$ and with $HOT([x-p])/[x-p]$ tending to 0 as $x$ tends to $p$. As an ideal, sums and differences of any two $HOT_1([x-p])$ and $HOT_2([x-p])$ terms in $I_p$ remain in $I_p$. Multiplication of any $HOT_1([x-p])$ in $I_p$ by any function $g(x)$ defined and continuous around $p$ remains in $I_p$, thus $g(x) \cdot HOT_2([x-p])$ and, in particular, $HOT_1([x-p]) \cdot HOT_2([x-p])$ remain in $I_p$.

For instance both $[x-p]^2$ and $6[x-p]^3$ are higher-order terms at $p$ and fall in $I_p$. Clearly their sum $[x-p]^2 + 6[x-p]^3$ and their difference $[x-p]^2 - 6[x-p]^3$ remain a higher-order term and belong to $I_p$. Their product $[x-p]^2 \cdot 6[x-p]^3 = 6[x-p]^5$ is of higher order and is in $I_p$. The product $x^2 \cdot [x-p]^2$ is of higher order and belongs to $I_p$ even though $x^2$ is *not* in $I_p$ (for $p \neq 0$ ) as $x^2[x-p]^2/[x-p] = x^2[x-p]$ tends to 0 as $x$ tends to $p$. Alternatively, we can expand the term $x^2$ about $p$ to obtain $x^2[x-p]^2 = (p + [x-p])^2[x-p]^2 = (p^2 + 2p[x-p] + [x-p]^2)[x-p]^2 = p^2[x-p]^2 + 2p[x-p]^3 + [x-p]^4$, which is clearly of higher order at $p$ and then belongs to $I_p$.

To conclude that the linearization is uniquely represented and well defined, assume that $f(x) = c_0 + c_1[x-p] + HOT_1([x-p]) = b_0 + b_1[x-p] + HOT_2([x-p])$. Then by bringing all lower-order terms to one side, and all higher-order terms to the other side, we obtain $(c_0 - b_0) + (c_1 - b_1)[x-p] = HOT_2([x-p]) - HOT_1([x-p])$. The right-hand side, as the difference of higher-order terms at $p$, remains a higher-order term, but the only way the constant plus linear term on the left-hand side is a higher-order term is when $c_0 = b_0$ and $c_1 = b_1$ and the left-hand side vanishes. We conclude that the constant term $c_0$ and the slope term $c_1$ given by the coefficient of $[x-p]$ in the linearization are unique. So the linearization is uniquely represented, and as such, the derivative $f'(p)$ is unique and well defined as the coefficient of $[x-p]$ in the linearization of a differentiable function $f(x)$ at $p$.

We mention that

### ■ Theorem 5.3

If $f(x)$ is differentiable at $p$ then $f$ must be continuous at $p$.

**Proof** If $f$ is differentiable at $p$ then $f(x) = f(p) + f'(p)[x-p] + HOT([x-p])$ as in Definition (5.1), with $HOT([x-p])/[x-p]$ tending to 0 as $x$ tends to $p$. This is equivalent to saying $\lim_{x \to p}(HOT([x-p])/[x-p]) = 0$. This tells us that $\lim_{x \to p} HOT([x-p]) = \lim_{x \to p}(\{HOT([x-p])/[x-p]\}\{[x-p]\}) = 0 \cdot 0 = 0$, which gives that

$$\lim_{x \to p} f(x) = \lim_{x \to p}\{f(p) + f'(p)[x-p] + HOT([x-p])\} = f(p) + 0 + 0 = f(p),$$

from which we conclude that $f$ is continuous at $p$. □

### ■ Exercises

1. Given the function $f(x) = x^2 + 3x$, use the substitution $x = (p + [x-p])$ and expand to find:
   (a) the linearization of $f(x)$ at $p$;
   (b) $f'(p)$ as the coefficient of $[x-p]$; and
   (c) the higher-order terms $HOT([x-p])$.

2. Use Theorem (5.3) to evaluate $\lim_{x \to p} x^2 + 3x$

3. Given $f(x) = x^2$:
   (a) find the linearization at the point $p = 1$;
   (b) find the derivative $f'(1)$;
   (c) find the equation of the tangent line at the point $p = 1$; and
   (d) graph $f(x)$ and the line in (c) and zoom in on the point $(1,1)^T$.

### ■ 5.4 Derivatives of Basic Functions

We begin compiling a list of functions with their derivatives, concentrating in this section on the first, most basic functions. We already have the Power Rule for monomials with positive integer powers $n$:

$$(x^n)' = n\, x^{n-1}.$$

## 5.4 Derivatives of Basic Functions

Our next goal is to be able to find the derivative of any polynomial. Our strategy will be to figure out the derivative of each term in the polynomial, and then we will be able to add the derivatives of each of the terms to obtain the derivative of the whole polynomial. The simplest term is the constant term.

■ **Theorem 5.4**

Let $c$ be any constant in $\mathbb{R}$, and let $f(x) = c$ be the constant function, then

$$f'(p) = (c)' = 0. \qquad (5.11)$$

**Proof** Since $f(x) = c = c + 0[x - p] + HOT([x - p])$ where $HOT([x - p]) = 0$ and $0/[x - p] = 0$ tends to 0 as $x$ tends to $p$, we have the coefficient of $[x - p]$ is 0, so $f'(p) = (c)' = 0$. □

This theorem is really no surprise, if we think geometrically, in that $f(x) = c$ is a line with slope 0. So $f(x) = c$ serves as its own tangent line in that $f(x)$ is best approximated by itself. Since this is a level line with slope 0, the derivative must be 0.

Knowing how to find the derivative of the constant term in a polynomial, we proceed next to the other terms of form $cx^n$ in the polynomial.

■ **Theorem 5.5**

Let $c$ be any constant in $\mathbb{R}$. If $f(x)$ is differentiable at $p$, then $cf(x)$ is differentiable at $p$, and

$$(cf)'(p) = c \, (f'(p)). \qquad (5.12)$$

One verbalizes this by saying: constants pull out of derivatives.

**Proof** Since $f(x)$ is differentiable at $p$, we have

$$f(x) = f(p) + (f'(p)) \cdot [x - p] + HOT([x - p]).$$

Multiplying both sides by $c$ gives

$$c\,f(x) = c\,f(p) + c\,(f'(p))\cdot[x-p] + c\cdot HOT([x-p]), \tag{5.13}$$

and, since $c\cdot HOT([x-p])$ is still a higher-order term, we see that the derivative of $c\,f(x)$ at $p$ must be the coefficient of $[x-p]$, namely $c\cdot f'(p)$. □

### EXAMPLE 5.3

Find the derivatives of the monomials $7x^5$, $-3x^4$, and $cx^n$ for $n$ a positive integer.
**Answer:** Pulling constants out of each derivative, and then using the Power Rule, we obtain

$$(7\,x^5)' = 7\,(x^5)' = 7\,(5x^4) = 35\,x^4$$
$$(-3\,x^4)' = (-3)(x^4)' = (-3)(4x^3) = -12x^3$$
$$(c\cdot x^n)' = c\,(x^n)' = c\,(n\,x^{n-1}) = (c\cdot n)x^{n-1}.$$

Now that we know how to differentiate each term in a polynomial we combine them via a differentiation rule for sums.

### ■ Theorem 5.6

Let $f(x)$ and $g(x)$ both be differentiable at $p$, then their sum $f(x)+g(x)$ and their difference $f(x)-g(x)$ are each differentiable at $p$, with $(f+g)'(p) = f'(p)+g'(p)$ and $(f-g)'(p) = f'(p)-g'(p)$. Since this holds for all $p$, we have:

$$(f+g)'(x) = (f(x)+g(x))' = f'(x)+g'(x)$$
$$(f-g)'(x) = (f(x)-g(x))' = f'(x)-g'(x).$$

One says: the derivative of a sum is the sum of the derivatives, and the derivative of a difference is the difference of the derivatives.

**Proof** Since $f(x)$ and $g(x)$ are differentiable at $p$ we have:

$$f(x) = f(p) + (f'(p))\,[x-p] + HOT_1([x-p])$$
$$g(x) = g(p) + (g'(p))\,[x-p] + HOT_2([x-p]).$$

First adding the two equations yields

$$\begin{aligned}(f+g)(x) &= f(x) + g(x) \\ &= f(p) + g(p) + (f'(p) + g'(p))\,[x-p] \\ &\quad + HOT_1([x-p]) + HOT_2([x-p]).\end{aligned} \quad (5.14)$$

Then subtracting yields

$$\begin{aligned}(f-g)(x) &= f(x) - g(x) \\ &= f(p) - g(p) + (f'(p) - g'(p))\,[x-p] \\ &\quad + HOT_1([x-p]) - HOT_2([x-p]).\end{aligned} \quad (5.15)$$

Since $HOT_1([x-p]) \pm HOT_2([x-p])$ remain higher-order terms, by reading off the coefficients of $[x-p]$ in Eq. (5.14) we obtain $(f+g)'(p) = f'(p) + g'(p)$. From Eq. (5.15) we see $(f-g)'(p) = f'(p) - g'(p)$. Since these both hold for all $p$, we have

$$\begin{aligned}(f+g)'(x) &= (f(x) + g(x))' = f'(x) + g'(x) \\ (f-g)'(x) &= (f(x) - g(x))' = f'(x) - g'(x).\end{aligned}$$

$\square$

By applying the preceding theorem repeatedly to sums of more than two terms we realize we can differentiate any finite sum of differentiable functions by first differentiating the individual summands and then adding the results. For instance, to differentiate the threefold sum $f(x) + g(x) + h(x)$ of differentiable functions, we proceed in steps:

$$\begin{aligned}(f(x) + g(x) + h(x))' &= ([f(x) + g(x)] + h(x))' \\ &= [f(x) + g(x)]' + h'(x) = f'(x) + g'(x) + h'(x).\end{aligned}$$

We are now well-equipped to differentiate any polynomial.

### EXAMPLE 5.4

Differentiate the polynomial $5x^3 - 4x^2 + 7x + 11$.

Using the sum rule we have

$$(5x^3 - 4x^2 + 7x + 11)' = (5x^3)' + (-4x^2)' + (7x)' + (11)'.$$

Next we pull out constants and differentiate the constant term 11 to get 0, obtaining

$$(5x^3 - 4x^2 + 7x + 11)' = 5(x^3)' + (-4)(x^2)' + 7(x)' + 0.$$

Applying the power rule to each monomial completes the computation

$$(5x^3 - 4x^2 + 7x + 11)' = 5(3x^{3-1}) + (-4)(2x^{2-1}) + 7(1x^0) = 15x^2 - 8x + 7.$$

We now have a way to differentiate any polynomial, via the following Polynomial Rule.

■ **Theorem 5.7**

The derivative of any polynomial is given by

$$\left(\sum_{p=0}^{n} c_p \, x^p\right)' = \sum_{p=0}^{n} p \, c_p \, x^{p-1}. \tag{5.16}$$

**Proof** We proceed as in Example (5.4). Distribute the derivatives over the summands, pull out constants, and use the Power Rule.

$$\left(\sum_{p=0}^{n} c_p \, x^p\right)' = \sum_{p=0}^{n} (c_p \, x^p)' = \sum_{p=0}^{n} c_p \, (x^p)' = \sum_{p=0}^{n} c_p \, (p \, x^{p-1}) = \sum_{p=0}^{n} p \, c_p \, x^{p-1}.$$

This uses the fact that the derivative is a linear operation. □

We next show the Product Rule.

## Theorem 5.8
### Product Rule

If $f(x)$ and $g(x)$ are both differentiable at $p$, then their product $f(x)g(x)$ is differentiable at $p$, with

$$(f\ g)'(p)\ =\ f'(p)\ g(p)\ +\ f(p)\ g'(p). \tag{5.17}$$

Since this holds for all $p$, we have

$$(f\ g)'(x)\ =\ (f(x)\ g(x))'\ =\ f'(x)\ g(x)\ +\ f(x)\ g'(x). \tag{5.18}$$

---

**Proof** By differentiability of $f(x)$ and $g(x)$ at $p$, we have

$$\begin{aligned} f(x) &= f(p)\ +\ f'(p)\ [x-p]\ +\ HOT_1([x-p]) \\ g(x) &= g(p)\ +\ g'(p)\ [x-p]\ +\ HOT_2([x-p]). \end{aligned}$$

Their product is given by

$$\begin{aligned} f(x) \cdot g(x) &= \{f(p)\ +\ f'(p)\ [x-p]\ +\ HOT_1([x-p])\} \\ &\quad \cdot \{g(p)\ +\ g'(p)\ [x-p]\ +\ HOT_2([x-p])\} \\ &= f(p)\ g(p)\ +\ f(p)\ g'(p)\ [x-p]\ +\ f(p)\ HOT_2([x-p]) \\ &\quad +\ f'(p)\ g(p)\ [x-p]\ +\ f'(p)\ g'(p)\ [x-p]^2 \\ &\quad +\ f'(p)\ [x-p]\ HOT_2([x-p]) \\ &\quad +\ g(p)\ HOT_1([x-p])\ +\ g'(p)\ [x-p]\ HOT_1([x-p]) \\ &\quad +\ HOT_1([x-p])\ HOT_2([x-p]) \\ &= f(p)\ g(p)\ +\ (f'(p)\ g(p)\ +\ f(p)\ g'(p))\ [x-p]\ +\ HOT([x-p]), \end{aligned}$$

where any term having a $[x-p]^2$, $HOT_1([x-p])$, or $HOT_2([x-p])$ factor has been subsumed into the higher-order term $HOT([x-p])$. Reading off the coefficient of $[x-p]$ gives the derivative to be

$$(f\ g)'(p)\ =\ (f'(p)\ g(p)\ +\ f(p)\ g'(p)),$$

proving the Product Rule. □

### EXAMPLE 5.5

Find the derivative $\{(x^5 + 7)(x^3 + x)\}'$. **Answer:** When we don't need to simplify, the Product Rule proves easier to apply:

$$\{(x^5 + 7)\,(x^3 + x)\}' = (x^5 + 7)'\,(x^3 + x) \;+\; (x^5 + 7)\,(x^3 + x)'$$
$$= (5x^4)\,(x^3 + x) \;+\; (x^5 + 7)\,(3x^2 + 1).$$

**Warning:** a common error is to assume the Product Rule behaves like its predecessors the Sum and Difference Rules. It is *not* true that the derivative of a product is the product of the derivatives. The Product Rule tells us that we must differentiate each factor, leaving all other factors alone, and then add the results.

### EXAMPLE 5.6

Use the Product Rule to verify the Power Rule for $x^2$.

Correct method: $(x^2)' = (x \cdot x)' = (x)'(x) + (x)(x)' = (1)(x) + (x)(1) = 2x^1$ agreeing with the Power Rule.

It is false to assume that the derivative of a product is the product of the derivatives:

$$(x^2)' \;=\; (x \cdot x)' \;\neq\; (x)' \cdot (x)' \;=\; (1)(1) \;=\; 1 \neq 2x^1.$$

### EXAMPLE 5.7

Find the derivative of the product $f(x)g(x)h(x)$ of three differentiable functions $f(x)$, $g(x)$, $h(x)$.
**Answer:** We successively employ the Product Rule for two functions:

$$\begin{aligned}(f(x)\,g(x)\,h(x))' &= ([f(x)\,g(x)]\,h(x))' = [f(x)\,g(x)]'\,h(x) + [f(x)\,g(x)]\,h'(x) \\ &= [f'(x)\,g(x) + f(x)\,g'(x)]\,h(x) + [f(x)\,g(x)]\,h'(x) \\ &= f'(x)\,g(x)\,h(x) + f(x)\,g'(x)\,h(x) + f(x)\,g(x)\,h'(x).\end{aligned}$$

Note the pattern in the final result. We differentiate each factor one by one, leaving the remaining factors alone, and add up the results. This pattern generalizes to products of any finite number of differentiable functions.

For now it may seem that the Product Rule is superfluous, as we can multiply out the product of polynomials and rely on the Polynomial Rule to differentiate the resulting large polynomial. However, as we shall see after learning about derivatives of trigonometric functions for instance, that the Product Rule is essential to differentiate a product such as $x\cos(x)$, which cannot be expanded algebraically. We come back to the derivative of this product later.

We next visit the Reciprocal Rule for differentiating the function $1/x$.

■ **Theorem 5.9**

**Reciprocal Rule**

The reciprocal function $f(x) = 1/x$ is differentiable at any $p \neq 0$, with

$$f'(p) = \frac{-1}{p^2} = -p^{-2}.$$

Since this is true for any $p \neq 0$ we have $f'(x) = (x^{-1})' = -x^{-2}$, or

$$(x^{-1})' = -x^{-2}.$$

**Proof** By obtaining a common denominator, we see that the difference between $1/x$ and $1/p$ can be expressed as

$$\frac{1}{x} - \frac{1}{p} = \frac{p-x}{xp},$$

or equivalently,

$$\frac{1}{x} = \frac{1}{p} + \frac{p-x}{xp}. \tag{5.19}$$

We can factor out a $\frac{1}{x}$ expression from the rightmost term in the right-hand side to obtain

$$\frac{1}{x} = \frac{1}{p} + \frac{1}{x}\frac{p-x}{p}, \tag{5.20}$$

and then we replace the $\frac{1}{x}$ in the last term of Eq. (5.20) by Eq. (5.19), getting

$$\frac{1}{x} = \frac{1}{p} + \left(\frac{1}{p} + \frac{p-x}{xp}\right)\frac{p-x}{p}.$$

Multiplying out yields

$$\frac{1}{x} = \frac{1}{p} + \frac{-1}{p^2}[x-p] + \frac{[x-p]^2}{xp^2}.$$

Now $[x-p]^2/xp^2$ is a higher-order term since $([x-p]^2/xp^2)/[x-p] = [x-p]/xp^2$, which tends to 0 for $p \neq 0$ and for $x$ near $p$ with $x \neq 0$. Thus,

$$[x-p]^2/(xp^2) = HOT([x-p]),$$

and the derivative is the coefficient of $[x-p]$ or

$$f'(p) = \frac{-1}{p^2} = -p^{-2}. \tag{5.21}$$

□

We are now in a position to show that the Power Rule holds for negative integer powers.

■ **Theorem 5.10**

Let $-n$ be a negative integer, then for all $x \neq 0$, we have:

$$\left(x^{-n}\right)' = -n\, x^{-n-1}. \tag{5.22}$$

**Proof** We write out $x^{-n} = (x^{-1})^n = x^{-1} \cdot x^{-1} \cdot x^{-1} \cdot \ldots \cdot x^{-1}$ as an $n$-fold product of the factor $x^{-1}$ and apply the Product Rule, differentiating each factor one by one, leaving all other factors alone, and adding the results.

$$\begin{aligned}(x^{-n})' &= ((x^{-1})^n)' = (x^{-1} \cdot x^{-1} \cdot x^{-1} \cdot \ldots \cdot x^{-1})' \\ &= (x^{-1})' \cdot (x^{-1}) \cdot (x^{-1}) \cdot \ldots \cdot (x^{-1}) \\ &\quad + (x^{-1}) \cdot (x^{-1})' \cdot (x^{-1}) \cdot \ldots \cdot (x^{-1}) \\ &\quad + (x^{-1}) \cdot (x^{-1}) \cdot (x^{-1})' \cdot \ldots \cdot (x^{-1}) \\ &\quad \vdots \\ &\quad + (x^{-1}) \cdot (x^{-1}) \cdot (x^{-1}) \cdot \ldots \cdot (x^{-1})' & (5.23)\\ &= n\,(x^{-1})' \cdot (x^{-1})^{n-1} & (5.24)\\ &= n\,(-1\,x^{-2}) \cdot x^{-n+1} & (5.25)\\ &= -n\,(x^{-2-n+1}) \\ &= -n\,(x^{n-1}).\end{aligned}$$

Since there are $n$ equal factors to differentiate, there are $n$ summands in Eq. (5.23) leading to the coefficient $n$ in Eq. (5.24). We have also applied the Reciprocal Rule to obtain Eq. (5.25). □

An alternate proof for $(x^{-n})'$ can be given by examining $1/x^n - 1/p^n$, obtaining common denominators, and paralleling the proof we gave for $(x^{-1})'$ by producing the linearization for $x^{-n}$ directly.

We now know the Power Rule holds for all integer powers. Indeed, by the fact that $x^0 = 1$, we observe $(x^0)' = (1)' = 0 = 0 x^{-1}$ since the derivative of any constant is 0.

---

### ■ Exercises

1. Use the Polynomial Rule in Theorem (5.7) as applied to $f(x) = x^2 + 3x$ to find $f'(x)$, and $f'(p)$, and compare your answers with Exercise 1 in Section 5.3.

2. Given $f(x) = 5x^{10} + 3x^7 + 2x^4 + 5x + 2$,
   (a) Find $f'(x)$; and
   (b) Find the equation of the tangent line to $f(x)$ at the point $p = 1$, by relying on Part (a) to find $f'(1)$.

3. Given $f(x) = 1/x$,
   (a) Find $f'(x)$;

(b) Find the equation of the tangent line to $f(x)$ at the point $p = 1$, by relying on Part (a) to find $f'(1)$; and
(c) Graph $f(x) = 1/x$ along with your tangent line in (b) and zoom in on the point $(1, 1)$.

4. Use the Product Rule in Theorem (5.8) to find the derivative of
   (a) $(x^3 + 4x^2 + 1)(5x^2 - 3)$; and
   (b) $(x^2 + x + 2)(x^{-3} + x^{-1} + 3)$.

## ■ 5.5 The Chain and Quotient Rules

Our first result in this section is a very powerful and versatile differentiation rule, called the Chain Rule, which holds for compositions of differentiable functions.

### ■ Theorem 5.11
**Chain Rule**

If $f(x)$ is differentiable at $p$, and if $g(u)$ is differentiable at $q = f(p)$, then the composition $g \circ f(x) = g(f(x))$ is differentiable at $p$, with

$$(g \circ f)'(p) = g'(f(p)) \cdot f'(p).$$

Since this is true for all $p$ with $(g \circ f)(p)$ defined, we have

$$(g \circ f)'(x) = (g(f(x)))' = g'(f(x)) \cdot f'(x). \tag{5.26}$$

To differentiate a composition of functions, we (1) differentiate the outside function, (2) leave the inside alone, and (3) multiply by the derivative of the inside.

**Proof** Let $q = f(p)$. Since $g(u)$ is differentiable at $q$, we have

$$g(u) = g(q) + g'(q)[u - q] + HOT_g(u - q), \tag{5.27}$$

where $HOT_g(u - q)/[u - q]$ tends to 0 as $u$ tends to $q$. Since $f(x)$ is differentiable at $p$, we have

$$f(x) = f(p) + f'(p)[x - p] + HOT_f(x - p), \tag{5.28}$$

where $HOT_f(x-p)/[x-p]$ tends to 0 as $x$ tends to $p$. Composing the left- and right-hand sides of Eq. (5.28) into the respective left- and right-hand sides of Eq. (5.27), we obtain

$$\begin{aligned}
g(f(x)) &= g(q) + g'(q)[\{f(p) + f'(p)[x-p] + HOT_f(x-p)\} - q] \\
&\quad + HOT_g(\{f(p) + f'(p)[x-p] + HOT_f(x-p)\} - q) \\
&= g(f(p)) + g'(f(p))[\{f(p) + f'(p)[x-p] + HOT_f(x-p)\} - f(p)] \\
&\quad + HOT_g(\{f(p) + f'(p)[x-p] + HOT_f(x-p)\} - f(p)) \\
&= g(f(p)) + g'(f(p))[\{f'(p)[x-p] + HOT_f(x-p)\}] \\
&\quad + HOT_g(\{f'(p)[x-p] + HOT_f(x-p)\}) \\
&= g(f(p)) + g'(f(p))f'(p)[x-p] + g'(f(p))HOT_f(x-p) \\
&\quad + HOT_g(\{f'(p)[x-p] + HOT_f(x-p)\}).
\end{aligned}$$

This reveals the linearization of $g \circ f$ to be

$$g(f(p)) + g'(f(p))\ f'(p)\ [x-p], \tag{5.29}$$

and the higher-order terms to be

$$HOT_{g \circ f}(x) = g'(f(p))HOT_f(x-p) + HOT_g(\{f'(p)[x-p] + HOT_f(x-p)\}). \tag{5.30}$$

We will come back to the higher-order terms shortly. Our main point is to notice that the coefficient $g'(f(p))f'(p)$ of $[x-p]$ in the linearization Eq. (5.29) gives us our derivative $(g \circ f)'(p)$. Thus we have shown our result

$$(g \circ f)'(p) = g'(f(p))\ f'(p). \tag{5.31}$$

There are some subtleties in recognizing the expression Eq. (5.30) as the higher-order term. We outline them here for the interested reader. Clearly the first term $g'(f(p))HOT_f(x-p)$ in Eq. (5.30) is a higher-order term because $HOT_f(x-p)$ is higher order. So we examine the second term in Eq. (5.30). Since $HOT_g(0) = 0$, we have that $HOT_g(\{f'(p)[x-p] + HOT_f(x-p)\}) = 0$ whenever $\{f'(p)[x-p] + HOT_f(x-p)\} = 0$, including the case when $x = p$. When $x \neq p$ and $\{f'(p)[x-p] + HOT_f(x-p)\} = 0$, we have $HOT_g(\{f'(p)[x-p] + HOT_f(x-p)\})/[x-p] = 0/[x-p] = 0$ and this certainly tends toward 0 since it is 0. When $x \neq p$ and

$\{f'(p)[x-p] + HOT_f(x-p)\} \neq 0$, we have

$$\frac{HOT_g(\{f'(p)[x-p] + HOT_f(x-p)\})}{[x-p]} \qquad (5.32)$$
$$= \frac{HOT_g(\{f'(p)[x-p] + HOT_f(x-p)\})}{\{f'(p)[x-p] + HOT_f(x-p)\}} \cdot \frac{\{f'(p)[x-p] + HOT_f(x-p)\}}{[x-p]}.$$

This expression also tends to 0 as $[x-p]$ tends to 0, since the left-hand factor of Eq. (5.32)

$$\frac{HOT_g(\{f'(p)[x-p] + HOT_f(x-p)\})}{\{f'(p)[x-p] + HOT_f(x-p)\}}$$
$$= \frac{HOT_g(\{f(p) + f'(p)[x-p] + HOT_f(x-p)\} - f(p))}{\{f(p) + f'(p)[x-p] + HOT_f(x-p)\} - f(p)}$$

tends to 0 as $[x-p]$ tends to 0 (by the hypothesis that $g(x)$ is differentiable at $q = f(p)$), and since the right-hand factor of Eq. (5.32) tends to $f(p)$ as $[x-p]$ tends to 0. □

We employ the Chain Rule immediately to obtain the derivative of a power of a function.

### ■ Theorem 5.12

For any integer $n$ and any differentiable function $f(x)$,

$$((f(x))^n)' = n\,(f(p))^{n-1} \cdot f'(x).$$

**Proof** Let $g(u) = u^n$ then $g'(u) = nu^{n-1}$ by the Power Rule. By the Chain Rule we have

$$(g(f(x))' = g'(f(x)) \cdot f'(x) = n(f(x))^{n-1} \cdot f'(x).$$

□

## EXAMPLE 5.8

Find the derivatives of: (a) $(x^2+1)^{100}$ and (b) $(x^2+1)^{-1}$. **Answer (a):**

$$((x^2+1)^{100})' = 100(x^2+1)^{100-1} \cdot (x^2+1)' = 100(x^2+1)^{99} \cdot 2x;$$

**Answer (b):**

$$((x^2+1)^{-1})' = -1(x^2+1)^{-1-1} \cdot (x^2+1)' = -1(x^2+1)^{-2} \cdot 2x.$$

Find the derivative of $(g(x))^{-1}$ for $g(x)$ differentiable. **Answer:**

$$((g(x))^{-1})' = -1(g(x))^{-1-1} \cdot g'(x) = -1(g(x))^{-2} \cdot g'(x).$$

Our last example allows us to differentiate a quotient of differentiable functions via the so-called Quotient Rule.

### ■ Theorem 5.13
**Quotient Rule**

Let $f(x)$ and $g(x)$ be differentiable with $g(x) \neq 0$. Then

$$\left(\frac{f(x)}{g(x)}\right)' = \frac{f'(x)g(x) - f(x)g'(x)}{g^2(x)}.$$

**Proof** We first write the quotient as a product $f(x)(g(x))^{-1}$ and then apply the Product Rule and the Reciprocal Rule, recombining the resulting expression by obtaining a common denominator of the form $g^2(x)$:

$$\begin{aligned}
\left(\frac{f(x)}{g(x)}\right)' &= (f(x)\,(g(x))^{-1})' \\
&= (f(x))'\,(g(x))^{-1} + f(x)\,((g(x))^{-1})' \\
&= f'(x)\,(g(x))^{-1} + f(x)\,((-1)(g(x))^{-2}\,g'(x)) \\
&= \frac{f'(x)}{g(x)} - \frac{f(x)\,g'(x)}{g^2(x)} \\
&= \frac{f'(x)g(x)}{g^2(x)} - \frac{f(x)\,g'(x)}{g^2(x))} \\
&= \frac{f'(x)\,g(x) - f(x)\,g'(x)}{g^2(x)}.
\end{aligned}$$

This function is defined as long as $g(x) \neq 0$. □

> **EXAMPLE 5.9**
>
> Find the derivative of $(x^3 + x)/(x^2 + 1)$, and of $x^{-n}$ for $-n$ a negative integer.
>
> $$\left(\frac{x^3 + x}{x^2 + 1}\right)' = \frac{(x^3 + x)'(x^2 + 1) - (x^3 + x)(x^2 + 1)'}{(x^2 + 1)^2}$$
> $$= \frac{(3x^2 + 1)(x^2 + 1) - (x^3 + x)(2x)}{(x^2 + 1)^2}.$$
>
> $$(x^{-n})' = \left(\frac{1}{x^n}\right)' = \frac{(1)'(x^n) - (1)(x^n)'}{(x^n)^2}$$
> $$= \frac{(0)(x^n) - (1)(nx^{n-1})}{(x^{2n})} = -nx^{n-1-2n} = -n\,x^{-n-1},$$
>
> and the Power Rule for negative integers is verified via the Quotient Rule.

### ■ Exercises

1. Find the derivative of each $f(x)$ as follows:

    (a) $f(x) = (x^3 + 4x^2 - 5x + 1)^{71}$;
    (b) $f(x) = (x^2 - 7x + 4)^{-1}$;
    (c) $f(x) = (x^3 + 4x^2 - 5x + 1)^{71} (3x^2 - 2x + 6)^5$; and
    (d) $f(x) = (3x^2 + x - 4)/(x^2 - 7x + 4)$.

2. Take the second derivative of the composition $(f \circ g)(x)$, that is, take another derivative of Eq. (5.26).

## ■ 5.6 Derivatives of Trigonometric Functions

From the diagram in Figure 5.4 of a sector of a circle of radius 1 subtended by the angle $\theta \geq 0$, we see that the shortest distance between the points A and B lies along the straight line formed by the hypothenuse of the right triangle with legs $\sin(\theta)$ and $1 - \cos(\theta)$.

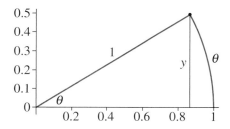
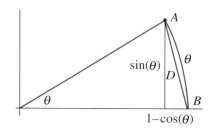

**FIGURE 5.4** ■ Arclength $\theta$ for a right triangle is greater than $D$.

This hypothenuse has length $D = \sqrt{\sin^2(\theta) + (1 - \cos(\theta))^2}$. Since the linear distance $D$ between A and B is less than or equal to the distance along the circle, $s = 1 \cdot \theta = \theta$ for $\theta$ in radians, we obtain $0 \leq D \leq \theta$ or equivalently $D^2 = \sin^2(\theta) + (1 - \cos(\theta))^2 \leq \theta^2$. Expanding the squared term yields

$$\begin{aligned} \sin^2(\theta) + (1 - \cos(\theta))^2 &= \sin^2(\theta) + 1 - 2\cos(\theta) + \cos^2(\theta) \\ &= 2 - 2\cos(\theta) \leq \theta^2. \end{aligned}$$

Dividing both sides by 2 yields our main estimate of this section:

$$0 \leq 1 - \cos(\theta) \leq \frac{1}{2}\theta^2, \qquad (5.33)$$

which holds for all angles $\theta$ by evenness of $\cos(\theta)$ and $\theta^2$.

If we examine the graph of $\cos(x)$ near 0, it is not hard to guess that the tangent line should have equation $y = 1$ and that the linearization of $\cos(x)$ at the point $p = 0$ can be given by $L(\cos(x); 0; x) = 1 + 0[x - 0] = 1$.

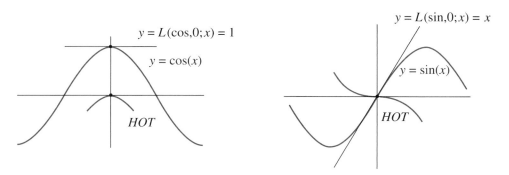

**FIGURE 5.5** ■ The cosine (left) and sine (right) functions with linearizations.

To see this, let $\cos(x) = 1 + 0[x-0] + E(x) = 1 + E(x)$ where $E(x)$ is the error between $\cos(x)$ and 1. Then $-E(x) = 1 - \cos(x)$ and our estimate in Eq. (5.33) gives that $0 \leq -E(x) = 1 - \cos(x) \leq x^2/2$, hence $E(0) = 0$ and

$$|E(x)/x| \leq |x|^2/(2|x|) = |x|/2,$$

which tends to 0 as $x$ tends to 0. Thus $E(x)$ is a higher-order term at 0, $E(x) = HOT(x-0)$, and the linearization of $\cos(x)$ at $p = 0$ is $1 + 0[x-0]$. We conclude, by examining the coefficient of $[x-0]$ in our linearization, that

■ **Lemma 5.1**

$$\cos'(0) = 0. \tag{5.34}$$

We have used an important idea in our reasoning that bears clarification. If an expression $E(x)$ satisfies $|E(x)| \leq HOT([x-p])$, then $E(x)$ is also a higher-order term at $p$, since $|E(p)| \leq HOT(p-p) = 0$, and $|E(x)/[x-p]| \leq HOT([x-p])/[x-p]$, which tends to 0 forcing $E(x)/[x-p]$ to tend to 0.

We turn next to the sine function to search for a linearization at 0. By the diagram in Figure 5.6, we have that for $\pi/2 > \theta > 0$ the smaller triangle has area less than or equal to the area of the circular sector, which has area less than or equal to the larger triangle. Thus

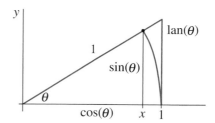

**FIGURE 5.6** ■ Using approximations for the area within a sector to estimate the sine function.

$$\frac{1}{2}\sin(\theta) \cdot \cos(\theta) \leq \frac{\theta}{2} \leq \frac{1}{2}\frac{\sin(\theta)}{\cos(\theta)}.$$

Dividing by $\theta$ and multiplying by 2 yields

$$\frac{\sin(\theta)}{\theta} \cdot \cos(\theta) \leq 1 \leq \frac{\sin(\theta)}{\theta} \cdot \frac{1}{\cos(\theta)},$$

which implies

$$\cos(\theta) \leq \frac{\sin(\theta)}{\theta} \leq \frac{1}{\cos(\theta)}. \tag{5.35}$$

Applying our key estimate Eq. (5.33) to both the far left and the far right sides of Eq. (5.35) gives

$$1 - \frac{\theta^2}{2} \leq \cos(\theta) \leq \frac{\sin(\theta)}{\theta} \leq \frac{1}{\cos(\theta)} \leq \frac{1}{1 - \theta^2/2} = \frac{1 - \theta^2/2 + \theta^2/2}{1 - \theta^2/2} = 1 + \frac{\theta^2}{2 - \theta^2}.$$

We conclude that

$$-\frac{\theta^2}{2} \leq \frac{\sin(\theta)}{\theta} - 1 = \frac{\sin(\theta) - \theta}{\theta} \leq \frac{\theta^2}{2 - \theta^2}. \qquad (5.36)$$

For $\theta^2 < 1$ we have $2 - \theta^2 > 2 - 1 = 1$, which gives that $1/(2 - \theta^2) < 1$ and then $\theta^2/(2 - \theta^2) < \theta^2$. Applying this last bound to the right-hand side of Eq. (5.36) yields

$$-\frac{\theta^2}{2} \leq \frac{\sin(\theta) - \theta}{\theta} \leq \theta^2.$$

We conclude that

$$\left| \frac{\sin(\theta) - \theta}{\theta} \right| \leq \theta^2 \qquad (5.37)$$

for $|\theta| < 1$. This is sufficient to deduce that the linearization of $\sin(x)$ at 0 is $x$. Let

$$\sin(x) = x + E(x), \qquad (5.38)$$

where $E(x)$ is the error between $\sin(x)$ and $x$. Then $E(0) = 0$, and

$$\left| \frac{E(x)}{x} \right| = \left| \frac{\sin(x) - x}{x} \right| \leq x^2,$$

which tends to 0 as $x$ tends to 0. Thus $E(x)$ is a higher-order term at 0, and Eq. (5.38) gives that the linearization of $\sin(x)$ at 0 is $x = 0 + 1[x - 0]$. We conclude that the coefficient 1 of $[x - 0]$ is the derivative $\sin'(0)$.

---

■ **Lemma 5.2**

$$\sin'(0) = 1.$$

---

Our two fundamental limits are the tool that lets us compute all trigonometric derivatives. First, for sine and cosine we have:

■ **Theorem 5.14**

For all $x$ in $\mathbb{R}$

$$\sin'(x) = \cos(x), \tag{5.39}$$
$$\cos'(x) = -\sin(x). \tag{5.40}$$

**Proof** To find the linearization of $\sin(x)$ at $p$ we expand about $p$ in the following way:

$$\begin{aligned} \sin(x) &= \sin(p + [x-p]) \\ &= \sin(p) \cdot \cos([x-p]) + \cos(p) \cdot \sin([x-p]) \quad (5.41) \\ &= \sin(p) \,\{1 + 0[x-p] + HOT_1([x-p])\} \\ &\quad + \cos(p) \,\{[x-p] + HOT_2([x-p])\} \\ &= \sin(p) + \cos(p)\,[x-p] \\ &\quad + \sin(p) \cdot HOT_1([x-p]) + \cos(p) \cdot HOT_2([x-p]) \\ &= \sin(p) + \cos(p) \cdot [x-p] + HOT([x-p]). \tag{5.42} \end{aligned}$$

We have used the angle addition formula for sine to obtain Eq. (5.41). To obtain the next line we observe that since $\cos(u) = 1 + HOT_1(u)$ we have $\cos([x-p]) = 1 + HOT_1([x-p])$, and since $\sin(u) = u + HOT_2(u)$ we have $\sin([x-p]) = [x-p] + HOT_2([x-p])$. Line (5.42) tells us that the linearization of $\sin(x)$ at $p$ is $\sin(p) + \cos(p)[x-p]$. The coefficient $\cos(p)$ of $[x-p]$ must be the derivative $\sin'(p)$. We conclude $\sin'(p) = \cos(p)$. Since this is true for all $p$, we obtain Eq. (5.39).

Finding the linearization of $\cos(x)$ at $p$ proceeds in an analogous manner.

$$\begin{aligned}
\cos(x) &= \cos(p + [x-p]) \\
&= \cos(p) \cdot \cos([x-p]) - \sin(p) \cdot \sin([x-p]) \quad (5.43) \\
&= \cos(p) \cdot \{1 + 0[x-p] + HOT_1([x-p])\} \\
&\quad - \sin(p)\{[x-p] + HOT_2([x-p])\} \\
&= \cos(p) - \sin(p) \cdot [x-p] \\
&\quad + \cos(p) \cdot HOT_1([x-p]) - \sin(p) \cdot HOT_2([x-p]) \\
&= \cos(p) - \sin(p) \cdot [x-p] + HOT([x-p]). \quad (5.44)
\end{aligned}$$

We have used the angle addition formula for cosine to obtain Eq. (5.43). We have again used the facts that $\cos([x-p]) = 1 + HOT_1([x-p])$ and $\sin([x-p]) = [x-p] + HOT_2([x-p])$. Line (5.44) tells us that the linearization of $\cos(x)$ at $p$ is $\cos(p) - \sin(p)[x-p]$. The coefficient $-\sin(p)$ of $[x-p]$ must be the derivative $\cos'(p)$. We conclude $\cos'(p) = -\sin(p)$. Since this is true for all $p$, we obtain Eq. (5.40). $\square$

Once the derivatives of $\sin(x)$ and $\cos(x)$ are known, we can compute the derivatives of all other basic trigonometric functions by converting them to expressions in sine and cosine and using the Quotient Rule.

### ■ Theorem 5.15

For $x$ in domain of the differentiated trigonometric function, we have:

$$\begin{aligned}
\tan'(x) &= \sec^2(x); \\
\cot'(x) &= -\csc^2(x); \\
\sec'(x) &= \sec(x) \cdot \tan(x); \text{ and} \\
\csc'(x) &= -\csc(x) \cdot \cot(x).
\end{aligned}$$

**Proof** In the case of $\tan(x)$ we convert to a ratio of $\sin(x)$ over $\cos(x)$ and proceed with the Quotient Rule.

$$\begin{aligned}
\tan'(x) &= \left(\frac{\sin(x)}{\cos(x)}\right)' \\
&= \frac{\sin'(x) \cdot \cos(x) - \sin(x) \cdot \cos'(x)}{\cos^2(x)} \\
&= \frac{\cos(x) \cdot \cos(x) - \sin(x)(-\sin(x))}{\cos^2(x)} \\
&= \frac{1}{\cos^2(x)} = \sec^2(x).
\end{aligned}$$

In the case of $\sec(x) = 1/\cos(x)$ we have:

$$\begin{aligned}
\sec'(x) &= \left(\frac{1}{\cos(x)}\right)' \\
&= \frac{1' \cos(x) - 1 \cdot \cos'(x)}{\cos^2(x)} \\
&= \frac{0 \cdot \cos(x) - 1 \cdot (-\sin(x))}{\cos^2(x)} \\
&= \frac{\sin(x)}{\cos^2(x)} = \frac{1}{\cos(x)}\frac{\sin(x)}{\cos(x)} = \sec(x) \cdot \tan(x).
\end{aligned}$$

$\cot(x) = \cos(x)/\sin(x)$ and $\csc(x) = 1/\sin(x)$ are handled similarly. $\square$

We return to a previous example, to find the derivative of $x\cos(x)$.

$$(x \cdot \cos(x))' = (x)'\cos(x) + x(\cos(x))' = (1)\cos(x) + x(-\sin(x))$$
$$= \cos(x) - x \cdot \sin(x).$$

Note that, unlike in the polynomial case where we had the option to expand the product of two polynomials and avoid the Product Rule, here we have to use the Product Rule in tandem with our knowledge of the derivative of the cosine function.

## EXAMPLE 5.10

Find the derivative of $\sin^4(x)$. **Answer:**

$$(\sin^4(x))' = ((\sin(x))^4)' = 4(\sin(x))^3 \cdot (\sin(x))' = 4(\sin(x))^3 \cdot \cos(x).$$

The Chain Rule was used in combination with our trigonometric rule.

Find the derivative of $\tan(x)\sin^2(x)$. **Answer:**

$$\begin{aligned}(\tan(x) \cdot \sin^2(x))' &= (\tan(x))'(\sin^2(x)) + \tan(x)(\sin^2(x))' \\ &= \sec^2(x) \cdot \sin^2(x) + \tan(x) 2\sin^1(x)(\sin(x))' \\ &= \sec^2(x) \cdot \sin^2(x) + 2\tan(x) \cdot \sin(x) \cdot \cos(x).\end{aligned}$$

The Product Rule was used as the overall rule; within the Product Rule we used trigonometric rules and the Power Rule in conjunction with the Chain Rule.

Find the derivative of $\sin(x^2)$ **Answer:**

$$\begin{aligned}(\sin(x^2))' &= \sin'(x^2) \cdot (x^2)' \\ &= \cos(x^2) \, 2x.\end{aligned}$$

Here the Chain Rule was used.

## ■ Exercises

1. Find the derivative $f'(x)$ for each function $f(x)$:

    (a) $\sin(x) \cdot \tan(x)$;
    (b) $(x^2 + 1) \cdot \sec(x)$; and
    (c) $x^3/(2 + \cos(x))$.

2. Find the derivative $f'(x)$ for each function $f(x)$:

    (a) $\cos^5(x)$;
    (b) $(\cos(x^2 + 1))^5$; and
    (c) $\cos(x^2 + 3x + 2)$.

## 5.7 Derivatives of Exponential Functions

The exponential function is defined by:

$$\begin{aligned}
\exp(x) &= \sum_{n=0}^{\infty} \frac{x^n}{n!} \\
&= 1 + x + \frac{x^2}{2!} + \frac{x^3}{3!} + \frac{x^4}{4!} \cdots \\
&= 1 + x + x^2 \left\{ \frac{x^0}{2!} + \frac{x^1}{3!} + \frac{x^2}{4!} \cdots \right\} \\
&= 1 + x + x^2 \left( \sum_{k=0}^{\infty} \frac{x^k}{(k+2)!} \right) \\
&= 1 + x + HOT([x-0])
\end{aligned} \qquad (5.45)$$

for any $x$ in $\mathbb{R}$, from which we see that the linearization of $\exp(x)$ at 0 is $1 + x = 1 + 1[x - 0]$, and thus the coefficient 1 of $[x - 0]$ is the derivative

$$\exp'(0) = 1.$$

We need the following additivity property of $\exp(x)$:

### Lemma 5.3

$$\exp(a) \cdot \exp(b) = \exp(a + b) \qquad (5.46)$$

## Proof

$$\begin{aligned}
\exp(a) \cdot \exp(b) &= \left(\sum_{k=0}^{\infty} \frac{a^k}{k!}\right)\left(\sum_{p=0}^{\infty} \frac{b^p}{p!}\right) = \sum_{k=0}^{\infty}\sum_{p=0}^{\infty} \frac{a^k}{k!}\frac{b^p}{p!} \\
&= \sum_{n=0}^{\infty} \sum_{k+p=n} \frac{a^k}{k!}\frac{b^p}{p!} &(5.47)\\
&= \sum_{n=0}^{\infty} \sum_{k=0}^{n} \frac{a^k}{k!}\frac{b^{n-k}}{(n-k)!} = \sum_{n=0}^{\infty}\sum_{k=0}^{n} \frac{1}{k!(n-k)!} a^k b^{n-k} &(5.48)\\
&= \sum_{n=0}^{\infty} \frac{1}{n!} \sum_{k=0}^{n} \frac{n!}{k!(n-k)!} a^k b^{n-k} &(5.49)\\
&= \sum_{n=0}^{\infty} \frac{1}{n!} \sum_{k=0}^{n} \binom{n}{k} a^k b^{n-k} &(5.50)\\
&= \sum_{n=0}^{\infty} \frac{1}{n!}(a+b)^n = \exp(a+b). &(5.51)
\end{aligned}$$

We distributed the multiplication through the sums in the first row. Then we moved from summing first over the $k$ and then the $p$ in the $k,p$ index plane, to summing over the northwest-southeast diagonals $k+p=n$ in the $k,p$ index plane to obtain Line (5.47). We next relabeled $p = n - k$, since we have $k + p = n$ on the $n$th diagonal, to get Line (5.48). We then multiplied up and down by $n!$ in Line (5.49) to obtain the binomial coefficients in Line (5.50). Finally we used the binomial expansion to obtain Line (5.51). □

We are now in a position to obtain the linearization of $\exp(x)$ at the point $p$, and find the derivative $\exp'(p)$.

### ■ Theorem 5.16

The linearization of $\exp(x)$ at $p$ is $\exp(p) + \exp(p)[x - p]$. We conclude $\exp'(p) = \exp(p)$. Since this is true for all $p$ in $\mathbb{R}$, we have

$$\exp'(x) = \exp(x).$$

**Proof**

$$\begin{aligned}
\exp(x) &= \exp(p + [x-p]) = \exp(p) \cdot \exp([x-p]) \\
&= \exp(p)\{1 + [x-p] + HOT([x-p])\} \quad (5.52)\\
&= \exp(p) + \exp(p) \cdot [x-p] + \exp(p) \cdot HOT([x-p]).
\end{aligned}$$

Here we used the previous lemma to re-express $\exp(x)$ as $\exp(p) \cdot \exp([x-p])$. We then relied on the linearization of $\exp(x)$ in Eq. (5.45) at 0 to re-express $\exp([x-p])$ to get Eq. (5.52). We then multiplied through to conclude that the linearization is $\exp(p) + \exp(p) \cdot [x-p]$, and the derivative at $p$ is the coefficient $\exp(p)$ of $[x-p]$, yielding $\exp'(p) = \exp(p)$ as claimed. $\square$

The number $e$ is defined as

$$e = \exp(1) = \sum_{n=0}^{\infty} \frac{1}{n!} = 1 + 1 + \frac{1}{2!} + \frac{1}{3!} + \frac{1}{4!} + \cdots \approx 2.7182818\ldots.$$

By our property $\exp(a)\exp(b) = \exp(a+b)$, we have that for $m$ a positive integer, the $m$-fold product of the exponential of $1/m$ converts to the exponential of the $m$-fold sum of $1/m$. Since the $m$-fold sum of $1/m$ is 1, we have

$$\begin{aligned}
(\exp(1/m))^m &= \exp(1/m) \cdot \exp(1/m) \cdots \exp(1/m) \\
&= \exp(1/m + 1/m + \cdots + 1/m) = \exp(1) = e.
\end{aligned}$$

Since $(\exp(1/m))^m = e$, we conclude by taking $m$th roots that $\exp(1/m)$ is the $m$th root of $e$, or $\exp(1/m) = e^{1/m}$. For $n$ a positive integer, the $n$-fold product of $\exp(1/m)$ converts to exponential of the $n$-fold sum of $1/m$ to give

$$\begin{aligned}
(e^{1/m})^n &= (\exp(1/m))^n = \exp(1/m) \cdot \exp(1/m) \cdots \exp(1/m) \\
&= \exp(1/m + 1/m + \cdots + 1/m) = \exp(n/m) = e^{n/m}.
\end{aligned}$$

We have concluded that $\exp(n/m) = e^{n/m}$ for all positive rational numbers $n/m$. We observe $\exp(0) = 1$ from the definition of $\exp$, and thus $\exp(0) = e^0$. For negative rational powers $-n/m$ we have $1 = \exp(0) = \exp(n/m + (-n/m)) = \exp(n/m)\exp(-n/m)$, from which we conclude that $\exp(-n/m) = 1/\exp(n/m) = 1/e^{n/m} = e^{-n/m}$. We conclude that

$$\exp(x) = e^x, \quad (5.53)$$

for every rational number $x$. In this sense, $\exp(x)$ is a differentiable function defined on all of $\mathbb{R}$ extending $e^x$ defined for rational $x$. Until this point, it was unclear how to take a number to an irrational exponent. But Eq. (5.53) tells us how we can do this in a consistent way. For *any* real number $x$ we *define* $e^x$ to be $\exp(x)$. We conclude that:

■ **Theorem 5.17**

For all real numbers $x$, we take $\exp(x) = e^x$. Since $\exp'(x) = \exp(x)$ we conclude

$$(e^x)' = e^x.$$

The function $e^x$ should be one of your favorite functions to differentiate. There is no work to do, just leave it alone!

By the Chain Rule, the composition of exp with differentiable function $f(x)$ is differentiable, and we have

$$\exp(f(x))' = \exp'(f(x)) \cdot f'(x) = \exp(f(x)) \cdot f'(x). \quad (5.54)$$

Equivalently,

$$(e^{f(x)})' = e^{f(x)} \cdot f'(x).$$

Exponential functions are key functions in modeling many phenomena, including population growth in the early stages, continuously compounded investments, radioactive decay, and carbon dating, to name a few. We examine a few examples that will be helpful in this context.

### EXAMPLE 5.11

Differentiate each of: $1000 \cdot e^{.04x}$ ; $C \cdot e^{rx}$ ; $e^{-x^2/2}$. **Answer:**

$$(1000 \cdot e^{.04x})' = 1000 \cdot (e^{.04x})' = 1000 \cdot e^{.04x} \cdot (.04x)' = 1000 \cdot e^{.04x} \cdot (.04) = 40 e^{.04x}.$$

We have used Eq. (5.54) to complete our computation. This example could be used to model a population with initial population 1000 at time $x = 0$ years, with a continuous growth rate of .04 or 4% per year.

$$(C \cdot e^{rx})' = C \cdot (e^{rx})' = C \cdot e^{rx} \cdot (rx)' = C e^{rx} r.$$

> We again used the Chain Rule in the form of Eq. (5.54) here. This example could also be used in population modeling if the initial population at $x = 0$ were $C$ and the growth rate were $r$.
>
> $$(e^{-x^2/2})' = (e^{-x^2/2}) \cdot (-x^2/2)' = (e^{-x^2/2}) \cdot (-2x^1/2) = (e^{-x^2/2}) \cdot (-x).$$
>
> Again, Eq. (5.54) was used. This function is a constant times the standard normal distribution (or bell curve) in statistics, and is used to model probabilities. Later on we will be able to conclude from this computation that the mean (or center) of the standard normal distribution is 0, which can be also be seen from graphing this function on your calculator.

We point out that $\exp(x) = e^x > 0$ for all $x$. Because the derivative is positive, $(e^x)' = e^x > 0$, we shall soon see that $e^x$ is an increasing function. As such, it will have an inverse function denoted $\ln(x)$, defined for positive $x$, and called the natural logarithm. We conclude

$$e^{\ln(x)} = x,$$

for $x > 0$. Similarly,

$$\ln(e^x) = x,$$

for all $x$ in $\mathbb{R}$. We can also convert from an arbitrary base $b > 0$ to base $e$ via

$$b^x = e^{\ln(b^x)} = e^{x \cdot \ln(b)}.$$

We then obtain derivative formulas for arbitrary bases

$$(b^x)' = (e^{\ln(b^x)})' = (e^{x \cdot \ln(b)})' = e^{x \cdot \ln(b)} \cdot (x \cdot \ln(b))' = b^x \cdot \ln(b).$$

### ■ Exercises

1. Find the derivative of each function $f(x)$ as follows:
    (a) $100 \, e^{.02 \, x}$;
    (b) $x \, e^x$;
    (c) $10^x$;
    (d) $e^{-x^2}$; and
    (e) $\sin(x) \, e^{-2x}$.

2. Verify that for given constants $C, r > 0$ the function $P(t) = C\, e^{rt}$ satisfies the differential equation

$$P'(t) = r \cdot P(t),$$

by computing $P'(t)$ and simplifying it, through substitution, to the right-hand side. One concludes that the rate of change of $P$ is proportional to $P$, with constant of proportionality $r$. Discuss why this statement would be relevant to modeling the size of a population.

3. Given the function $A(t) = 1000\, e^{.04t}$, from Exercise (2) one knows that $A(t)$ satisfies the differential equation

$$A'(t) = .04\, A(t).$$

This function models how a bank would continuously compound interest on an investment of $1000 at a 4% annual return. In actuality 4% compounded continuously is slightly better than 4%, as you can see by computing $A(1) = 1000\, e^{.04}$. Evaluate $A(1)$ on your calculator and compare with 4% of $1000. This is due to the fact that one receives interest upon your accrued interest under continuous compounding.

## 5.8 The Derivative of an Inverse Function

If $f^{-1}(x)$ is the inverse function to the differentiable function $f(x)$, we know that $f(f^{-1}(x)) = (f \circ f^{-1})(x) = x$. If we assume for the moment that $f^{-1}(x)$ is differentiable, then by the Chain Rule we have

$$\left( f\left[f^{-1}(x)\right] \right)' = x'$$
$$f'\left[f^{-1}(x)\right] \cdot (f^{-1})'(x) = 1.$$

We conclude in this case, that when $f'(f^{-1}(x)) \neq 0$, that we know how to differentiate an inverse function $f^{-1}$ when we know how to differentiate $f$, by the following formula:

$$(f^{-1})'(x) = \frac{1}{f'\left[f^{-1}(x)\right]}. \tag{5.55}$$

We see that $f^{-1}$ is differentiable at $p$ in the following way. We know $q = f^{-1}(p)$ is equivalent to $f(q) = f(f^{-1}(p)) = p$. The tangent line to $f$ at $q$ is $y = f(q) +$

$f'(q)[x-q]$. Solving for $x$ yields the linearization for $f^{-1}$ in terms of $y$. That is, if we reflect $f$ about the line $y = x$ the tangent line to $f$ at $q$ reflects about $y = x$ to be the tangent line to $f^{-1}$ at $f(q) = p$. In solving for $x$ we obtain $x = q + (1/f'(q))[y - f(q)] = f^{-1}(p) + (1/f'(f^{-1}(p)))[y - p]$, and conclude that $f^{-1}$ is differentiable at $p$ with derivative $1/f'(f^{-1}(p))$, provided this denominator is not 0.

As an immediate application of Eq. (5.55) we can find the derivative of the inverse function $f^{-1}(x) = x^{1/n}$ the $n$th root of $x$ for $f(x) = x^n$ (where we pick the positive $n$th root when $n$ is even and then require $x > 0$ in the even case). Then Eq. (5.55) becomes

$$\begin{aligned}(f^{-1})'(x) &= (x^{1/n})' = \frac{1}{f'(f^{-1}(x))} = \frac{1}{n\,(f^{-1}(x))^{n-1}} \\ &= \frac{1}{n(x^{1/n})^{n-1}} = \frac{1}{n(x^{1-1/n})} = \frac{1}{n}x^{(1/n)-1},\end{aligned} \qquad (5.56)$$

and the Power Rule holds for fractional powers $1/n$.

■ **Theorem 5.18**

The Power Rule holds for rational exponents

$$\begin{aligned}(x^{1/n})' &= \frac{1}{n}x^{\frac{1}{n}-1} \\ (x^{m/n})' &= \frac{m}{n}x^{\frac{m}{n}-1}.\end{aligned}$$

**Proof** We know $(x^{1/n})' = (1/n)x^{(1/n)-1}$ holds by the remarks leading up to our theorem. To handle the case that the exponent is $m/n$ we rely on the fact that $x^{m/n} = (x^{1/n})^m$ and use our Chain Rule in combination with our Root Rule.

$$\begin{aligned}(x^{m/n})' &= ((x^{1/n})^m)' = m \cdot (x^{1/n})^{m-1} \cdot (x^{1/n})' \\ &= m \cdot x^{(m/n)-(1/n)} \cdot (1/n)x^{(1/n)-1} \\ &= (m/n)\,x^{(m/n)-(1/n)+(1/n)-1} = (m/n)\,x^{(m/n)-1}.\end{aligned}$$

□

We apply our technique when $f(x) = e^x$ and $f^{-1}(x) = \ln(x)$.

## 5.8 The Derivative of an Inverse Function

■ **Theorem 5.19**

$$(\ln(x))' = \frac{1}{x}.$$

**Proof** $x = e^{\ln(x)}$ implies that $x' = 1 = e^{\ln(x)} \cdot (\ln(x))' = x \cdot (\ln(x))'$. Thus $1 = x \cdot (\ln(x))'$, which gives $1/x = (\ln(x))'$. □

For $r$ any real number, we make the definition

$$x^r = e^{r \ln(x)},$$

in which case we have the most general form of the Power Rule:

■ **Theorem 5.20**
**General Power Rule**
For $r$ any real number

$$(x^r)' = r\, x^{r-1}.$$

**Proof** We have:

$$\begin{aligned}
(x^r)' &= (e^{r \ln(x)})' \\
&= (e^{r \ln(x)})\, (r \ln(x))' \\
&= (e^{r \ln(x)})\, r\, (1/x) \\
&= r\, (e^{r \ln(x)})\, (e^{-\ln(x)}) \\
&= r\, (e^{(r-1) \ln(x)}) = r\, x^{r-1}.
\end{aligned}$$

□

For the two most important inverse trigonometric functions, we have

■ **Theorem 5.21**

$$(\sin^{-1}(x))' = \frac{1}{\sqrt{1-x^2}}$$
$$(\tan^{-1}(x))' = \frac{1}{1+x^2}.$$

**Proof** For the arcsine function $\sin^{-1}(x)$, Eq. (5.55) becomes our starting point as Line (5.57):

$$(\sin^{-1}(x))' = \frac{1}{\sin'((\sin^{-1}(x))} \qquad (5.57)$$
$$= \frac{1}{\cos((\sin^{-1}(x))}$$
$$= \frac{1}{\sqrt{1-x^2}}, \qquad (5.58)$$

where we relied on the Pythagorean Theorem in Line (5.58) to obtain the identity $\cos(\sin^{-1}(x)) = \sqrt{1-x^2}$.

The arctangent function $\tan^{-1}(x)$ is handled similarly. To begin, Line (5.55) becomes Line (5.59):

$$(\tan^{-1}(x))' = \frac{1}{\tan'((\tan^{-1}(x))} \qquad (5.59)$$
$$= \frac{1}{\sec^2((\tan^{-1}(x))}$$
$$= \frac{1}{(\sqrt{1+x^2})^2} \qquad (5.60)$$
$$= \frac{1}{1+x^2},$$

where we relied on the Pythagorean Theorem in Line (5.60) to obtain the identity $\sec(\tan^{-1}(x)) = \sqrt{1+x^2}$. □

More generally, a differentiable function is invertible around $x = x_0$ if $f'(x_0) \neq 0$. In this case, the linear approximation of the inverse, around the point $y_0 \equiv f(x_0)$ is

$$L(f^{-1}; y_0; y) \equiv x_0 + \frac{1}{f'(x_0)} \cdot (y - y_0). \tag{5.61}$$

---

**Definition 5.3** For a differentiable function $f : \mathcal{C}^1(\mathcal{D}; \mathbb{R})$, the points $x \in \mathbb{R}$ where $f'(x) = 0$ or where $f'$ is undefined, or is $\pm\infty$, are called *critical points*. The complement of these points are *regular* points of $f(x)$.

---

A function $f(x)$ is invertible at its regular points. In general, unique invertibility requires only monotonicity. For example $x^3$ has the inverse $\sqrt[3]{x}$, even though $x = 0$ is a critical point for both functions. This is because $x^3$ is monotonically increasing on $\mathbb{R}$. Conversely, $x^2$ does not have a unique inverse and depending on the application, one must choose between the positive branch $\sqrt{x}$ or the negative branch $-\sqrt{x}$, on $\mathbb{R}_0^+$.

The exponential function $\exp(x) = e^x$ is monotonically increasing on all of $\mathbb{R}$ and has the unique inverse $\ln(x)$, which is defined for $x \in \mathbb{R}^+$. The derivatives of both functions are positive on their domains.

The sine function $f(x) = \sin(x)$ has the critical points $Crit(\sin(x)) \equiv \{n \cdot \pi + \pi/2\}_{n \in \mathbb{Z}}$. The *principle branch* of $\sin(x)$ is defined as a differentiable function on the open interval

$$\mathcal{D}(\arcsin(x)) = I_{\arcsin} \equiv (\sin(-\pi/2), \sin(\pi/2)) = (-1, 1),$$

using the convergent series

$$\arcsin(x) = \sum_{n=0}^{\infty} \frac{(2 \cdot n)!}{(n!)^2} \cdot \frac{x^{2 \cdot n+1}}{4^n (2 \cdot n + 1)}, \quad \forall |x| < 1. \tag{5.62}$$

The intervals between adjacent critical points represents a different *branch* of the inverse function

$$\arcsin |_{I_{\arcsin}+2\cdot n}(x) = n \cdot \pi + (-1)^n \cdot \arcsin(x - 2 \cdot n). \tag{5.63}$$

Similar complications arise in trying to define the other inverse trigonometric functions. However, if a linear approximation will suffice, then only the derivative of the inverse function is needed. This can be found using the formula

$$\frac{df^{-1}(x)}{dx} = \frac{1}{f'\left[f^{-1}(x)\right]}, \tag{5.64}$$

which seems to require knowledge of the inverse function. Often this can be obtained by modifying the form of $f(x)$ so that the identity $f\left[f^{-1}(x)\right] = x$ can be used.

### ■ Exercises

1. Find the derivative of each function $f(x)$ as follows:

    (a) $x \, \sin^{-1}(x)$;
    (b) $\sin^{-1}(2x)$;
    (c) $\ln(x) \, \tan^{-1}(x)$;
    (d) $\ln(x^4 + 5x^2 + 1)$; and
    (e) $(3x^4 + 5x + 7)^{4/9}$.

## 5.9 A Table of Derivatives

The following is a compilation of derivative formulas:

$$(x^r)' = rx^{r-1}$$
$$((f(x))^r)' = r(f(p))^{r-1} \cdot f'(x)$$
$$\sin'(x) = \cos(x)$$
$$\cos'(x) = -\sin(x)$$
$$\tan'(x) = \sec^2(x)$$
$$\cot'(x) = -\csc^2(x)$$
$$\sec'(x) = \sec(x)\tan(x)$$
$$\csc'(x) = -\csc(x)\cot(x)$$
$$(e^x)' = e^x$$
$$(b^x)' = b^x \ln(b)$$
$$(\ln(x))' = \frac{1}{x}$$
$$(\sin^{-1}(x))' = \frac{1}{\sqrt{1-x^2}}$$
$$(\tan^{-1}(x))' = \frac{1}{1+x^2}$$
$$(f(x)g(x))' = f'(x)g(x) + f(x)g'(x)$$
$$\left(\frac{f(x)}{g(x)}\right)' = \frac{f'(x)g(x) - f(x)g'(x)}{g^2(x)}$$
$$(g(f(x)))' = g'(f(x)) \cdot f'(x)$$

## 5.10 The Differential of a Function

If we move along the $x$-axis from the point $x$ to the point $\hat{x}$, the symbol $dx$ represents the change in the independent variable $dx = \Delta x = \hat{x} - x$, for $\hat{x}$ near $x$.

The symbol $df$ represents the rise of the tangent line at the point $x$ corresponding to a run of $dx$. Since our slope at $x$ is $f'(x)$ and our run is $dx = \hat{x} - x$, the "rise = slope times run" formula gives our *differential of f* at $x$,

$$df = f'(x) \cdot dx,$$

due to the change in the independent variable $dx$.

If we let $\Delta f = f(\hat{x}) - f(x)$ be the corresponding change in the dependent variable, $f$, we have that for small $dx$,

$$\Delta f = f(\hat{x}) - f(x) \simeq df.$$

This is equivalent to saying that the linear approximation of $f$ is close to $f$ near $x$: $\Delta f = f(\hat{x}) - f(x) \simeq df = f'(x)dx = f'(x)[\hat{x} - x]$, which is equivalent to $f(\hat{x}) \simeq f(x) + f'(x)[\hat{x} - x] = L(f; x; \hat{x})$.

This discussion results in an alternative expression for the derivative of $f$, which is due to the German mathematician Leibniz.

$$\frac{df}{dx} = f'(x) \iff df = f'(x) \cdot dx, \tag{5.65}$$

where $df$ is the *differential* of the function $f$ at $x$. Use of the differential will be important in area computations because it has elegantly built the Chain Rule into the differential notation in a consistent manner. We will come back to this point in our study of integral calculus.

The linearization of $f(\hat{x})$ at the point $x$ is $f(\hat{x}) = f(x) + f'(x)[\hat{x} - x] + HOT([\hat{x} - x])$, which gives

$$\frac{f(\hat{x}) - f(x)}{[\hat{x} - x]} = f'(x) + \frac{HOT([\hat{x} - x])}{[\hat{x} - x]}.$$

In this setting, since $HOT([\hat{x} - x])/[\hat{x} - x]$ tends to 0 as $\hat{x}$ tends to $x$, we conclude

$$\lim_{\Delta x \to 0} \frac{\Delta f}{\Delta x} = \lim_{\hat{x} \to x} \frac{f(\hat{x}) - f(x)}{[\hat{x} - x]} = \lim_{\hat{x} \to x} \left( f'(x) + \frac{HOT([\hat{x} - x])}{[\hat{x} - x]} \right)$$
$$= f'(x) + \lim_{\hat{x} \to x} \frac{HOT([\hat{x} - x])}{[\hat{x} - x]} = f'(x) + 0 = f'(x).$$

This says that the change in $f$ per change in $x$ approaches $\frac{df}{dx}(x) = f'(x)$, and one sees that we *convert from Greek to English* in the Leibniz notation when computing the derivative:

$$\lim_{\Delta x \to 0} \frac{\Delta f}{\Delta x} = \frac{df}{dx}(x) = f'(x). \tag{5.66}$$

The simpler $f'(x)$ is inherited from Newton, while the more detailed Leibniz notation $\frac{df}{dx}(x)$ indicates the derivative as a limit of change of the dependent variable $f$ per change in the independent variable $x$.

## EXAMPLE 5.12

Find the differential of the function $u(x) = x^n$. **Answer:** For $u(x) = x^n$, we have $du = d(u(x)) = d(x^n) = (x^n)'dx = nx^{n-1}dx$.

Find the differential of the function $g(u) = \sin(u)$. **Answer:** For $g(u) = \sin(u)$, we have $dg = d(g(u)) = d(\sin(u)) = (\sin(u))'du = \cos(u)du$.

Find the differential of $g(u(x)) = \sin(x^n)$, and compare it to the preceding expressions. **Answer:**

$$d(g(u(x))) = (g(u(x)))'dx = g'(u(x))u'(x)dx = \cos(x^n) \cdot nx^{n-1}dx = \cos(u)du = d(g(u)), \tag{5.67}$$

where we substituted $u = u(x)$, with corresponding differential $du = u'(x)dx$, to see the last equations of (5.67).

We note that the differential with respect to $x$ is consistent with the differential with respect to $u$ under substitution via the Chain Rule, and $d(g(u(x))) = d(g(u))$, so we refer to each of these as $dg$.

## Exercises

1. Find the differential $df$ of each function $f(x)$ as follows:

   (a) $x^5$;
   (b) $e^{2x}$;
   (c) $\sec(x)$;
   (d) $\sin(x)$; and
   (e) $e^{x^3}$.

2. Find $\frac{df}{dx}(x)$ for the following functions, and compare these answers with your answers to the previous exercise.

   (a) $x^5$;
   (b) $e^{2x}$;
   (c) $\sec(x)$;
   (d) $\sin(x)$; and
   (e) $e^{x^3}$.

## 5.11 Continuous Differentiability and Higher Derivatives

If a function $f : I \to \mathbb{R}$ is continuous, this is expressed as $f \in \mathcal{C}^0(I;\mathbb{R})$. If the slope function, or derivative, $f' : I \to \mathbb{R}$ associated with $f$ is also continuous, then the expression $f \in \mathcal{C}^1(I;\mathbb{R})$ is used.

If a function $f$ can be continuously differentiated twice in succession then we say it has a second derivative and write $f \in \mathcal{C}^2(I;\mathbb{R})$.

**Definition 5.4** The *second derivative* of a continuous function $f \in \mathcal{C}^0(I;\mathbb{R})$ has several alternative expressions

$$(f')'(x) = f''(x) = f^{(2)}(x) = \left(\frac{d^2 f}{dx^2}\right)_x, \tag{5.68}$$

and if $f'' : I \to \mathbb{R}$ is continuous, then $f \in \mathcal{C}^2(I;\mathbb{R})$. For any $k \in \mathbb{N}$, the space of $k$th *continuously differentiable functions* is

$$\mathcal{C}^k(I;\mathbb{R}) = \{f \in \mathcal{C}^{k-1}(I;\mathbb{R}) \mid f^{(k)} \in \mathcal{C}^0(I;\mathbb{R})\}. \tag{5.69}$$

### EXAMPLE 5.13

Find the second derivative $f''(x) = f^{(2)}(x)$ and the third derivative $f^{(3)}(x)$ of the function $f(x) = x^4$. **Answer:** The (first) derivative is $f'(x) = (x^4)' = 4x^3$. The second derivative is $f''(x) = (f'(x))' = ((x^4)')' = (4x^3)' = 12x^2$. And the third derivative is $(f''(x))' = (12x^2)' = 24x$.

Higher derivatives, especially second derivatives, have an important role in the understanding of functions and in applications in physics. If we let $x = t$ be time, and let our function $s(t)$ be the position of a moving object at time $t$, then the derivative $s'(t)$ as the slope of the tangent line is the instantaneous rate of change of $s$ with respect to $t$. We then see that the first derivative of position is the instantaneous velocity $v(t) = s'(t)$ of the object at time $t$. The second derivative of $s$ is $s'' = (s')' = v'$, which is our instantaneous rate of change of velocity, better known as acceleration. Thus acceleration and force = mass × acceleration can both be expressed in terms of second derivatives of position.

## 5.11 Continuous Differentiability and Higher Derivatives

### EXAMPLE 5.14

If the height above the ground of a falling ball is given by $s(t) = -4.9t^2 + 40t + 60$ meters at time $t$ seconds, find the velocity $v(t)$ and the acceleration $a(t)$ of the ball at time $t$.

**Answer:** The height above ground gives us the position of the ball. As such the velocity of the ball is given by $v(t) = s'(t) = (-4.9t^2 + 40t + 60)' = -9.8t + 40$ m/sec. And the acceleration of the ball is given by $a(t) = s''(t) = v'(t) = (-9.8t + 40)' = -9.8$ m/sec$^2$. Acceleration due to gravity on Earth is $-9.8$ m/sec$^2$, and, as we shall see when we learn to reverse the process of computing a derivative, the height $s(t) = -4.9t^2 + 40t + 60$ corresponds to the height of a ball thrown straight up at 40 m/sec from a height of 60 m above ground (on Earth).

In examining the graph of a function $f$, the first derivative $f'(x)$ gives the slope of the tangent line to $f$ at $x$, or the instantaneous rate of change of $f$ with respect to $x$. The second derivative $f''$ of $f$ is the derivative of $f'$, which is the derivative of the slope. Thus $f''$ is the the instantaneous rate of change of the slope of the curve, and in turn this tells us the rate of bending or the concavity of our graph.

We return to our space of $k$ continuously differentiable functions to mention their mathematical properties.

### ■ Theorem 5.22

For each $k \in \mathbb{N}_0$ and interval $I = [a,b]$ for $a < b$, the spaces $\mathcal{C}^k(I;\mathbb{R})$ is a vector space over the field $\mathbb{R}$. Furthermore, for each $\ell > k$, $\mathcal{C}^\ell(I;\mathbb{R})$ is a subspace of $\mathcal{C}^k(I;\mathbb{R})$.

**Proof** Since, by definition $\mathcal{C}^\ell(I;\mathbb{R}) \subset \mathcal{C}^k(I;\mathbb{R})$, it is sufficient to show that $\mathcal{C}^\ell(I;\mathbb{R})$ is closed under linear combination. This requires the following:

### ■ Lemma 5.4
For every $f, g \in \mathcal{C}^1(I;\mathbb{R})$ and $a, b \in \mathbb{R}$, the *Linear Combination Rule* holds for differentiation, in that

$$(a \cdot f(x) + b \cdot g(x))' = a \cdot f'(x) + b \cdot g'(x),$$

and this new function is in $\mathcal{C}^0(I;\mathbb{R})$.

The remaining vector space properties follow from those of $\mathcal{C}^0(I;\mathbb{R})$. □

## Exercises

1. Given the polynomial $f(x) = x^3 + 4x^2 + 9x + 2$,

    (a) find the $k$th derivatives $f^{(k)}(x)$ for $k = 1, 2, 3, 4$.
    (b) decide what $f^{(k)}(x)$ is for $k \geq 4$. Is $f(x)$ infinitely differentiable?

2. Given the height in meters above the ground of an object falling on Earth is $s(t) = -4.9t^2 + v_0 t + s_0$ at time $t$ seconds,

    (a) find the position $s(0)$ at time 0;
    (b) find the velocity $v(t) = s'(t)$;
    (c) find the velocity $v(0) = s'(0)$ at time 0; and
    (d) find the acceleration $a(t) = v'(t) = s''(t)$ for all times $t$.

## 5.12 Optimization

A function $f(x)$ defined near $x_0$ has a *local maximum* at $x_0$ if $f(x) \leq f(x_0)$ for all $x$ near $x_0$. For instance $f(x) = 1 - x^2$ has a local maximum at 0 since $f(x) = 1 - x^2 \leq 1 = f(0)$ for all $x$ near 0. The function $f(x)$ defined on a set $D$ has a *global maximum* at $x_0$ if $f(x) \leq f(x_0)$ for all $x$ in $D$. In our example, $f(x) = 1 - x^2$, $f$ has a global maximum at 0 (in fact, every global maximum is a local maximum).

A function $f(x)$ defined near $x_1$ has a *local minimum* at $x_1$ if $f(x) \geq f(x_1)$ for all $x$ near $x_1$. For instance, $f(x) = x^2$ has a local minimum at 0 since $f(x) = x^2 \geq 0 = f(0)$ for all $x$ near 0. The function $f(x)$ defined on a set $D$ has a *global minimum* at $x_1$ if $f(x) \geq f(x_1)$ for all $x$ in $D$. In our example, $f(x) = x^2$, $f$ has a global minimum at 0 (in fact, every global minimum is a local minimum).

Either a local maximum or a local minimum is referred to as a *local extremum*, and a global maximum or global minimum is referred to as a *global extremum*. The process of finding global extrema is known as *optimization*.

If $f$ has a local extremum at a point $p$, then there are two possibilities. Either $f$ is not differentiable at $p$ and then $f'(p)$ does not exist, or $f$ is differentiable at $p$ and then $f'(p)$ does exist and must be 0. Points $q$ with $f'(q) = 0$ or $f'(q)$ undefined are called *critical points*.

### Theorem 5.23

Let $f(x)$ be defined for all $x$ in an open interval about $p$. If $f$ has a local extremum at $p$, then either $f'(p)$ does not exist, or $f'(p) = 0$. Thus $p$ is a critical point of $f$.

**Proof** Without any loss of generality, let's assume we have a local maximum at $p$. If $f'(p)$ does not exist, we are done, as this is one of the cases in our theorem. If $f'(p)$ exists at this local maximum, we rule out every case except $f'(p) = 0$.

Assume that $f'(p)$ were not 0. Using the linearization of $f$ at $p$ we have $f(x) = f(p) + f'(p)[x - p] + HOT([x - p])$. Since the higher-order terms satisfy that ratio $HOT([x-p])/[x-p]$ approaches 0, for $x$ near $p$, we have that $|HOT([x-p])/[x-p]| \leq |f'(p)|/2 \neq 0$ for all $x$ sufficiently close to $p$. We conclude that $|HOT([x-p])| \leq |f'(p)||[x-p]|/2$ for $x$ in an interval $I$ about $p$.

If $f'(p) > 0$, then for $x > p$ in $I$, we have $f'(p)[x-p] > 0$ and $HOT([x-p]) > -f'(p)[x-p]/2$. Adding $f'(p)[x-p]$ to both sides of the last inequality gives that $f'(p)[x-p] + HOT([x-p]) > f'(p)[x-p]/2 > 0$. Adding $f(p)$ to all sides gives $f(x) = f(p) + f'(p)[x-p] + HOT([x-p]) > f(p) + f'(p)[x-p]/2 > f(p)$ contradicting that we have a local maximum at $p$.

If $f'(p) < 0$, then for $x < p$ in $I$, we have $f'(p)[x-p] > 0$ and $HOT([x-p]) > -f'(p)[x-p]/2$. Adding $f'(p)[x-p]$ to both sides of the last inequality gives that $f'(p)[x-p] + HOT([x-p]) > f'(p)[x-p]/2 > 0$. Adding $f(p)$ to all sides gives $f(x) = f(p) + f'(p)[x-p] + HOT([x-p]) > f(p) + f'(p)[x-p]/2 > f(p)$ contradicting that we have a local maximum at $p$.

So it must be true that $f'(p) = 0$ as was claimed. □

Our preceding theorem gives us a strategy for optimizing a function on a closed bounded interval $I = [a, b]$: find all critical points $p_1, p_2, \ldots, p_k$ in the interval, add in the endpoints $a, b$, compute and compare the values of $f(p_1), f(p_2), \ldots, f(p_k), f(a), f(b)$ choosing the largest to maximize and the smallest to minimize. Thus we have used calculus to reduce optimization to the algebraic process of solving $f'(x) = 0$ or undefined.

### EXAMPLE 5.15

Find the largest and smallest values of $y = f(x) = x^3 - 3x$ on the interval $[0, 2]$.

**Answer:** We find $f'(x) = 3x^2 - 3$ and we ask ourselves where this is undefined or 0. Since a polynomial is always defined, we only need solve $0 = 3x^2 - 3 = 3(x-1)(x+1)$ to obtain the critical points $x = \pm 1$. Since $-1$ is not in our interval $[0, 2]$ we disregard it and check our values of $f$ at the critical point(s) 1 in the interval and at the endpoints 0, 2. We have $f(1) = -2, f(0) = 0,$ and $f(2) = 2$. We conclude that the maximum value of $f(x) = x^3 - 3x$ is $y = 2$ when $x = 2$ and the minimum value is $y = -2$ when $x = 1$.

Find the largest and smallest values of $y = f(x) = x^{2/3}$ on the interval $[-1, 2]$.

**Answer:** We find $f'(x) = (2/3)x^{-1/3} = 2/\{3x^{1/3}\}$ and we ask ourselves where this is undefined or 0. Since in simplest form, the numerator is always non-vanishing (since it is 2), we have no solutions to $f'(x) = 0$ and we must decide when $f'(x)$ is undefined. This will correspond to the case when our denominator vanishes, so we solve $x^{1/3} = 0$ to obtain $x = 0$ as our only critical point in our interval $[-1, 2]$. We now check our values of $f$ at the critical point(s) 0 in the interval and at the endpoints $-1, 2$. We have $f(0) = 0, f(-1) = 1$, and $f(2) = 2^{2/3}$. We conclude that the maximum value of $f(x) = x^{2/3}$ is $y = 2^{2/3}$ when $x = 2$ and the minimum value is $y = 0$ when $x = 0$.

It is possible to now deploy optimization techniques to solve any problem concerned with a largest or smallest outcome. The goal is to model the problem with a differentiable function $f$ on a closed interval $[a, b]$, and then find critical points in $[a, b]$, comparing the value of $f$ at these critical points with $f(a), f(b)$ to obtain optimal values in applied settings.

### EXAMPLE 5.16

A gardener wishes to build a small rectangular garden with one side lying along her house. She has 40 m of flexible fencing to place on three sides of her garden to keep out rabbits, with the house protecting the fourth side. What are the dimensions of the garden she can place along her house with largest area?

**Answer:** We let $x$ be the length of the sides of the garden perpendicular to the house, and $y$ be the length of the side of the garden parallel to the house. The limit of 40 m of fencing gives $2x + y = 40$, from which one sees that $y = 40 - 2x$. The area of the garden is $A = xy = x(40 - 2x) = 40x - 2x^2$. Since $0 \leq x$ and $0 \leq 40 - 2x$, the constraints on $x$ are $0 \leq x \leq 20$. Thus the problem is posed on the interval $[0, 20]$. The endpoints are then 0 and 20. Critical points are obtained by solving $A'(x) = 40 - 4x = 0$ (there is no need to search for undefined derivatives as $A$ is a polynomial differentiable everywhere). Hence $x = 10$ is the only critical point in $[0, 20]$. Comparing $A(0) = 0, A(10) = 10(40 - 20) = 200, A(20) = 0$, one concludes that the garden has maximal area of 200 m$^2$ with dimensions $x = 10$ m by $y = 20$ m.

Continuity is a desired property to establish for a function because it can be used to guarantee the existence of an absolute maximum point and an absolute minimum point for a continuous function on a closed bounded interval $I = [a, b]$, as is proven in advanced calculus [59]. Assuming the existence of absolute extrema of a continuous function on a closed bounded interval $I$, we can prove Rolle's theorem:

### ■ Theorem 5.24

#### Rolle's Theorem

Let $f(x)$ be continuous on $[a, b]$ and differentiable on $(a, b)$, with $f(a) = f(b)$. Then there is a point $c$ with $a < c < b$ with $f'(c) = 0$.

**Proof** If $f(x)$ is a constant function, then we are done, as the derivative of $f$ is 0 throughout $(a,b)$ and we pick any $c \in (a,b)$ and $f'(c) = 0$. If $f$ is not a constant function, then there is a point $d \in (a,b)$ with $f(d) \neq f(a) = f(b)$.

If $f(d) < f(a)$, then the absolute minimum occurs at a point $p \neq a,b$, as we know that at the minimum $f(p) \leq f(d) < f(a) = f(b)$. Thus $p \in (a,b)$. We conclude $f$ is differentiable at $p$ by our hypotheses, and then we must have $f'(p) = 0$ by the preceding theorem. We then take $c = p$.

If $f(d) > f(a)$, then the absolute maximum occurs at a point $q \neq a,b$, as we know that at the maximum $f(q) \geq f(d) > f(a) = f(b)$. Thus $q \in (a,b)$. We conclude $f$ is differentiable at $q$ by our hypotheses, and then we must have $f'(q) = 0$ by the preceding theorem. We then take $c = q$. □

### Exercises

1. Given the function $f(x) = x^3 - 3x$ on the interval $[0, 2]$,

   (a) find the maximum value that $f$ attains on $[0, 2]$; and
   (b) find the minimum value that $f$ attains on $[0, 2]$.

2. An object is thrown straight up in the air. It satisfies the equation of motion that the height above ground in meters is $s(t) = -4.9t^2 + 29.4t + 25.7$ after $t$ seconds.

   (a) Find the time $t$ at which $s(t)$ reaches a maximum value.
   (b) How high does the object get?
   (c) How long is the object in the air?

3. Revenue from the sale of a small appliance is computed as the price times the number sold. Let $x$ be the number sold. Economic forecast says that the price will drop as more of the appliances come on the market according to the predicted formula $p(x) = \$(100 - .2x)$. Find the number of appliances to be sold that gives the largest revenue.

## 5.13 The Mean Value Theorem and its Consequences

We use Rolle's theorem to prove an equivalent theorem called the Mean Value theorem. The Mean Value theorem says that a continuous function $f$, which is also differentiable between two points $a$ and $b$, has at least one point on the graph where the tangent line is parallel to the secant line connecting the point $(a, f(a))$ and $(b, f(b))$. Since parallel lines have the same slope, we conclude that there is some intermediate point $c$ where the tangent slope $f'(c)$ must equal the secant slope $(f(b) - f(a))/(b-a)$.

### ∎ Theorem 5.25
### The Mean Value Theorem

Let $f(x)$ be continuous on $[a,b]$ and differentiable on $(a,b)$, then there is a point $c$ with $a < c < b$ with $f'(c) = (f(b) - f(a))/(b-a)$.

**Proof** Let $L(x)$ be the line through the points $(a, f(a))$ and $(b, f(b))$. The slope of $L$ is then $(f(b) - f(a))/(b-a)$. Observe that $L(a) = f(a)$ and $L(b) = f(b)$. Now $L(x)$ is differentiable and continuous on all of $\mathbb{R}$. This gives that $g(x) = f(x) - L(x)$ is continuous on $[a,b]$ and differentiable on $(a,b)$. Now $g(a) = f(a) - L(a) = 0$ and $g(b) = f(b) - L(b) = 0$. Apply Rolle's theorem to $g$ to obtain a point $c \in (a,b)$ with $g'(c) = 0$. But $g'(c) = f'(c) - L'(c) = 0$. Thus $f'(c) = L'(c) = (f(b) - f(a))/(b-a)$, as the derivative of the line $L$ is always the slope of $L$. □

The Mean Value theorem has three important consequences. The first is a relation between the sign of the derivative and the rising or falling of a function. We say $f(x)$ is *increasing* on an interval $I$ if for all choices of $x_1 < x_2$ in $I$ we have $f(x_1) < f(x_2)$. That is $f$ rises as one moves to the right. We say $f(x)$ is *decreasing* on an interval $I$ if for all choices of $x_1 < x_2$ in $I$ we have $f(x_1) > f(x_2)$. In this case, $f$ falls as one moves to the right.

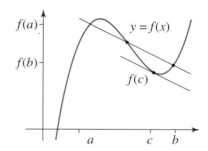

**FIGURE 5.7** ∎ A smooth function $f(x)$ with slope $f'(c) = (f(b) - f(a))/(b-a)$.

If a differentiable function $f$ has a positive derivative, then its linearization has positive slope and is an increasing function, and the tangent line increases to the right. Since the function $f$ hugs its tangent line, $f$ should also be increasing. This intuition is verified in the next theorem.

### ∎ Theorem 5.26

If $f'(x) > 0$ for all $x$ in an interval $I$, then $f$ is increasing on $I$. If $f'(x) < 0$ for all $x$ in an interval $I$, then $f$ is decreasing on $I$.

## 5.13 The Mean Value Theorem and its Consequences

**Proof** Assume $f'(x) > 0$ for all $x$ in an interval $I$, and pick any $x_1 < x_2$ in $I$. Apply the Mean Value theorem to $f$ on the interval $[x_1, x_2]$ to find a point $c \in (x_1, x_2)$ with $(f(x_2) - f(x_1))/(x_2 - x_1) = f'(c)$. Since our derivative is positive on $I$, $f'(c) > 0$. We conclude that since $(x_2 - x_1) > 0$, we must have the numerator $(f(x_2) - f(x_1)) > 0$ too. Thus $f(x_2) > f(x_1)$ and $f$ is increasing. A similar approach shows $f$ is decreasing when $f' < 0$. □

### EXAMPLE 5.17

Find the intervals of increase and of decrease of $f(x) = x^3 - 3x$.

**Answer:** We compute $f'(x) = 3x^2 - 3 = 3(x-1)(x+1)$. We obtain the critical points by solving $f'(x) = 0 \iff x = \pm 1$. We check the sign of $f'$ on each of the intervals $(-\infty, -1)$, $(-1, 1)$, and $(1, \infty)$ determined by our critical points. By a sign analysis, or by substituting test points from each interval, we determine that $f'(x) > 0$ on $(-\infty, -1)$, $f'(x) < 0$ on $(-1, 1)$, and $f'(x) > 0$ on $(1, \infty)$. For instance, the factors $(x - 1)$ and $(x + 1)$ of $f$ are both negative on $(-\infty, -1)$, so their product and $f'$ are positive on $(-\infty, -1)$. Alternatively, if we pick a point, say $-2$ in $(-\infty, -1)$, then $f'(-2) = 9 > 0$ and $f'$ remains positive on this interval. This approach works for a continuously differentiable function $f$, such as our current polynomial. We do a similar factor sign analysis, or test point computation, on all the remaining intervals to determine the sign of $f'$ there. We conclude from the sign of $f'$ on each interval, that $f$ is increasing on the intervals $(-\infty, -1)$ and $(1, \infty)$, and $f$ is decreasing on the interval $(-1, 1)$.

From the preceding theorem, we deduce an important test to classify the behavior of a function $f$ at a critical point. This test is called the First Derivative Test.

### ■ Theorem 5.27
### First Derivative Test

Let $p$ be a critical point of a continuous function $f$, where $f(x)$ is differentiable for all points $x$ near $p$ except possibly at $p$ itself. If $f'(x) > 0$ for all $x < p$ near $p$ and if $f'(x) < 0$ for all $x > p$ near $p$, then $f$ has a local maximum at $p$. If $f'(x) < 0$ for all $x < p$ near $p$ and if $f'(x) > 0$ for all $x > p$ near $p$, then $f$ has a local minimum at $p$.

**Proof** In the first case, since $f'(x) > 0$ for $x < p$ we have $f$ is increasing to the left of $p$ until we reach $p$, then because $f'(x) < 0$ for $x > p$ we decrease after we pass through $p$. This tells us that we have a local maximum for $f$ at $p$. The reasoning in the second case is similar: we decrease to the left of $p$ and increase to the right of $p$, and so we must have a local minimum at $p$. $\square$

A second result that follows from the Mean Value theorem is the fact that any function with 0 derivative on an interval is constant on the interval.

### ■ Theorem 5.28

If $f'(x) = 0$ for all $x$ in an interval $I$, then $f(x) = k$ is a constant function on $I$.

**Proof** Let $x_0$ be fixed in $I$ and let $k = f(x_0)$. For any other point $x$ in $I$, apply the Mean Value theorem on the interval with $J$ endpoints $x_0$ and $x$ to obtain a point $c$ in $J$ with $(f(x) - f(x_0))/(x - x_0) = f'(c) = 0$ where the vanishing of the derivative is by hypothesis. Thus our numerator $(f(x) - f(x_0)) = 0$ and $f(x) = f(x_0) = k$ for all $x$ in $I$. $\square$

A third result following from the Mean Value theorem will be a cornerstone fact in reversing the process of computing a derivative. This will be crucial in the arena of integral calculus. It states that any two functions with the same derivative on an interval $I$ differ by a constant on $I$.

### ■ Theorem 5.29

Let $f(x)$ and $g(x)$ be differentiable on an interval $I$, with $f'(x) = g'(x)$ for all $x$ in $I$. Then there is a constant $C$ with $f(x) = g(x) + C$ for all $x$ in $I$.

**Proof** Let $H(x) = f(x) - g(x)$. Then $H$ is differentiable on $I$ by the Difference Rule, and $H'(x) = f'(x) - g'(x) = 0$, where the vanishing is by hypothesis. By our previous theorem there is a constant $C$ with $H(x) = f(x) - g(x) = C$ for all $x$ in $I$. We conclude that $f(x) = g(x) + C$ for all $x$ in $I$. $\square$

### EXAMPLE 5.18

Find all functions $f(x)$ having derivative $2x$. **Answer:** Since $g(x) = x^2$ has derivative $2x$ on $\mathbb{R}$, the preceding theorem tells us that all other functions $f$ with the same derivative $2x$ differ by a constant from $g$. Thus $f(x) = x^2 + C$ for any choice of $C$ gives the complete family of functions having derivative $f'(x) = 2x$ on $\mathbb{R}$.

### ■ Exercises

1. Find all functions $f(x)$ with $f'(x) = 3x^2$.
2. Given $f(x) = x^4 - 2x^2$ on $\mathbb{R}$,

   (a) find the intervals of increase of $f$;
   (b) find the intervals of decrease of $f$; and
   (c) find all local maxima and minima for $f$.

3. Given $f(x) = e^{-x^2/2}$ on $\mathbb{R}$,

   (a) find the intervals of increase of $f$;
   (b) find the intervals of decrease of $f$;
   (c) find all local maxima and minima for $f$.

### ■ 5.14 Newton's Method for the Zeros of a Function

In optimizing $f(x)$, one needs to find the zeros of a function $f'(x)$, as these will lie among our critical points. If solving $f'(x) = 0$ is not easy algebraically, then we resort to approximating these roots. This alone is a sufficient reason for finding roots of functions $g(x) = 0$.

Another place where finding roots is important is in finding equilibrium points for systems defined by the differential equation $f'(x) = g(x)$, which will be addressed later.

Suppose $g$ is continuous on $[a, b]$ with $g(a) < 0 < g(b)$. Then 0 is intermediate to $g(a), g(b)$, and by the Intermediate Value theorem there is a $c \in (a, b)$ where $g(c) = 0$. Thus finding a difference in the sign of $g$ at the endpoints implies a root between the endpoints. By cutting our interval in half repeatedly and checking signs at the resulting endpoints, we can home in on our root. Since we are cutting the interval in half at each stage we are bisecting.

The *Bisection Method* is an algorithm for finding $c$, which starts by checking the midpoint $(a+b)/2$. If $g((a+b)/2) = 0$, then $c = (a+b)/2$ gives our sought root. Otherwise, one of the two intervals $[a,(a+b)/2]$ or $[(a+b)/2,b]$ has endpoints with $g$ having opposite signs. Call the interval with opposite signs $[a_1, b_1]$ and repeat the process, always cutting the interval in half and picking a new interval of half the length with opposite signs for $g$ at each endpoint.

After $n$ steps the resulting interval $[a_n, b_n]$ must contain a zero $c$ of $g(x)$ and length of $[a_n, b_n]$ is $b_n - a_n = (b-a)/2^n$. Thus $a_n$ can be used to approximate the root $c$ to within an error of at most $(b-a)/2^n$. Although this method is slow, it is guaranteed to converge to a solution to $g(x) = 0$.

Alternatively, for *Newton's Method* we assume that $g$ is continuously differentiable on an interval $[a,b]$ with $g(a) < 0 < g(b)$. We will also assume that $g'(c) \neq 0$ at the root $c$ where $g(c) = 0$.

We start Newton's method with an $x_0$ chosen in $[a,b]$ where we guess $x_0$ as near $c$. The idea is to produce the tangent line to $g$ at $x_0$, and since the tangent line well approximates $g$ for points near $x_0$, the point where the tangent line crosses the $x$-axis should be very near the point where our curve $g$ crosses the $x$-axis. Since our tangent line is given by our linearization at $x_0$ we have $y = g(x_0) + g'(x_0)[x - x_0]$ is the equation of our tangent line. And this tangent line crosses the $x$-axis when $y = 0$, so we solve

$$0 = g(x_0) + g'(x_0)[x - x_0],$$

for $x$ to obtain

$$x = x_0 - \frac{g(x_0)}{g'(x_0)} = x_1,$$

where we have labeled $x_1$ as the crossing value of $x$. We then repeat the process using $x_1$ as our guess, to find an $x_2$ even closer to $c$, and continue the process producing a sequence of points $x_n$ homing in on the root $c$ very rapidly in general. If we reach $x_n$, we find the tangent line at $x_n$ as $y = g(x_n) + g'(x_n)[x - x_n]$, then set $y = 0 = g(x_n) + g'(x_n)[x_{n+1} - x_n]$ for the new point of crossing $x_{n+1}$ to obtain

$$x_{n+1} = x_n - \frac{g(x_n)}{g'(x_n)}.$$

The sequence $x_n$ approaches the root in most cases, and it does so very rapidly. However, if $g'(x_n)$ becomes close to 0 our small denominator can send us far away from $c$, sometimes near another root. This is why we presupposed that $g'(c) \neq 0$.

In applying Newton's Method, algebraic simplification of the term $x_n - (g(x_n))/(g'(x_n))$ is recommended. In practice, we truncate the process when the $x_n$ remain sufficiently close to each other.

### EXAMPLE 5.19

Approximate $c = \sqrt{2}$ using Newton's Method. **Answer:** Squaring both sides gives $c^2 - 2 = 0$, and we see $c$ is a root of the polynomial $g(x) = x^2 - 2$. We compute the derivative $g'(x) = 2x$, and find the expression

$$x - \frac{g(x)}{g'(x)} = x - \frac{x^2 - 2}{2x} = x - \left(\frac{x}{2} - \frac{1}{x}\right) = \frac{x}{2} + \frac{1}{x}.$$

If we let $x_0 = 1$ be our seed or first guess, then $x_1 = x_0/2 + 1/x_0 = 1/2 + 1 = 3/2$. We find $x_2 = x_1/2 + 1/x_1 = (3/2)/2 + 1/(3/2) = 3/4 + 2/3 = 17/12 \approx 1.4166\overline{6}$. We compare with $\sqrt{2} = 1.414213562\ldots$ and see we are within $.0024\ldots$. If we want even more accuracy, we compute $x_3 = x_2/2 + 1/x_2 = 17/24 + 12/17 = 577/408 = 1.414215686\ldots$ and we are now within $.00000212\ldots$. We have very rapid convergence indeed!

### ■ Exercises

1. Use two steps of the Bisection Method on the interval $[1, 2]$ to approximate $c = \sqrt{3}$. (*Hint:* Use the fact that $c^2 - 3 = 0$ to deduce the polynomial $g(x)$ in the Bisection Method.)

2. Use two steps of Newton's Method with initial guess $x_0 = 2$ (the right-hand endpoint of $[1, 2]$) to approximate $c = \sqrt{3}$. (*Hint:* use the fact that $c^2 - 3 = 0$ to deduce the polynomial $g(x)$ in Newton's Method.)

3. Use a calculator to approximate $\sqrt{3}$ to within eight decimal places, and compare your accuracy in Exercises (1) and (2).

### ■ 5.15 Linearization and L'Hopital's Rule

Suppose both $f(x)$ and $g(x)$ are continuously differentiable at $p$, then

$$f(x) = f(p) + f'(p)[x - p] + HOT_1([x - p]),$$

and

$$g(x) = g(p) + g'(p)[x-p] + HOT_2([x-p]),$$

where $\lim_{x \to p} HOT_i([x-p])/[x-p] = 0$ for $i = 1, 2$.

If we wish to examine the limiting value of the ration $f(x)/g(x)$ as $x$ approaches $p$, we are fine if $g(p) \neq 0$, as then $\lim_{x \to p} f(x)/g(x) = f(p)/g(p)$.

However if both $f(p) = 0 = g(p)$, we say the limit is in *indeterminate form* $0/0$. In this case, we are still able to understand the limiting behavior of $f(x)/g(x)$ at $p$ in terms of the derivatives $f'(p)$ and $g'(p)$ if $g'(p) \neq 0$. We utilize our linearizations to see this:

$$\frac{f(x)}{g(x)} = \frac{f(p) + f'(p)[x-p] + HOT_1([x-p])}{g(p) + g'(p)[x-p] + HOT_2([x-p])} = \frac{0 + f'(p)[x-p] + HOT_1([x-p])}{0 + g'(p)[x-p] + HOT_2([x-p])}$$
$$= \frac{[x-p](f'(p) + HOT_1([x-p])/[x-p])}{[x-p](g'(p) + HOT_2([x-p])/[x-p])} = \frac{(f'(p) + HOT_1([x-p])/[x-p])}{(g'(p) + HOT_2([x-p])/[x-p])}.$$

Since the higher-order term ratios in the last rightmost expression approach $0$ as $x$ approaches $p$, that is since $\lim_{x \to p} HOT_i([x-p])/[x-p] = 0$ for $i = 1, 2$, we have the last ratio approaches $f'(p)/g'(p)$. We conclude that when

$$\lim_{x \to p} f(x)/g(x) \text{ is in } \textit{indeterminate } \text{form} \frac{0}{0},$$

that

$$\lim_{x \to p} f(x)/g(x) = f'(p)/g'(p) = \lim_{x \to p} f'(x)/g'(x).$$

We conclude with a rule for limits of ratios in indeterminate form $0/0$ known as L'Hopital's Rule:

$$\lim_{x \to p} \frac{f(x)}{g(x)} = \lim_{x \to p} \frac{f'(x)}{g'(x)},$$

when $\lim_{x \to p} f'(x)/g'(x)$ exists.

### EXAMPLE 5.20

Find
$$\lim_{x \to 0} \frac{\sin(x)}{x}. \tag{5.70}$$

**Answer:** First, we observe that we are in indeterminate form $0/0$ since $\lim_{x \to 0} \sin(x) = 0$ and $\lim_{x \to 0} x = 0$. We then recall that $(\sin(x))' = \cos(x)$ and $x' = 1$. Thus by L'Hopital's Rule

$$\lim_{x \to 0} \frac{\sin(x)}{x} = \lim_{x \to 0} \frac{(\sin(x))'}{x'} = \lim_{x \to 0} \frac{\cos(x)}{1} = \frac{\cos(0)}{1} = 1.$$

### ■ Exercises

1. Use L'Hopital's Rule to evaluate each of the following indeterminate $0/0$ form limits:

   (a) $\lim_{x \to 0} (\cos(x) - 1)/x$.
   (b) $\lim_{x \to 1} (e^x - 1)/(x - 1)$.
   (c) $\lim_{x \to 9} (\sqrt{x} - 3)/(x - 9)$.

### ■ 5.16 Antiderivatives

If we differentiate a function $F(x)$ to get $F'(x) = f(x)$, we consider differentiation a forward moving process starting at $F$ and ending at $f$. We will concern ourselves with reversing this process, starting at $f$ and asking ourselves, "What functions differentiate to give $f$?" One answer is then clearly $F$, since $F' = f$. However there are also other answers; for instance, any function $G$ of form $G(x) = F(x) + C$ is also an answer because $G' = (F + C)' = F' + C' = f + 0 = f$. We know that any two functions $G$ and $F$ having the same derivative must differ by a constant, so all answers to the reverse process form the family of functions $G(x) = F(x) + C$. We call this reverse process of finding all functions with derivative a given function $f(x)$ the process of *antidifferentiation*. Our notation is

$$\int f(x)\,dx = F(x) + C,$$

and we emphasize that the following statements are equivalent:

$$F'(x) = f(x) \iff \int f(x)\,dx = F(x) + C.$$

In applying the antidifferentiation process to $f$ to get $\int f(x)dx = F(x)+C$, we call any such $F$ an *antiderivative* of $f$. We also call $\int f(x)dx$ the *indefinite integral* of $f(x)$.

Since every derivative formula now "clones" itself in the form of an equivalent antiderivative formula, and we obtain from our table of derivatives the following corresponding table of antiderivatives (or indefinite integrals). This is called an integral table.

$$(x^r)' = r\, x^{r-1} \iff \int x^r\, dx = \frac{x^{r+1}}{r+1} + C \text{ for } r \neq -1$$

$$((f(x))^r)' = r\,(f(p))^{r-1} \cdot f'(x) \iff \int (f(x))^r f'(x)\, dx = \frac{(f(x))^{r+1}}{r+1} + C$$

$$\sin'(x) = \cos(x) \iff \int \cos(x)\, dx = \sin(x) + C$$

$$\cos'(x) = -\sin(x) \iff \int \sin(x)\, dx = -\cos(x) + C$$

$$\tan'(x) = \sec^2(x) \iff \int \sec^2(x)\, dx = \tan(x) + C$$

$$\cot'(x) = -\csc^2(x) \iff \int \csc^2(x)\, dx = -\cot(x) + C$$

$$\sec'(x) = \sec(x)\,\tan(x) \iff \int \sec(x)\,\tan(x)\, dx = \sec(x) + C$$

$$\csc'(x) = -\csc(x)\,\cot(x) \iff \int \csc(x)\,\cot(x)\, dx = -\csc(x) + C$$

$$(e^x)' = e^x \iff \int e^x\, dx = e^x + C$$

$$(b^x)' = b^x\,\ln(b) \iff \int b^x\, dx = \frac{b^x}{\ln(b)} + C$$

$$(\ln(x))' = \frac{1}{x} \iff \int \frac{1}{x}\, dx = \ln(x) + C$$

$$(\sin^{-1}(x))' = \frac{1}{\sqrt{1-x^2}} \iff \int \frac{1}{\sqrt{1-x^2}}\, dx = \sin^{-1}(x) + C$$

$$(\tan^{-1}(x))' = \frac{1}{1+x^2} \iff \int \frac{1}{1+x^2}\, dx = \tan^{-1}(x) + C$$

$$(f(x) \cdot g(x))' =$$
$$f'(x) g(x) + f(x) g'(x) \iff \int f(x) g'(x) dx$$
$$= f(x) g(x) - \int f'(x) g(x) dx$$

$$(c_1 \cdot F(x) + c_2 \cdot G(x))' =$$
$$c_1 F'(x) + c_2 G'(x) =$$
$$c_1 f(x) + c_2 g(x) \iff \int c_1 f(x) + c_2 g(x) dx = c_1 \cdot F(x) + c_2 \cdot G(x)$$
$$= c_1 \int f(x) dx + c_2 \int g(x) dx$$

$$(F(u(x)))' = F'(u(x)) \cdot u'(x)$$
$$= f(u(x)) \cdot u'(x) \iff \int f(u(x)) \cdot u'(x) dx = F(u(x)) + C.$$

Differentiation is a linear operation in that the derivative of a sum is the sum of the derivatives and in that constants pull out of differentiation. Our next to last rule in our table of integrals says that because differentiation is a linear operation, so is antidifferentiation a linear operation. Integrals of sums are sums of integrals; and constants pull out of integrals.

Thus we know how to find antiderivatives of polynomials and other algebraic combinations of functions in the preceding integral table. For instance $\int x^3 + 5x \, dx = \int x^3 dx + 5 \int x dx = x^4/4 + 5x^2/2 + C$, and $\int \sin(x) - 3\cos(x) \, dx = \int \sin(x) dx - 3 \int \cos(x) dx = -\cos(x) - 3\sin(x) + C$. While most of the preceding integral formulas are self-evident, we explicitly confirm the Power Rule.

Since, by use of the Power Rule, we have $((x^{r+1}/(r+1))' = (x^{r+1})'/(r+1) = ((r+1)x^r)/(r+1) = x^r$ when $r \neq -1$, we conclude, by reversing our steps, that $\int x^r \, dx = x^{r+1}/(r+1) + C$. We also point out that the missing link $r = -1$ in the Power Rule is handled by the fact that $\int 1/x \, dx = \int x^{-1} \, dx = \ln(x) + C$.

We also point out that the last rule in our integral table is equivalent to the Chain Rule and it is consistent with our differential notation under substitution. In this setting it provides for the *Method of Substitution*. The Method of Substitution proceeds as follows. Given an integral of form $\int f(u(x)) \cdot u'(x) dx$ where we know the antiderivative of $f$ to be $F$, we define a new variable $u = u(x)$ to be the inside function, and we compute the differential of $u$ to be $du = d(u(x)) = u'(x)dx$. We make these substitutions into our integral to obtain

$$\int f(u(x)) \cdot u'(x)\, dx = \int f(u)\, du$$
$$= F(u) + C$$
$$= F(u(x)) + C,$$

where we have substituted back to an expression in $x$ in the last equation by replacing $u$ by $u(x)$. This substitution method is correct because it is consistent with the Chain Rule, but the reader is cautioned to keep track of all differentials $dx$ and $du$ in the substitution process. Sometimes this is also called a *change of variables* in that we replaced an expression in $x$ by one in $u$, at least in the intermediate stages.

### EXAMPLE 5.21

Find $\int (\sin(x))^{30} \cdot \cos(x) dx$. **Answer:** We find an inner function, in this case $\sin(x)$ is inside the 30th power function, and make the substitution $u = \sin(x)$ with $du = \cos(x)dx$. Our integral then becomes

$$\int (\sin(x))^{30} \cdot \cos(x)\, dx = \int u^{30} \cdot du = u^{31}/31 + C = \sin^{31}(x)/31 + C.$$

Notice that the more complicated appearing integral $\int (\sin(x))^{30} \cdot \cos(x)dx$ has been converted into the easier integral $\int u^{30} \cdot du$, which is evaluated by a simple power rule. Our answer is reconverted into an expression in $x$ and not left as an expression in $u$, because our original problem was posed in terms of $x$.

The preceding example illustrates the power of the Method of Substitution. If we can find the right inside function $u$, we convert to a simpler problem via substitution, perform our simpler integral, and convert back to $x$ when done to obtain our result.

When taken in concert, our integral table and our Method of Substitution allow us to find antiderivatives of a wide category of functions.

Before proceeding we note that $(|x|)' = 1$ if $x > 0$ and $(|x|)' = -1$ if $x < 0$. From this we obtain $(\ln(|x|))' = (1/|x|)(|x|)'$, which tells us $(\ln(|x|))' = (1/x) \cdot 1 = 1/x$ if $x > 0$, and $(\ln(|x|))' = (1/(-x)) \cdot (-1) = 1/x$ if $x < 0$. We conclude that $(\ln(|x|))' = 1/x$ for all $x \neq 0$, including for negative $x$. We can rely on this to extend our reciprocal integral to include nonzero values of $x$:

$$\int 1/x\, dx = \ln(|x|) + C.$$

A few further examples are in order.

### EXAMPLE 5.22

Find $\int \tan(x)\, dx$. **Answer:** We convert $\tan(x)$ into an expression of sine and cosine, $\tan(x) = \sin(x)/\cos(x)$, and make the substitution $u = \cos(x)$ inside the denominator, with $du = -\sin(x)dx$.

$$\begin{aligned}\int \tan(x)\, dx &= \int \frac{\sin(x)}{\cos(x)}\, dx \\ &= \int \frac{-1}{u}\, du = -\ln(|u|) + C = -\ln(|\cos(x)|) + C.\end{aligned}$$

Find $\int \sec(x)\, dx$. **Answer:** We multiply $\sec(x)$ up and down by $\sec(x) + \tan(x)$ and make the substitution $u = \sec(x) + \tan(x)$ inside the denominator, with $du = (\sec(x)\tan(x) + \sec^2(x))dx$.

$$\begin{aligned}\int \sec(x)\, dx &= \int \frac{\sec(x)(\sec(x) + \tan(x))}{(\sec(x) + \tan(x))}\, dx = \int \frac{\sec^2(x) + \sec(x)\tan(x)}{\sec(x) + \tan(x)}\, dx \\ &= \int \frac{1}{u} du = \ln(|u|) + C = \ln(|\sec(x) + \tan(x)|) + C.\end{aligned}$$

In a similar way we see

$$\int \cot(x)\, dx = \ln(|\sin(x)|) + C,$$

and

$$\int \csc(x)\, dx = -\ln(|\csc(x) + \cot(x)|) + C.$$

### EXAMPLE 5.23

As a final example in this section we find $\int \sqrt{1+x^2}\, x\, dx$. **Answer:** We let $u = 1 + x^2$ and $du = 2x dx$. We also note that upon dividing by 2 we have $du/2 = x\, dx$

$$\begin{aligned}\int \sqrt{1+x^2}\, x\, dx &= \int \sqrt{u}\, \frac{du}{2} = \frac{1}{2}\int u^{1/2}\, du \\ &= \frac{1}{2}\frac{u^{3/2}}{3/2} + C = \frac{1}{3}(1+x^2)^{3/2} + C.\end{aligned}$$

## Exercises

1. Find the following antiderivatives, or indefinite integrals:

   (a) $\int (x^5 - 4x^2 + 3x + 7) \, dx$.
   (b) $\int \cos^4(x) \, \sin(x) \, dx$.
   (c) $\int \cos(2x) \, dx$.
   (d) $\int e^{x^4} \, x^3 \, dx$.
   (e) $\int (\sec(x) + \tan(x)) \, dx$.
   (f) $\int 1/\sqrt{1-x^2} \, dx$.
   (g) $\int 1/(1+x^2) \, dx$.
   (h) $\int e^x/(1+e^{2x}) \, dx$.

## 5.17 Initial Value Problems and Euler's Method

An *Initial Value Problem*, or IVP, is a differential equation of form $f'(x) = g(x)$, together with an *initial condition* $f(x_0) = y_0$. Here $g$ is a given function and $f$ is an unknown function to be found. The initial condition gives us a point on the graph of the solution $f$. To solve an initial value problem we first antidifferentiate the differential equation to get

$$f(x) = \int g(x) \, dx = G(x) + C, \tag{5.71}$$

which gives us a family of curves satisfying the differential equation. However only one of these curves passes through the point $(x_0, y_0)$. So we substitute the initial value point $(x_0, y_0)$ into Eq. (5.71) to determine $C$

$$y_0 = f(x_0) = G(x_0) + C, \quad \text{which gives} \quad C = y_0 - G(x_0), \tag{5.72}$$

and we obtain a unique solution

$$f(x) = G(x) + y_0 - G(x_0), \tag{5.73}$$

to our initial value problem.

## EXAMPLE 5.24

Solve the IVP

$$f'(x) = 2x \quad \text{with initial condition} \quad f(1) = 2. \tag{5.74}$$

**Answer:** We have $g(x)$ is given to be $2x$, and $x_0 = 1$ with $y_0 = 2$. Antidifferentiating we have $f(x) = \int g(x)dx = \int 2xdx = x^2 + C$. Then initial condition $2 = f(1) = (1)^2 + C$ determines $C$ as $C = 2 - 1^2 = 1$. We conclude that the solution to our IVP is $f(x) = x^2 + 1$.

## EXAMPLE 5.25

Acceleration due to gravity: If acceleration $a(t)$ due to gravity on a planet is the constant $-g$, find the velocity $v(t)$ of a falling object acted upon only by gravity (thus ignoring any air resistance), if the initial velocity at time 0 is $v_0$ m/sec.

**Answer:** We know that $v'(t) = a(t)$, so we have the initial value problem

$$v'(t) = -g \quad \text{with initial condition} \quad v(0) = v_0.$$

Then by antidifferentiating we obtain $v(t) = \int -g dt = -gt + C$. Now $v_0 = v(0) = -g \times 0 + C$ implies $C = v_0$. Our solution is then $v(t) = -gt + v_0$.

If our velocity is now $v(t) = -gt + v_0$, and the initial position of the falling object at time 0 is $s_0$, determine the height $s(t)$ at any time $t$.

**Answer:** Since $s'(t) = v(t)$, we have been asked to solve the IVP

$$s'(t) = -gt + v_0 \quad \text{with initial condition} \quad s(0) = s_0.$$

We antidifferentiate the differential equation to obtain $s(t) = \int(-gt+v_0)dt = -gt^2/2 + v_0t + C$. The initial condition gives $s_0 = s(0) = -g0^2 + v_00 + C$, implying that $C = s_0$. Thus $s(t) = -gt^2/2 + v_0t + s_0$. We can now predict the height of a falling object acted upon only by constant gravity, provided we know our starting position $s_0$ and starting velocity $v_0$. On Earth $-g = -9.8$ m/sec$^2$, and the height $s(t)$ in meters at $t$ seconds of a falling object is modeled by $s(t) = -4.9t^2 + v_0t + s_0$.

If one is unable to find the antiderivative $\int g(x)dx$ of a more complicated given function $g(x)$ in an IVP, all is not lost. We can use linear approximation. A given fixed step size $\Delta x$ is agreed upon, and the process is as follows. Since the curve passes through the point $(x_0, y_0)$ given by the initial condition, we know $f(x_0) = y_0$.

We know the tangent line through this point has equation

$$y = f(x_0) + f'(x_0) \cdot [x - x_0] = y_0 + g(x_0) \cdot [x - x_0], \quad (5.75)$$

where we used the fact $f'(x) = g(x)$ from our differential equation. We move to a new point $x_1$ by incrementing by the step size $\Delta x$, so $x_1 = x_0 + \Delta x$. Plugging $x_1$ into Eq. (5.75) we obtain the tangent line height

$$y_1 = y_0 + g(x_0) \cdot [x_1 - x_0] = y_0 + g(x_0) \cdot \Delta x, \quad (5.76)$$

which essentially says that the new height $y_1$ should be the old height $y_0$ plus the rise (= slope $\times$ run = $g(x_0) \times \Delta x$). This gives us a good approximation $y_1 \approx f(x_1)$, by linearization. We then proceed as if $(x_1, y_1)$ were on the graph of the solution $f$, to produce an $x_2 = x_1 + \Delta x$ and a $y_2 = y_1 + g(x_1)\Delta x$, and repeat until we have covered an agreed upon number of steps $N$.

At each stage we have

$$x_{n+1} = x_n + \Delta x \quad \text{and} \quad y_{n+1} = y_n + g(x_n) \cdot \Delta x. \quad (5.77)$$

This process is called *Euler's Method*. Computing the points $(x_n, y_n)$ for $n = 0, 1, \ldots, N$ and *connecting the dots* yields a broken line segment approximation called Euler's Approximation that converges to the true solution $f(x)$ as $\Delta x$ approaches 0 when the given function $g(x)$ is continuous.

### ■ Exercises

1. Solve the IVP $f'(x) = 4x + 6$ with $f(0) = 5$.
2. Use Euler's Method on the IVP $f'(x) = 4x + 6$ with $f(0) = 5$, by fixing $\Delta x = 1/2$ and observing $x_0 = 0$, $y_0 = 5$. Find the points $(x_1, y_1)$ and $(x_2, y_2)$ obtained from Euler's Method.

### ■ 5.18 Area and the Fundamental Theorem of Calculus

We wish to find the area under a curve $y = f(x)$ over a given interval $[a, b]$. Two geometric examples enlighten us as to how we can use calculus to do this.

As a first example, let $f(x) = k$ be a constant height function. For now let $k > 0$. Then the region that falls under $k$ above $[a, b]$ is a rectangle of area height

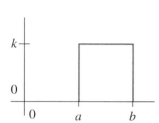 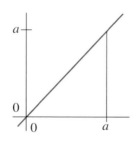 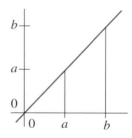

**FIGURE 5.8** ■ A constant and linear function. Area using a trapezoid.

times width $= k(b-a) = kb - ka$. We observe that the antiderivatives of $f(x) = k$ are $F(x) = \int k\,dx = kx + C$. We set $C = 0$ to obtain one such antiderivative $F(x) = kx$, and we observe that $F(b) - F(a) = kb - ka$ is our area! In fact, any of the antiderivatives of $k$ gives the same result. If $F(x)$ is taken to be $F(x) = kx + C$, then $F(b) - F(a) = kb + C - (ka + C) = kb - ka$, which is still our area.

As a second example, we rely on the fact that the area of a right triangle is one half the base times the height. In this example we let $f(x) = x$. Then the region under $x$ from 0 to $a$ is an equilateral right triangle with base $a$ and height $a$, so the area of this region is $(1/2)a \cdot a = a^2/2$. The area of the region under $x$ from 0 to $b$ is likewise $b^2/2$. The area of the trapezoidal region under $x$ over $[a,b]$ is the area of the triangle to $b$ minus the area of the triangle to $a$, or more succinctly the area of our trapezoid is $b^2/2 - a^2/2$. If we compare this to our antiderivative $F(x) = \int x\,dx = x^2/2 + C$, we see that for any choice of $C$, we have

$$F(b) - F(a) = b^2/2 + C - (a^2/2 + C) = b^2/2 - a^2/2,$$

which is indeed our area.

Could this be a coincidence that the area under a curve over an interval appears to be the difference of antiderivatives at the endpoints? The next goal is to explore and answer this question. It is a central tenet of integral calculus that the area under a reasonable curve over a given interval is indeed related to the difference of antiderivatives at the endpoints.

First we examine the process of approximating the area under a curve $y = f(x)$ over an interval $[a,b]$. For now, we think of the case that $f(x) \geq 0$ over our interval, but all our computations that follow work even if $f$ becomes negative. To estimate our area, we subdivide or *partition* $[a,b]$ into $N$ subintervals each of length $\Delta x = (b-a)/N$. From each subinterval we select any point, say $x_i^*$ from the $i$th subinterval. We approximate the area over the $i$th subinterval by the area $A_i$ of a rectangle of height $f(x_i^*)$ and width $\Delta x$. That is $A_i = f(x_i^*)\Delta x$. The actual area under $f$ should then be close to the sum of all the areas $A_i$. So we form an approximate area

$$\sum_{i=1}^{N} A_i = \sum_{i=1}^{N} f(x_i^*) \cdot \Delta x,$$

called a *Riemann Sum* in honor of the famous German mathematician Riemann.

As the widths $\Delta x$ get smaller and approach 0, we should get a more accurate estimate, at least if $f$ is a reasonable function. When *all* Riemann Sums for $f$ over $[a, b]$ approach a common value, $DI \in \mathbb{R}$, as our widths $\Delta x$ approach 0, we say

$$\lim_{\Delta x \to 0} \sum_{i=1}^{N} f(x_i^*) \, \Delta x = DI,$$

and we call the number $DI$ the *definite integral* of $f$ over the interval $[a, b]$. To indicate that the limiting value $DI$ came from approximating Riemann Sums of form $\sum f(x_i^*) \Delta x$, we *translate* from Greek to English replacing the Greek letter sigma $\sum$ by its English counterpart S, where we use the old English S, $\int$, and we also replace the Greek letter delta $\Delta$ by it English counterpart d. We also indicate the interval $[a, b]$ by lower and upper *limits of integration* $a$ and $b$, and we denote

$$DI = \int_a^b f(x) \, dx.$$

Equivalently we have

$$\lim_{\Delta x \to 0} \sum_{i=1}^{N} f(x_i^*) \, \Delta x = \int_a^b f(x) \, dx.$$

If this limit exists we say that $f(x)$ is *integrable* on $[a, b]$. It is a fact proven in advanced calculus [59] that if $f$ is continuous on the interval $[a, b]$, then $f$ is integrable and the definite integral $\int_a^b f(x)dx$ is a finite number that is approached by all approximating Riemann Sums, the smaller $\Delta x$ gets the better the approximation gets. We then have

$$\sum_{i=1}^{N} f(x_i^*) \, \Delta x \approx \int_a^b f(x) \, dx,$$

for a sufficiently small $\Delta x$. It is often the case in applications that we do not know how to compute a given quantity. However if we can approximate it by an expression having the form of a Riemann Sum, then we can *translate from Greek to English* to see the form of the exact answer expressed as a definite integral. We will repeat this type of approximation and translation several times in our own applications here, but you should be learning how to perform this approximation/translation process in any new or specialized setting perhaps not covered in a general text. We will return again to this theme later.

As an example, we observe that any increasing function $f$ on an interval $[a, b]$ is integrable, and we estimate the error for any Riemann Sum for $f$ on this interval from above. Partition the interval into $N$ equal subintervals each of length $\Delta x = (b-a)/N$ and let the endpoints be denoted $x_i$, for $i = 0, 1, \ldots, n$. Because $f$ is increasing $f(x_{i-1}) \leq f(x) \leq f(x_i)$ for all $x$ in the $i$th subinterval. We estimate the net area under $f$ from below with the Riemann Sums of form $\sum_{i=1}^{N} f(x_{i-1}) \Delta x$ and from above with the Riemann Sums of form $\sum_{i=1}^{N} f(x_i) \Delta x$. The difference in these two estimates gives an upper bound on the error in estimating the associated definite integral $DI$. This difference is

$$\sum_{i=1}^{N} f(x_i) \Delta x - \sum_{i=1}^{N} f(x_{i-1}) \Delta x \qquad (5.78)$$

$$= \left\{ \sum_{i=1}^{N} [f(x_i) - f(x_{i-1})] \right\} \cdot \Delta x = [F(b) - F(a)] \Delta x,$$

and any other Riemann Sum for this partition has error less than the amount in Eq. (5.78). The amount in Eq. (5.78) can be visualized as the rectangular stack in Figure 5.9.

By letting $N$ approach infinity, which then forces $\Delta x$ toward 0, we see that our total possible error in estimation via a Riemann Sum is also forced to 0. A similar argument works for decreasing functions by placing an absolute value on Eq. (5.78). For any function $f$ that can be broken up into a finite number of pieces that are increasing or decreasing, a combination of the preceding methods works to show $f$ is integrable.

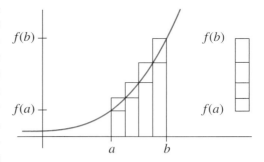

FIGURE 5.9 ■ Rectangular stack of errors for $f(x)$ on $[a,b]$.

At this point we are in a position to return to our original question of relating area computations, or definite integrals, to antiderivatives at endpoints. We have the first part of a remarkable theorem called the Fundamental Theorem of Calculus (FTC), due to Newton and Leibniz, independently.

### ■ Theorem 5.30

**Fundamental Theorem of Calculus (FTC) I**

Let $f(x)$ be continuous on the interval $[a,b]$, with $F(x)$ being any antiderivative of $f(x)$. Then the definite integral of $f$ over the interval $[a,b]$ is given by

$$\int_a^b f(x)\, dx = F(b) - F(a). \tag{5.79}$$

**Proof** Partition our interval $[a,b]$ into $N$ subintervals of width $\Delta x = (b-a)/N$, with the left and right endpoints of the $i$th subinterval denoted by $x_{i-1}$ and $x_i$ respectively. We apply the Mean Value theorem to $F$ on the interval $[x_{i-1}, x_i]$ to obtain a $c_i$ in $(x_{i-1}, x_i)$ with $f(c_i) = F'(c_i) = (F(x_i) - F(x_{i-1}))/(x_i - x_{i-1})$. We conclude by cross-multiplying by the denominator $(x_i - x_{i-1})$, that

$$f(c_i)\,(x_i - x_{i-1}) = (F(x_i) - F(x_{i-1})).$$

If we sum the differences $(F(x_i) - F(x_{i-1}))$ from 1 to $N$, we cancel out any point that is both a right endpoint of one interval and also a left endpoint of the next adjacent interval, due to opposite signs. We then have only a $F(b)$ term and a $-F(a)$ term left that did not cancel out:

$$\begin{aligned} F(b) - F(a) &= \sum_{i=1}^{N}(F(x_i) - F(x_{i-1})) \\ &= \sum_{i=1}^{N} f(c_i)\,(x_i - x_{i-1}) \\ &= \sum_{i=1}^{N} f(c_i)\,\Delta x. \end{aligned}$$

This last summation is an approximating Riemann Sum for the definite integral $DI = \int_a^b f(x)\,dx$ and so becomes arbitrarily close to $DI$ for small $\Delta x$, by the hypothesis that $f$ was continuous on $[a,b]$. However our Riemann Sum is seen to be the preceding constant $F(b) - F(a)$, and we conclude that if the constant $F(b) - F(a)$ is arbitrarily close to the constant $DI = \int_a^b f(x)\,dx$, they must be equal, so

$$F(b) - F(a) = \int_a^b f(x)\, dx$$

is proven. □

## EXAMPLE 5.26

Find the area $A$ under the curve $y = f(x) = x^2$ over the interval $[0, 1]$. **Answer:** We cannot easily use geometry to see our area as in our first rectangular and triangular examples in this section. The FTC I makes the computation easy though. It is exact and has no need to rely on approximations. First we find an antiderivative $F(x) = x^3/3$ by the Power Rule, and then we compute the area to be

$$A = \int_0^1 x^2 \, dx = F(1) - F(0) = 1^3/3 - 0^3/3 = 1/3.$$

Since it will often be convenient to shorthand our expression $F(b) - F(a)$, we make the following notational definition:

$$F(x) \Big|_a^b = F(b) - F(a).$$

In the previous example we would have

$$A = \int_0^1 x^2 \, dx = \frac{x^3}{3}\Big|_0^1 = \frac{1}{3} - \frac{0}{3} = \frac{1}{3}.$$

This shortens the process as we proceed along one line, without needing to stop and say elsewhere that the antiderivative was $x^3/3$.

We also pause to remark here that if $f$ were negative on the interval $[a, b]$ all the preceding computations would proceed in the same manner. After examining the approximating Riemann Sums in this case, we conclude that the definite integral gives us the negative of the area of the region between $f$ and the interval $[a, b]$. In general $\int_a^b f(x)dx$ gives us the net area, counting positively when $f > 0$ and counting negatively when $f < 0$.

## EXAMPLE 5.27

Find $\int_0^1 (-x^2)dx$. **Answer:** Clearly $-x^2$ lies below the $x$-axis. We compute

$$\int_0^1 (-x^2) \, dx = -x^3/3 \Big|_0^1 = -1/3 - (-0/3) = -1/3.$$

Thus $-1/3$ is now minus the area of the region between $-x^2$ and $[0, 1]$, which of course should be clear geometrically upon reflecting $x^2$ about the $x$-axis to get $-x^2$.

When $a > b$ we define $\int_a^b f(x)dx$ to be $-\int_b^a f(x)dx$, consistent with the FTC. Thus interchanging the limits of integration changes the sign of the resulting definite integral.

Before proceeding, we record several properties of the definite integral $\int_a^b f(x)dx$.

---

### ■ Lemma 5.5

$$\int_a^b f(x)\,dx = -\int_b^a f(x)\,dx$$

$$\int_a^a f(x)\,dx = 0$$

$$\int_a^b (c_1 f(x) + c_2 g(x))\,dx = c_1 \int_a^b f(x)\,dx + c_2 \int_a^b g(x)\,dx$$

$$\int_a^c f(x)\,dx = \int_a^b f(x)\,dx + \int_b^c f(x)\,dx,$$

and finally, if $A \leq f(x) \leq B$ on the interval $[a,b]$, we obtain

$$A \cdot (b-a) \leq \int_a^b f(x)\,dx \leq B \cdot (b-a).$$

---

The first line of Lemma 5.5 follows from the the preceding discussion, by definition. The second line says that there can only be 0 net area above the interval $[a,a]$. The third line says that the definite integral is a linear operation. The fourth line says that net area is additive: if we add the net area under $f$ from $a$ to $b$ with the net area under $f$ from $b$ to $c$, then we obtain the total net area from $a$ to $c$. The last line says that if we place a rectangle of height $A$ underneath the graph of $f$ over $[a,b]$ and a rectangle of height $B$ above the graph of $f$ over $[a,b]$, then we have underestimated the net area under $f$ by the area of the lower rectangle, and we have overestimated by the area of the upper rectangle.

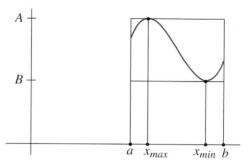

**FIGURE 5.10** ■ Range of $f(x)$ on $[a,b]$.

We next have the second part of the FTC.

■ **Theorem 5.31**

**Fundamental Theorem of Calculus (FTC) II**

Let $f(x)$ be continuous on the interval $[a,b]$. Then for $t \in (a,b)$ the function

$$F(x) = \int_a^x f(t)\,dt, \qquad (5.80)$$

is differentiable for all $x \in [a,b]$, and

$$F'(x) = f(x). \qquad (5.81)$$

---

**Proof** Let $p$ be a point in $[a,b]$, then

$$\begin{aligned}
F(x) &= \int_a^x f(t)\,dt = \int_a^p f(t)\,dt + \int_p^x f(t)\,dt = F(p) + \int_p^x f(t)\,dt \\
&= F(p) + f(p)\cdot[x-p] + \int_p^x (f(t)-f(p))\,dt. \qquad (5.82)
\end{aligned}$$

We used additivity of areas in the first line to split up the first integral into a pair of integrals.

We then observed that, for $x$ near $p$, $\int_p^x f(t)dt$ is the area of a very thin nearly rectangular region with height approximately $f(p)$ and base $[x-p]$. As such $\int_p^x f(t)dt$ very close to $f(p)[x-p]$ with error term $\int_p^x (f(t)-f(p))dt$. We obtain the error term by noting that $f(t) = f(p) + [f(t)-f(p)]$ and integrating to get

$$\begin{aligned}
\int_p^x f(t)\,dt &= \int_p^x (f(p) + (f(t)-f(p)))\,dt = \int_p^x f(p)\,dt + \int_p^x (f(t)-f(p))\,dt \\
&= f(p)\cdot[x-p] + \int_p^x (f(t)-f(p))\,dt,
\end{aligned}$$

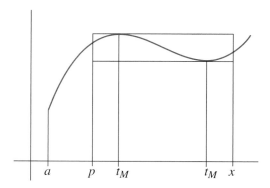

**FIGURE 5.11** ■ Using linearization of $f(x)$ at a point $p$ to estimate integral.

which gives Eq. (5.82). Now we claim $F(p) + f(p) \cdot [x - p]$ in Eq. (5.82) is the linearization of $F$ at $p$, with $E(x) = \int_p^x (f(t) - f(p))dt$ the higher-order terms at $p$.

Clearly $E(p) = \int_p^p (f(t) - f(p))dt = 0$. We next show $|E(x)/[x - p]|$ tends to 0 as $x$ tends to $p$. We handle the case $x > p$ here (with the case $x < p$ being similar). On the interval $[p, x]$, there are $t_m, t_M \in [p, x]$ with $f(t_m) \leq f(t) \leq f(t_M)$ serving as absolute extrema for $f$ on $[p, x]$, by continuity of $f$. Then $f(t_m)[x-p] \leq \int_p^x f(t)dt \leq f(t_M)[x - p]$ from the previous lemma. We subtract $f(p)[x - p]$ from all sides to obtain

$$(f(t_m) - f(p)) \cdot [x - p] \leq \int_p^x (f(t) - f(p))\, dt = E(x) \leq (f(t_M) - f(p)) \cdot [x - p].$$

Thus,

$$(f(t_m) - f(p)) \leq E(x)/[x - p] \leq (f(t_M) - f(p)),$$

and since $t_m, t_M \in [p, x]$ as $x$ tends to $p$ so do $t_m, t_M$ tend to $p$. We conclude $(f(t_m) - f(p))$ tends to 0 and $(f(t_M) - f(p))$ tends to 0 as $x$ tends to $p$ by continuity of $f$ at $p$, so $E(x)/[x-p]$ also tends to 0 as $x$ tends to $p$, and we now know $E(x)$ is a higher-order term. Thus we have confirmed that $F(p) + f(p)[x-p]$ is the linearization at $p$, and so $F'(p) = f(p)$, the slope of the linearization. Since $F'(p) = f(p)$ holds for all $p$ in $[a, b]$, we have $F'(x) = f(x)$, and the theorem now follows. □

### EXAMPLE 5.28

Find the derivative of

$$F(x) = \int_5^x \frac{1}{1+t^2+t^4}\,dt.$$

**Answer:** Since $F(x) = \int_5^x f(t)\,dt$ for $f(t) = 1/(1+t^2+t^4)$, we have by the FTC that

$$F'(x) = f(x) = 1/(1+x^2+x^4),$$

directly, without the need to compute an integral and then differentiate the result.

## ■ Exercises

1. Evaluate each definite integral as follows, using the FTC I:

    (a) $\int_1^2 x^3\,dx$.
    (b) $\int_0^1 e^x\,dx$.
    (c) $\int_0^{\pi/2} \sin(x)\,\cos(x)\,dx$.

2. Evaluate each derivative as follows, using the FTC II:

    (a) $\frac{d}{dx}\left(\int_0^x \cos(t)\,dt\right)$.
    (b) $\frac{d}{dx}\left(\int_0^x 1/(1+t^4+t^{10}+\cos(t))\,dt\right)$.
    (c) $\frac{d}{dx}\left(\int_0^{x^2} 1/e^{\sqrt{t}}\,dt\right)$.

## ■ 5.19 Applications of the Definite Integral

We return to our process of approximating a computation with a Riemann Sum and then seeing the form of the exact answer as a definite integral:

$$\sum_{i=1}^N f(x_i^*)\,\Delta x \approx \int_a^b f(x)\,dx,$$

a process we informally called *translating from Greek to English* in that the Sigma became an old English S (which we recognize as an integral sign now) and the Delta became a "d." All other subscripts and superscripts were dropped in this translation, and the limits of integration $a$ and $b$ told us which interval was appropriate. Our advantage now is that we know how to evaluate our definite interval exactly as $F(b) - F(a)$ for any antiderivative $F$ of $f$.

### ■ 5.19.1 Volumes by Cross-Sectional Area & Volumes of Revolution

Set a solid region on the $x$-axis and take a plane perpendicular to the $x$-axis at a point $x$. Slice the solid by this plane to get a *cross-section* with *cross-sectional area* $A(x)$. If we know the formula for $A(x)$, we can recover the volume $V$ of the solid. Assuming the solid sits over the interval $[a, b]$, partition $[a, b]$ into $N$ subintervals of length $\Delta x = (b-a)/N$. Let $V_i$ be the volume of our solid over the $i$th subinterval. At a point $x_i$ in the $i$th subinterval, take the cross-sectional area $A(x_i)$ at $x_i$ and multiply by $\Delta x$ to approximate $V_i \approx A(x_i)\Delta x$. Summing over $i$ yields

$$V = \sum_{i=1}^{N} V_i \approx \sum_{i=1}^{N} A(x_i)\,\Delta x \approx \int_a^b A(x)\,dx,$$

where we have recognized the rightmost approximating sum to be in the form of a Riemann Sum, and then *translated from Greek to English* to see the definite integral. We conclude that the volume $V$ of a solid with cross-sectional area $A(x)$ at $x$ is

$$V = \int_a^b A(x)\,dx.$$

Notice that we are no longer computing areas, but rather volumes. But the volume Riemann Sum had the same format as an area Riemann Sum, and we still had a definite integral as an exact answer in this volume setting.

We employ our volume by cross-section technique to obtain the volume of a solid of revolution obtained by revolving a function $f(x)$ over $[a, b]$ around the $x$-axis. Since $f(x)$ is the radius at $x$, and the cross-section is a circle, we know the cross-sectional area is $A(x) = \pi f^2(x)$. Hence our volume of revolution is

$$V = \int_a^b A(x)\,dx = \int_a^b \pi f^2(x)\,dx = \pi \int_a^b f^2(x)\,dx.$$

### EXAMPLE 5.29

Find the volume of the sphere of radius $R$. **Answer:** We take the circle of radius $R$ centered at the origin in the plane. The formula is $x^2 + y^2 = R^2$, and solving for $y$ gives $y = \pm\sqrt{R^2 - x^2}$. We select as $f(x)$ the upper semicircle $f(x) = +\sqrt{R^2 - x^2}$ as our function to revolve around the $x$-axis over the interval $[-R, R]$. Our volume is computed then as:

$$\begin{aligned} V &= \pi \int_a^b f^2(x)\, dx = \pi \int_a^b (\sqrt{R^2 - x^2})^2\, dx = \pi \int_a^b (R^2 - x^2)\, dx \\ &= \pi(R^2 x - x^3/3)|_{-R}^{R} = \pi(R^3 - R^3/3) - \pi(-R^3 + R^3/3) = 4\pi R^3/3. \end{aligned}$$

We have computed a difficult classical volume from geometry in just two lines by relying on the power of the definite integral.

### 5.19.2 Mass and Probability

We will first compute the mass $M$ of a thin rigid wire of radius $r$ by harnessing definite integrals. We place the wire along the $x$-axis over an interval $[a, b]$. We assume that the wire is of varying density, $D(x)$, where density is mass per volume. From the definition of density we have that mass = density × volume. If we partition our interval $[a, b]$ into $N$ equal subintervals of length $\Delta x = (b-a)/N$, we obtain the volume of the $i$th piece of the wire to be $V_i = \pi r^2 \Delta x$. Then we approximate the mass $M_i$ of the $i$th piece by $M_i \approx D(x_i) V_i = D(x_i) \pi r^2 \Delta x$. We then have a good approximation for our mass as

$$M = \sum_{i=1}^{N} M_i \approx \sum_{i=1}^{N} D(x_i)\, \pi\, r^2\, \Delta x \approx \int_a^b (\pi\, r^2\, D(x))\, dx.$$

We have once again translated from Greek to English to see the form of the integral for the exact answer, $M = \int_a^b \pi r^2 D(x)\, dx$. Some texts refer to $\delta(x) = \pi r^2 D(x)$ as the *linear density*, from which we obtain

$$M = \int_a^b \delta(x)\, dx. \tag{5.83}$$

> **EXAMPLE 5.30**
>
> Find the mass of a thin wire over an interval from $[0,2]$ m, if the linear density is $\delta(x) = 3x$ grams/meter. **Answer:** $M = \int_0^2 3x\,dx = (3x^2/2)|_0^2 = 3(2^2)/2 - 3(0^2)/2 = 6$ g.

We mention that Eq. (5.83) has an analogue in statistics, where the *mass over* $[a,b]$ corresponds to the *probability over* $[a,b]$, denoted $P([a,b])$. The *linear density* $\delta(x)$ corresponds to a *probability density*, denoted $\rho(x)$. In this setting Eq. (5.83) becomes

$$P([a,b]) = \int_a^b \rho(x)\,dx. \tag{5.84}$$

The reader is undoubtedly familiar with the standard normal probability density (commonly called the bell curve). Since the integrals associated with the normal probability density are complicated, we content ourselves with using an exponential probability density $\rho(x) = \lambda e^{-\lambda x}$ over the interval $[0, \infty)$ to illustrate the connection to statistics.

> **EXAMPLE 5.31**
>
> Given the exponential probability density $\rho(x) = \lambda e^{-\lambda x}$, for $\lambda > 0$, find the probability over the interval $[0, A]$. **Answer:** We first observe that $P([0, A]) = \int_0^A \rho(x)\,dx = \int_0^A \lambda e^{-\lambda x}\,dx$. From this we realize that we need to evaluate the indefinite integral $\int \lambda e^{-\lambda x}\,dx$. We do so by substituting $u = -\lambda x$ as our inside function, with $du = -\lambda dx$ and thus $-du/\lambda = dx$. We obtain
>
> $$\int \lambda e^{-\lambda x}\,dx = \int \lambda e^u (-du/\lambda) = \int -e^u\,du = -e^u + C = -e^{-\lambda x} + C.$$
>
> Thus,
>
> $$P([0, A]) = \int_0^A \lambda e^{-\lambda x}\,dx = -e^{-\lambda x}\Big|_0^A = -e^{-\lambda A} - (-e^{-\lambda 0}) = -e^{-\lambda A} + 1.$$
>
> Notice that as $A$ tends to $\infty$, $e^{-\lambda A}$ tends to 0, and then $P([0, A]) = 1 - e^{-\lambda A}$ tends to 1. This is true for any probability density as the total probability is always 1.

The statement $F'(u) = f(u)$ has equivalent integral form $\int f(u)\,du = F(u) + C$. For the same $F$ and $f$ the Chain Rule $(F(u(x)))' = F'(u(x))u'(x) = f(u(x))u'(x)$ then has equivalent integral form $\int f(u(x))u'(x)\,dx = F(u(x)) + C$. It follows that

$\int_a^b f(u(x))u'(x)\,dx = F(u(x))|_a^b = F(u(b)) - F(u(a))$, and also $\int_{u(a)}^{u(b)} f(u)\,du = F(u)|_{u(a)}^{u(b)} = F(u(b)) - F(u(a))$. So the two definite integrals are equal, and there is a change of variable formula for definite integrals.

$$\int_a^b f(u(x))u'(x)\,dx = \int_{u(a)}^{u(b)} f(u)\,du. \tag{5.85}$$

This is utilized in many settings, including computation of probabilities in terms of the standard normal density function.

### EXAMPLE 5.32

The normal density function with mean $\mu$ and standard deviation $\sigma$ is given by

$$\rho(\mu, \sigma; x) = \frac{1}{\sqrt{2\pi}\sigma} e^{-\frac{1}{2}\left(\frac{x-\mu}{\sigma}\right)^2},$$

while the standard normal density function has mean 0 and standard deviation 1 and takes the form

$$\rho(0, 1; z) = \frac{1}{\sqrt{2\pi}} e^{-\frac{1}{2}(z)^2}.$$

Use the $z$-score substitution $z = (x - \mu)/\sigma$ to convert the probability that $x \in [a,b]$ for $x$ normal with mean $\mu$ and standard deviation $\sigma$, to the probability that $z \in [z(a), z(b)]$ for $z$ having standard normal density.

Letting $u(x) = z(x) = (x-\mu)/\sigma$ and $dz = dx/\sigma$ in Eq. (5.85) one obtains

$$\begin{aligned} P(x \in [a,b]) &= \int_a^b \frac{1}{\sqrt{2\pi}\sigma} e^{-\frac{1}{2}\left(\frac{x-\mu}{\sigma}\right)^2} dx = \int_{z(a)}^{z(b)} \frac{1}{\sqrt{2\pi}\sigma} e^{-\frac{1}{2}(z)^2} \sigma\,dz \\ &= \int_{z(a)}^{z(b)} \frac{1}{\sqrt{2\pi}} e^{-\frac{1}{2}(z)^2} dz = P(z \in [z(a), z(b)]). \end{aligned}$$

Thus every normal distribution probability can be computed in terms of the standard normal density function by the use of $z$-scores.

### ■ 5.19.3 The Center of Mass and the Mean of a Probability Density

The balance point of a thin rigid wire is called the *center of mass*, and it corresponds to the location of the point where the wire could be supported and balanced. We rely on the fact that the torque of a mass acting on a lever of a given length is proportional to mass × length. That is torque = C × mass × length. If we denote the position of our center of mass by $\bar{x}$, then the torque $T$ of a mass $m$ at position $x$ would be $T = C \times m \times (x - \bar{x})$. The torque is positive if $x$ is to the right of $\bar{x}$, indicating a tendency to rotate clockwise. The torque is negative if $x$ is to the left of $\bar{x}$, indicating a tendency to rotate counterclockwise. The total torque should be 0 if balance is achieved. We compute the total torque by adding the torque of individual masses at their respective distances from $\bar{x}$. We partition our wire along $[a, b]$ as before into $N$ pieces over subintervals of length $\Delta x = (b-a)/N$. The mass $M_i$ of the $i$th piece can be approximated by $M_i \approx D(x_i)V_i = D(x_i)\pi r^2 \Delta x$. The torque of the $i$th piece about $\bar{x}$ is then approximated by $C(x_i - \bar{x})M_i \approx C(x_i - \bar{x})D(x_i)\pi r^2 \Delta x$. We sum over $i$ to obtain the total torque. Since the total torque should be 0 we have

$$0 \approx \sum_{i=1}^{N} C(x_i - \bar{x})D(x_i)\pi r^2 \, \Delta x = \sum_{i=1}^{N} C(x_i - \bar{x}) \, \delta(x_i) \, \Delta x \approx \int_a^b C(x - \bar{x}) \, \delta(x) \, dx,$$

where we have again translated from Greek to English to see the corresponding definite integral. We conclude that to balance at $\bar{x}$, we must have, after dividing both sides by C,

$$0 = \int_a^b (x - \bar{x}) \, \delta(x) \, dx.$$

We then solve for $\bar{x}$ in the following manner:

$$0 = \int_a^b (x - \bar{x})\delta(x)dx = \int_a^b x\delta(x)dx - \int_a^b \bar{x}\delta(x)dx = \int_a^b x\delta(x)dx - \bar{x}\int_a^b \delta(x)dx$$

gives our formula for the center of mass as

$$\bar{x} = \frac{\int_a^b x \, \delta(x) \, dx}{\int_a^b \delta(x) \, dx}. \tag{5.86}$$

We recognize our denominator in Eq. (5.86) as the mass of the thin wire.

### EXAMPLE 5.33

Find the center of mass of the thin wire over $[0, 2]$ m, with linear density $\delta(x) = 3x$ g/m. **Answer:** We computed the mass of the wire in the previous example to be $M = \int_0^2 3x\,dx = 6$ g. We now compute the numerator of Eq. (5.86) to be $\int_0^2 x\delta(x)dx = \int_0^2 x(3x)dx = \int_0^2 3x^2\,dx = x^3|_0^2 = 8$ gm. Our center of mass is now seen to be

$$\bar{x} = (8 \text{ gm})/(6\text{g}) = (4/3)\text{m}.$$

Notice that, because the wire has a greater density at the right near 2 than it does at the left near 0, the center of mass 4/3 is closer to 2 than to 0.

We again return to our statistical example. The center of mass is the balance point for a linear density $\delta(x)$, and in some sense measures where the weighted center of the wire is located. The *center of mass* for a probability density over $[a,b]$ would be given by the same formula in an attempt to obtain the weighted center of the probability distribution. As such we would have

$$\bar{x} = \frac{\int_a^b x\, \rho(x)\, dx}{\int_a^b \rho(x)\, dx}. \tag{5.87}$$

We call $\bar{x}$ the mean, or average, of the probability distribution. It is indeed the weighted center. Since the total probability (corresponding to the total mass) is always 1 in a probability distribution, we have our denominator $\int_a^b \rho(x)dx = 1$. We conclude that our mean of the probability distribution is given by

$$\bar{x} = \int_a^b x\, \rho(x)\, dx. \tag{5.88}$$

## EXAMPLE 5.34

Find the mean of the exponential probability distribution $\rho(x) = \lambda e^{-\lambda x}$ on $[0, \infty)$. **Answer:** We have

$$\bar{x} = \int_0^\infty x \cdot (\lambda e^{-\lambda x})\, dx.$$

This example creates two new interesting questions. The first is: how do we antidifferentiate $\lambda x e^{-\lambda x}$? If we differentiate $x e^{-\lambda x}$, we use the Product Rule to obtain $(x e^{-\lambda x})' = (x)' e^{-\lambda x} + x(e^{-\lambda x})' = 1 e^{-\lambda x} + x e^{-\lambda x}(-\lambda)$. We conclude, by solving for $\lambda x e^{-\lambda x}$ on the right-hand side, that

$$\lambda\, x\, e^{-\lambda x} = e^{-\lambda x} - (x\, e^{-\lambda x})'.$$

Antidifferentiating both sides yields

$$\int \lambda x\, e^{-\lambda x} = \int e^{-\lambda x} - \int (x\, e^{-\lambda x})' = (-1/\lambda)\, e^{-\lambda x} - x\, e^{-\lambda x} + C.$$

We observe that the Product Rule helped us antidifferentiate a more difficult function. We will formalize this process later and call it integration by parts.

The second question is: how do we find a definite integral over the infinite interval $[0, \infty)$. Our answer is to find our integral on $[0, A]$ and then let $A$ tend to $\infty$ and see what we get. Thus,

$$\begin{aligned}\int_0^A x\, (\lambda e^{-\lambda x})\, dx &= ((-1/\lambda) e^{-\lambda x} - x\, e^{-\lambda x})\big|_0^A \\ &= (-1/\lambda) e^{-\lambda A} - A e^{-\lambda A} - \{(-1/\lambda) e^{-\lambda 0} - 0 e^{-\lambda 0}\} \\ &= (-1/\lambda) e^{-\lambda A} - A e^{-\lambda A} + (1/\lambda).\end{aligned}$$

As $A$ tends to $\infty$ the the left and middle terms also tend to 0 (see L'Hopital's Rule for the middle term), and we then say that

$$\int_0^\infty x\, (\lambda e^{-\lambda x})\, dx = 1/\lambda. \tag{5.89}$$

Integrating over an interval of infinite length falls under a topic called improper integrals.

Thus our statistics example points us in the pursuit of three new directions: integration by parts, improper integrals, and L'Hopital's Rule.

# Exercises

1. Find the volume of revolution of the solid obtained by revolving the region underneath $f(x) = \sqrt{x}$ over the interval $[1, 2]$ around the $x$-axis.
2. Find the mass of the thin wire over the interval $[0, 4]$ m with linear density $\delta(x) = \sqrt{x}$ g/m.
3. Find the center of mass of the thin wire over the interval $[0, 4]$ m with linear density $\delta(x) = \sqrt{x}$ g/m.
4. Find the probability $P([1, 2])$ for the probability density function $\rho(x) = e^{-x}$.

## 5.20 Improper Integrals and L'Hopital's Rule

As we saw when computing means in statistics, we often find the need to integrate over an infinite interval. In order to understand this phenomenon better, we first examine the notion of a *limit at infinity*.

If a function $f(x)$ levels off and tends to a finite value $L$ as $x$ gets arbitrarily large, we say that $f(x)$ approaches $L$ as $x$ approaches $\infty$, and we write

$$\lim_{x \to \infty} f(x) = L.$$

As an example we point out that $\lim_{x \to \infty} 1/x = 0$; we also note $\lim_{x \to \infty} e^{-x} = 0$; and a third example is $\lim_{x \to \infty} (x+1)/x = 1$. One can also compute limits as $x$ approaches $-\infty$ in an analogous manner.

If $f(x)$ grows arbitrarily large as $x$ gets arbitrarily large, we say that $f(x)$ approaches $\infty$ as $x$ approaches $\infty$, and we write

$$\lim_{x \to \infty} f(x) = \infty.$$

For instance $\lim_{x \to \infty} x^2 = \infty$. In general,

$$\lim_{x \to \infty} x^p = \begin{cases} \infty & \text{if } p > 0 \\ 1 & \text{if } p = 0 \\ 0 & \text{if } p < 0. \end{cases} \tag{5.90}$$

Furthermore the limit at infinity of any rational function can be understood by taking the limit at infinity of the ratio of the leading terms. That is

$$\lim_{x \to \infty} \frac{a_n x^n + a_{n-1} x^{n-1} + \ldots}{b_m x^m + b_{m-1} x^{m-1} + \ldots} = \lim_{x \to \infty} \frac{a_n x^n}{b_m x^m}. \qquad (5.91)$$

For instance,

$$\lim_{x \to \infty} (3x^2 + 4x + 2)/(5x^2 + x + 6) = \lim_{x \to \infty} (3x^2)/(5x^2) = \lim_{x \to \infty} (3)/(5) = 3/5.$$

The reasoning behind the statement in Eq. (5.91) runs this way: If we factor out leading terms, the remaining ratio approaches 1. To illustrate this point we utilize our example.

$$\lim_{x \to \infty} \frac{3x^2 + 4x + 2}{5x^2 + x + 6} = \lim_{x \to \infty} \frac{3x^2(1 + 4x^{-1}/3 + 2x^{-2}/3)}{5x^2(1 + x^{-1}/5 + 6x^{-2}/5)}$$

$$= \lim_{x \to \infty} \frac{3x^2}{5x^2} \frac{(1 + 4x^{-1}/3 + 2x^{-2}/3)}{(1 + x^{-1}/5 + 6x^{-2}/5)} = \lim_{x \to \infty} \frac{3x^2}{5x^2} \cdot 1$$

$$= \lim_{x \to \infty} \frac{3x^2}{5x^2} \cdot 1 = \lim_{x \to \infty} \frac{3}{5} = \frac{3}{5}.$$

We employ limits at infinity to define the following common type of *improper integral*:

$$\int_a^\infty f(x)dx = \lim_{A \to \infty} \int_a^A f(x)\, dx. \qquad (5.92)$$

The definite integral on the left is new to us, but we approximate it with an integral over the finite interval $[a, A]$, which is familiar to us, and then take a limit as $A$ approaches $\infty$. If the limit at infinity on the right-hand side of Eq. (5.92) is finite, we say the integral $\int_a^\infty f(x)dx$ *converges*, and if the right-hand side of Eq. (5.92) is not finite or fails to exist we say $\int_a^\infty f(x)dx$ *diverges*.

## EXAMPLE 5.35

Find $\int_0^\infty e^{-x} dx$. **Answer:** As a preliminary we compute the antiderivative $\int e^{-x} dx$ by making the substitution $u = -x$ with corresponding $du = -1 dx$. Then $\int e^{-x} dx = \int -e^u du = -e^u + C = -e^{-x} + C$. We then compute the improper integral:

$$\int_0^\infty e^{-x} dx = \lim_{A \to \infty} \int_0^A e^{-x} dx = \lim_{A \to \infty} -e^{-x} \Big|_0^A$$
$$= \lim_{A \to \infty} -e^{-A} - (-e^{-0}) = \lim_{A \to \infty} -e^{-A} + 1 = 1.$$

It turns out, as in the case of finding the mean of the exponential density in Example (5.34), that we are required to evaluate a limit at $\infty$ of type $\lim_{A \to \infty} A e^{-\lambda A}$, which we claimed was 0.

There is an extension of L'Hopital's Rule that governs this type of limit. Let $\alpha$ represent either a finite point or $\pm \infty$. We say

$$\lim_{x \to \alpha} \frac{f(x)}{g(x)} \quad \text{is in } \textit{indeterminate} \text{ form } \quad \frac{0}{0} \quad \text{or} \quad \frac{\infty}{\infty},$$

provided either $\lim_{x \to \alpha} f(x) = 0 = \lim_{x \to \infty} g(x)$ together, or $\lim_{x \to \alpha} f(x) = \pm \infty = \lim_{x \to \infty} g(x)$ together. For continuously differentiable functions $f$ and $g$, L'Hopital's Rule [59] tells us that these limits of indeterminate form 0/0 or $\infty/\infty$ also satisfy

$$\lim_{x \to \alpha} \frac{f(x)}{g(x)} = \lim_{x \to \alpha} \frac{f'(x)}{g'(x)}.$$

Thus according to L'Hopital's Rule we have, for $\lambda > 0$,

$$\lim_{x \to \infty} x e^{-\lambda x} = \lim_{x \to \infty} \frac{x}{e^{\lambda x}} = \lim_{x \to \infty} \frac{(x)'}{(e^{\lambda x})'} = \lim_{x \to \infty} \frac{1}{\lambda e^{\lambda x}} = 0,$$

and this computation now completes our computation of the mean in Example (5.34).

It turns out that one needs a multiple application of L'Hopital's Rule.

### EXAMPLE 5.36

Find $\lim_{x \to \infty} \dfrac{x^2}{e^x}$.

If we use the symbol $\sim$ to mean *behaves like in the limit*, then at $\infty$ we have

$$\frac{x^2}{e^x} \sim \frac{(x^2)'}{(e^x)'} = \frac{2x}{e^x} \sim \frac{(2x)'}{(e^x)'} = \frac{2}{e^x},$$

and we conclude since both

$$\lim_{x \to \infty} \frac{x^2}{e^x} \quad \text{and} \quad \lim_{x \to \infty} \frac{2x}{e^x} \quad \text{are in indeterminate form} \quad \frac{\infty}{\infty},$$

that

$$\lim_{x \to \infty} \frac{x^2}{e^x} = \lim_{x \to \infty} \frac{2x}{e^x} = \lim_{x \to \infty} \frac{2}{e^x} = 0,$$

by a double application of L'Hopital's Rule.

One can also use L'Hopital's Rule to find the limit of rational functions at infinity.

### EXAMPLE 5.37

Use L'Hopital's Rule to find $\lim_{x \to \infty} (3x^2 + 4x + 5)/(7x^2 + 2x + 11)$. **Answer:**

$$\lim_{x \to \infty} \frac{3x^2 + 4x + 5}{7x^2 + 2x + 11} = \lim_{x \to \infty} \frac{6x + 4}{14x + 2} = \lim_{x \to \infty} \frac{6}{14} = \frac{6}{14} = \frac{3}{7}.$$

Here L'Hopital's Rule was applied successively to the limits in indeterminate form $\infty/\infty$.

### ■ Exercises

1. Evaluate the improper integral $\int_0^\infty e^{-2x} dx$.
2. Use L'Hopital's Rule to evaluate each limit at infinity:
   (a) $\lim_{x \to \infty} x^3/e^x$.
   (b) $\lim_{x \to \infty} (x^2 + x)/(3x^2 + 5x + 2)$.

(c) $\lim_{x \to \infty}(5x)/e^{2x}$.

## 5.21 Integration Techniques

In this section we familiarize ourselves with a few further basic integration techniques. Our aim here is not to be comprehensive. If the reader finds an indefinite integral $\int f(x)dx$ for an unfamiliar function $f(x)$, a reasonable option is to consult an integral table either on the Internet (search for *integral table pdf* or a similar phrase) or in a handbook of mathematical equations. Here you can find numerous antiderivatives worked out in detail.

However there are a few further techniques with which you should be familiar. First and foremost of these is *integration by parts*, which is an integral clone of our product rule for the derivative of a product.

### 5.21.1 Integration by Parts

The derivative of the product $u(x)v(x)$ is computed as

$$(u(x)v(x))' = (u(x))'v(x) + u(x)(v(x))'.$$

Antidifferentiating both sides gives

$$u(x)v(x) = \int (u(x))'v(x)dx + \int u(x)(v(x))' \, dx,$$

where the arbitrary constant of integration on the left-hand side is moved to the right and placed within the constants of integration implicit in the antiderivatives of the right-hand side. Solving for the right-most integral we obtain

$$\int u(x)(v(x))' \, dx = u(x)v(x) - \int (u(x))'v(x) \, dx.$$

In differential notation we have for $u = u(x)$, $v = v(x)$ and $du = u'(x)dx$, $dv = v'(x)dx$, that

$$\int u\,dv = uv - \int v\,du, \tag{5.93}$$

which is our integration technique called *integration by parts*.

The power of integration by parts comes from the fact that it moves the derivative off of the $v$ term in the left-hand side of Eq. (5.93) and moves the derivative to the $u$

term on the right-hand side of Eq. (5.93). This often greatly simplifies our integral computation. In practice one may have many options for choosing $u$ and $dv$ on the left side of Eq. (5.93). The following mnemonic for the choice of $u$ often helps pare down the number of possibilities:

$$\text{LIATE} \leftrightarrow \text{Logarithm} \quad \text{Inverse} \quad \text{Algebraic} \quad \text{Trigonometric} \quad \text{Exponential.}$$

LIATE is a hierarchy telling us to pick $u$ to be a logarithm function over an inverse function over an algebraic function over a trigonometric function over an exponential function in selecting $u$. Once $u$ is selected, the remaining terms, including the $dx$ term, comprise $dv$.

### EXAMPLE 5.38

Find $\int x e^x dx$. **Answer:** Since $x$ is an algebraic function it outranks the exponential function $e^x$ in the LIATE hierarchy and so we choose $u = x$. Then $du = dx$, and the remaining terms $e^x dx$ give $dv = e^x dx$. Now $v = \int dv = \int e^x dx = e^x + C_1$. We conclude that

$$\begin{aligned}\int x\, e^x\, dx &= uv - \int v\, du = x\,(e^x + C_1) - \int (e^x + C_1)\, dx \\ &= x\, e^x + C_1\, x - e^x - C_1\, x + C \\ &= x\, e^x - e^x + C.\end{aligned}$$

Notice that the constant of integration $C_1$ completely canceled out, as will always be the case in integration by parts. Thus when we compute $v = \int dv$ we will always take the constant of integration $C_1$ to be 0.

### EXAMPLE 5.39

When we computed the mean of our exponential distribution, we had to evaluate $\int_0^\infty x e^{-\lambda x} dx$. To do this systematically now, we employ integration by parts, picking the algebraic function $u = x$ and $dv = e^{-\lambda x} dx$. Then $du = dx$ and $v = \int dv = \int e^{-\lambda x} dx = \int e^w (-1/\lambda) dw = (-1/\lambda) e^w = (-1/\lambda) e^{-\lambda x}$, where we have made the substitution $w = -\lambda x$ and $(-1/\lambda) dw = dx$. Next we employ integration by parts to get

$$\int xe^{-\lambda x}\,dx = \int u\,dv = uv - \int v\,du$$
$$= x\,(-1/\lambda)\,e^{-\lambda x} - \int (-1/\lambda)\,e^{-\lambda x}\,dx$$
$$= x\,(-1/\lambda)\,e^{-\lambda x} - (-1/\lambda)\int e^{-\lambda x}\,dx$$
$$= x\,(-1/\lambda)\,e^{-\lambda x} - (1/\lambda^2)\,e^{-\lambda x} + C.$$

Now we use our antiderivative to evaluate our definite integral

$$\int_0^\infty xe^{-\lambda x}\,dx = \lim_{A\to\infty}\int_0^A xe^{-\lambda x}\,dx = \lim_{A\to\infty}\left(x(-1/\lambda)e^{-\lambda x} - (1/\lambda^2)e^{-\lambda x}\right)\Big|_0^A$$
$$= \lim_{A\to\infty}\left(A(-1/\lambda)e^{-\lambda A} - (1/\lambda^2)e^{-\lambda A} - (0(-1/\lambda)e^{-\lambda 0} - (1/\lambda^2)e^{-\lambda 0})\right) = 1/\lambda^2,$$

where

$$\lim_{A\to\infty} A(-1/\lambda)e^{-\lambda A} = (-1/\lambda)\lim_{A\to\infty} A/e^{\lambda A}$$
$$= (-1/\lambda)\lim_{A\to\infty}(A)'/(e^{\lambda A})' = (-1/\lambda)\lim_{A\to\infty} 1/(\lambda e^{\lambda A}) = 0,$$

by an application of L'Hopital's Rule.

### ■ Exercises

1. Evaluate each indefinite integral using integration by parts:

    (a) $\int x\,e^x\,dx$.
    (b) $\int x^2\,e^x\,dx$. (*Hint:* Try a double application of integration by parts.)
    (c) $\int x\,\cos(x)\,dx$.
    (d) $\int \ln(x)\,dx$.
    (e) $\int \sin^{-1}(x)\,dx$.
    (f) $\int \sin(x)e^x\,dx$. (*Hint:* A double application of integration by parts is needed here; and then solve for your indefinite integral with algebra.)

### ■ 5.21.2 Integrals of Trigonometric Expressions

An expression that needs to be integrated can often be converted via a change of variables into a trigonometric expression that is more easily antidifferentiated. It will therefore become important to be able to integrate trigonometric expressions. We start with a pair of trigonometric identities:

$$\cos^2(x) = \frac{1+\cos(2x)}{2}, \quad \sin^2(x) = \frac{1-\cos(2x)}{2}. \tag{5.94}$$

The utility of the identities in Eq. (5.94) is that they reduce higher even powers of sine or cosine to lower powers with higher frequency coefficients, where the latter are not difficult to integrate via substitution. For instance,

$$\begin{aligned}
\int \cos^2(x)\,dx &= \int \frac{1+\cos(2x)}{2}\,dx = \frac{1}{2}\int dx + \frac{1}{2}\int \cos(2x)\,dx \\
&= \frac{x}{2} + \frac{1}{2}\int \cos(u)\,\frac{du}{2} = \frac{x}{2} + \frac{1}{4}\sin(u) + C \\
&= \frac{x}{2} + \frac{1}{4}\sin(2x) + C,
\end{aligned}$$

where we made the substitution $u = 2x$, $du = 2dx$. One shows similarly that

$$\int \sin^2(x)\,dx = \frac{x}{2} - \frac{1}{4}\sin(2x) + C.$$

Alternatively, we could have used integration by parts combined with a trigonometric identity. If we let $u = \cos(x)$ and $dv = \cos(x)dx$, then $du = -\sin(x)dx$ and $v = \sin(x)$. Then

$$\begin{aligned}
\int \cos^2(x)\,dx &= \int \cos(x)\cos(x)\,dx = uv - \int v\,du \\
&= \cos(x)\sin(x) + \int \sin(x)\sin(x)\,dx \\
&= \cos(x)\sin(x) + \int (1-\cos^2(x))\,dx \\
&= \cos(x)\sin(x) + x + C - \int \cos^2(x)\,dx.
\end{aligned}$$

We have now reached the point that

$$\int \cos^2(x)\,dx = \cos(x)\sin(x) + x + C - \int \cos^2(x)\,dx,$$

and moving the rightmost integral to the left side lets us algebraically solve for our integral: $2 \int \cos^2(x) dx = \cos(x)\sin(x) + x + C$. So we have

$$\int \cos^2(x)\, dx = \cos(x)\sin(x)/2 + x/2 + C.$$

The trigonometric identity $\sin(2x) = 2\sin(x)\cos(x)$ shows our two results to be equivalent answers.

We use Eq. (5.94) to successively reduce integrals of products of even powers of sine and cosine to integrals of lower powers, eventually reaching simpler antiderivatives. For instance,

$$\begin{aligned}
\int \sin^2(x)\, \cos^2(x)\, dx &= \int \frac{1-\cos(2x)}{2}\frac{1+\cos(2x)}{2}\, dx = \int \frac{1-\cos^2(2x)}{4}\, dx \\
&= \int \frac{1}{4} - \frac{1+\cos(4x)}{8}\, dx \\
&= \int \frac{1}{8}\, dx - \frac{1}{8}\int \cos(u)\, \frac{du}{4} = \frac{x}{8} - \frac{1}{32}\sin(u) + C \\
&= \frac{x}{8} - \frac{1}{32}\sin(4x) + C,
\end{aligned}$$

where we used the substitution $u = 4x$. Other products of even powers are handled similarly.

On the other hand, given a product of powers of sine and cosine where one of them has an odd power, the process is even simpler. Keep one copy from the odd power, convert all other terms using $\sin^2(x) + \cos^2(x) = 1$, and then convert via substitution to an integral of a polynomial. We illustrate this process with the following example:

$$\begin{aligned}
\int \sin^2(x)\, \cos^3(x)\, dx &= \int \sin^2(x)\, \cos^2(x)\, (\cos(x)\, dx) \\
&= \int \sin^2(x)(1 - \sin^2(x))(\cos(x)\, dx) \\
&= \int u^2(1-u^2)\, (du) = \int (u^2 - u^4)\, du \\
&= \frac{u^3}{3} - \frac{u^5}{5} + C = \frac{\sin^3(x)}{3} - \frac{\sin^5(x)}{5} + C.
\end{aligned}$$

We now know how to integrate integrals of form $\int \sin^n(x)\cos^m(x) dx$ for integers $n, m \geq 0$.

We next turn to integrals of form $\int \tan^n(x)\sec^m(x) dx$ for integers $n, m \geq 0$, and mention a few examples from a range of techniques here.

If $m$ is even: keep $\sec^2 dx$, convert the remaining $\sec^{m-2}(x)$ to tangents via $\sec^2(x) = \tan^2(x) + 1$, then make the substitution $u = \tan(x), du = \sec^2(x)dx$, and integrate the resulting polynomial. As an example we compute

$$\begin{aligned}
\int \tan^2(x) \sec^4(x) \, dx &= \int \tan^2(x) \sec^2(x)(\sec^2(x) \, dx) \\
&= \int \tan^2(x)(1 + \tan^2(x)) \sec^2(x) \, dx \\
&= \int u^2(1 + u^2) \, du = \int (u^2 + u^4) \, du \\
&= \frac{u^3}{3} + \frac{u^5}{5} + C = \frac{\tan^3(x)}{3} + \frac{\tan^5(x)}{5} + C,
\end{aligned}$$

where we made the substitution $u = \tan(x)$.

If $m \geq 1$ and if $n \geq 1$ is odd: keep one power each of $\sec(x)$ and $\tan(x)$ with $dx$, convert the remaining terms to a polynomial in $\sec(x)$ via the identity $\tan^2(x) = \sec^2(x) - 1$, then substitute $u = \sec(x)$, and reduce to an integral of a polynomial. As an example we compute

$$\begin{aligned}
\int \tan^3(x) \sec^3(x) \, dx &= \int \tan^2(x) \sec^2(x)(\sec(x) \tan(x) \, dx) \\
&= \int (\sec^2(x) - 1) \sec^2(x)(\sec(x) \tan(x) \, dx) \\
&= \int (u^2 - 1)(u^2) \, du = \int u^4 - u^2 \, du \\
&= \frac{u^5}{5} - \frac{u^3}{3} + C = \frac{\sec^5(x)}{5} - \frac{\sec^3(x)}{3} + C,
\end{aligned}$$

where we made the substitution $u = \sec(x)$.

An interesting integral that does not fall into either preceding category, but often appears in computations, is $\int \sec^3(x) dx$. This is handled via integration by parts, a trigonometric identity, and use of algebra to solve for the integral. We start off by letting $u = \sec(x), dv = \sec^2(x)dx$, and $du = \sec(x)\tan(x)dx, v = \tan(x)$. We obtain

$$\int \sec^3(x)\, dx = \int \sec(x)\sec^2(x)\, dx = \int u\, dv = uv - \int v\, du$$
$$= \sec(x)\tan(x) - \int \tan(x)\sec(x)\tan(x)\, dx$$
$$= \sec(x)\tan(x) - \int \sec(x)(\sec^2(x) - 1)\, dx$$
$$= \sec(x)\tan(x) + \int \sec(x)\, dx - \int \sec^3(x)\, dx.$$

We conclude, by placing both integrals involving $\sec^3(x)$ on the left side, that

$$2\int \sec^3(x)\, dx = \sec(x)\tan(x) + \int \sec(x)\, dx$$
$$= \sec(x)\tan(x) + \ln|\sec(x) + \tan(x)| + C,$$

which gives

$$\int \sec^3(x)\, dx = (1/2)\sec(x)\tan(x) + (1/2)\ln|\sec(x) + \tan(x)| + C.$$

The preceding cases should handle most integrals of trigonometric expressions that you might encounter in applications. An integral table serves as a backup to handle any unfamiliar cases not yet handled.

### Exercises

1. Evaluate each indefinite integral of a trigonometric expression:

   (a) $\int \sin^3(x)\cos^2(x)\, dx$.
   (b) $\int \sin^2(x)\cos^4(x)\, dx$.
   (c) $\int \tan(x)\sec^4(x)\, dx$.
   (d) $\int \tan(x)\sec^3(x)\, dx$.

## 5.21.3 Trigonometric Substitutions

When integrating any expressions involving powers of $\sqrt{r^2 - x^2}$, $\sqrt{r^2 + x^2}$, $\sqrt{x^2 - r^2}$, we replace $x$ by an appropriate trigonometric expression, with the goal of converting the square root term to a square root of a perfect square, and simplifying.

The technique for $\sqrt{r^2 - x^2}$ is to: use the trigonometric substitution $x = r\sin(\theta)$, $dx = r\cos(\theta)d\theta$ for $\theta \in [-\pi/2, \pi/2]$; then $\sqrt{r^2 - x^2} = \sqrt{r^2 - r^2\sin^2(\theta)} = \sqrt{r^2(1 - \sin^2(\theta))} = \sqrt{r^2 \cos^2(\theta)} = r\cos(\theta)$. We then integrate the resulting trigonometric integral and convert our answer in $\theta$ to an answer in terms of $x$ via the right triangle with $\theta$ as angle, $x$ as opposite side, $r$ as hypothenuse, and $\sqrt{r^2 - x^2}$ as adjacent side. Then $\theta = \sin^{-1}(x/r)$.

The technique for $\sqrt{r^2 + x^2}$ is to: use the trigonometric substitution $x = r \cdot \tan(\theta)$, $dx = r\sec^2(\theta)d\theta$ for $\theta \in (-\pi/2, \pi/2)$; then $\sqrt{r^2 + x^2} = \sqrt{r^2 + r^2\tan^2(\theta)} = \sqrt{r^2(1 + \tan^2(\theta))} = \sqrt{r^2 \sec^2(\theta)} = r\sec(\theta)$. We then integrate the resulting trigonometric integral and convert our answer in $\theta$ to an answer in terms of $x$ via the right triangle with $\theta$ as angle, $x$ as opposite side, $r$ as adjacent side, and $\sqrt{r^2 + x^2}$ as hypothenuse. Here $\theta = \tan^{-1}(x/r)$.

The technique for $\sqrt{x^2 - r^2}$ is to: use the trigonometric substitution $x = r\sec(\theta)$, $dx = r\sec(\theta)\tan(\theta)d\theta$; then $\sqrt{x^2 - r^2} = \sqrt{r^2\sec^2(\theta) - r^2} = \sqrt{r^2(\sec^2(\theta) - 1)} = \sqrt{r^2 \tan^2(\theta)} = r\tan(\theta)$. We then integrate the resulting trigonometric integral and convert our answer in $\theta$ to an answer in terms of $x$ via the right triangle with $\theta$ as angle, $x$ as hypothenuse, $r$ as adjacent side, and $\sqrt{x^2 - r^2}$ as opposite side. Here $\theta = \sec^{-1}(x/r)$.

As a first example, we evaluate $\int \sqrt{r^2 - x^2}\, dx$ by letting $x = r\sin(\theta)$, $dx = r\cos(\theta)d\theta$. We obtain

$$\int \sqrt{r^2 - x^2}\, dx = \int \sqrt{r^2 - r^2\sin^2(\theta)}\, r\cos(\theta)\, d\theta$$
$$= \int \sqrt{r^2(1 - \sin^2(\theta))}\, r\cos(\theta)d\theta = \int \sqrt{r^2\cos^2(\theta)}\, r\cos(\theta)d\theta = \int r^2\cos^2(\theta)d\theta$$
$$= r^2 \int \frac{1 + \cos(2\theta)}{2} d\theta = \frac{r^2}{2}\left(\theta + \frac{\sin(2\theta)}{2}\right) + C = \frac{r^2}{2}(\theta + \sin(\theta)\cos(\theta)) + C$$
$$= \frac{r^2}{2}\left(\sin^{-1}(x/r) + \frac{x}{r}\frac{\sqrt{r^2 - x^2}}{r}\right) + C = \frac{r^2}{2}\sin^{-1}(x/r) + \frac{1}{2}\left(x\sqrt{r^2 - x^2}\right) + C.$$

We use the preceding computation in the next example to find the area of a circle.

### EXAMPLE 5.40

Find the area $A$ of the circle of radius $r$. **Answer:** We find the area under the semicircle $y = \sqrt{r^2 - x^2}$ by computing $\int_{-r}^{r} \sqrt{r^2 - x^2}\, dx$ and then doubling.

$$\begin{aligned} A &= 2\int_{-r}^{r} \sqrt{r^2 - x^2}\, dx = 2\left(\frac{r^2}{2}\sin^{-1}(x/r) + \frac{1}{2}\left(x\sqrt{r^2 - x^2}\right)\right)\Big|_{-r}^{r} \\ &= 2\frac{r^2}{2}(\sin^{-1}(r/r) - \sin^{-1}(-r/r)) = r^2(\pi/2 - (-\pi/2))) = \pi r^2. \end{aligned}$$

Here is another example of the use of a trigonometric substitution. We evaluate $\int 1/(r^2 + x^2)\, dx$ by our new techniques. In this case, we let $x = r\tan(\theta)$ with $dx = r\,\sec^2(\theta)\, d\theta$.

$$\begin{aligned} \int \frac{1}{r^2 + x^2}\, dx &= \int \frac{1}{r^2 + r^2 \tan^2(\theta)}\, r\,\sec^2(\theta)\, d\theta \\ &= \int \frac{1}{r^2 \sec^2(\theta)}\, r\,\sec^2(\theta)\, d\theta \\ &= \int \frac{1}{r}\, d\theta = \frac{\theta}{r} + C = \frac{1}{r}\tan^{-1}(x/r) + C, \end{aligned}$$

where we replaced $1 + \tan^2(\theta)$ by $\sec^2(\theta)$ in the denominator. Notice that when $r = 1$ we are in agreement with our previous formula that $\int 1/(1 + x^2)\, dx = \tan^{-1}(x) + C$.

Here is a final illustration of the technique of trigonometric substitution. We evaluate $\int dx/\sqrt{r^2 - x^2}$. In this case, we let $x = r\,\sin(\theta)$ with $dx = r\,\cos(\theta)\, d\theta$.

$$\begin{aligned} \int \frac{dx}{\sqrt{r^2 - x^2}} &= \int \frac{r\,\cos(\theta)\, d\theta}{\sqrt{r^2 - r^2 \sin^2(\theta)}} = \int \frac{r\,\cos(\theta)\, d\theta}{\sqrt{r^2 \cos^2(\theta)}} \\ &= \int d\theta = \theta + C = \sin^{-1}(x/r) + C, \end{aligned}$$

where $1 - \sin^2(\theta)$ in our square root was replaced by $\cos^2(\theta)$. If $r = 1$ our formula reduces to the case that $\int 1/\sqrt{1 - x^2}\, dx = \sin^{-1}(x) + C$, which we have previously seen.

## Exercises

1. Find each indefinite integral using trigonometric substitution:
   (a) $\int \sqrt{9 - x^2}\, dx$.
   (b) $\int 1/\sqrt{1 + x^2}\, dx$.
   (c) $\int (4 - x^2)^{3/2}\, dx$.
   (d) $\int \sqrt{x^2 - 1}\, dx$. (*Hint:* Rely on the computation of $\int \sec^3(x)\, dx$ from the text.)

### 5.21.4 Examples Illustrating the Method of Partial Fractions

In studying population growth, one starts with the law of natural growth, which says that the rate of change of the population is proportional to the population. This is expressed as

$$P'(t) = r\, P(t),$$

or in differential notation as $dP = rP dt$. If one divides both sides by $P$ and integrates, one obtains

$$\int \frac{dP}{P} = \int r\, dt,$$

which gives $\ln(P) = rt + C$. Taking exponentials of both sides gives $P = e^{\ln(P)} = e^{rt+C} = e^C e^{rt}$. Evaluating the population at time $t = 0$ gives the initial population, say $P_0$, so $P(0) = P_0 = e^C e^0 = e^C$. These combine to give the population at future times:

$$P(t) = P_0 e^{rt}.$$

Thus the law of natural growth is equivalent to exponential growth.

The law of natural growth also applies to investments compounded continuously. $P(t)$ is considered in this case to be the value of the investment at time $t$, where $r$ is the rate of return. The preceding formulas again hold and give exponential growth of the investment.

The preceding has essentially been a study of the IVP

$$P'(t) = r\, P(t), \quad P(0) = P_0.$$

To obtain an expression for $P(t)$ it was necessary to be able to antidifferentiate the rational function $1/P$. In fact there are many similar models where antidifferentiating rational functions is a necessity. In doing so, the rational function is decomposed into component functions that are simpler to antidifferentiate. This process is called *partial fraction decomposition*.

Here is an illustration of partial fraction decomposition within the context of constrained population growth, called the *logistic population model*. Here it is assumed that the population has a *carrying capacity*, $M$, thought of as the maximal supportable population. The previous initial value problem is modified to account for a slower rate of growth as the population nears $M$ in size in the following way:

$$P'(t) = r\, P(t)\, (M - P(t)), \quad P(0) = P_0.$$

In the early stages when $P$ is very small compared to $M$ the factor $M - P \approx M$ and the differential equation becomes $P' \approx (rM)P$, which says this model is close to exponential growth in the early stages. On the other hand, when $P$ nears $M$, the factor $M - P \approx 0$ and then $P' \approx 0$, which says that population growth slows down to zero growth rate as the carrying capacity is attained.

In differential notation this gives $dP = rP(M - P)dt$, or equivalently

$$\frac{dP}{P\,(M - P)} = r\, dt,$$

which integrates to give

$$\int \frac{dP}{P\,(M - P)} = \int r\, dt = r\, t + C. \tag{5.95}$$

The utility of antidifferentiating rational functions such as $1/(P(M - P))$ is immediately seen. This example is now used as our gateway to the method of partial fractions.

First, a partial fraction decomposition of the integrand $1/(P(M - P))$, is found by writing

$$\frac{1}{P\,(M - P)} = \frac{A}{P} + \frac{B}{M - P},$$

where $A, B$ are constants to be determined. Obtaining a common denominator gives

$$\frac{1}{P\,(M - P)} = \frac{A\,(M - P) + B\,P}{P\,(M - P)},$$

from which the numerators must be equal

$$1 = A(M - P) + BP. \tag{5.96}$$

$A$ and $B$ are now determined algebraically: first letting $P = 0$ yields $A = 1/M$, and then letting $P = M$ yields $B = 1/M$. The strategy here was to pick values of $P$ that let each factor in the sum of Eq. (5.96) vanish. Now

$$\frac{1}{P(M-P)} = \frac{1/M}{P} + \frac{1/M}{M-P},$$

which allows for integration of the rational function on the left in terms of the antiderivatives of the simpler functions on the right.

$$\int \frac{1}{P(M-P)} dP = \int \frac{1/M}{P} dP + \int \frac{1/M}{M-P} dP$$
$$= (1/M)\ln(P) - (1/M)\ln(M-P) + C.$$

The rightmost integral was obtained from the substitution $u = M - P, du = -dP$. Now Eq. (5.95) becomes

$$(1/M)\ln(P) - (1/M)\ln(M-P) = rt + C,$$

or $\ln(P/(M-P)) = rMt + C$. Exponentiating both sides gives

$$(P/[M-P]) = C_1 \exp(rMt), \tag{5.97}$$

or $P = (M-P)C_1 \exp(rMt)$. Solving for $P$ gives

$$P = \frac{M C_1 e^{rMt}}{1 + C_1 e^{rMt}} = \frac{M}{1(C_1)^{-1} e^{-rMt} + 1}. \tag{5.98}$$

If one examines Eq. (5.97) at time 0, one sees that $P$ becomes $P_0$ and Eq. (5.97) is $P_0/[M - P_0] = C_1$, and then Eq. (5.98) becomes

$$P(t) = \frac{MP_0}{(M-P_0)e^{-rMt} + P_0}.$$

At time 0, $P(0) = MP_0/[M - P_0 + P_0] = P_0$, and the initial condition is met. As $t$ approaches $\infty$ the term $e^{-rMt}$ approaches 0 and then $P(t)$ approaches $MP_0/P_0 = M$, and in the long run the population approaches the carrying capacity $M$.

In general the method of partial fractions decomposes a rational function $p(x)/q(x)$ with the degree of $p$ less than the degree of $q$ into pieces based on the irreducible factors of the denominator $q(x)$, with each such factor creating a set of terms in a partial fractions decomposition. If $q(x) = \ldots (Ax+B)^N \ldots (ax^2+bx+c)^M \ldots$ then the partial fraction decomposition takes the form:

$$\frac{p(x)}{q(x)} = \ldots \tag{5.99}$$

$$\frac{A_1}{(Ax+B)^1} + \frac{A_2}{(Ax+B)^2} + \frac{A_3}{(Ax+B)^3} + \cdots + \frac{A_N}{(Ax+B)^N} \tag{5.100}$$

$$+ \ldots$$

$$\frac{B_1 x + C_1}{(ax^2+bx+c)^1} + \frac{B_2 x + C_2}{(ax^2+bx+c)^2} + \cdots + \frac{B_M x + C_M}{(ax^2+bx+c)^M} \tag{5.101}$$

$$+ \ldots$$

where the terms in Eq. (5.100) are associated to the linear irreducible factor $(Ax+B)$ raised to the $N$th power in the factorization of $q$, and the terms in Eq. (5.101) are associated to the quadratic irreducible factor $(ax^2+bx+c)$ raised to the $M$th power in the factorization of $q$. In Eq. (5.99) one then obtains a common denominator, and sets the two numerators on the left and right equal. By matching the coefficients of like powers of $x^k$ in each numerator, we obtain sufficiently many conditions to determine all the unknown constants, $A_i$, $B_i$, and $C_i$. To this point, only algebraic techniques have been utilized.

To integrate $p(x)/q(x)$, it is now necessary to be able to integrate each summand in the partial fractions decomposition.

The terms associated to the linear factor $(Ax + B)$ in $q$ are handled via the substitution $u = Ax + B$:

$$\int \frac{A_k}{(Ax+B)^k} \, dx = \int \frac{A_k}{(u)^k} \frac{du}{A},$$

which yields either a logarithm when $k = 1$, or a power $u^{-k+1}/(-k+1)$ when $k \geq 2$. The terms associated to the quadratic factor $(ax^2 + bx + c)$ are handled in the following way.

$$\int \frac{B_k x + C_k}{(ax^2+bx+c)^k} \, dx = \int \frac{D_k \{2ax + b\} + E_k}{(ax^2+bx+c)^k} \, dx,$$

gives two terms to integrate. The first is handled via the substitution $u = ax^2 + bx + c$, which gives

$$\int \frac{D_k \{2ax + b\}}{(ax^2 + bx + c)^k} \, dx = \int \frac{D_k \, du}{(u)^k},$$

resulting in a logarithmic antiderivative when $k = 1$, or a power $D_k u^{-k+1}/(-k+1)$ when $k \geq 2$. In the second integral one completes the square in the denominator obtaining

$$\int \frac{E_k}{(ax^2 + bx + c)^k} \, dx = \int \frac{E_k}{(a[x + b/(2a)]^2 + c - b^2/(4a))^k} \, dx,$$

and sets $r^2 = c - b^2/(4a)$ and $u = \sqrt{a}[x + b/(2a)], du = \sqrt{a} \, dx$ to get

$$\int \frac{E_k}{(ax^2 + bx + c)^k} dx = \int \frac{E_k}{(u^2 + r^2)^k} \frac{du}{\sqrt{a}}.$$

In this last form we see that the trigonometric substitution $u = r\tan(\theta)$ converts to an integrable format using powers of secants and tangents; or in some cases one obtains an arctangent directly from the integral.

An illustration of this last type of integral is

$$\int \frac{4x}{4x^2 + 8x + 5} \, dx = \int \frac{(1/2)(8x + 8) - 4}{4x^2 + 8x + 5} \, dx$$
$$= \frac{1}{2} \int \frac{(8x + 8)}{4x^2 + 8x + 5} \, dx + \int \frac{-4}{4x^2 + 8x + 5} \, dx.$$

Now $u = 4x^2 + 8x + 5, \ du = (8x + 8)dx$ gives the first integral as

$$\frac{1}{2} \int \frac{(8x + 8)}{4x^2 + 8x + 5} \, dx = \frac{1}{2} \int \frac{du}{u}$$
$$= \frac{1}{2} \ln(|u|) + C = \frac{1}{2} \ln(|4x^2 + 8x + 5|) + C.$$

The second integral relies first on a completion of squares, and then on the substitution $u = 2(x+1)$, $du = 2dx$:

$$\int \frac{-4}{4x^2 + 8x + 5} \, dx = -4 \int \frac{dx}{4(x+1)^2 + 1}$$
$$= -4 \int \frac{du/2}{u^2 + 1} = -2 \int \frac{du}{u^2 + 1}$$
$$= -2\tan^{-1}(u) + C = -2\tan^{-1}(2(x+1)) + C.$$

Piecing these together yields

$$\int \frac{4x}{4x^2 + 8x + 5} \, dx = \frac{1}{2}\ln(|4x^2 + 8x + 5|) - 2\tan^{-1}(2(x+1)) + C.$$

### ■ Exercises

1. (a) Find the partial fraction decomposition of $1/(x(x-5))$.
   (b) Evaluate the indefinite integral $\int 1/(x(x-5)) \, dx$.
2. (a) Find the partial fraction decomposition of $1/(x(x^2+1))$.
   (b) Evaluate the indefinite integral $\int 1/(x(x^2+1)) \, dx$.
3. Evaluate each indefinite integral:

   (a) $\int 5/(3x+4)^3 \, dx$; and
   (b) $\int 2/(3x+4) \, dx$.

4. Evaluate each indefinite integral:

   (a) $\int 1/(x^2 + 2x + 2) \, dx$; and
   (b) $\int x/(x^2 + 2x + 2) \, dx$.

## ■ 5.22 Series and Taylor Series

By performing the following subtraction of polynomials

$$\begin{array}{cccccc}
 & 1 + & x & + \cdots + & x^n & \\
- & & x & - \cdots - & x^n - & x^{n+1} \\
\hline
 & 1 + & 0 & + \cdots + & 0 - & x^{n+1}
\end{array}$$

one sees that

$$(1-x) \cdot \sum_{k=0}^{n} x^k = \sum_{k=0}^{n} x^k - x \cdot \sum_{k=0}^{n} x^k = 1 - x^{n+1}.$$

This gives that for $x \neq 1$

$$\sum_{k=0}^{n} x^k = \frac{1-x^{n+1}}{1-x} = \frac{1}{1-x} - \frac{x^{n+1}}{1-x}. \qquad (5.102)$$

Precisely when $|x| < 1$ one has $\lim_{n \to \infty} x^{n+1} = 0$, giving

$$\lim_{n \to \infty} \sum_{k=0}^{n} x^k = \frac{1}{1-x} - \lim_{n \to \infty} \frac{x^{n+1}}{1-x} = \frac{1}{1-x}. \qquad (5.103)$$

The *infinite geometric series* is defined to be

$$\sum_{k=0}^{\infty} x^k = \lim_{n \to \infty} \sum_{k=0}^{n} x^k = \frac{1}{1-x} \quad \text{for } |x| < 1, \qquad (5.104)$$

and this series is said to *converge* when $|x| < 1$ and *diverge* when $|x| \geq 1$.
For instance when $x = .1$ then Eq. (5.104) becomes

$$\sum_{k=0}^{\infty} (.1)^k = \lim_{n \to \infty} \sum_{k=0}^{n} (.1)^k = 1.11111\ldots = 1/(1-(.1)) = 10/9,$$

which is recognized as the repeating decimal expansion for $10/9$.

In general, an *infinite series* expanded about the point $p$ is taken to be

$$\sum_{k=0}^{\infty} a_k \, [x-p]^k = \lim_{n \to \infty} \sum_{k=0}^{n} a_k \, [x-p]^k \quad \text{should this limit exist,} \qquad (5.105)$$

and one sets about determining the values of $x$ for which the infinite series *converges*, or makes sense. Surprisingly, the geometric series can often be harnessed to decide when a given infinite series converges. As a first example of this, examine the infinite series $\sum_{k=0}^{\infty} 2^k [x-3]^k$. Because $\sum_{k=0}^{\infty} 2^k [x-3]^k = \sum_{k=0}^{\infty} (2[x-3])^k$ is the geometric series evaluated at $2[x-3]$, it converges for $|2[x-3]| < 1$, or equivalently for $|x-3| < 1/2$. One says the *radius of convergence* of $\sum_{k=0}^{\infty} 2^k[x-3]^k$ is $1/2$ about the point of expansion 3.

In general, the infinite series $\sum_{k=0}^{\infty} a_k[x-p]^k$ *converges absolutely* when $\sum_{k=0}^{\infty} |a_k[x-p]^k|$ converges. It can be shown that if a series converges absolutely, then the series converges and does so in a very strong sense (allowing for rearrangements of the summands). A first result in deciding convergence is the *Root test*.

### ■ Theorem 5.32

**Root Test**

Given an infinite series $\sum_{k=0}^{\infty} a_k[x-p]^k$, if

$$\lim_{k \to \infty} \sqrt[k]{|a_k|} = L$$

then the series converges absolutely precisely when $|x-p| < 1/L$.

We remark that if $L = 0$ we take $1/L$ to be $\infty$ and understand that the series converges for all $x$. Also, $R = 1/L$ is said to be the *radius of convergence* of the series.

**Proof** Let $\lim_{k \to \infty} \sqrt[k]{|a_k|} = L$. Take a small value $\epsilon > 0$ and let $\tilde{L} = L + \epsilon$. Then there is a $k_0$ so that for all $k \geq k_0$ we have $\sqrt[k]{|a_k|} < \tilde{L}$. For these $k$ we then know $|a_k| < \tilde{L}^k$ and $|a_k[x-p]^k| < (\tilde{L}|x-p|)^k$. Then $\sum_{k=0}^{\infty}(\tilde{L}|x-p|)^k$ is the geometric series evaluated at $\tilde{L}|x-p|$, so it converges when $\tilde{L}|x-p| < 1$, or equivalently when $|x-p| < 1/\tilde{L}$. Thus $\sum_{k=k_0}^{\infty} |a_k[x-p]^k| \leq \sum_{k=k_0}^{\infty}(\tilde{L}|x-p|)^k$ gives that the smaller sum converges when $|x-p| < 1/\tilde{L}$. One concludes by adding the terms $\sum_{k=0}^{k_0-1} |a_k[x-p]^k|$ that $\sum_{k=0}^{\infty} |a_k[x-p]^k|$ converges for $|x-p| < 1/\tilde{L}$. Letting $\epsilon$ approach 0 forces $\tilde{L}$ to $L$, and now $\sum_{k=0}^{\infty} |a_k[x-p]^k|$ converges for $|x-p| < 1/L$. Thus the series converges absolutely. It is true, but not shown here, that $|x-p| > 1/L$ gives divergence. □

As a quick example, examine the series $\sum_{k=0}^{\infty} C^k[x-p]^k$ for $C \neq 0$. Then $\lim_{k \to \infty} \sqrt[k]{|a_k|} = \lim_{k \to \infty} \sqrt[k]{|C^k|} = \lim_{k \to \infty} |C| = |C|$, and the Root test tells us that the series converges for $|x-p| < 1/|C|$. The radius of convergence is seen to be $R = 1/|C|$.

There is another test called the Ratio test, similar in flavor to the Root test, that is also obtained by comparing with a geometric series.

### ■ Theorem 5.33
**Ratio Test**

Let $\sum_{k=0}^{\infty} a_k[x-p]^k$ be an infinite series with $a_k \neq 0$ for all $k$. If

$$\lim_{k \to \infty} \left| \frac{a_{k+1}}{a_k} \right| = L,$$

then the series converges absolutely precisely when $|x-p| < 1/L$.

**Proof** Let $\lim_{k\to\infty} |a_{k+1}/a_k| = L$. Take a small value $\epsilon > 0$ and let $\tilde{L} = L + \epsilon$. Then there is a $k_0$ so that for all $k \geq k_0$ $|a_{k+1}/a_k| < \tilde{L}$. This is equivalent to $|(a_{k+1}[x-p]^{k+1})/(a_k[x-p]^k)| < \tilde{L}|x-p|$ for all $k \geq k_0$.

For these $k$ we then know $|a_{k_0+1}||x-p|^{k_0+1} < \{\tilde{L}|x-p|\}|a_{k_0}||x-p|^{k_0}$; and $|a_{k_0+2}||x-p|^{k_0+2} < (\tilde{L}|x-p|)|a_{k_0+1}||x-p|^{k_0+1} < (\tilde{L}|x-p|)^2|a_{k_0}||x-p|^{k_0}$; and $|a_{k_0+3}||x-p|^{k_0+3} < (\tilde{L}|x-p|)|a_{k_0+2}||x-p|^{k_0+2} < (\tilde{L}|x-p|)^3|a_{k_0}||x-p|^{k_0}$. Proceeding in this manner indefinitely for all integers $m \geq 0$ it is true that $|a_{k_0+m}||x-p|^{k_0+m} < (\tilde{L}|x-p|)^m|a_{k_0}||x-p|^{k_0}$.

Thus $\sum_{m=0}^{\infty} |a_{k_0+m}||x-p|^{k_0+m} < \sum_{m=0}^{\infty} (\tilde{L}|x-p|)^m |a_{k_0}||x-p|^{k_0} = |a_{k_0}||x-p|^{k_0} \sum_{m=0}^{\infty} (\tilde{L}|x-p|)^m$ where the last sum is a geometric series evaluated at $\tilde{L}|x-p|$. Then $\sum_{m=0}^{\infty} (\tilde{L}|x-p|)^m$ converges when $\tilde{L}|x-p| < 1$, or equivalently when $|x-p| < 1/\tilde{L}$. Thus

$$\sum_{k=k_0}^{\infty} |a_k[x-p]^k| = \sum_{m=0}^{\infty} |a_{k_0+m}||x-p|^{k_0+m}$$

$$\leq |a_{k_0}||x-p|^{k_0} \sum_{m=0}^{\infty} (\tilde{L}|x-p|)^m,$$

gives that the smaller sum converges when $|x-p| < 1/\tilde{L}$. By adding the terms $\sum_{k=0}^{k_0-1} |a_k[x-p]^k|$ one sees that $\sum_{k=0}^{\infty} |a_k[x-p]^k|$ converges for $|x-p| < 1/\tilde{L}$. Letting $\epsilon$ approach 0 forces $\tilde{L}$ to $L$, and gives that $\sum_{k=0}^{\infty} |a_k[x-p]^k|$ converges for $|x-p| < 1/L$. Thus the series converges absolutely. It is not shown here that $|x-p| > 1/L$ gives divergence. □

Returning to the earlier series $\sum_{k=0}^{\infty} C^k[x-p]^k$ for $C \neq 0$, one has $\lim_{k\to\infty} |a_{k+1}/a_k| = \lim_{k\to\infty} |C^{k+1}/C^k| = \lim_{k\to\infty} |C| = |C|$, and the Ratio test now says that the series converges for $|x-p| < 1/|C|$. The radius of convergence is again seen to be $R = 1/|C|$.

Examine the series $\sum_{k=0}^{\infty}(2^k + 1)[x - p]^k$. Then $\lim_{k\to\infty}|a_{k+1}/a_k| = \lim_{k\to\infty}|(2^{k+1} + 1)/(2^k + 1)| = \lim_{k\to\infty}|2^{k+1}/2^k||(1 + 1/2^{k+1})/(1 + 1/2^k)| = \lim_{k\to\infty}2|(1 + 1/2^{k+1})/(1 + 1/2^k)| = 2 \cdot 1 = 2$, and the Ratio test gives that the series converges for $|x - p| < 1/2$. The radius of convergence is $R = 1/2$.

Examine the series $\sum_{k=0}^{\infty} x^k/k!$. Then $\lim_{k\to\infty}|a_{k+1}/a_k| = \lim_{k\to\infty}|(1/(k+1)!)/(1/k!)| = \lim_{k\to\infty} 1/(k+1) = 0$, and the Ratio test now shows that the series converges for $|x| < \infty$. The radius of convergence is $R = \infty$. Of course this series represents our function $\exp(x)$ and it is formally proven that this converges for all $x$ now.

Infinite series have one large advantage in that they behave like polynomials with respect to both differentiation and integration. It can be shown that

$$f(x) = \sum_{k=0}^{\infty} a_k [x - p]^k \implies f'(x) = \sum_{k=0}^{\infty} k \, a_k [x - p]^{k-1}$$

and

$$\int f(x)\, dx = C + \sum_{k=0}^{\infty} \frac{a_k}{k + 1} [x - p]^{k+1},$$

and the new series for $f'(x)$ and $\int f(x)dx$ have the same radius of convergence as does $f(x)$.

Suppose a given function $f(x)$ is infinitely differentiable. The tangent line for $f(x)$ has equation $y(x) = f(p) + f'(p)[x-p]$. And this line satisfies that $y(p) = f(p)$ and $y'(p) = f'(p)$. One says $y(x)$ matches 0th and 1st derivatives with $f(x)$. The question becomes, "Is there a series $\sum_{k=0}^{\infty} a_k[x - p]^k = g(x)$ that has the same derivatives of all orders as does $f(x)$ at the point $p$?" We proceed to solve for such a series by finding its coefficients $a_k$. First $f(p) = g(p) = \sum_{k=0}^{\infty} a_k[0]^k = a_0$ since only the constant term is nonzero in this case. The value of $a_0$ is now known. The following is a table of the behavior of the higher derivatives of the series $g(x)$ and their evaluations at $p$.

$$g'(x) = \sum_{k=1}^{\infty} k a_k [x - p]^{k-1} \qquad g'(p) = a_1 = f'(p)$$

$$g^{(2)}(x) = \sum_{k=2}^{\infty} k(k - 1) a_k [x - p]^{k-2} \qquad g^{(2)}(p) = 2! a_2 = f^{(2)}(p)$$

$$g^{(3)}(x) = \sum_{k=3}^{\infty} k(k - 1)(k - 2) a_k [x - p]^{k-3} \qquad g^{(3)}(p) = 3! a_3 = f^{(3)}(p)$$

$$g^{(4)}(x) = \sum_{k=4}^{\infty} k(k - 1)(k - 2)(k - 3) a_k [x - p]^{k-4} \qquad g^{(4)}(p) = 4! a_4 = f^{(4)}(p)$$

$$\vdots$$

For all integers $n \geq 0$ we have

$$g^{(n)}(x) = \sum_{k=n}^{\infty} \frac{k!}{(k-n)!} a_k [x-p]^{k-n} \quad g^{(n)}(p) = n! \, a_n = f^{(n)}(p).$$

The coefficients are seen to be $a_n = f^{(n)}(p)/n!$ and one concludes that the power series with derivatives of all orders matching those of $f(x)$ takes the form:

$$\sum_{k=0}^{\infty} \frac{f^k(p)}{k!} [x-p]^k, \tag{5.106}$$

which is called the *Taylor Series* for $f$ expanded about the point $p$. The next concern is the relation of the Taylor Series $g$ to the original function $f$. In particular, the question becomes where they are equal.

To consider this question, we define the *Taylor Polynomial* $P_N(x)$ of $f$ expanded about the point $p$ to be the first terms up to degree $N$ of the Taylor series:

$$P_N(x) = \sum_{k=0}^{N} \frac{f^k(p)}{k!} [x-p]^k.$$

Notice that the first degree Taylor Polynomial is just the linearization for $f(x)$ at $p$:

$$P_1(x) = f(p) + f'(p) [x-p],$$

which is the line (degree one polynomial) best approximating $f(x)$ near $p$. The degree 2 Taylor Polynomial for $f(x)$ at $p$ is:

$$P_2(x) = f(p) + f'(p) [x-p] + \frac{f^{(2)}(p)}{2!} [x-p]^2,$$

which is the quadratic best approximating $f(x)$ near $p$. $P_2(x)$ shares the same second derivative with $f$ at $p$, so it is curving in sync with $f$ at $p$, which lets $P_2$ better approximate $f$ on a wider interval about $p$ than does $P_1(x)$. In general, $P_N(x)$ sharing higher derivatives with $f$ at $p$ allows $P_N(x)$ to maintain a better approximation over an even larger interval about $p$.

Given the Taylor Polynomial $P_N(x)$, there is a corresponding *Taylor Remainder*:

$$R_N(x) = f(x) - \sum_{k=0}^{N} \frac{f^k(p)}{k!} [x-p]^k.$$

Thus the remainder term $R_N(x)$ measures how far the Taylor polynomial $P_N(x)$ is from $f(x)$, and $f(x) = P_N(x) + R_N(x)$. The Taylor Polynomial $P_N(x)$ will exist even if $f$ has only $N$ continuous derivatives at $p$. Furthermore, one obtains equality of $f(x)$ with the Taylor series precisely when

$$f(x) = \sum_{k=0}^{\infty} \frac{f^k(p)}{k!} [x-p]^k \iff \lim_{N \to \infty} R_N(x) = 0.$$

Thus the goal will be to show the Taylor Remainder $R_N(x)$ approaches 0 as $N$ approaches $\infty$.

It can be shown that, for $f$ having $N+1$ continuous derivatives, the Taylor Remainder takes the form:

$$R_N(x) = \frac{f^{N+1}(c)}{(N+1)!} [x-p]^{N+1},$$

for some point $c$ lying between $p$ and $x$. This form for the Taylor Remainder is very similar to the $(N+1)$st term of the Taylor series, except that the derivative is evaluated at the undetermined point $c$ instead of the expansion point $p$. This form is also the most convenient to analyze when taking the limit $\lim_{N \to \infty} R_N(x)$.

### EXAMPLE 5.41

Given the function $f(x) = e^x$, find the Taylor Series, the Taylor Polynomial $P_N(x)$ of degree $N$, the corresponding remainder term $R_N(x)$ for $e^x$ expanded about $p = 0$, and decide on which interval $e^x$ equals its Taylor Series. **Answer:** The derivatives of $f(x) = e^x$ are easily now seen to be $f^{(k)}(x) = e^x$. Thus $f^{(k)}(0) = e^0 = 1$ and $a_k = f^{(k)}(0)/k! = 1/k!$. The Taylor Series about 0 now takes the form

$$\sum_{k=0}^{\infty} \frac{1}{k!} [x-0]^k = \sum_{k=0}^{\infty} \frac{x^k}{k!}.$$

As such the Taylor Polynomial of degree $N$ is

$$P_N(x) = \sum_{k=0}^{N} \frac{x^k}{k!},$$

and the corresponding remainder term is

$$R_N(x) = \frac{f^{N+1}(c)}{(N+1)!} [x-0]^{N+1} = \frac{e^c}{(N+1)!} [x]^{N+1}.$$

Eventually $(N+1) > |x|$ from which one concludes $1 > |x|/(N+1) > |x|/(N+2) > \dots$. Thus, $|x|^N/N! > |x|^{N+1}/(N+1)! > |x|^{N+2}/(N+2)! > \dots$ and these terms decrease and tend to 0. This shows that

$$|R_N(x)| = \left| \frac{e^c}{(N+1)!} [x]^{N+1} \right| \leq \max\{1, e^x\} \frac{|x|^{N+1}}{(N+1)!},$$

must approach 0 as $N$ approaches $\infty$. Since $\lim_{N \to \infty} R_N(x) = 0$, we know

$$e^x = \sum_{k=0}^{\infty} \frac{x^k}{k!},$$

for all $x \in \mathbb{R}$. This has been seen from other perspectives earlier in the book. One checks, as in an earlier example, that the radius of convergence is $R = \infty$. Then $|a_{k+1}/a_k| = |1/(k+1)!/1/(k)!| = k!/(k+1)! = 1/(k+1)$ approaches 0 as $k$ approaches $\infty$, so the radius of convergence of our Taylor series is $\infty$.

The Taylor Series at 0 is also called the *Maclaurin Series*.

### EXAMPLE 5.42

Given the function $f(x) = \sin(x)$, find the Taylor Series, the Taylor Polynomial $P_N(x)$ of degree $N$, the corresponding remainder term $R_N(x)$ for $\sin(x)$ expanded about $p = 0$, and decide on which interval $\sin(x)$ equals its Taylor Series. **Answer:** First compute the derivatives of $f(x) = \sin(x)$. The first derivative $f^{(1)}(x) = \sin'(x) = \cos(x)$, and $f^{(2)}(x) = \cos'(x) = -\sin(x)$, while $f^{(3)}(x) = -\sin'(x) = -\cos(x)$. At the fourth derivative a repetitive cycle emerges, because $f^{(4)}(x) = \sin(x)$. Thus $f(0) = 0$, $f^{(1)}(0) = \cos(0) = 1$, $f^{(2)}(0) = -\sin(0) = 0$, while $f^{(3)}(0) = -\cos(0) = -1$. The next four derivatives at 0 are: $0, 1, 0, -1, \dots$ with the pattern continuing indefinitely: $f^{2k}(0) = 0$, $f^{2k+1}(0) = (-1)^k$. Thus the even indexed coefficients vanish: $a_{2k} = f^{(2k)}(0)/(2k)! = 0$, while $a_{2k+1} = f^{(2k+1)}(0)/(2k+1)! = (-1)^k/(2k+1)!$. The Taylor Series about 0 now takes the form

$$\sum_{k=0}^{\infty} \frac{(-1)^k}{(2k+1)!} x^{2k+1}.$$

As such, the Taylor Polynomial of degree $2N+1$ is

$$P_{2N+1}(x) = \sum_{k=0}^{N} \frac{(-1)^k}{(2k+1)!} x^{2k+1},$$

and the corresponding remainder term is

$$R_{2N+1}(x) = \frac{f^{2N+2}(c)}{(2N+2)!}[x]^{2N+2} = \frac{(-1)^{N+1}\sin(c)}{(2N+2)!}[x]^{2N+2}.$$

It is now evident that

$$|R_{2N+2}(x)| = \left|\frac{(-1)^{N+1}\sin(c)}{(2N+2)!}[x]^{2N+2}\right| \leq 1\frac{|x|^{2N+2}}{(2N+2)!},$$

which must approach 0 as $N$ approaches $\infty$. Since $\lim_{N\to\infty} R_N(x) = 0$, we know

$$\sin(x) = \sum_{k=0}^{\infty} \frac{(-1)^k x^{2k+1}}{(2k+1)!},$$

for all $x \in \mathbb{R}$. We confirm directly that our radius of convergence is $R = \infty$ by factoring out one power of $x$ out of the Taylor Series and then making the following substitution $u = x^2$ in the remaining series to get

$$\sum_{k=0}^{\infty} \frac{(-1)^k x^{2k+1}}{(2k+1)!} = x \cdot \sum_{k=0}^{\infty} \frac{(-1)^k u^k}{(2k+1)!}.$$

The resulting series in $u$ has no vanishing coefficients, and one now employs the Ratio test. The coefficient of $u^k$ is now $a_k = (-1)^k/(2k+1)!$ and then $|a_{k+1}/a_k| = |(-1)^{k+1}/(2k+3)!/((-1)^k/(2k+1)!| = (2k+1)!/(2k+3)! = 1/[(2k+3)(2k+2)]$, which approaches 0 as $k$ approaches $\infty$, so the radius of convergence of our Taylor Series in $u$ is $\infty$. Upon substitution back for $x^2 = u$, the radius of convergence in $x$ is also $\infty$.

The interested reader is encouraged to confirm that $\cos(x)$ equals its Maclaurin series on $\mathbb{R}$ and that this series has radius of convergence $\infty$. That is, the equality

$$\cos(x) = \sum_{k=0}^{\infty} \frac{(-1)^k x^{2k}}{(2k)!},$$

holds for all $x \in \mathbb{R}$.

### EXAMPLE 5.43

Given the function $f(x) = 1/(1-x)$, find the Taylor Series, the Taylor Polynomial $P_N(x)$ of degree $N$, the corresponding remainder term $R_N(x)$ for $1/(1-x)$ expanded about $p = 0$, and decide on which interval $1/(1-x)$ equals its Taylor Series. **Answer:** First, the derivatives of $f(x) = (1-x)^{-1}$ are computed and evaluated at 0.

$$f(x) = (1-x)^{-1} \qquad f(0) = 1$$
$$f^{(1)}(x) = (1-x)^{-2} \qquad f^{(1)}(0) = 1$$
$$f^{(2)}(x) = 2\,(1-x)^{-3} \qquad f^{(2)}(0) = 2$$
$$f^{(3)}(x) = 3!\,(1-x)^{-4} \qquad f^{(3)}(0) = 3!$$
$$f^{(k)}(x) = k!\,(1-x)^{-k-1} \qquad f^{(k)}(0) = k!$$

The Maclaurin coefficients are $a_k = f^{(k)}(0)/k! = k!/k! = 1$ and our Maclaurin Series is

$$\frac{1}{(1-x)} = \sum_{k=0}^{\infty} 1 \cdot x^k.$$

Since $\lim_{k\to\infty} |a_{k+1}/a_k| = \lim_{k\to\infty} |a_{k+1}/a_k| = 1$, the radius of convergence is 1. Thus the goal is to gain equality of $1/(1-x) = \sum_{k=0}^{\infty} x^k$ on the interval $(-1, 1)$.

At the beginning of this section it was shown that

$$P_N(x) = \sum_{k=0}^{N} x^k = \frac{1}{1-x} - \frac{x^{N+1}}{1-x},$$

which explicitly gives our remainder term as

$$R_N(x) = \frac{x^{N+1}}{1-x}.$$

Thus $R_N(x)$ approaches 0 as $N$ approaches $\infty$. So

$$\frac{1}{1-x} = \sum_{k=0}^{\infty} x^k,$$

on the interval for $x \in (-1, 1)$, and the radius of convergence of the Taylor Series is $R = 1$. However, even though $1/(1-x)$ is defined on $\mathbb{R}\setminus\{1\}$ beyond the interval $(-1, 1)$, its Taylor Series does not even converge there and the two expressions are not equal beyond $(-1, 1)$.

In statistics it is important to know the area under the standard normal density function

$$f(x) = \frac{1}{\sqrt{2\pi}} e^{-x^2/2},$$

in particular over an interval of form $[0, z]$.

However when one attempts to antidifferentiate $f(x)$, it becomes clear that no current technique applies. This is where Taylor Series comes into play. The Maclaurin Series gives the exponential function as $e^u = \sum_{k=0}^{\infty} u^k/k!$, and, upon making the substitution $u = -x^2/2$, it becomes

$$f(x) = e^{-x^2/2} = \sum_{k=0}^{\infty} \frac{(-1)^k \, x^{2k}}{2^k \, k!},$$

which is amenable to integration under term-by-term application of the Power Rule:

$$F(x) = \int e^{-x^2/2} \, dx = \sum_{k=0}^{\infty} \int \frac{(-1)^k \, x^{2k}}{2^k \, k!} \, dx = \sum_{k=0}^{\infty} \frac{(-1)^k \, x^{2k+1}}{2^k \, (2k+1) \, (k!)} + C. \tag{5.107}$$

Setting $C = 0$, $F(x)$ is utilized to obtain the area over the interval $[0, z]$ as

$$F(z) - F(0) = F(z) - 0 = \sum_{k=0}^{\infty} \frac{(-1)^k \, z^{2k+1}}{2^k \, (2k+1) \, (k!)}.$$

**EXAMPLE 5.44**

Find the area $A([0, z])$ under the standard normal density over $[0, z]$. **Answer:**

$$A([0, z]) = \frac{1}{\sqrt{2\pi}} \int_0^z e^{-x^2/2} \, dx = \frac{1}{\sqrt{2\pi}} F(z) = \frac{1}{\sqrt{2\pi}} \sum_{k=0}^{\infty} \frac{(-1)^k \, z^{2k+1}}{2^k \, (2k+1) \, (k!)}. \tag{5.108}$$

This is the basis for constructing standard normal tables based on $z$-scores. For each $z > 0$ the corresponding probability in the table is $A([0, z])$. The probability over $[z_1, z_2]$ is then $A([z_1, z_2]) = A([0, z_2]) - A([0, z_1])$ for $0 < z_1 < z_2$. These can all be approximated to any accuracy by relying on Eq. (5.108) and truncating when one reaches the desired number of decimal places of accuracy.

Other topics on sequences, series, and tests for convergence can be found in [59].

## Exercises

1. (a) Show that the Taylor Series for $\sin(x)$ expanded about $p = \pi/2$ is given by $\sum_{k=0}^{\infty} (-1)^k [x - \pi/2]^{2k}/(2k)!$.
   (b) Substitute $u = [x - \pi/2]$ in the Maclaurin Series for $\cos(u) = \sum_{k=0}^{\infty} (-1)^k u^{2k}/(2k)!$.
   (c) Find a relation between $\sin(x)$ and $\cos(x - \pi/2)$.

2. Evaluate $\int e^{x^3} dx$ by substituting $u = x^3$ in the Maclaurin Series for $e^u = \sum_{k=0}^{\infty} u^k/k!$ and integrating the result term by term.

3. Use the 7th degree = 8th degree Taylor polynomial $P_8(x) = x - x^3/3! + x^5/5! - x^7/7!$ for $\sin(x)$ with an associated remainder term estimated by $|R_8(x)| \leq 1|x|^9/9!$ to estimate $\sin(1) \approx P_8(1)$ to within $1/9!$.

4. Use the Maclaurin Series for the exponential function $e^x = \sum_{k=0}^{\infty} x^k/k!$ to estimate $e = e^1 = \sum_{k=0}^{\infty} 1/k!$ to within four decimal places by appropriate truncation of the series.

# 6 Vector Calculus

## Introduction

Vector calculus gives the framework to understand geometric structures in higher dimensions. It also provides a basis for applications in the sciences. These applications occur in a number of settings, where vector calculus allows for precise computation of forces, equations of motion, the geometric properties of curves and surfaces, flux through a boundary, rates of expansion and contraction, the partial differential equations associated with gradient flow, and several variable optimization and estimation.

## ■ 6.1 Algebra and Geometry in $\mathbb{R}^n$

The set of $n$-tuples of real numbers is denoted $\mathbb{R}^n$. A point $\vec{x}$ in $\mathbb{R}^n$ is a column of $n$ real numbers, often called a *vector* in $\mathbb{R}^n$. $\vec{x} = (x_1, x_2, x_3, \ldots, x_n)^T$, for $x_i \in \mathbb{R}$ for $i = 1, \ldots, n$. The $i$th *coordinate* of $\vec{x}$ is $x_i$. Also, the $T$ denotes the *transpose* operation that converts a row of numbers into a column of numbers, and vice versa. (The use of the transpose $T$ here is basically as a space saver.) While $\mathbb{R}^1 = \mathbb{R}$ is the real line, $\mathbb{R}^2$ is the plane, and $\mathbb{R}^3$ is space or 3-space, in principal there are many uses for higher dimensional Euclidean spaces. One such example is *space-time*, thought of as $\mathbb{R}^4$, with coordinates $(x, y, z, t)^T$, the first three denoting position in space, and the fourth denoting the time.

While $\vec{x}$ is at times considered to be the point in $\mathbb{R}^n$ given by its coordinates $(x_1, x_2, x_3, \ldots, x_n)^T$, it will often be convenient in applications to also visualize $\vec{x}$ as an arrow emanating from the origin $\vec{0} = (0, 0, \ldots, 0)^T$ to the point $(x_1, x_2, x_3, \ldots, x_n)^T$. For instance a force acting at the origin can be represented by an arrow emanating from the origin to the point $(x_1, x_2, x_3, \ldots, x_n)^T$. Such an arrow gives both a direction of the force and a magnitude, or length.

### 6.1.1 The Algebra of $\mathbb{R}^n$ as a Vector Space

There are two basic operations on vectors in $\mathbb{R}^n$: adding and scaling (or magnifying) vectors. Two vectors $\vec{x} = (x_1, x_2, x_3, \ldots, x_n)^T$, $\vec{y} = (y_1, y_2, y_3, \ldots, y_n)^T$ are added by adding componentwise: $\vec{x} + \vec{y} = (x_1 + y_1, x_2 + y_2, x_3 + y_3, \ldots, x_n + y_n)^T$. When two vector forces are added in physics according to this addition rule, their sum is the resolvent force, which is the resulting combined force.

A vector is scaled by a number $\lambda \in \mathbb{R}$ by multiplying each component by $\lambda$. *Scalar multiplication* is given by: $\lambda \vec{x} = (\lambda x_1, \lambda x_2, \lambda x_3, \ldots, \lambda x_n)^T$. There is a *zero vector* $\vec{0} = (0, 0, 0, \ldots, 0)^T \in \mathbb{R}^n$. And for each vector $\vec{x} = (x_1, x_2, x_3, \ldots, x_n)^T$ there is an additive inverse $-\vec{x} = (-x_1, -x_2, -x_3, \ldots, -x_n)^T$. The two operations, addition and scalar multiplication, form the paradigm for any algebraic structure called a *vector space*. That is, $\mathbb{R}^n$, together with the operations of addition and scalar multiplication, satisfies the following properties: for all $\vec{x}, \vec{y}, \vec{w} \in \mathbb{R}^n$, and for all $\lambda, \mu \in \mathbb{R}$ we have

$$\vec{x} + (\vec{y} + \vec{w}) = (\vec{x} + \vec{y}) + \vec{w}$$
$$\vec{x} + \vec{y} = \vec{y} + \vec{x}$$
$$\vec{0} + \vec{x} = \vec{x}$$

for each $\vec{x}$ there is a $-\vec{x}$ with $\vec{x} + (-\vec{x}) = \vec{0}$

$$\lambda(\vec{x} + \vec{y}) = (\lambda \vec{x}) + (\lambda \vec{y})$$
$$(\lambda + \mu)\vec{x} = (\lambda \vec{x}) + (\mu \vec{x})$$
$$1\vec{x} = \vec{x}$$
$$(\lambda \mu)\vec{x} = \lambda(\mu \vec{x})$$

Subtraction of two vectors is now given by $\vec{x} - \vec{y} = \vec{x} + (-1)\vec{y}$, which is computed by subtracting componentwise. $\vec{x} - \vec{y}$ can be visualized as the arrow, or vector, starting at $\vec{y}$ and ending at $\vec{x}$.

The distance between any two points in $\mathbb{R}^n$ is defined by the Pythagorean theorem, in conjunction with the assumption that the coordinate axes are at right angles, to be

$$d(\vec{x}, \vec{y}) = \left( \sum_{i=1}^{n} (x_i - y_i)^2 \right)^{1/2}.$$

This distance formula lets us do familiar geometric computations, so with this additional notion of distance our space $\mathbb{R}^n$ is called *n-dimensional Euclidean space*.

In this setting, the length of a vector $\vec{x}$ is given by

$$|\vec{x}| = d(\vec{0}, \vec{x}) = \left( \sum_{i=1}^{n} (x_i)^2 \right)^{1/2}.$$

### Exercises

1. Perform the given algebraic operation on the given vector(s):

   (a) $(1, 2, -1)^T + (4, 5, 0)^T$
   (b) $3(5, 4)^T$
   (c) $(4, 6, -3)^T - (3, 7, 5)^T$

2. Find the distance between the points:

   (a) $(1, 5)^T$ and $(4, 9)^T$
   (b) $(2, 5, 3)^T$ and $(4, 6, 1)^T$

3. Find the absolute value, or length, of each vector:

   (a) $|(1, 2)^T|$
   (b) $|(1, 2, 3)^T|$

### 6.1.2 The Dot Product and its Geometry in $\mathbb{R}^n$

There is an operation called *dot product* on Euclidean space that lets us multiply two vectors and obtain a number. The dot product of two vectors $\vec{x} = (x_1, x_2, x_3, \ldots, x_n)^T$ and $\vec{y} = (y_1, y_2, y_3, \ldots, y_n)^T$ is given by

$$\vec{x} \cdot \vec{y} = x_1 y_1 + x_2 y_2 + x_3 y_3 + \cdots + x_n y_n = \sum_{i=1}^{n} x_i y_i.$$

The distance formula combined with the dot product operation tells us many important geometric and trigonometric facts.

The dot product operation has several properties. For all $\vec{x}, \vec{y}, \vec{w} \in \mathbb{R}^n$, and $\lambda \in \mathbb{R}$

$$\vec{x} \cdot \vec{y} = \vec{y} \cdot \vec{x}$$
$$\vec{x} \cdot \vec{x} = |\vec{x}|^2$$
$$\vec{0} \cdot \vec{x} = 0$$
$$\vec{x} \cdot (\vec{y} + \vec{w}) = (\vec{x} \cdot \vec{y}) + (\vec{x} \cdot \vec{w})$$
$$\lambda(\vec{x} \cdot \vec{y}) = (\lambda\vec{x}) \cdot \vec{y} = \vec{x} \cdot (\lambda\vec{y}).$$

Dot product can compute the length of a vector as $|\vec{x}| = (\vec{x} \cdot \vec{x})^{1/2}$. But $\vec{x} \cdot \vec{y}$ also gives trigonometric information about the angle $\theta$ between two vectors $\vec{x}, \vec{y}$. To see this, first examine two-dimensional Euclidean space $\mathbb{R}^2$. The claim will be that rotating two vectors by the same angle leaves their dot product alone. Rotation by an angle $\psi$ will be denoted $R_\psi$. Let $\vec{x} = (r_1 \cos(\alpha), r_1 \sin(\alpha))^T$ and $\vec{y} = (r_2 \cos(\beta), r_2 \sin(\beta))^T$, where $r_1 = |\vec{x}|, r_2 = |\vec{y}|$, and $\alpha, \beta$ are the angles that $\vec{x}$ and $\vec{y}$ make with the positive $x$-axis, respectively. Let $\theta$ be the angle between $\vec{x}, \vec{y}$.

If one rotates $\vec{x} = (x_1, x_2)^T$ through an angle $\psi$, one sees that

$$R_\psi(\vec{x}) = (r_1 \cos(\psi + \alpha), r_1 \sin(\psi + \alpha))^T$$
$$= (r_1\{\cos(\psi)\cos(\alpha) - \sin(\psi)\sin(\alpha)\}, r_1\{\sin(\psi)\cos(\alpha) + \cos(\psi)\sin(\alpha)\})^T$$
$$= (\{\cos(\psi)r_1\cos(\alpha) - \sin(\psi)r_1\sin(\alpha)\}, \{\sin(\psi)r_1\cos(\alpha) + \cos(\psi)r_1\sin(\alpha)\})^T$$
$$= (\{\cos(\psi)x_1 - \sin(\psi)x_2\}, \{\sin(\psi)x_1 + \cos(\psi)x_2\})^T$$

by the angle addition formulas of trigonometry. Similarly, rotation by $\psi$ gives

$$R_\psi(\vec{y}) = (\{\cos(\psi)y_1 - \sin(\psi)y_2\}, \{\sin(\psi)y_1 + \cos(\psi)y_2\})^T.$$

■ **Lemma 6.1**

Rotation by any angle $\psi$ leaves dot product invariant. That is, for any two vectors $\vec{x}, \vec{y} \in \mathbb{R}^2$

$$(R_\psi(\vec{x})) \cdot (R_\psi(\vec{y})) = \vec{x} \cdot \vec{y}.$$

**Proof**

$$(R_\psi(\vec{x})) \cdot (R_\psi(\vec{y})) = (\{\cos(\psi)x_1 - \sin(\psi)x_2\}, \{\sin(\psi)x_1 + \cos(\psi)x_2\})^T$$
$$\cdot (\{\cos(\psi)y_1 - \sin(\psi)y_2\}, \{\sin(\psi)y_1 + \cos(\psi)y_2\})^T.$$

Hence,

$$\begin{aligned}
&(R_\psi(\vec{x})) \cdot (R_\psi(\vec{y})) \\
&= \{\cos(\psi)x_1 - \sin(\psi)x_2\}\{\cos(\psi)y_1 - \sin(\psi)y_2\} \\
&\quad + \{\sin(\psi)x_1 + \cos(\psi)x_2\}\{\sin(\psi)y_1 + \cos(\psi)y_2\} \\
&= \cos^2(\psi)x_1 y_1 - \cos(\psi)\sin(\psi)x_1 y_2 - \sin(\psi)\cos(\psi)x_2 y_1 + \sin^2(\psi)x_2 y_2 \\
&\quad + \sin^2(\psi)x_1 y_1 + \sin(\psi)\cos(\psi)x_1 y_2 + \cos(\psi)\sin(\psi)x_2 y_1 + \cos^2(\psi)x_2 y_2 \\
&= (\cos^2(\psi) + \sin^2(\psi))x_1 y_1 + (\sin^2(\psi) + \cos^2(\psi))x_2 y_2 = x_1 y_1 + x_2 y_2 \\
&= \vec{x} \cdot \vec{y}.
\end{aligned}$$

$\square$

### ■ Lemma 6.2
For $\vec{x}, \vec{y} \in \mathbb{R}^2$ and $\theta$ the angle between $\vec{x}$ and $\vec{y}$, the dot product satisfies

$$\vec{x} \cdot \vec{y} = |\vec{x}||\vec{y}|\cos(\theta). \tag{6.1}$$

**Proof** Let $\alpha$ be the angle that $\vec{x}$ makes with the positive $x$-axis in $\mathbb{R}^2$. Then rotate both $\vec{x}$ and $\vec{y}$ by the angle $-\alpha$ to obtain

$$\vec{x} \cdot \vec{y} = (R_{-\alpha}(\vec{x})) \cdot (R_{-\alpha}(\vec{y})) = (|\vec{x}|, 0)^T \cdot (|\vec{y}|\cos(\theta), |\vec{y}|\sin(\theta))^T = |\vec{x}||\vec{y}|\cos(\theta),$$

because rotation in the Euclidean plane leaves lengths and angles unchanged. $\square$

As an important consequence, one sees the Law of Cosines:

### ■ Theorem 6.1
**The Law of Cosines**

In the Euclidean plane $\mathbb{R}^2$, let $\vec{x}$ and $\vec{y}$ have angle $\theta$ between them. They form a triangle having third side $\vec{x} - \vec{y}$, which is opposite $\theta$. Then the lengths of the three sides of the triangle are related by:

$$|\vec{x} - \vec{y}|^2 = |\vec{x}|^2 - 2|\vec{x}||\vec{y}|\cos(\theta) + |\vec{y}|^2. \tag{6.2}$$

## Proof

$$|\vec{x} - \vec{y}|^2 = (\vec{x} - \vec{y}) \cdot (\vec{x} - \vec{y}) = \vec{x} \cdot \vec{x} - \vec{x} \cdot \vec{y} - \vec{y} \cdot \vec{x} + \vec{y} \cdot \vec{y} = |\vec{x}|^2 - 2\vec{x} \cdot \vec{y} + |\vec{y}|^2$$
$$= |\vec{x}|^2 - 2|\vec{x}||\vec{y}|\cos(\theta) + |\vec{y}|^2.$$

Equation (6.1) from the previous Lemma has been applied to obtain the last equation. □

The Law of Cosines in Eq. (6.2) holds in any Euclidean plane, beyond $\mathbb{R}^2$, as it is stated in terms of lengths and angles only. Equation (6.2) can be thought of as a generalization of the Pythagorean theorem, where the side $\vec{x} - \vec{y}$ opposite $\theta$ plays the role of the hypotenuse, and $\vec{x}$, $\vec{y}$ play the role of the legs. The term $-2|\vec{x}||\vec{y}|\cos(\theta)$ is a correction term, which vanishes when $\theta = \pi/2$ in the case of a right triangle, yielding the Pythagorean theorem.

The fact that the Law of Cosines holds in any Euclidean plane lets us conclude that the Law of Cosines holds in $\mathbb{R}^n$.

### ■ Theorem 6.2
**Law of Cosines in $\mathbb{R}^n$**

Let $\vec{x}, \vec{y} \in \mathbb{R}^n$, with $\theta$ the angle between them, then

$$|\vec{x} - \vec{y}|^2 = |\vec{x}|^2 - 2|\vec{x}||\vec{y}|\cos(\theta) + |\vec{y}|^2. \tag{6.3}$$

**Proof** The plane $P$ spanned by $\vec{x}$ and $\vec{y}$ is a Euclidean subplane of $\mathbb{R}^n$. As such the Law of Cosines holds on $P$, and the result follows immediately.

Alternatively one could find higher-dimensional rotations moving $\vec{x}, \vec{y}$ back to the $x_1 x_2$ plane preserving the dot product, and use the fact that one is in a coordinate $\mathbb{R}^2$ to see that the result must hold. The study of such motions preserving the dot product forms the beginning of a topic called *Lie Groups*, which is beyond the scope of this text. □

The Law of Cosines for $\mathbb{R}^n$ now converts to its equivalent expression in $\mathbb{R}^n$.

### Theorem 6.3

For $\vec{x}, \vec{y} \in \mathbb{R}^n$ and $\theta$ the angle between $\vec{x}$ and $\vec{y}$, the dot product satisfies

$$\vec{x} \cdot \vec{y} = |\vec{x}||\vec{y}| \cos(\theta). \tag{6.4}$$

### Proof

$$\begin{aligned} |\vec{x} - \vec{y}|^2 &= (\vec{x} - \vec{y}) \cdot (\vec{x} - \vec{y}) = \vec{x} \cdot \vec{x} - \vec{x} \cdot \vec{y} - \vec{y} \cdot \vec{x} + \vec{y} \cdot \vec{y} = |\vec{x}|^2 - 2\vec{x} \cdot \vec{y} + |\vec{y}|^2 \\ &= |\vec{x}|^2 - 2|\vec{x}||\vec{y}| \cos(\theta) + |\vec{y}|^2, \end{aligned}$$

where the Law of Cosines in Eq. (6.3) gave the last equality. One concludes by subtracting equal expressions from both sides of the last equality, and then dividing by $-2$, that

$$\vec{x} \cdot \vec{y} = |\vec{x}||\vec{y}| \cos(\theta).$$

$\square$

There are three corollaries yielding geometric consequences in $\mathbb{R}^n$.

### Corollary 6.1 Cauchy-Schwarz Inequality

For any $\vec{x}, \vec{y} \in \mathbb{R}^n$, the inequality

$$|\vec{x} \cdot \vec{y}| \leq |\vec{x}||\vec{y}| \tag{6.5}$$

holds, with equality if and only if one of $\vec{x}, \vec{y}$ is a multiple of the other.

### Proof

$$|\vec{x} \cdot \vec{y}| = |\vec{x}||\vec{y}||\cos(\theta)| \leq |\vec{x}||\vec{y}|,$$

where the first equality holds by Eq. (6.4). If $|\vec{x} \cdot \vec{y}| = |\vec{x}||\vec{y}|$, then $\cos(\theta) = \pm 1$ and then $\vec{x}, \vec{y}$ are collinear, giving that one is a multiple of the other. $\square$

## Corollary 6.2 Triangle Inequality

For any $\vec{x}, \vec{y} \in \mathbb{R}^n$, the inequality

$$|\vec{x} + \vec{y}| \leq |\vec{x}| + |\vec{y}| \tag{6.6}$$

holds.

**Proof**

$$|\vec{x} + \vec{y}|^2 = (\vec{x} + \vec{y}) \cdot (\vec{x} + \vec{y}) = \vec{x} \cdot \vec{x} + 2\vec{x} \cdot \vec{y} + \vec{y} \cdot \vec{y} \leq |\vec{x}|^2 + 2|\vec{x}||\vec{y}| + |\vec{y}|^2 = (|\vec{x}| + |\vec{y}|)^2$$

where the inequality holds by the Cauchy-Schwarz inequality in Eq. (6.5). Taking square roots of both sides yields the triangle inequality. □

## Corollary 6.3

For two nonzero vectors $\vec{x}, \vec{y} \in \mathbb{R}^n$, one has $\vec{x} \perp \vec{y} \iff \vec{x} \cdot \vec{y} = 0$.

**Proof** Letting $\theta$ be the angle between $\vec{x}$ and $\vec{y}$, one has $\vec{x} \perp \vec{y} \iff \theta = \pi/2 \iff \cos(\theta) = 0 \iff |\vec{x}||\vec{y}|\cos(\theta) = \vec{x} \cdot \vec{y} = 0$. □

### ■ Exercises

1. Let $\vec{x} = (1, 0, 1)^T$ and $\vec{y} = (1, 1, 0)^T$.

   (a) Find $\cos(\theta)$ the angle between $\vec{x}$ and $\vec{y}$.
   (b) Find $\theta$ the angle between $\vec{x}$ and $\vec{y}$, using your knowledge of special angles.

2. Which pair(s) of vectors on the following list are perpendicular? $(1, 4)^T$, $(2, 3)^T$, $(8, -2)^T$?

3. Let $R_{-\pi/2}$ be rotation by $-\pi/2$ (equivalently, rotation by $\pi/2$ in the clockwise direction). Show that $R_{-\pi/2}$ acting on the vector $(A, B)^T$ gives $R_{-\pi/2}(A, B)^T = (B, -A)^T$.

### 6.1.3 The Cross Product in $\mathbb{R}^3$

Euclidean 3-space $\mathbb{R}^3$ plays a special role in modeling our spacial surroundings. In the setting of $\mathbb{R}^3$ there is another type of vector product called the cross product that depends on familiarity with the computation of a determinant. A determinant of a $2 \times 2$ matrix is given by

$$\det \begin{pmatrix} a & b \\ c & d \end{pmatrix} = ad - bc,$$

while the determinant of a $3 \times 3$ matrix is given by

$$\det \begin{pmatrix} a_{11} & a_{12} & a_{13} \\ a_{21} & a_{22} & a_{23} \\ a_{31} & a_{32} & a_{33} \end{pmatrix}$$
$$= a_{11}\det \begin{pmatrix} a_{22} & a_{23} \\ a_{32} & a_{33} \end{pmatrix} - a_{12}\det \begin{pmatrix} a_{21} & a_{23} \\ a_{31} & a_{33} \end{pmatrix} + a_{13}\det \begin{pmatrix} a_{21} & a_{22} \\ a_{31} & a_{32} \end{pmatrix}$$
$$= a_{11}\{a_{22}a_{33} - a_{32}a_{23}\} - a_{12}\{a_{21}a_{33} - a_{31}a_{23}\} + a_{13}\{a_{21}a_{32} - a_{31}a_{22}\}$$
$$= a_{11}a_{22}a_{33} + a_{12}a_{23}a_{31} + a_{13}a_{21}a_{32} - a_{11}a_{23}a_{32} - a_{12}a_{21}a_{33} - a_{13}a_{22}a_{31}.$$

The *standard basis* vectors in $\mathbb{R}^3$ are denoted $\vec{i} = (1,0,0)^T$, $\vec{j} = (0,1,0)^T$ $\vec{j} = (0,0,1)^T$. Then any vector $(a,b,c)^T \in \mathbb{R}^3$ can be written uniquely as $(a,b,c)^T = a\vec{i} + b\vec{j} + c\vec{k}$. The *cross product* of any two vectors $\vec{x}$ and $\vec{y}$ in $\mathbb{R}^3$ is given by

$$\vec{x} \times \vec{y} = \det \begin{pmatrix} \vec{i} & \vec{j} & \vec{k} \\ x_1 & x_2 & x_3 \\ y_1 & y_2 & y_3 \end{pmatrix}$$
$$= \vec{i} \det \begin{pmatrix} x_2 & x_3 \\ y_2 & y_3 \end{pmatrix} - \vec{j} \det \begin{pmatrix} x_1 & x_3 \\ y_1 & y_3 \end{pmatrix} + \vec{k} \det \begin{pmatrix} x_1 & x_2 \\ y_1 & y_2 \end{pmatrix}$$
$$= \vec{i}\{x_2y_3 - y_2x_3\} - \vec{j}\{x_1y_3 - y_1x_3\} + \vec{k}\{x_1y_2 - y_1x_2\}$$
$$= (x_2y_3 - y_2x_3, -x_1y_3 + y_1x_3, x_1y_2 - y_1x_2)^T.$$

Note that the cross product $\vec{x} \times \vec{y}$ is another vector, while the dot product $\vec{x} \cdot \vec{y}$ is a number.

The operation of cross product has several properties, including the following:

$$\vec{x} \times \vec{y} = -\vec{y} \times \vec{x}$$
$$\vec{x} \cdot (\vec{x} \times \vec{y}) = 0$$
$$\vec{y} \cdot (\vec{x} \times \vec{y}) = 0$$
$$\vec{a} \cdot (\vec{b} \times \vec{c}) = \det \begin{pmatrix} a_1 & a_2 & a_3 \\ b_1 & b_2 & b_3 \\ c_1 & c_2 & c_3 \end{pmatrix}.$$

The last property follows from the fact that

$$\begin{aligned} \vec{a} \cdot (\vec{b} \times \vec{c}) &= (a_1, a_2, a_3)^T \cdot (b_2 c_3 - c_2 b_3, -b_1 c_3 + c_1 b_3, b_1 c_2 - c_1 b_2)^T \\ &= a_1 \{b_2 c_3 - c_2 b_3\} - a_2 \{+b_1 c_3 - c_1 b_3\} + a_3 \{b_1 c_2 - c_1 b_2\} \\ &= \det \begin{pmatrix} a_1 & a_2 & a_3 \\ b_1 & b_2 & b_3 \\ c_1 & c_2 & c_3 \end{pmatrix}. \end{aligned}$$

The first three properties now follow from the fourth along with the fact that interchanging any two rows (or any two columns) of the determinant changes only the sign of the result.

Another related result is the following:

### ■ Lemma 6.3

$$|\vec{x} \times \vec{y}|^2 + (\vec{x} \cdot \vec{y})^2 = |\vec{x}|^2 |\vec{y}|^2. \tag{6.7}$$

**Proof**

$$\begin{aligned} &|\vec{x} \times \vec{y}|^2 + (\vec{x} \cdot \vec{y})^2 \\ &= (x_2 y_3 - y_2 x_3)^2 + (-x_1 y_3 + y_1 x_3)^2 + (x_1 y_2 - y_1 x_2)^2 + (x_1 y_1 + x_2 y_2 + x_3 y_3)^2 \\ &= x_2^2 y_3^2 - 2 x_2 y_2 x_3 y_3 + y_2^2 x_3^2 + x_1^2 y_3^2 - 2 x_1 y_1 x_3 y_3 + y_1^2 x_3^2 + x_1^2 y_2^2 - 2 x_1 y_1 x_2 y_2 \\ &\quad + y_1^2 x_2^2 + x_1^2 y_1^2 + x_2^2 y_2^2 + x_3^2 y_3^2 + 2 x_1 y_1 x_2 y_2 + 2 x_1 y_1 x_3 y_3 + 2 x_2 y_2 x_3 y_3 \\ &= x_2^2 y_3^2 + y_2^2 x_3^2 + x_1^2 y_3^2 + y_1^2 x_3^2 + x_1^2 y_2^2 + y_1^2 x_2^2 + x_1^2 y_1^2 + x_2^2 y_2^2 + x_3^2 y_3^2 \\ &= (x_1^2 + x_2^2 + x_3^2)(y_1^2 + y_2^2 + y_3^2) = |\vec{x}|^2 |\vec{y}|^2. \end{aligned}$$

□

## Corollary 6.4

For $\theta$ the angle between two vectors $\vec{x}, \vec{y} \in \mathbb{R}^3$:

$$|\vec{x} \times \vec{y}| = |\vec{x}||\vec{y}|\sin(\theta) \tag{6.8}$$

**Proof** By Lemma (6.3) one has

$$\begin{aligned}|\vec{x} \times \vec{y}|^2 &= |\vec{x}|^2|\vec{y}|^2 - (\vec{x} \cdot \vec{y})^2 = |\vec{x}|^2|\vec{y}|^2 - (|\vec{x}||\vec{y}|\cos(\theta))^2 \\ &= |\vec{x}|^2|\vec{y}|^2(1 - \cos^2(\theta)) = |\vec{x}|^2|\vec{y}|^2 \sin^2(\theta).\end{aligned}$$

Taking square roots of both sides, together with the fact that $0 \leq \theta \leq \pi$ implies $\sin(\theta) \geq 0$, gives the result. □

## Corollary 6.5

The length of the cross product $\vec{x} \times \vec{y}$ is the area of the parallelogram spanned by $\vec{x}$ and $\vec{y}$.

**Proof** The area $A$ of the parallelogram is the base $= |\vec{x}|$ times the altitude $= |\vec{y}|\sin(\theta)$, for $\theta$ the angle between the two sides $\vec{x}$ and $\vec{y}$. So $A = |\vec{x}||\vec{y}|\sin(\theta) = |\vec{x} \times \vec{y}|$, where the last equality holds by Corollary (6.4). □

Thus the length of the cross product of two vectors tells us the area of the parallelogram having the two vectors as sides. Since $\vec{x} \times \vec{y}$ is perpendicular to both $\vec{x}$ and $\vec{y}$, with specified length the area of the parallelogram, one only needs determine the orientation of the cross product to represent it geometrically. This is accomplished with the *Right Hand Rule*: if one lays the index finger of one's right hand along $\vec{x}$ with the second finger along $\vec{y}$, then the thumb will be pointing in the direction of the cross product $\vec{x} \times \vec{y}$.

As a direct consequence of Corollary (6.5), one knows that the determinant of a $2 \times 2$ matrix now gives $\pm$ the area of the parallelogram spanned by the rows (or the columns) of the matrix.

## Corollary 6.6

The determinant

$$\det \begin{pmatrix} a & b \\ c & d \end{pmatrix} = ad - bc$$

is $\pm$ the area of the parallelogram spanned by $(a,b)^T$ and $(c,d)^T$ in $\mathbb{R}^2$.

**Proof** Embed $\mathbb{R}^2$ in $\mathbb{R}^3$ as the $(x,y,0)^T$ coordinate plane. So $(a,b)^T$ corresponds to $(a,b,0)^T$ and $(c,d)^T$ corresponds to $(c,d,0)^T$. The area of the parallelogram spanned by $(a,b)^T$ and $(c,d)^T$ is then the same as the area of the parallelogram spanned by $(a,b,0)^T$ and $(c,d,0)^T$, which is the length of the cross product $|(a,b,0)^T \times (c,d,0)^T| = |(ad-bc,0,0)^T| = \sqrt{(ad-bc)^2} = \pm(ad-bc)$. This gives the result. $\square$

## Corollary 6.7

The determinant

$$\det \begin{pmatrix} a_1 & a_2 & a_3 \\ b_1 & b_2 & b_3 \\ c_1 & c_2 & c_3 \end{pmatrix} = \vec{a} \cdot (\vec{b} \times \vec{c})$$

is $\pm$ the volume of the parallelepiped spanned by $\vec{a}$, $\vec{b}$, and $\vec{c}$ in $\mathbb{R}^3$, where $\vec{a} = (a_1, a_2, a_3)^T$, $\vec{b} = (b_1, b_2, b_3)^T$, and $\vec{c} = (c_1, c_2, c_3)^T$.

**Proof** First, by sliding the vector $\vec{b}$ parallel to itself along the vector $\vec{c}$ one has the parallelogram spanned by $\vec{b}$ and $\vec{c}$. This parallelogram has area $|\vec{b} \times \vec{c}|$, and we take this parallelogram to be the base of the parallelepiped. Sliding this parallelogram parallel to itself along the vector $\vec{a}$ fills out the so-called parallelepiped spanned by $\vec{a}$, $\vec{b}$, and $\vec{c}$. To find an altitude, we project $\vec{a}$ along a unit normal $\vec{N}$ to the base parallelogram. We choose this normal to be $\vec{N} = (\vec{b} \times \vec{c})/|\vec{b} \times \vec{c}|$. The altitude given by $\vec{a}$ is then $|\vec{a}||\cos(\alpha)|$ where $\alpha$ is the angle between $\vec{a}$ and $\vec{N}$. Thus the volume $V$

of the parallelepiped is given by the altitude $|\vec{a}||\cos(\alpha)|$ times the area of the base parallelogram $|\vec{b} \times \vec{c}|$.

$$\begin{aligned} V &= |\vec{a}| \, |\cos(\alpha)| \, |\vec{b} \times \vec{c}| = |\vec{a}| \, |\cos(\alpha)| \, |\vec{N}| \, |\vec{b} \times \vec{c}| \\ &= |\,|\vec{a}|\cos(\alpha)|\vec{N}|\,|\vec{b} \times \vec{c}|\,| = |\,(\vec{a} \cdot \vec{N})|\vec{b} \times \vec{c}|\,| \\ &= |\,(\vec{a} \cdot (\vec{b} \times \vec{c})/|\vec{b} \times \vec{c}|)\,|\vec{b} \times \vec{c}|\,| = |\,(\vec{a} \cdot (\vec{b} \times \vec{c}))\,| \\ &= \pm(\vec{a} \cdot (\vec{b} \times \vec{c})) = \pm \det \begin{pmatrix} a_1 & a_2 & a_3 \\ b_1 & b_2 & b_3 \\ c_1 & c_2 & c_3 \end{pmatrix}. \end{aligned}$$

This gives the result. $\square$

The moral of this story is that the determinant of a $2 \times 2$ matrix measures $\pm$ area (or 2-volume), while the determinant of a $3 \times 3$ matrix measures $\pm$ volume (or 3-volume). It is also true that in higher dimensions the determinant measures $n$-volume up to a sign. The purely algebraic construction, determinant, is a geometric measurer of $n$-volume. As a further fact, the length of the cross product in $\mathbb{R}^3$ measures the surface area of the parallelogram spanned by the two vectors in the cross product.

### ■ Exercises

1. Find the area of the parallelogram spanned by the vectors $(1,2)^T$ and $(6,7)^T$ in $\mathbb{R}^2$.
2. Find the volume of the parallelepiped spanned by the vectors $(1,2,0)^T$, $(4,2,1)^T$, and $(1,1,3)^T$ in $\mathbb{R}^3$.
3. (a) Find $(1,2,4)^T \times (1,1,1)^T$.
   (b) Find the area of the parallelogram spanned by the vectors $(1,2,4)^T$ and $(1,1,1)^T$ in $\mathbb{R}^3$.
4. Find the following determinants:

   (a) $\det \begin{pmatrix} 4 & 5 \\ 1 & 2 \end{pmatrix}$

   (b) $\det \begin{pmatrix} 1 & 2 \\ 4 & 5 \end{pmatrix}$

   (c) $\det \begin{pmatrix} 1 & 2 & 4 \\ -1 & 1 & 2 \\ 1 & 0 & 1 \end{pmatrix}$

5. Verify directly that switching the rows (or switching the columns) of the matrix $\begin{pmatrix} a & b \\ c & d \end{pmatrix}$ changes the sign of the determinant.

### ■ 6.1.4 Flat Sets in $\mathbb{R}^n$: Lines, Planes, and Hyperplanes

There are various types of subsets in $\mathbb{R}^n$ for $n \geq 2$. These can be defined in terms of inequalities, as in the case of half spaces, or in terms of equalities, as in the case of a plane or sphere. The first focus will be on the flat sets. One reason for this is that flat sets will be good approximations for differentiable functions, just as the line given by a linearization to a differentiable function $f$ at $p$ is a good approximation to the curve given by the graph of $f$ in $\mathbb{R}^2$.

The general equation of a line in $\mathbb{R}^2$ is given by $n_1 x_1 + n_2 x_2 = C$ for constants $n_1, n_2, C \in \mathbb{R}$. If the line is to pass through a given point $(p_1, p_2)^T$ then $n_1 p_1 + n_2 p_2 = C$. Subtracting these two equalities gives:

$$n_1(x_1 - p_1) + n_2(x_2 - p_2) = 0 \tag{6.9}$$

as the equation of a line through a point $(p_1, p_2)^T$ in $\mathbb{R}^2$. Take $\vec{n} = (n_1, n_2)^T$ to be the coefficient vector, $\vec{x} = (x_1, x_2)^T$ to be an arbitrary point on the line, and $\vec{p} = (p_1, p_2)^T$ to be the fixed point on the line. Then Eq. (6.9) becomes

$$\vec{n} \cdot (\vec{x} - \vec{p}) = 0, \tag{6.10}$$

from which one observes that $\vec{n}$ is perpendicular to any vector $\vec{x} - \vec{p}$ lying completely inside the line. For this reason $\vec{n}$ is called a *normal* to the line.

The process for obtaining the (flat) lines in $\mathbb{R}^2$ generalizes to all dimensions, giving flat planes in $\mathbb{R}^3$ and so called *hyperplanes* in $\mathbb{R}^n$. In particular, in $\mathbb{R}^3$, we proceed by analogy to the preceding process for $\mathbb{R}^2$.

Assume the general equation of a plane in $\mathbb{R}^3$ is given by $n_1 x_1 + n_2 x_2 + n_2 x_3 = C$ for constants $n_1, n_2, n_3, C \in \mathbb{R}$. If the plane is to pass through a given point $(p_1, p_2, p_3)^T$ then $n_1 p_1 + n_2 p_2 + n_3 p_3 = C$. Subtracting these two equalities gives:

$$n_1(x_1 - p_1) + n_2(x_2 - p_2) + n_3(x_3 - p_3) = 0 \tag{6.11}$$

as the equation of a plane through the point $(p_1, p_2, p_3)^T$ in $\mathbb{R}^3$. Now take $\vec{n} = (n_1, n_2, n_3)^T$ to be the coefficient vector, $\vec{x} = (x_1, x_2, x_3)^T$ to be an arbitrary point on the plane, and $\vec{p} = (p_1, p_2, p_3)^T$ to be the fixed point on the plane. Then Eq. (6.11) becomes

$$\vec{n} \cdot (\vec{x} - \vec{p}) = 0. \tag{6.12}$$

Again $\vec{n}$ is a perpendicular to any vector $\vec{x} - \vec{p}$ lying completely inside the plane, and so it is called a normal to the plane. Since the set of vectors perpendicular to a given normal $\vec{n}$ does form a plane, and since Eq. (6.12) can be rewritten as $n_1 x_1 + n_2 x_2 + n_2 x_3 = C$, we were justified in our first assumption that the equation of a plane could be taken to be a sum of coefficients times variables set equal to a constant.

The process is completely analogous in higher dimensions. The equation of the hyperplane in $\mathbb{R}^n$ through the point $\vec{p} \in \mathbb{R}^n$ is given by

$$\vec{n} \cdot (\vec{x} - \vec{p}) = 0. \tag{6.13}$$

To illustrate the power of recognizing the equations of the hyperplanes, we turn to $\mathbb{R}^3$ and a simple curved surface in $\mathbb{R}^3$, the sphere. Let $S$ be the sphere of radius $R$ with center $\vec{0} = (0,0,0)^T$ in $\mathbb{R}^3$. Then the distance from a point $\vec{x} = (x_1, x_2, x_3)^T$ on the sphere to the center $\vec{0}$ is $R$. More succinctly: $|\vec{x} - \vec{0}| = |\vec{x}| = R$. Squaring both sides gives $|\vec{x}|^2 = \vec{x} \cdot \vec{x} = R^2$. Since $\vec{x} \cdot \vec{x} = x_1^2 + x_2^2 + x_3^2$, we conclude that the sphere of radius $R$ centered at $\vec{0}$ in $\mathbb{R}^3$ has equation:

$$x_1^2 + x_2^2 + x_3^2 = R^2. \tag{6.14}$$

Suppose $\vec{p} = (p_1, p_2, p_3)^T$ is a point on the sphere, then Eq. (6.14) gives $p_1^2 + p_2^2 + p_3^2 = R^2$. Next expand $x_1$ about $p_1$, $x_2$ about $p_2$, $x_3$ about $p_3$ in Eq. (6.14) to obtain

$$\begin{aligned} R^2 &= x_1^2 + x_2^2 + x_3^2 = (p_1 + [x_1 - p_1])^2 + (p_2 + [x_2 - p_2])^2 + (p_3 + [x_3 - p_3])^2 \\ &= p_1^2 + 2p_1[x_1 - p_1] + [x_1 - p_1]^2 + p_2^2 + 2p_2[x_2 - p_2] + [x_2 - p_2]^2 \\ &\quad + p_3^2 + 2p_3[x_3 - p_3] + [x_3 - p_3]^2 \\ &= \{p_1^2 + p_2^2 + p_3^2\} + 2p_1[x_1 - p_1] + 2p_2[x_2 - p_2] + 2p_3[x_3 - p_3] \\ &\quad + [x_1 - p_1]^2 + [x_2 - p_2]^2 + [x_3 - p_3]^2 \\ &= \{R^2\} + \{2p_1[x_1 - p_1] + 2p_2[x_2 - p_2] + 2p_3[x_3 - p_3]\} \\ &\quad + \{[x_1 - p_1]^2 + [x_2 - p_2]^2 + [x_3 - p_3]^2\}. \end{aligned}$$

Subtracting $R^2$ from both sides gives a linear expression plus higher degree (higher-order) terms $HOT = \{[x_1 - p_1]^2 + [x_2 - p_2]^2 + [x_3 - p_3]^2\}$:

$$0 = 2p_1[x_1 - p_1] + 2p_2[x_2 - p_2] + 2p_3[x_3 - p_3] + HOT.$$

For $\vec{x}$ near $\vec{p}$, we linearize the equation by dropping higher-order terms because square powers of small numbers become negligible compared to the linear part.

Thus the linearized equation is

$$0 = 2p_1[x_1 - p_1] + 2p_2[x_2 - p_2] + 2p_3[x_3 - p_3] = (2p_1, 2p_2, 2p_3)^T \cdot (\vec{x} - \vec{p}),$$

which is recognizable as a plane through the point $\vec{p} = (p_1, p_2, p_3)^T$ with normal vector $\vec{n} = (2p_1, 2p_2, 2p_3)^T$. We have just seen our first tangent plane to a curved surface! Note this confirms that the normal vector to a tangent plane at a point $\vec{p}$ on a sphere can be taken to be the point $\vec{p} = (1/2)\vec{n}$. In other words, the radial vector is perpendicular to the tangent plane of a sphere.

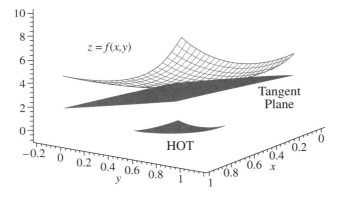

**FIGURE 6.1** ■ Surface, tangent, and *HOT*.

■ **Exercises**

1. Find the equation of the line in $\mathbb{R}^2$ through the point $\vec{p} = (1, 2)^T$ and perpendicular to $\vec{n} = (4, 7)^T$.
2. Find the equation of the plane in $\mathbb{R}^3$ through the point $\vec{p} = (2, 7, 1)^T$ and perpendicular to $\vec{n} = (4, -2, 6)^T$.
3. Use linearization at the point $\vec{p} = (1, 1, 1)^T$ to find the tangent plane of $xyz = 1$ at $\vec{p}$.

**4.** Find the equation of the plane containing the three points $(2,0,2)^T$, $(1,2,1)^T$, and $(1,1,1)^T$.

## 6.2 Examples of Surfaces in $\mathbb{R}^3$

We next encounter a series of examples of surfaces in $\mathbb{R}^3$. The goal will be to gain intuition and familiarity with such higher-dimensional sets. In this section the variables $(x_1, x_2, x_3)^T$ will be denoted $(x, y, z)^T$, as is common.

The *plane* through the point $(p_1, p_2, p_3)^T$ with normal $(n_1, n_2, n_3)^T$ has been seen in the previous section to have equation

$$n_1(x - p_1) + n_2(y - p_2) + n_3(z - p_3) = 0.$$

The *sphere* with center $(p_1, p_2, p_3)^T$ and radius $R$ is also familiar, with equation

$$(x - p_1)^2 + (y - p_2)^2 + (z - p_3)^2 = R^2.$$

When $\vec{p} = \vec{0}$ we have $x^2 + y^2 + z^2 = R^2$ is the sphere of radius $R$ centered at $\vec{0}$. From here on out the various surfaces will be centered at $\vec{0}$.

The sphere can be morphed to an oval shape called an *ellipsoid* resulting in the following equation:

$$\frac{x^2}{A^2} + \frac{y^2}{B^2} + \frac{z^2}{C^2} = 1.$$

When $A = B = C = R$ the preceding equation becomes that of the sphere. The longer one parameter is relative to the others, the more elongated is the ellipsoid. For instance if $A$ is double $B$ and $C$, the ellipsoid has stretched the sphere and doubled in the $x$ direction.

Another surface familiar to anyone with a satellite dish is the *elliptic paraboloid*

$$z = \frac{x^2}{A^2} + \frac{y^2}{B^2}.$$

One way of studying surfaces in higher dimensions is to slice the surface by taking cross sections to obtain a curve. As one slides the cross sections the resulting curve moves with the plane of slicing, and one can see these moving cross sectional curves as filling out the surface. For instance in the case of the elliptic paraboloid with $A = B$, slicing by the plane at constant height $z = z_0$ yields $z_0 = x^2/A^2 + y^2/A^2$, or after multiplying by $A^2$, $z_0 A^2 = x^2 + y^2$, which is seen to be a circle of radius

$\sqrt{z_0}A$ in the $x$-$y$ plane. In the $x$-$y$ plane this is called a level curve. When this level curve is slid up to height $z_0$ one obtains the cross section. As we move the height $z_0$ of the slice upward, the cross sections are seen to be expanding circles. On the other hand if one sets $y = y_0$ one gets a parabolic cross section $z = x^2/A^2 + y_0^2/A^2$, in the $x$-$z$ plane. As one increases $y_0$ the parabolas slide upward. In all, one begins to see these moving cross sections piece together a surface that resembles a satellite dish or the reflective mirror of a flashlight or headlight of a car. Paraboloids are common, in that they have useful reflective properties, including: focusing incoming satellite beams to a receiver point (called a *focus*); or reflecting a lightbulb's beam out the front of the headlight in a controlled manner.

Here are a few more surfaces given by degree two polynomials in three variables. Such surfaces are called *conics*. The following surfaces can be morphed, as the sphere was morphed, by inserting an $A, B, C$ in the denominator of $x, y, z$, respectively.

The *cone*: $z^2 = x^2 + y^2$.
The *hyperbolic paraboloid*: $z = x^2 - y^2$.
The *hyperboloid of one sheet*: $1 = x^2 + y^2 - z^2$.
The *hyperboloid of two sheets*: $1 = x^2 - y^2 - z^2$.
The *cylinder*: $x^2 + y^2 = R^2$.

These surfaces are logically named. Just as the elliptic paraboloid had cross sectional ellipses and parabolas, so do the other named surfaces have the indicated cross sections. For instance, the hyperbolic paraboloid has cross sections of hyperbolas and parabolas: the cross section $z = z_0$ is given by the hyperbola $z_0 = x^2 - y^2$ in the $x$-$y$ plane; the cross section $x = x_0$ is given by the upside-down parabola $z = x_0^2 - y^2$ in the $z$-$y$ plane; and the cross section $y = y_0$ is given by the upward parabola $z = x^2 - y_0^2$ in the $z$-$x$ plane.

Linearization techniques can be harnessed to see the tangent planes to each of the conics, except at the point $\vec{0}$ on the cone, as the cone is not a regular surface at $\vec{0}$. See Theorem 6.7.

For instance, suppose $\vec{p} = (p_1, p_2, p_3)^T$ is a point on the elliptic paraboloid $z = x^2 + y^2$. Then $p_3 = p_1^2 + p_2^2$. Expand $x$ about $p_1$, $y$ about $p_2$, and $z$ about $p_3$ to obtain

$$(p_3 + [z - p_3]) = (p_1 + [x_1 - p_1])^2 + (p_2 + [x_2 - p_2])^2$$
$$= p_1^2 + 2p_1[x_1 - p_1] + [x_1 - p_1]^2 + p_2^2 + 2p_2[x_2 - p_2] + [x_2 - p_2]^2$$
$$= \{p_1^2 + p_2^2\} + 2p_1[x_1 - p_1] + 2p_2[x_2 - p_2] + [x_1 - p_1]^2 + [x_2 - p_2]^2$$
$$= \{p_3\} + 2p_1[x_1 - p_1] + 2p_2[x_2 - p_2] + [x_1 - p_1]^2 + [x_2 - p_2]^2.$$

Subtracting $p_3$ from both sides gives a linear expression plus higher-degree (higher-order) terms $HOT = \{[x_1 - p_1]^2 + [x_2 - p_2]^2\}$:

$$[z - p_3] = 2p_1[x_1 - p_1] + 2p_2[x_2 - p_2] + HOT.$$

For $\vec{x}$ near $\vec{p}$, we linearize the equation by dropping higher-order terms because square powers of small numbers become negligible compared to the linear part. Thus the linearized equation is

$$0 = 2p_1[x - p_1] + 2p_2[y - p_2] - 1[z - p_3] = (2p_1, 2p_2, -1)^T \cdot (\vec{x} - \vec{p}),$$

which is recognizable as a plane through the point $\vec{p} = (p_1, p_2, p_3)^T$ with normal vector $\vec{n} = (2p_1, 2p_2, -1)^T$. We take this plane to be the tangent plane. For instance when $\vec{p} = \vec{0}$ one sees that the tangent plane at $\vec{0}$ is $0 = 0x + 0y - z$, or more simply $z = 0$.

## Exercises

1. Given the surface $z = x^2 - y^2$ in $\mathbb{R}^3$,
    (a) find the cross sections for $z = 1$, $z = 0$, $z = -1$.
    (b) find the cross sections for $x = 0$, $x = 1$.
    (c) find the cross sections for $y = 0$, $y = 1$.

2. Use linearization to find the tangent plane to the surface $1 = x^2 + y^2 - z^2$ at the point $(1, 1, 1)^T$.

## 6.3 Functions of One Variable: Curves in $\mathbb{R}^n$

Let $[a, b]$ be an interval in $\mathbb{R}$ and consider the function given by

$$\vec{f}(t) = (f_1(t), f_2(t), \cdots, f_n(t))^T, \qquad (6.15)$$

where each $f_i$ is a function from $[a, b]$ to $\mathbb{R}$ for $i = 1, 2, \ldots, n$. The function $\vec{f}$ is called a *vector-valued function* on $[a, b]$.

The independent variable $t$ is called the *parameter*. The set of points in $\mathbb{R}^n$ in the range of $\vec{f}$ is denoted

$$\mathcal{C} \equiv \{\vec{f}(t) \mid t \in [a, b]\} = \vec{f}([a, b]),$$

and is called the *curve*, or sometimes called the *orbit*,[1] of the *parameterization* given by $\vec{f}: [a,b] \to \mathcal{C} \subset \mathbb{R}^n$.

Often one is given the curve $\mathcal{C}$ and a parameterization has to be found or constructed. That is, one must find a variable $t$ and an $n$-tuple of functions $f_i(t)$ that form the coordinates of $\vec{f}$, with $\vec{f}$ having range $\mathcal{C}$. For example, to parameterize the upper unit semicircle, $SC$, given by

$$SC \equiv \{(x,y)^T \mid x^2 + y^2 = 1, \ y \geq 0\},$$

consider these two choices of parameterizations. The first parameterization of $SC$ is $\vec{s}(t) = (-t, \sqrt{1-t^2})^T$, for $t \in [-1,1]$. The second parameterization of $SC$ is $\vec{r}(\theta) = (\cos(\theta), \sin(\theta))^T$, for $\theta \in [0, \pi]$.

A limitation of using the parameterization $\vec{s}(t)$ is that it is not possible to extend the parameterization beyond its end points $t = \pm 1$. However $\vec{r}(\theta)$ extends beyond $\theta = 0, \pi$, making sense for any $\theta \in \mathbb{R}$, and so one can parameterize the entire unit circle. Hence $\vec{r}(\theta)$ is often a better choice of parameterization.

In many cases the parameter, or variable, $t$ will correspond to time. For example, consider a fly in a room whose position is seen to follow the path $\vec{r}([a,b])$ described by $\vec{r}(t) \equiv (x(t), y(t), z(t))^T$ in $\mathbb{R}^3$. The quantity $\vec{r}(t)$ is called the *position vector* of the fly.

If each of the coordinate functions in the position function is twice differentiable, then the velocity and acceleration functions can be defined as follows.

---

**Definition 6.1** The derivative of the position vector $\vec{r}(t)$ is the *velocity vector* $\vec{v}(t) \equiv \vec{r}\,'(t) = (x'(t), y'(t), z'(t))^T$, and the derivative of the velocity is the *acceleration vector*

$$\vec{a}(t) \equiv \vec{v}\,'(t) = \vec{r}\,''(t) = (x''(t), y''(t), z''(t))^T.$$

If a particle of mass $m$ with position $\vec{r}(t)$ moves through a force field $\vec{F}(\vec{x})$ then the acceleration solves the *equations of motion*

$$m\vec{a}(t) \equiv m\frac{d^2\vec{r}(t)}{dt^2} = \vec{F}(\vec{r}(t)). \tag{6.16}$$

---

The equations of motion can be difficult to solve, however there are some simple examples that lead to an understanding of well-known phenomena. For instance,

---

[1]The notion of an orbit usually implies that the end points agree.

if the force field is $\vec{0}$ in $\mathbb{R}^3$, then $\vec{a}(t) = \vec{0}$. By antidifferentiating each component function, one obtains

$$\int \vec{a}(t)dt = \vec{v}(t) = \int \vec{0} \, dt = \left( \int 0 \, dt, \int 0 \, dt, \int 0 \, dt \right)^T = (v_1, v_2, v_3)^T,$$

a constant vector, as the velocity. Antidifferentiating again recovers the position function as

$$\int \vec{v}(t) \, dt = \vec{x}(t) = \int (v_1, v_2, v_3)^T \, dt$$
$$= \left( \int v_1 \, dt, \int v_2 \, dt, \int v_3 \, dt \right)^T = (v_1 t + p_1, v_2 t + p_2, v_3 t + p_3)^T.$$

If there is no force, motion is in a straight line. The equation

$$\vec{x}(t) = (v_1 t + p_1, v_2 t + p_2, v_3 t + p_3)^T = (v_1, v_2, v_3)^T t + (p_1, p_2, p_3)^T = \vec{v} \, t + \vec{p}$$

is that of the line in $\mathbb{R}^3$ through the point $\vec{p}$ in the direction $\vec{v}$. In general, in $\mathbb{R}^n$ the *line through a point $\vec{p}$ in the direction $\vec{d}$* has equation $\vec{x}(t) = \vec{p} + t \, \vec{d}$.

We return to the concept of the derivative of a curve $\vec{f}$ in $\mathbb{R}^n$. $\vec{f}(t) = (f_1(t), f_2(t), \ldots, f_n(t))^T$ is differentiable at a point $t = p$ precisely when each of its coordinate functions $f_i(t)$ is differentiable at $p$. This is equivalent to saying that $f_i(t) = f_i(p) + f'_i(p)[t - p] + HOT_i([t - p])$ with $HOT_i([0]) = 0$ and $\lim_{t \to p} HOT_i([t - p])/[t - p] = 0$. If these equations are written out in vector form, they are equivalent to the vector equation:

$$\begin{aligned}
\vec{f}(t) &= (f_1(p), f_2(p), \ldots, f_n(p))^T + (f'_1(p), f'_2(p), \ldots, f'_n(p))^T [t - p] \\
&\quad + (HOT_1([t - p]), HOT_2([t - p]), \ldots, HOT_n([t - p]))^T \\
&= \vec{f}(p) + \vec{f}\,'(p)[t - p] + \vec{HOT}([t - p])
\end{aligned}$$

where $\vec{HOT}([0]) = \vec{0}$ and $\lim_{t \to p} \vec{HOT}([t - p])/[t - p] = \vec{0}$. This matches exactly the single variable format for the Definition (5.1) of a function to be differentiable at $p$, with vector modification.

This leads to the following definition of the derivative of a vector-valued function.

---

**Definition 6.2** The vector-valued function $\vec{f}(t)$ is *differentiable* at $p$ if there is a constant vector, called $\vec{f}\,'(p)$, and a linear function $\vec{f}(p) + \vec{f}\,'(p)[t - p]$ such that

$$\vec{f}(t) = \vec{f}(p) + \vec{f}\,'(p)[t - p] + \vec{HOT}([t - p])$$

with $\vec{HOT}([0]) = \vec{0}$ and $\lim_{t \to p} \vec{HOT}([t - p])/[t - p] = \vec{0}$.

The function $\vec{f}(p) + \vec{f}\,'(p)[t-p]$ is called the *linearization* of $\vec{f}$ at $p$.

The curve $\vec{r}(t) = \vec{f}(p) + \vec{f}\,'(p)[t-p]$ is called the *tangent line* to $\vec{f}$ at $p$. The vector $\vec{f}\,'(p)$ is called the *tangent vector*.

In general, the vector-valued function is differentiable precisely when each of its coordinate functions is differentiable.

### EXAMPLE 6.1

Find the derivative of the vector-valued function $\vec{f}(t) = (t^3, \sin(t), e^t)^T$. Find the equation of the tangent line to $\vec{f}$ at time $t = 1$.

One only needs to differentiate component-wise to obtain $\vec{f}\,'$:

$$\vec{f}\,'(t) = ((t^3)', \sin'(t), (e^t)')^T = (3t^2, \cos(t), e^t)^T.$$

The equation of the tangent line at 1 is

$$\vec{r}(t) = \vec{f}(1) + \vec{f}\,'(1)[t-1] = (1, \sin(1), e^1)^T + (3, \cos(1), e^1)^T[t-1].$$

### ■ Exercises

1. Given $\vec{r}(t) = (t^2, t^3, t^4)^T$,

    (a) find the velocity vector $\vec{v}(t) = \vec{r}\,'(t)$;
    (b) find the acceleration vector $\vec{a}(t) = \vec{r}\,''(t)$;
    (c) find the linearization of $\vec{r}(t)$ at $t = 1$; and
    (d) find the equation of the tangent line to $\vec{r}(t)$ at $t = 1$.

2. Parameterize the line $\ell = \{(x,y)^T | y = 3x - 2,\ |x| < 3\}$.
3. Parameterize the parabola $P = \{(x,y)^T | y = x^2 - 2x + 1,\ |x| < 2\}$.
4. Parameterize the parabola $\ell = \{(x,y)^T | x = y^2 - 2y + 1,\ |y| < 1\}$.
5. Parameterize the ellipse $\ell = \{(x,y)^T | x^2 + 4y^2 = 16\}$.
6. Parameterize the hyperbola $\ell = \{(x,y)^T | 9x^2 - y^2 = 1,\ x > 0\}$.

### ■ 6.4 Functions of Several Variables

For many reasons one may desire to study functions of several variables. For instance, if 3-space has position coordinates $(x, y, z)^T$ and time is given by $t$, an atmospheric scientist or a weather person would be interested in the temperature $T(x, y, z, t)$ at any given position at any given time. Knowing this data, they would be interested in predicting the temperature at various locations at future times. This would involve understanding how heat tends to redistribute, or flow. Another example of a function of several variables of interest to this scientist would be the atmospheric pressure $P(x, y, z, t)$ at a position at a given time. A related concept would be to find the places and times where the temperature or the pressure is constant. For instance, for a fixed time, $t = t_0$, and a fixed pressure value $P_0$, the relation

$$P(x, y, z, t_0) = P_0 \quad \text{for} \quad P_0 \in \mathbb{R}^+$$

defines a pressure *isobar*, or a surface of constant pressure in 3-space. As time evolves the isobars will move, usually so the high-pressure region decreases and the low-pressure increases, both toward an equilibrium value. The same would be true for temperatures. For a fixed time, $t = t_0$, and a fixed temperature value $T_0$, the relation

$$T(x, y, z, t_0) = T_0 \quad \text{for} \quad T_0 \in \mathbb{R}$$

defines a temperature *isobar*, or a surface of constant temperature in 3-space. As time evolves the isobars will move, usually so the high-temperature region decreases and the low temperature increases, both toward an equilibrium value. Again, temperature flows from high-value locations toward low-value locations.

With these examples in mind, we turn to a real-valued function $f(x_1, x_2, \ldots, x_n) = f(\vec{x})$ of several variables $(x_1, x_2, \ldots, x_n)^T = \vec{x} \in \mathbb{R}^n$.

Functions of two variables can be visualized by defining a type of surface in $\mathbb{R}^3$ called a *graph*, e.g., $z = f(x, y)$. A particular example of a function of two variables is given by $f(x, y) = x^2 + y^2$. It has graph

$$z = f(x, y) = x^2 + y^2,$$

defined for all values of $(x, y)^T$ in $\mathbb{R}^2$. This is seen to be the surface in $\mathbb{R}^3$ given by the elliptic paraboloid, which might model a type of satellite dish. The graph of a function satisfies a vertical line test, that is, above each $(x, y)^T$ in the domain of $f$ there is only one value for $z \in \mathbb{R}$ on the graph of the function.

Functions $f(x, y, z)$ of three variables can be graphed only in $\mathbb{R}^4$. This is accomplished by taking a fourth variable $w$ and setting the graph to be $w = f(x, y, z)$. This cannot in general be easily visualized, given our familiar three-dimensional perspective. However, choosing a constant $w_0$ and obtaining a cross section in $\mathbb{R}^4$ by setting $w = w_0$, one has a *level surface* $f(x, y, z) = w_0$ in our domain in $\mathbb{R}^3$. One visualizes this level surface as being transported to a height of $w_0$ in the fourth, $w$, dimension to obtain an image of the $w = w_0$ cross section of the graph of $f$. By varying $w_0$ one reconstructs the whole graph in $\mathbb{R}^4$ as a collection of moving cross sections. For instance, given the function $f(x, y, z) = x^2 + y^2 + z^2$, a level surface given by the cross section $w = w_0$ would be

$$w_0 = x^2 + y^2 + z^2 = f(x, y, z),$$

which is seen to be a sphere of radius $\sqrt{w_0}$ centered at $\vec{0}$. As $w_0$ increases, one obtains a cross section at height $w = w_0$ to be this sphere of radius $\sqrt{w_0}$. The whole graph is now visualized as a collection of these moving surfaces in $\mathbb{R}^4$.

We turn our focus now to the concept of a differentiable function $f(\vec{x})$ of several variables. Just as in the one variable case we want a linearization at a point $\vec{p}$ that well approximates $f(\vec{x})$ up to higher-order terms that vanish rapidly compared to how far $\vec{x}$ is from $\vec{p}$, that is compared to $|\vec{x} - \vec{p}|$. In this setting, one needs $HOT([\vec{x} - \vec{p}])/|\vec{x} - \vec{p}|$ to approach 0 as $\vec{x}$ approaches $\vec{p}$. Thus $f$ should satisfy that

$$f(\vec{x}) = f(\vec{p}) + \vec{n} \cdot (\vec{x} - \vec{p}) + HOT([\vec{x} - \vec{p}])$$

for the linearization $f(\vec{p}) + \vec{n} \cdot (\vec{x} - \vec{p})$, with the higher-order terms, $HOT([\vec{x} - \vec{p}])$, satisfying $HOT(\vec{0}) = 0$ and $\lim_{\vec{x} \to \vec{p}} HOT([\vec{x} - \vec{p}])/|\vec{x} - \vec{p}| = 0$. Note that when we had a function of one variable we could divide HOT by $[x - p]$ in a meaningful way. We cannot divide $HOT$ by a vector $[\vec{x} - \vec{p}]$. However, if one considers $[x - p]$ in the one variable case to be measuring how far $x$ is from $p$, then one can use the distance $|\vec{x} - \vec{p}|$ from $\vec{x}$ to $\vec{p}$ as the denominator in the several variable case. The vector $\vec{n}$ in the linearization is given a name. It is the *gradient* of $f$ at the point $\vec{p}$, and is denoted

$$\vec{\nabla} f(\vec{p}) = grad\ f(\vec{p}) = \vec{n}$$

and each component $n_i$ of $\vec{n}$ is called the $i$th *partial derivative* of $f$ at $\vec{p}$ and is denoted $f_{x_i}(\vec{p}) = n_i$.

---

**Definition 6.3** The real-valued function $f(\vec{x}) = f(x_1, x_2, \ldots, x_n)$ of several variables is *differentiable* at $\vec{p}$ when there is a linearization

$$f(\vec{p}) + \vec{\nabla} f(\vec{p}) \cdot [\vec{x} - \vec{p}] = f(\vec{p}) + (f_{x_1}(\vec{p}), f_{x_2}(\vec{p}), \ldots, f_{x_n}(\vec{p}))^T \cdot [\vec{x} - \vec{p}]$$

such that

$$f(\vec{x}) = f(\vec{p}) + \vec{\nabla} f(\vec{p}) \cdot [\vec{x} - \vec{p}] + HOT([\vec{x} - \vec{p}])$$

with

$$HOT(\vec{0}) = 0 \text{ and } \lim_{\vec{x} \to \vec{p}} HOT([\vec{x} - \vec{p}])/\|[\vec{x} - \vec{p}]\| = 0.$$

The *gradient* of $f$ at $\vec{p}$ is denoted

$$\vec{\nabla} f(\vec{p}) = (f_{x_1}(\vec{p}), f_{x_2}(\vec{p}), \ldots, f_{x_n}(\vec{p}))^T,$$

and is the vector of coefficients of the linear terms. Each component of the gradient, $f_{x_i}(\vec{p})$ is the $i$th *partial derivative* of $f$ at $\vec{p}$.

---

As an example, consider the function $f(x_1, x_2) = x_1^2 + 3x_2^2$. Let $\vec{p} = (p_1, p_2)^T$ be a point in $\mathbb{R}^2$. Expand $x_1$ about $p_1$, and $x_2$ about $p_2$, to obtain

$$\begin{aligned} f(x_1, x_2) &= (p_1 + [x_1 - p_1])^2 + 3(p_2 + [x_2 - p_2])^2 \\ &= p_1^2 + 2p_1[x_1 - p_1] + [x_1 - p_1]^2 + 3p_2^2 + 6p_2[x_2 - p_2] + 3[x_2 - p_2]^2 \\ &= \{p_1^2 + 3p_2^2\} + 2p_1[x_1 - p_1] + 6p_2[x_2 - p_2] + [x_1 - p_1]^2 + 3[x_2 - p_2]^2. \end{aligned}$$

From this one sees that the linearization of $f$ is

$$\{p_1^2 + 3p_2^2\} + 2p_1[x_1 - p_1] + 6p_2[x_2 - p_2] = f(p_1, p_2) + (2p_1, 6p_2)^T \cdot [x_1 - p_1, x_2 - p_2]^T,$$

and the higher-order terms are

$$HOT([\vec{x} - \vec{p}]) = \{[x_1 - p_1]^2 + 3[x_2 - p_2]^2\}.$$

The gradient is $\vec{\nabla} f(\vec{p}) = (2p_1, 6p_2)^T$ and the partial derivative of $f$ with respect to $x_1$ is $f_{x_1}(\vec{p}) = 2p_1$ with the partial derivative of $f$ with respect to $x_2$ given by $f_{x_2}(\vec{p}) = 6p_2$.

To examine the higher-order terms, denote $\Delta x_1 = [x_1 - p_1], \Delta x_2 = [x_2 - p_2]$, then $HOT([\vec{x} - \vec{p}]) = (\Delta x_1)^2 + 3(\Delta x_2)^2$, and $|\vec{x} - \vec{p}| = \sqrt{(\Delta x_1)^2 + (\Delta x_2)^2}$. Then

$$\begin{aligned} HOT([\vec{x} - \vec{p}])/|\vec{x} - \vec{p}| &= \frac{(\Delta x_1)^2 + 3(\Delta x_2)^2}{\sqrt{(\Delta x_1)^2 + (\Delta x_2)^2}} \\ &= (\Delta x_1)\frac{(\Delta x_1)}{\sqrt{(\Delta x_1)^2 + (\Delta x_2)^2}} + 3(\Delta x_2)\frac{(\Delta x_2)}{\sqrt{(\Delta x_1)^2 + (\Delta x_2)^2}}. \end{aligned} \quad (6.17)$$

Note that

$$\left| \frac{(\Delta x_i)}{\sqrt{(\Delta x_1)^2 + (\Delta x_2)^2}} \right| \leq 1 \text{ for each } i = 1, 2. \quad (6.18)$$

Applying Eq. (6.18) to Eq. (6.17) we get $|HOT([\vec{x} - \vec{p}])/|\vec{x} - \vec{p}|| \leq |(\Delta x_1)| \cdot 1 + 3|(\Delta x_2)| \cdot 1 = |x_1 - p_1| + 3|x_2 - p_2|$, which clearly approaches 0 as $\vec{x}$ approaches $\vec{p}$.

From this example we can see that the analysis of the higher-order terms $HOT$ generalizes to any expression $HOT$ consisting of sums of powers of $\Delta x_i$ with each term having total degree greater than 1. One simply factors out one power of say $\Delta x_j$ from a summand, divide by $|\vec{x} - \vec{p}| = \sqrt{(\Delta x_1)^2 + (\Delta x_2)^2 + \cdots + (\Delta x_n)^2}$, use the bound in Eq. (6.18), and let the remaining powers of the $\Delta x_i$ take the summand to 0, which shows the limit of the sum to be 0 as well. For this reason, whenever the higher-order terms are polynomials of total degree two or more, one will never need to repeat this analysis on $HOT([\vec{x}-\vec{p}])/|\vec{x}-\vec{p}|$ to show it approaches 0 when $\vec{x}$ approaches $\vec{p}$. In this case one is now justified in simply ignoring the higher-order terms.

It bears pointing out that there is an alternate, more geometric, way of obtaining partial derivatives by cross-sectional techniques. For the function $f(x_1, x_2) = x_1^2 + 3x_2^2$ one can find the first partial derivative $f_{x_1}$ in the following way: consider $x_2$ to be a constant with only $x_1$ varying (this is where the cross sectioning occurs) and differentiate $f$ with respect to the variable $x_1$ (considering $3x_2^2$ to now be constant). Then $f_{x_1} = 2x_1 + 0$ using the Power Rule, and at $\vec{p}$ this becomes $f_{x_1}(p_1, p_2) = 2p_1$, as was just shown. If one holds $x_1$ constant and lets $x_2$ vary, then $f_{x_2} = 3 \cdot 2(x_2)^1 = 6x_2$, which at $\vec{p}$ becomes $f_{x_2}(p_1, p_2) = 6p_2$, again agreeing with the preceding technique. The notation is $\partial f/\partial x_1 = f_{x_1} = 2x_1$ and $\partial f/\partial x_2 = f_{x_2} = 6x_2$. This leads us into the next section.

## ■ Exercises

1. For the function $f(x, y) = x^2 - y^2 - 2y$ describe the three curves $f(x, y) = 1, 0, -1$.

2. Given $f(x_1, x_2) = x_1^2 + x_2^3$,
   
   (a) find the linearization of $f$ at $(2, 1)^T$;
   (b) find the gradient $\vec{\nabla} f(2, 1)$; and
   (c) find the partial derivatives $\partial f/\partial x_1$, $\partial f/\partial x_2$, and evaluate them at the point $(2, 1)^T$.

## ■ 6.5 The Chain Rule; Partial and Directional Derivatives

Given a real-valued function $f(\vec{x}) = f(x_1, x_2, \ldots, x_n)$ that is differentiable at the point $\vec{p} = (p_1, p_2, \ldots, p_n)^T$ and a vector-valued function $\vec{x}(t) = (x_1(t), x_2(t), \ldots, x_n(t))^T$ that is differentiable at $t = t_0$ where $\vec{x}(t_0) = \vec{p}$, then the composed function $f(\vec{x}(t)) = f(x_1(t), x_2(t), \ldots, x_n(t))$ is a real-valued function of one variable $t$, and is differentiable at $t = t_0$. This derivative at $t_0$ is computed in the next theorem.

## ■ Theorem 6.4

### The Chain Rule for a Function of Several Variables

Let the real-valued function $f(\vec{x}) = f(x_1, x_2, \ldots, x_n)$ be differentiable at the point $\vec{p} = (p_1, p_2, \ldots, p_n)^T$; and let $\vec{x}(t) = (x_1(t), x_2(t), \ldots, x_n(t))^T$ be a vector-valued function of one variable that is differentiable at $t = t_0$ with $\vec{x}(t_0) = \vec{p}$. Then the derivative of the composed function is computed by:

$$\begin{aligned} (f \circ \vec{x})'(t_0) &= \frac{d(f \circ \vec{x})}{dt}(t_0) = \vec{\nabla} f(\vec{x}(t_0)) \cdot \vec{x}\,'(t_0) \qquad (6.19) \\ &= \frac{\partial f}{\partial x_1}(\vec{x}(t_0)) \frac{dx_1}{dt}(t_0) + \frac{\partial f}{\partial x_2}(\vec{x}(t_0)) \frac{dx_2}{dt}(t_0) + \cdots + \frac{\partial f}{\partial x_n}(\vec{x}(t_0)) \frac{dx_n}{dt}(t_0). \end{aligned}$$

**Proof** Because $f$ is differentiable at $p$, one knows

$$f(\vec{x}) = f(\vec{p}) + \vec{\nabla} f(\vec{p}) \cdot [\vec{x} - \vec{p}] + HOT_f([\vec{x} - \vec{p}]) \qquad (6.20)$$

with

$$HOT_f(\vec{0}) = 0 \quad \text{and} \quad \lim_{\vec{x} \to \vec{p}} HOT_f([\vec{x} - \vec{p}])/\|[\vec{x} - \vec{p}]\| = 0.$$

Since $\vec{x}(t)$ is differentiable at $t_0$ one also knows

$$\vec{x}(t) = \vec{x}(t_0) + \vec{x}\,'(t_0)(t - t_0) + \vec{HOT}_{\vec{x}}([t - t_0]) \qquad (6.21)$$

with $\vec{HOT}_{\vec{x}}(0) = \vec{0}$ and $\vec{HOT}_{\vec{x}}([t - t_0])/[t - t_0]$ approaching $\vec{0}$ as $t$ approaches $t_0$. Substituting $\vec{x}(t)$ into $f$ in Eq. (6.20) then yields

$$f(\vec{x}(t)) = f(\vec{p}) + \vec{\nabla} f(\vec{p}) \cdot [\vec{x}(t) - \vec{p}] + HOT_f([\vec{x}(t) - \vec{p}]). \qquad (6.22)$$

Now Eq. (6.21), along with the fact that $\vec{x}(t_0) = \vec{p}$, gives $[\vec{x}(t) - \vec{p}] = \vec{x}(t) - \vec{x}(t_0) = \vec{x}\,'(t_0)(t - t_0) + \vec{HOT}_{\vec{x}}([t - t_0])$, which is substituted into Eq. (6.22) to obtain

$$f(\vec{x}(t)) = f(\vec{p}) + \vec{\nabla} f(\vec{p}) \cdot [\vec{x}\,'(t_0)(t - t_0) + \vec{HOT}_{\vec{x}}([t - t_0])] + HOT_f([\vec{x}(t) - \vec{p}])$$
$$= f(\vec{p}) + \{\vec{\nabla} f(\vec{p}) \cdot \vec{x}\,'(t_0)\}(t - t_0)$$
$$+ \{\vec{\nabla} f(\vec{p}) \cdot \vec{HOT}_{\vec{x}}([t - t_0]) + HOT_f([\vec{x}(t) - \vec{p}])\}. \tag{6.23}$$

Now the linearization of $f(\vec{x}(t))$ at $t_0$ is seen to be $f(\vec{p}) + \{\vec{\nabla} f(\vec{p}) \cdot \vec{x}\,'(t_0)\}(t - t_0)$, and then the derivative $(f \circ \vec{x})'(t_0)$ is the coefficient $\{\vec{\nabla} f(\vec{p}) \cdot \vec{x}\,'(t_0)\} = \{\vec{\nabla} f(\vec{x}(t_0)) \cdot \vec{x}\,'(t_0)\}$ of $(t - t_0)$, which gives the main statement of the theorem.

The interested reader can continue on to ensure that the higher-order terms are $\{\vec{\nabla} f(\vec{p}) \cdot \vec{HOT}_{\vec{x}}([t - t_0]) + HOT_f([\vec{x}(t) - \vec{p}])\}$. Clearly $\vec{\nabla} f(\vec{p}) \cdot \vec{HOT}_{\vec{x}}(0) = \vec{\nabla} f(\vec{p}) \cdot \vec{0} = 0$ and $(\vec{\nabla} f(\vec{p}) \cdot \vec{HOT}_{\vec{x}}([t - t_0]))/(t - t_0)$ approaches $(\vec{\nabla} f(\vec{p}) \cdot \vec{0}) = 0$ as $t$ approaches $t_0$. Finally at $t_0$, $HOT_f([\vec{x}(t_0) - \vec{p}]) = HOT_f([\vec{p} - \vec{p}]) = HOT_f(\vec{0}) = 0$ (and this is true at any time $\vec{x}(t) = \vec{p}$); and

$$\frac{HOT_f([\vec{x}(t) - \vec{p}])}{(t_0 - t)} = \frac{HOT_f([\vec{x}(t) - \vec{p}])}{|\vec{x}(t) - \vec{p}|} \frac{|\vec{x}(t) - \vec{p}|}{(t - t_0)}$$
$$= \frac{HOT_f([\vec{x}(t) - \vec{p}])}{|\vec{x}(t) - \vec{p}|} \frac{|\vec{x}\,'(t_0)(t - t_0) + \vec{HOT}_{\vec{x}}([t - t_0])|}{(t_0 - t)} \tag{6.24}$$
$$= \pm \frac{HOT_f([\vec{x}(t) - \vec{p}])}{|\vec{x}(t) - \vec{p}|} \left| \vec{x}\,'(t_0) + \frac{\vec{HOT}_{\vec{x}}([t - t_0])}{t - t_0} \right|,$$

which approaches $0|\vec{x}\,'(t_0)| = 0$ as $t$ approaches $t_0$. Observe that Eq. (6.21) was used to obtain Eq. (6.24). This finishes the proof of the Chain Rule. □

If one slices the graph of $f$ by the plane formed by the vertical axis and by the line through $\vec{p}$ parallel to the $x_i$ axis, and then differentiates the resulting function of $x_i$ one obtains $f_{x_i}(\vec{p})$, and so the $i$th partial of $f$ is the rate of change of $f$ in the $x_i$ direction:

### Corollary 6.8

The partial derivative $f_{x_i}$ can be computed by setting all $x_j$ constant for $j \neq i$, letting $x_i$ vary, and differentiating the resulting function of $x_i$.

**Proof** Fix $i$, let $x_i(t) = t$, and let all $x_j(t) = p_j$ be constant for $j \neq i$. The resulting differentiable curve is $\vec{x}(t) = (p_1, \ldots, p_{j-1}, t, p_{j+1}, \ldots, p_n)^T$, which has derivative $\vec{x}'(t) = (0, \ldots, 0, 1, 0, \ldots, 0)^T$ with the 1 being in the $i$th coordinate.

Clearly $\vec{x}(p_i) = \vec{p}$. By the Chain Rule

$$(f \circ \vec{x})'(p_i) = \vec{\nabla} f(\vec{p}) \cdot (0, \ldots, 0, 1, 0, \ldots, 0)^T = f_{x_i}(\vec{p}).$$

□

The partial derivative of $f$ with respect to the $i$th-variable $x_i$ is expressed in many different ways,

$$\frac{\partial f}{\partial x_i}(x_1, \ldots, x_i, \ldots, x_n) = \partial_{x_i} f(\vec{x}) = \partial_i f = f_{,i} = f_{x_i}$$

$$\equiv \lim_{\Delta x \to 0} \frac{f(x_1, \ldots, x_i + \Delta x, \ldots, x_n)}{\Delta x}. \quad (6.25)$$

This last limit gives the partial derivative $\partial f/\partial x_i$ for all differentiable functions $f$, but the limit in Eq. (6.25) can hold even when $f$ fails to be differentiable, and so Eq. (6.25) will be the default computation of the partial derivative.

The last corollary, or equivalently, Eq. (6.25), allows for quick computation of partial derivatives, gradients, and linearizations without the need for algebraic expansion. This can be particularly helpful if $f$ is not an algebraic function.

### EXAMPLE 6.2

Find the partial derivatives and the gradient of $f(x_1, x_2) = \cos(x_1) e^{3x_2}$. Find the linearization of $f$ at $\vec{p} = (\pi/4, 1)$.

First compute $\partial f/\partial x_1$ by holding $x_2$ fixed and differentiating with respect to $x_1$ to get $\partial f/\partial x_1 = -\sin(x_1) e^{3x_2}$. Next compute $\partial f/\partial x_2$ by holding $x_1$ fixed and differentiating with respect to $x_2$ to get $\partial f/\partial x_2 = \cos(x_1)(e^{3x_2})3$. The gradient is now $\vec{\nabla} f = (-\sin(x_1) e^{3x_2}, \cos(x_1)(e^{3x_2})3)^T$. The linearization of $f$ at $\vec{p} = (p_1, p_2)^T$ is given by

$$f(\vec{p}) + \vec{\nabla} f(\vec{p}) \cdot [\vec{x} - \vec{p}]$$
$$= \cos(p_1) e^{3p_2} + (-\sin(p_1) e^{3p_2}, \cos(p_1)(e^{3p_2})3)^T \cdot [x_1 - p_1, x_2 - p_2]^T$$
$$= \cos(p_1) e^{3p_2} - \sin(p_1) e^{3p_2}[x_1 - p_1] + \cos(p_1)(e^{3p_2})3[x_2 - p_2].$$

Letting $p_1 = \pi/4$ and $p_2 = 1$ gives the linearization at the chosen point as

$$\cos(\pi/4) e^3 - \sin(\pi/4) e^3 [x_1 - \pi/4] + \cos(\pi/4) e^3 3 [x_2 - 1]$$
$$= (\sqrt{2}/2) e^3 - (\sqrt{2}/2) e^3 [x_1 - \pi/4] + (\sqrt{2}/2) e^3 3 [x_2 - 1].$$

This last linearization could have been done algebraically using Taylor Series, but the preceding approach is much simpler and should be the default approach.

It is true that if the partial derivatives $\partial f/\partial x_i$ exist in the sense of Eq. (6.25) and are continuous in an open set containing $\vec{p}$, then $f$ is differentiable at $p$.

If the differentiable function $\vec{x}(t) : [a,b] \to \mathbb{R}^n$ traces out a curve $\mathcal{C} = \vec{x}([a,b])$, then the rate of change of $f$ along $\mathcal{C}$, as computed by the Chain Rule, is taken to be

$$\begin{aligned}\frac{df \circ \vec{x}(t)}{dt} &= \frac{df\,(x_1(t),\,x_2(t),\ldots,x_n(t))}{dt} \\ &= \frac{\partial f\,(\vec{x})}{\partial x_1}\bigg|_{\vec{x}(t)} \cdot \frac{dx_1(t)}{dt} + \cdots + \frac{\partial f\,(\vec{x})}{\partial x_n}\bigg|_{\vec{x}(t)} \cdot \frac{dx_n(t)}{dt} \\ &= \vec{\nabla}f(\vec{x}(t)) \cdot \vec{x}\,'(t). \end{aligned} \qquad (6.26)$$

As such, Eq. (6.26) says that the gradient at a point $p = \vec{x}(t)$

$$grad(f) \;=\; \vec{\nabla}f(\vec{x}) \;\equiv\; \left(\frac{\partial f}{\partial x_1},\ldots,\frac{\partial f}{\partial x_i},\ldots,\frac{\partial f}{\partial x_n}\right)^T, \qquad (6.27)$$

has built into it all the rates of change along any curve through $p$, provided one knows the velocity vector of the curve at $p$.

Equation (6.26) also shows that the greater the velocity vector the greater the rate of change of $f$ along $\mathcal{C}$. To see this let $\vec{x}\,'(t) = |\vec{x}\,'(t)|(\vec{x}\,'(t)/|\vec{x}\,'(t)|) = |\vec{x}\,'(t)|\vec{u}$, where $\vec{u}$ is the unit vector pointing in the same direction as $\vec{x}\,'(t)$. Then Eq. (6.26) becomes

$$\frac{df \circ \vec{x}(t)}{dt} \;=\; \vec{x}\,'(t) \cdot \vec{\nabla}f(\vec{x}(t)) \;=\; |\vec{x}\,'(t)|\vec{u} \cdot \vec{\nabla}f(\vec{x}(t)), \qquad (6.28)$$

from which it is explicit that $\vec{u} \cdot \vec{\nabla}f(\vec{x}(t))$ remains unchanged at $\vec{p} = \vec{x}(t)$ while the speed $|\vec{x}\,'(t)|$ determines the magnitude of the rate of change along $\mathcal{C}$.

In light of Eq. (6.28), to fairly compare the rate of change of $f$ in different directions, one utilizes unit vectors $\vec{u}$ coming from unit speed curves. This gives the notion of a *directional derivative*, in the $\vec{u}$ direction, defined by

$$D_{\vec{u}}f \;\equiv\; \vec{u} \cdot \vec{\nabla}f \;=\; (\vec{x}\,'/|\vec{x}\,'|) \cdot \vec{\nabla}f. \qquad (6.29)$$

The directional derivative is the slope of the function $f$ in direction $\vec{u}$.

One is often concerned with proceeding forth from a parameter point $\vec{p}$ while trying to increase $f$ most rapidly. The directional derivatives in all unit directions

## 6.5 The Chain Rule; Partial and Directional Derivatives

$\vec{u}$ emanating from $\vec{p}$ allow for such comparison. Letting $\theta$ be the angle between the unit vector $\vec{u}$ and $\vec{\nabla}f$ one sees that

$$D_{\vec{u}}f = \vec{u} \cdot \vec{\nabla}f = |\vec{u}||\vec{\nabla}f|\cos(\theta) = |\vec{\nabla}f|\cos(\theta) \qquad (6.30)$$

is optimized when $\cos(\theta) = 1$ or when $\theta = 0$. Thus the special direction $\vec{u}$ of optimal increase in $f$ is $\vec{u} = \vec{\nabla}f/|\vec{\nabla}f|$. That is, the direction of the gradient is the direction of maximal ascent in $f$. The value of Eq. (6.30) is minimized when $\theta = \pi$ and $\vec{u}$ points in the same direction as $-\vec{\nabla}f$.

As an example, if $T(x, y, t)$ denotes the temperature of a plate at position $(x, y)^T$ and time $t$, then heat will tend to flow from hot to cold, and so it should flow along the negative of the spatial gradient of $T$ given by $(\partial T/\partial x, \partial T/\partial y)^T$, that is, it flows in the direction of maximal decrease in $T$. This idea will be utilized in studying how heat tends to distribute itself, leading to a differential equation known as the heat equation, governing temperature distribution over time.

Another observation to make is that the gradient $\vec{\nabla}f$ is always perpendicular to level sets $f(\vec{x}) = C$. To see this, let $\vec{x}(t)$ parameterize any curve through a point $\vec{p}$ in the level set. Then $f(\vec{x}(t)) = C$ for all $t$. Differentiating this equation gives $\vec{\nabla}f(\vec{x}(t)) \cdot \vec{x}\,'(t) = 0$. Thus $\vec{\nabla}f(\vec{x}(t)) \perp \vec{x}\,'(t)$, and the gradient is perpendicular to any tangent vector to the level set, and so is then perpendicular to the level set.

### EXAMPLE 6.3

Let $f(x, y) = x^2 + y^2$, which has graph the elliptic paraboloid. Find the level curves of $f$, and find the gradient at the point $(x, y)^T$, and find the direction of most rapid ascent in $f$ from $(x, y)^T$. Observe geometrically that the gradient is indeed perpendicular to the level sets.

The level curves are of form $f(x, y) = x^2 + y^2 = C$, which are circles of radius $\sqrt{C}$ for $C > 0$. The gradient is $\vec{\nabla}f = (\partial f/\partial x, \partial f/\partial y)^T = (2x, 2y)^T$, which is the direction of most rapid ascent. Note that the $\vec{\nabla}f$ is twice the radial vector $(x, y)^T$, and from geometry one knows that the radial vector is perpendicular to the circle centered at $(0, 0)^T$ containing $(x, y)^T$.

### ■ Exercises

1. Given $f(x, y) = \sin(x + y^2)$,

   (a) find $\partial f/\partial x$;
   (b) find $\partial f/\partial y$;
   (c) find the gradient $\vec{\nabla}f$; and
   (d) given a differentiable curve $\vec{r}(t)$ with $\vec{r}(0) = (1, 2)^T$ and $\vec{r}\,'(0) = (3, 4)^T$, find the derivative of $f(\vec{r}(t))$ at time 0.

2. Find the directional derivative of the function $f(x,y) = x^2y + xy^2$ at the point $\vec{p} = (1,2)^T$ in the direction of the unit vector $\vec{u} = (\cos(\theta), \sin(\theta))^T$.

## 6.6 Vector-Valued Functions of Several Variables

A vector-valued function of several variables takes the form

$$\begin{pmatrix} f_1(x_1, x_2, \ldots x_n) \\ f_2(x_1, x_2, \ldots x_n) \\ \vdots \\ f_m(x_1, x_2, \ldots x_n) \end{pmatrix} = \vec{f}(\vec{x}), \quad \vec{x} = \begin{pmatrix} x_1 \\ x_2 \\ \vdots \\ x_n \end{pmatrix}, \quad \vec{f} = \begin{pmatrix} f_1 \\ f_2 \\ \vdots \\ f_m \end{pmatrix}, \tag{6.31}$$

and sends vectors $\vec{x} \in \mathbb{R}^n$ to vectors $\vec{y} = \vec{f}(\vec{x}) \in \mathbb{R}^m$.

One says that $\vec{f}$ is differentiable at a point $\vec{p}$ if each component function $f_i$ is differentiable at $p$. Equivalently,

$$f_i(\vec{x}) = f_i(\vec{p}) + \vec{\nabla} f_i(\vec{p}) \cdot [\vec{x} - \vec{p}] + HOT_i([\vec{x} - \vec{p}]),$$

where $HOT_i([\vec{0}]) = 0$ and $\lim_{\vec{x} \to \vec{p}} HOT_i([\vec{x} - \vec{p}])/\|[\vec{x} - \vec{p}]\| = 0$. These equations for the individual $f_i$ can be encoded in one vector equation for $\vec{f}$:

$$\begin{pmatrix} f_1(\vec{x}) \\ f_2(\vec{x}) \\ \vdots \\ f_m(\vec{x}) \end{pmatrix} = \begin{pmatrix} f_1(\vec{p}) \\ f_2(\vec{p}) \\ \vdots \\ f_m(\vec{p}) \end{pmatrix} + \begin{pmatrix} \vec{\nabla} f_1(\vec{p}) \cdot [\vec{x} - \vec{p}] \\ \vec{\nabla} f_2(\vec{p}) \cdot [\vec{x} - \vec{p}] \\ \vdots \\ \vec{\nabla} f_m(\vec{p}) \cdot [\vec{x} - \vec{p}] \end{pmatrix} + \begin{pmatrix} HOT_1([\vec{x} - \vec{p}]) \\ HOT_2([\vec{x} - \vec{p}]) \\ \vdots \\ HOT_m([\vec{x} - \vec{p}]) \end{pmatrix},$$
(6.32)

or even more succinctly,

$$\vec{f}(\vec{x}) = \vec{f}(\vec{p}) + D\vec{f}(p)[\vec{x} - \vec{p}] + \vec{HOT}([\vec{x} - \vec{p}]),$$

where $\vec{HOT}(\vec{0}) = \vec{0}$ and $\lim_{\vec{x} \to \vec{p}} \vec{HOT}([\vec{x} - \vec{p}])/\|[\vec{x} - \vec{p}]\| = \vec{0}$, and where

$$D\vec{f}(p) = \begin{pmatrix} \vec{\nabla} f_1(\vec{p}) \\ \vec{\nabla} f_2(\vec{p}) \\ \vdots \\ \vec{\nabla} f_m(\vec{p}) \end{pmatrix} = \begin{pmatrix} \frac{\partial f_1}{\partial x_1}(\vec{p}), & \frac{\partial f_1}{\partial x_2}(\vec{p}), & \ldots & \frac{\partial f_1}{\partial x_n}(\vec{p}) \\ \frac{\partial f_2}{\partial x_1}(\vec{p}), & \frac{\partial f_2}{\partial x_2}(\vec{p}), & \ldots & \frac{\partial f_2}{\partial x_n}(\vec{p}) \\ \vdots & & & \\ \frac{\partial f_m}{\partial x_1}(\vec{p}), & \frac{\partial f_m}{\partial x_2}(\vec{p}), & \ldots & \frac{\partial f_m}{\partial x_n}(\vec{p}) \end{pmatrix}$$

is called the *Jacobian matrix* of $\vec{f}$, consisting of the matrix of all partial derivatives of $\vec{f}$.

One has an immediate consequence for compositions.

### ■ Theorem 6.5
**The Chain Rule for Vector-Valued Functions of Several Variables**

Let $\vec{f}: \mathbb{R}^n \to \mathbb{R}^m$ be differentiable at $\vec{p}$, and let $\vec{g}: \mathbb{R}^m \to \mathbb{R}^k$ be differentiable at $\vec{f}(\vec{p})$, then the composition function is differentiable at $\vec{p}$ with

$$D(\vec{g} \circ \vec{f})(\vec{p}) = Dg(\vec{f}(\vec{p})) \cdot Df(\vec{p}).$$

---

**Proof** Note that if $\vec{x}(t)$ is any differentiable curve to $\mathbb{R}^n$ through $\vec{p} = \vec{x}(t_0)$, then the composition $\vec{f}(\vec{x}(t))$ is a differentiable curve to $\mathbb{R}^m$, as is justified by the Chain Rule for a function of several variables applied to each component function that yields

$$(f_i \circ \vec{x})'(t_0) = \vec{\nabla} f_i(\vec{x}(t_0)) \cdot \vec{x}\,'(t_0).$$

In vector format one expresses these equations as the derivative

$$D\vec{f}(\vec{x}(t_0)) \cdot \vec{x}\,'(t_0).$$

By this same version of the Chain Rule, the composition $\vec{g}(\vec{f}(\vec{x}(t)))$, is also a differentiable curve to $\vec{R}^k$ with derivative $D\vec{g}(\vec{f}(\vec{x}(t_0))) \cdot (\vec{f} \circ \vec{x})'(t_0) = D\vec{g}(\vec{f}(\vec{x}(t_0))) \cdot D\vec{f}(\vec{x}(t_0)) \cdot (\vec{x}\,'(t_0)) = D(g \circ f)(\vec{x}(t_0)) \cdot (\vec{x}\,'(t_0))$. Since this is true for all $\vec{x}\,'(t_0)$ one concludes $D\vec{g}(\vec{f}(\vec{x}(t_0))) \cdot D\vec{f}(\vec{x}(t_0)) = D(g \circ f)(\vec{x}(t_0))$ or $D\vec{g}(\vec{f}(\vec{p})) \cdot D\vec{f}(\vec{p}) = D(g \circ f)(\vec{p})$. One can also construct the linearization directly by composition, and conclude

$$\vec{g}(\vec{f}(\vec{x})) = \vec{g}(\vec{f}(\vec{p})) + D\vec{g}(\vec{f}(\vec{p})) \cdot D\vec{f}(\vec{p})[\vec{x} - \vec{p}] + HOT([\vec{x} - \vec{p}]).$$

□

Succinctly put, the Jacobian matrix of the composition is the product of the individual Jacobian matrices.

When $\vec{f}: \mathbb{R}^n \to \mathbb{R}^n$ and the number of independent and dependent variables agree, then the function $\vec{f}$ is often called a *transformation*. An important question is whether the transformation is invertible, answered by the following theorem [68].

## Theorem 6.6

**Inverse Function Theorem**

Let $\vec{f} : \Omega \to \mathbb{R}^n$ for $\Omega = \{\vec{x} \mid |\vec{x} - \vec{p}| < r\} \subset \mathbb{R}^n$ be continuously differentiable for all points $\vec{x} \in \Omega$, and let

$$\det\left(D\vec{f}(\vec{p})\right) \neq 0,$$

then there exist open[2] sets $U \subset \Omega$ and $V \subset \mathbb{R}^n$ with $\vec{p} \in U, \vec{f}(\vec{p}) \in V$ and a differentiable inverse function

$$\vec{f}^{-1} : V \to U$$

with $\vec{f}^{-1}(\vec{f}(\vec{x})) = \vec{x}$ for all $\vec{x} \in U$ and $\vec{f}(\vec{f}^{-1}(\vec{y})) = \vec{y}$ for all $\vec{y} \in V$. The Jacobian matrix of $\vec{f}^{-1}$ at $\vec{f}(\vec{p})$ is given by

$$D\vec{f}^{-1}(\vec{f}(\vec{p})) = \left(D\vec{f}(\vec{p})\right)^{-1}.$$

**Proof** The idea of the proof goes as follows. $\vec{f}$ is very nearly its linearization, up to very small higher-order terms. The linearization will be invertible precisely when $\det\left(D\vec{f}(\vec{p})\right) \neq 0$, as can be seen in the following way. Let $T_{\vec{v}}$ be translation by $\vec{v}$. That is $T_{\vec{v}}(\vec{x}) = \vec{x} + \vec{v}$ for all $\vec{x} \in \mathbb{R}^n$. $T_{\vec{v}}$ is differentiable with inverse $T_{-\vec{v}}$. The linearization $L(\vec{x}) = \vec{f}(\vec{p}) + D\vec{f}(\vec{p})[\vec{x} - \vec{p}]$, can be expressed as $T_{\vec{f}(\vec{p})} \circ D\vec{f}(\vec{p}) \circ T_{-\vec{p}}$. That is

$$T_{\vec{f}(\vec{p})} \circ D\vec{f}(\vec{p}) \circ T_{-\vec{p}}(\vec{x}) = T_{\vec{f}(\vec{p})} \circ D\vec{f}(\vec{p})([\vec{x} - \vec{p}])$$
$$= T_{\vec{f}(\vec{p})}\left(D\vec{f}(\vec{p}) \cdot ([\vec{x} - \vec{p}])\right) = \vec{f}(\vec{p}) + D\vec{f}(\vec{p}) \cdot ([\vec{x} - \vec{p}]).$$

The invertibility of the linearization is now equivalent to the invertibility of the matrix multiplication by $D\vec{f}(\vec{p})$ in the middle of the threefold composition $T_{\vec{f}(\vec{p})} \circ D\vec{f}(\vec{p}) \circ T_{-\vec{p}}$. But matrix multiplication by $D\vec{f}(\vec{p})$ is invertible precisely when $\det\left(D\vec{f}(\vec{p})\right) \neq 0$. So the linearization $L(\vec{x})$ is invertible when $\det\left(D\vec{f}(\vec{p})\right) \neq 0$. Now $\vec{f}(\vec{x}) \approx L(\vec{x})$ for $\vec{x}$ close to $\vec{p}$. So if $U, V$ are chosen small enough, $\vec{f}$ behaves sufficiently like $L(\vec{x})$ to also be invertible when $\det\left(D\vec{f}(\vec{p})\right) \neq 0$.

---

[2] A set $U$ in $\mathbb{R}^n$ is open if it is the union of open balls of form $\{\vec{x} \mid |\vec{x} - \vec{p}| < r\}$.

Since $\vec{f}^{-1}(\vec{f}(\vec{x})) = \vec{x}$, computing the Jacobian matrix of each side yields $D(\vec{f}^{-1} \circ \vec{f})(\vec{x}) = D(\vec{f}^{-1})(\vec{f}(\vec{x})) \cdot (D\vec{f})(\vec{x}) = Id$ where $Id$ denotes the $n \times n$ identity matrix. Evaluating this at $\vec{x} = \vec{p}$ and multiplying on the right by the inverse matrix $\left(D(\vec{f}(\vec{p}))\right)^{-1}$ gives

$$D(\vec{f}^{-1})(\vec{f}(\vec{p})) = \left((D\vec{f})(\vec{p})\right)^{-1}.$$

□

An example of the power of the Inverse Function theorem is obtained in showing regular surfaces are locally graphs of differentiable functions. A surface in $\mathbb{R}^3$ of form $g(x, y, z) = C$ is called regular if $\vec{\nabla} g(x, y, z) \neq \vec{0}$ at all points $(x, y, z)^T$ on the surface.

■ **Theorem 6.7**

**Implicit Function Theorem for Regular Surfaces**

If $g(x, y, z)$ is a continuously differentiable function defining a regular surface $g(x, y, z) = C$ in $\mathbb{R}^3$ then the surface is representable locally as the graph of a differentiable function.

**Proof** Since $\vec{\nabla} g(x, y, z) \neq \vec{0}$ for all points $(x, y, z)^T$ on the surface, assume that $\partial g / \partial z \neq 0$ near a point of interest on the surface. Define the function $\vec{f}(x, y, z) = (x, y, g(x, y, z))^T$ from $\mathbb{R}^3$ to $\mathbb{R}^3$. Then the Jacobian matrix $D\vec{f}$ is given by

$$D\vec{f}(x, y, z) = \begin{pmatrix} 1 & 0 & 0 \\ 0 & 1 & 0 \\ g_x & g_y & g_z \end{pmatrix},$$

from which one sees the determinant $\det\left(D\vec{f}\right) = g_z \neq 0$ at the point of interest on the surface (as well as for all nearby points). By the Inverse Function theorem $\vec{f}$ has a differentiable local inverse $\vec{f}^{-1}$, satisfying

$$(x, y, z)^T = \vec{f}^{-1}(\vec{f}(x, y, z)) = \vec{f}^{-1}(x, y, g(x, y, z)).$$

Restricted to the surface $g(x, y, z) = C$ this becomes

$$(x, y, z)^T = \vec{f}^{-1}(x, y, C),$$

for $(x, y, x)^T$ on the surface. From this one sees that $z = (\vec{f}^{-1})_3(x, y, C)$ is a differentiable function of $x$ and $y$ (where $(\vec{f}^{-1})_3$ denotes the differentiable third coordinate function of $\vec{f}^{-1}$) for $(x, y, z)^T$ on the surface. So on the surface near the point of interest $z$ is expressible as a differentiable function of $x$ and $y$, and is represented by the graph of $(\vec{f}^{-1})_3(x, y, C)$. □

This theorem says that one can study a most general class of surfaces simply by studying graphs of functions. It says that if the $z$-partial $g_z$ is nonzero, then $z$ is expressible differentiably in terms of the remaining variables $x, y$. Similarly if the $x$-partial $g_x$ is nonzero, then $x$ is expressible differentiably in terms of the remaining variables $y, z$; if the $y$-partial $g_y$ is nonzero, then $y$ is expressible differentiably in terms of the remaining variables $x, z$.

The Implicit Function theorem has many other forms and generalizations [68] that are not emphasized here.

### EXAMPLE 6.4

Show the surface given by the sphere $x^2 + y^2 + z^2 = R^2$ is a regular surface. Show the sphere can be represented as a graph near the point $(0, 0, R)^T$.

In this setting $g(x, y, z) = x^2 + y^2 + z^2$ and $C = R^2 > 0$. $\vec{\nabla} g(x, y, z) = (2x, 2y, 2z)^T \neq \vec{0}$ because if $(2x, 2y, 2z)^T = 2(x, y, z)^T = \vec{0}$ then $(x, y, z)^T = \vec{0}$ and $x^2 + y^2 + z^2 = 0 = R^2$, a contradiction. So the surface is regular. At the point $(0, 0, R)^T$, the gradient becomes $\vec{\nabla} g(0, 0, R) = (0, 0, 2R)^T$, and $g_z \neq 0$. The Implicit Function theorem now says that $z$ is expressible locally as the graph of a differentiable function in $x, y$. In this particular example one can even solve for $z$ algebraically:

$$x^2 + y^2 + z^2 = R^2 \iff z^2 = R^2 - x^2 - y^2 \iff z = \pm\sqrt{R^2 - x^2 - y^2}.$$

Since the concern was for the point $(0, 0, R)^T$ with $R > 0$, pick the positive square root to give the local graph $z = \sqrt{R^2 - x^2 - y^2}$. Notice that this local graph does *not* imply that the sphere is globally the graph of a function (as the sphere does not satisfy the vertical line test and cannot be the graph of a function). Near $(0, 0, R)^T$ one can only obtain a local graph of the upper hemisphere $z = \sqrt{R^2 - x^2 - y^2}$.

Another utilization of the Inverse Function theorem is linearization of the local inverse function. If one writes $\vec{y} = \vec{f}(\vec{x})$ and $\det\left(D\vec{f}(\vec{x}_0)\right) \neq 0$, then there is a differentiable local inverse $\vec{f}^{-1}$ defined near $\vec{y}_0 = \vec{f}(\vec{x}_0)$ with $\vec{f}^{-1}(\vec{y}) = \vec{f}^{-1}(\vec{y}_0) +$

$(D\vec{f}^{-1})(\vec{y}_0)[\vec{y} - \vec{y}_0] + HOT([\vec{y} - \vec{y}_0])$. Since $(D\vec{f}^{-1})(\vec{y}_0) = (D\vec{f}(\vec{x}_0))^{-1}$ by the Inverse Function theorem one sees that

$$\begin{aligned} \vec{f}^{-1}(\vec{y}) &= \vec{f}^{-1}(\vec{y}_0) + (D\vec{f})^{-1}(\vec{y}_0)[\vec{y} - \vec{y}_0] + HOT([\vec{y} - \vec{y}_0]) \\ &= \vec{x}_0 + (D\vec{f}(\vec{x}_0))^{-1}[\vec{y} - \vec{y}_0] + HOT([\vec{y} - \vec{y}_0]). \end{aligned}$$

Thus the linearization of the inverse $\vec{f}^{-1}$ at $\vec{y}_0 = \vec{f}(\vec{x}_0)$ is given by

$$L(\vec{f}^{-1}, \vec{y}_0, \vec{y}) = \vec{x}_0 + (D\vec{f}(\vec{x}_0))^{-1}[\vec{y} - \vec{y}_0] \approx \vec{f}^{-1}(\vec{y}). \tag{6.33}$$

■ **Exercises**

1. Given $\vec{f}(x, y) = (x^2 + xy + y^4, x^2 y)^T$, find
   (a) $D\vec{f}$;
   (b) $D\vec{f}(1, 1)$;
   (c) $(D\vec{f}(1, 1))^{-1}$;
   (d) the linearization of $\vec{f}$ at $\vec{p} = (1, 1)^T$; and
   (e) the linearization of the local inverse $\vec{f}^{-1}$ at $(3, 1)^T$. (*Hint:* Use Part (c))

2. Given the surface $x^2 + y^3 + z^4 = 2$,
   (a) decide if $z$ can be written as a differentiable function of $x, y$ near the point $(1, 1, 0)^T$; and
   (b) decide if $x$ can be written as a differentiable function of $y, z$ near the point $(1, 1, 0)^T$.

## ■ 6.7 Change of Variables

To perform integration and better understand surfaces and regions, it is often convenient to change variables, or to *parameterize* the surface or region. This will entail reexpressing the coordinates via a 1-1 onto vector-valued function of maximal rank, as described in the examples in this section.

### ■ 6.7.1 Parameterized Surfaces

The most basic example of a parameterized surface in $\mathbb{R}^3$ is the graph $z = f(x, y)$ of a differentiable function $f(x, y)$ for $(x, y)^T$ in some open set $\mathcal{D} \subset \mathbb{R}^2$. Given such a graph, the grid of lines parallel to the $x$- or $y$-coordinate axes in $\mathcal{D}$ are transported to the surface under $f$. One says that a *grid* on $\mathcal{D}$ is imaged onto the surface. Each gridline on the surface is a parameterized curve, or a *curvilinear coordinate*. The surface given by the graph of $f$ is said to be parameterized by the coordinates $(x, y)^T$. The Implicit Function theorem tells us that each regular surface in $\mathbb{R}^3$ can be taken locally to be the graph of a differentiable function. So each regular surface is locally parameterized by coordinate variables.

For any graph $z = f(x, y)$ of a function differentiable at $(x_0, y_0)^T$, one knows that $z = f(x, y) = f(x_0, y_0) + (\vec{\nabla} f)(x_0, y_0) \cdot [x - x_0, y - y_0]^T + HOT$, from which one uses the linearization to see the equation of the tangent plane:

$$z = f(x_0, y_0) + f_x(x_0, y_0)[x - x_0] + f_y(x_0, y_0)[y - y_0].$$

Rather than parameterize only one coordinate as a function of the other two coordinate parameters, it is possible to parameterize each coordinate in $(x, y, z)^T$ by parameters $(u, v)^T$ to obtain a parameterized surface $\vec{f}(u, v) = (x(u, v), y(u, v), z(u, v))^T$ where each coordinate is a differentiable function of $(u, v)^T$. If $\vec{f}$ is one-to-one and the Jacobian $D\vec{f}$ is of maximal rank, then the surface can be locally seen as the graph of coordinate variables.

Such a parameterization includes, but is more general than, the case of a graph $z = f(x, y)$. For the case of a graph one lets $x(u, v) = u$, $y(u, v) = v$, and $z(u, v) = f(u, v)$. Then define $\vec{f}(u, v) = (u, v, f(u, v))^T$. Computing the Jacobian matrix yields

$$D\vec{f} = \begin{pmatrix} \partial x/\partial u & \partial x/\partial v \\ \partial y/\partial u & \partial y/\partial v \\ \partial z/\partial u & \partial z/\partial v \end{pmatrix} = \begin{pmatrix} 1 & 0 \\ 0 & 1 \\ \partial f/\partial u & \partial f/\partial v \end{pmatrix}.$$

Since the determinant of the minor

$$\det \left( \frac{\partial(x, y)}{\partial(u, v)} \right) \equiv \det \begin{pmatrix} \partial x/\partial u & \partial x/\partial v \\ \partial y/\partial u & \partial y/\partial v \end{pmatrix} = \det \begin{pmatrix} 1 & 0 \\ 0 & 1 \end{pmatrix} = 1 \neq 0,$$

one knows the Jacobian matrix is of maximal rank for any graph.

A more explicit example of a parameterized surface that is not globally the graph of a function is given by the sphere of radius $r$ centered at $\vec{0}$, which has earlier been seen to be a regular surface, locally graphable in terms of coordinate variables. We parameterize the sphere away from the points $(0, 0, \pm r)^T$ by its spherical coordinates in the following manner. Let $u = \phi$ be the angle made with the positive $z$-axis and let $v = \theta$ be the angle made by the projection onto the $x$-$y$ plane with the positive $x$-axis. Then $x = r\cos(\theta)\sin(\phi)$, $y = r\sin(\theta)\sin(\phi)$, and $z = r\cos(\phi)$. Letting

## 6.7 Change of Variables

$\vec{f}(\theta, \phi) = (x(\theta, \phi), y(\theta, \phi), z(\theta, \phi))^T = (r\cos(\theta)\sin(\phi), r\sin(\theta)\sin(\phi), r\cos(\phi))^T$,
one has the Jacobian matrix

$$D\vec{f} = \begin{pmatrix} \partial x/\partial\theta & \partial x/\partial\phi \\ \partial y/\partial\theta & \partial y/\partial\phi \\ \partial z/\partial\theta & \partial z/\partial\phi \end{pmatrix} = \begin{pmatrix} -r\sin(\theta)\sin(\phi) & r\cos(\theta)\cos(\phi) \\ r\cos(\theta)\sin(\phi) & r\sin(\theta)\cos(\phi) \\ 0 & -r\sin(\phi) \end{pmatrix}.$$

Since the determinant of the minor

$$\det\left(\frac{\partial(x,y)}{\partial(\theta,\phi)}\right) \equiv \det\begin{pmatrix} \partial x/\partial\theta & \partial x/\partial\phi \\ \partial y/\partial\theta & \partial y/\partial\phi \end{pmatrix}$$
$$= \det\begin{pmatrix} -r\sin(\theta)\cdot\sin(\phi) & r\cos(\theta)\cdot\cos(\phi) \\ r\cos(\theta)\cdot\sin(\phi) & r\sin(\theta)\cdot\cos(\phi) \end{pmatrix}$$
$$= -r^2\sin(\phi)\cdot\cos(\phi) \neq 0$$

away from $\phi = 0, \pi, \pi/2$, (that is away from the north/south poles $(0, 0, \pm r)^T$ and the equator $(r\cos(\theta), r\sin(\theta), 0)^T$ the graph is regular. Also since the determinant of the minor

$$\det\left(\frac{\partial(x,z)}{\partial(\theta,\phi)}\right) \equiv \det\begin{pmatrix} \partial x/\partial\theta & \partial x/\partial\phi \\ \partial z/\partial\theta & \partial z/\partial\phi \end{pmatrix}$$
$$= \det\begin{pmatrix} -r\sin(\theta)\cdot\sin(\phi) & r\cos(\theta)\cdot\cos(\phi) \\ 0 & -r\sin(\phi) \end{pmatrix}$$
$$= -r^2\sin(\theta)\sin^2(\phi) \neq 0$$

for $\phi = \pi/2, \theta \neq 0, \pi$ the parameterization is regular now at equatorial points excepting the poles $(\pm r, 0, 0)^T$. However, at these poles the graph is seen to be regular since the determinant of the minor

$$\det\left(\frac{\partial(y,z)}{\partial(\theta,\phi)}\right) \equiv \det\begin{pmatrix} \partial y/\partial\theta & \partial y/\partial\phi \\ \partial z/\partial\theta & \partial z/\partial\phi \end{pmatrix}$$
$$= \det\begin{pmatrix} r\cos(\theta)\sin(\phi) & r\sin(\theta)\cos(\phi) \\ 0 & -r\sin(\phi) \end{pmatrix}$$
$$= -r^2\cos(\theta)\sin^2(\phi) \neq 0$$

for $\phi = \pi/2, \theta = 0, \pi$ giving regularity also at the points $(\pm r, 0, 0)^T$ as well. The parameterization is seen to be regular for all points away from the north/south poles $(0, 0, \pm r)^T$.

### ■ Exercises

1. Given $\vec{f}(u,v) = (u, v, u^2 + v^2)^T$, show that $D\vec{f}$ is of maximal rank (and thus $\vec{f}$ is regular).

2. Given $\vec{f}(u,v) = (u+v, u-v, u^2 - v^2)^T = (x, y, z)^T$,

   (a) show $\vec{f}$ is regular; and
   (b) find the equation of the surface in terms of $x, y, z$.

### ■ 6.7.2 Parameterized Regions

One can also parameterize a region in $\mathbb{R}^3$ with a continuously differentiable mapping $\vec{f}(u, v, w) = (x(u, v, w), y(u, v, w), z(u, v, w))^T$ with the Jacobian matrix $D\vec{f}$ of maximal rank, in this case with $\det D\vec{f} \neq 0$. For instance one can parameterize the ball centered at $\vec{0}$ of radius $r$ by $\vec{f}(\theta, \phi, w) = (x(\theta, \phi, w), y(\theta, \phi, w), z(\theta, \phi, w))^T = (w\cos(\theta)\sin(\phi), w\sin(\theta)\sin(\phi), w\cos(\phi))^T$, where $0 \leq \theta \leq 2\pi$, $0 \leq \phi \leq \pi$, and where letting $w$ vary from $0$ to $r$ fills out the ball centered at $0$ of radius $r$.

This parameterization of the ball is regular in that

$$\det D\vec{f} = \begin{pmatrix} \partial x/\partial \theta & \partial x/\partial \phi & \partial x/\partial w \\ \partial y/\partial \theta & \partial y/\partial \phi & \partial y/\partial w \\ \partial z/\partial \theta & \partial z/\partial \phi & \partial z/\partial w \end{pmatrix}$$

$$= \begin{pmatrix} -w\sin(\theta)\cdot\sin(\phi) & w\cos(\theta)\cdot\cos(\phi) & \cos(\theta)\cdot\sin(\phi) \\ w\cos(\theta)\cdot\sin(\phi) & w\sin(\theta)\cdot\cos(\phi) & \sin(\theta)\cdot\sin(\phi) \\ 0 & -w\sin(\phi) & \cos(\phi) \end{pmatrix}$$

$$= -w^2 \sin(\phi) \neq 0$$

away from $\phi = 0, \pi$ or equivalently, away from the polar axis $(0, 0, w)^T$ for $w \in [-r, r]$. Such a parameterization is said to be a change of coordinates when it is one-to-one, onto, and regular (when the Jacobian matrix $D\vec{f}$ is of maximal rank).

The ball of radius $r$ in $\mathbb{R}^3$ centered at $\vec{0}$ is a good example of a region with boundary a closed surface, in this case the sphere of radius $r$. The notion of a boundary, or edge, of a region in $\mathbb{R}^n$ is an important concept because many computations done over a region can be effectively computed on the boundary of the region. As a familiar example of this, if one considers $[a, b]$ to be a region in $\mathbb{R}$, then it has boundary the pair of points $a, b$. If one integrates a function $g$ over all of the region $[a, b]$ to obtain $\int_a^b g(x)dx$, one could have obtained the same answer by computing the antiderivative $G$ and evaluating over the boundary according to the formula $G(b) - G(a)$ via the Fundamental Theorem of Calculus. This process has many counterparts in higher dimensions. For this reason we return to the notion of a boundary of a region in $\mathbb{R}^n$.

### ■ 6.7.3 The Boundary of a Region in $\mathbb{R}^n$

Let $\mathcal{V} \subset \mathbb{R}^n$ be a region of interest, and define the boundary of $\mathcal{V}$ as

$$\partial \mathcal{V} \equiv \{\vec{x} \in \mathbb{R}^n | (\forall \epsilon) \implies (\exists \vec{x}_i \in \mathcal{V} \text{ and } \vec{x}_o \notin \mathcal{V} \text{ with } \|\vec{x} - \vec{x}_i\| < \epsilon, \|\vec{x} - \vec{x}_o\| < \epsilon)\}. \tag{6.34}$$

The basic idea in the preceding definition is that the boundary should be the edge of a region $\mathcal{V}$, and the edge should be near both the inside and the outside of $\mathcal{V}$. As such, any small ball around a point $\vec{x}$ in the boundary should meet both the inside and the outside of $\mathcal{V}$. So any ball with center $\vec{x}$ of arbitrary radius $\epsilon$ contains a point $\vec{x}_i$ inside $\mathcal{V}$ and also contains a point $\vec{x}_o$ outside $\mathcal{V}$.

There are two special cases:

---

**EXAMPLE 6.5**

An open ball in $\mathbb{R}^n$ that is centered at $\vec{p}$ with radius $a$, is defined to be

$$\mathcal{B}^n_{\vec{p},a} \equiv \{\vec{x} \in \mathbb{R}^n \mid \|\vec{x} - \vec{p}\| < a\}. \tag{6.35}$$

In three dimensions, the spherical polar parameterization of $\mathcal{B}^3_{\vec{p},a}$ is

$$\vec{\psi} : [0, a) \times [0, 2\pi) \times [-\pi/2, \pi/2] \to \mathcal{B}^3_{\vec{p},a} \tag{6.36}$$

$$\vec{\psi}(r, \theta, \phi) = \vec{p} + (r \cdot \cos(\theta) \cdot \cos(\phi), \ r \cdot \sin(\theta) \cdot \cos(\phi), \ r \cdot \sin(\phi))^T. \tag{6.37}$$

The boundary $\partial \mathcal{B}^n_{\vec{p},a}$ of the open ball $\mathcal{B}^n_{\vec{p},a}$ in $\mathbb{R}^n$ is the sphere

$$\partial \mathcal{B}^n_{\vec{p},a} = \mathcal{S}^{n-1}_{\vec{p}} \equiv \{\vec{x} \in \mathbb{R}^n \mid \|\vec{x} - \vec{p}\| = a\}. \tag{6.38}$$

In three dimensions, the spherical polar parameterization of $\partial \mathcal{B}^3_{\vec{p},a} = \mathcal{S}^2_{\vec{p},a}$ is

$$\vec{\delta} \equiv \vec{\psi}|_{r=a} : [0, 2\pi) \times [-\pi/2, \pi/2] \to \mathcal{S}^2_{\vec{p},a} \tag{6.39}$$

$$\vec{\delta}(\theta, \phi) = \vec{p} + (a \cdot \cos(\theta) \cdot \cos(\phi), \ a \cdot \sin(\theta) \cdot \cos(\phi), \ a \cdot \sin(\phi))^T. \tag{6.40}$$

These parameterizations can be extended to higher dimensions by introducing more angles about the center $\vec{p}$. Also, by setting $\phi = 0$ and $\vec{p} = \vec{0}$ one obtains the usual polar coordinates for a disc

$\mathcal{B}_{\vec{0},a}^2$ and a circle $\mathcal{S}_{\vec{0},a}^1$. When $a = 1$, the open unit ball in $\mathbb{R}^n$ is expressed as $\mathcal{B}_{\vec{0},1}^n = \mathcal{B}^n$, and the unit sphere is $\mathcal{S}_{\vec{0},1}^{n-1} = \mathcal{S}^{n-1}$.

### ■ Exercises

1. Find the boundary of the region $R = \{(x,y)^T | y \geq 0\}$.
2. Find the boundary of the region $R = \{(x,y)^T | 1 \leq x^2 + y^2 \leq 2\}$.

## ■ 6.8 Integration, Green's Theorem, and the Divergence Theorem

We examine double and triple integrals in their various settings in this section.

### ■ 6.8.1 Double Integrals

If one wants to find the area $\mathcal{A}$ of the region $R$ in $\mathbb{R}^2$ between an upper continuous curve $y = f_2(x)$ and a lower continuous curve $y = f_1(x)$ over the interval $[a, b]$, one would simply partition the interval into $N$ equal subintervals, pick $x_i^*$ in the $i$th subinterval, obtain rectangles of height $h_i = f_2(x_i^*) - f_1(x_i^*)$ and width $w_i = \Delta x = (b-a)/N$, add up the areas of these rectangles to get Riemann Sums of form $\sum_{i=1}^{n} h_i w_i = \sum_{i=1}^{n} (f_2(x_i^*) - f_1(x_i^*))\Delta x$, which are seen, by converting from Greek to English, to approach the definite integral $\int_a^b f_2(x) - f_1(x) dx = \mathcal{A}$.

An alternate approach would have been to partition the whole region $R$ by slicing it with an $N$ by $M$ grid of rectangles of widths $\Delta x_i$ and heights $\Delta y_j$. Letting $1 \leq i \leq N$ and $1 \leq j \leq M$ index the grid, we index each rectangle as the $(i, j)$ rectangle. Intersecting the $(i, j)$ rectangle with $R$ will form subregions $R_{i,j}$ (most of which are rectangles, except near the top and bottom edges) of area $A_{i,j}$. We approximate $A_{i,j}$ by the area $\Delta x_i \cdot \Delta y_j$ of the $(i, j)$ rectangle containing $R_{i,j}$, noting that in most cases $A_{i,j} = \Delta x_i \cdot \Delta y_j$. $\mathcal{A}$ is then approximated by the sum $\sum_{i=1}^{N} \sum_{j=1}^{M} 1 \Delta x_i \cdot \Delta y_j$ (where we agree not to count the area of any rectangles completely missing $R$). This sum is a Riemann Sum for a so-called double integral approximating $\mathcal{A}$, denoted $\int\int_R 1\, dxdy = \int\int_R 1\, dA$, where $dA$ is the area element $dxdy$. Note that one could have gridded $R$ with any geometrical shapes and summed their areas as a Riemann Sum for the area $\mathcal{A}$.

If one sums over a column of rectangles first (say holding $i$ fixed and summing over $j$) and then sums the areas of the columns over the $i$, one obtains an iterated sum of form

$$\sum_{i=1}^{N}\left(\sum_{j=1}^{M} 1\, \Delta y_j\right) \Delta x_i. \tag{6.41}$$

For a fixed $i$ the inner sum $\sum_{j=1}^{M} 1\, \Delta y_j$ is observed to be approximately the height of the rectangle $h_i = f_2(x_i^*) - f_1(x_i^*) = \int_{f_1(x_i^*)}^{f_2(x_i^*)} 1\, dy$. This is substituted into Eq. (6.41) to obtain the nearly equal expression:

$$\sum_{i=1}^{N}\left(\int_{f_1(x_i^*)}^{f_2(x_i*)} 1\, dy\right) \Delta x_i, \tag{6.42}$$

which is seen to be a Riemann Sum for the so-called iterated integral:

$$\int_a^b \left(\int_{f_1(x)}^{f_2(x)} 1\, dy\right) dx. \tag{6.43}$$

Thus the area $\mathcal{A}$ could be computed by iterated integrals with

$$\mathcal{A} = \int_a^b \left(\int_{f_1(x)}^{f_2(x)} 1\, dy\right) dx = \int_a^b \left(y|_{f_1(x)}^{f_2(x)}\right) dx = \int_a^b (f_2(x) - f_1(x))\, dx$$

where the last expression has already been seen to be $\mathcal{A}$. Thus we have three equivalent perspectives on computing $\mathcal{A}$ as:

$$\int\int_R 1\, dxdy = \int_a^b \left(\int_{f_1(x)}^{f_2(x)} 1\, dy\right) dx = \int_a^b (f_2(x) - f_1(x))\, dx.$$

The preceding discussion is the basis for a similarly constructed theory of double integrals. Taking a continuous function $f(x,y)$ defined over the same region $R$

in $\mathbb{R}^2$ as previously gridded, picking a point $(x_i, y_j)^T$ in the $(i,j)$ rectangle and computing the volume of the box with height $f(x_i, y_j)$ and base $R_{i,j}$ of area $A_{i,j}$ to be $V_{i,j} = f(x_i, y_j) A_{i,j}$, summing these volumes over the grid gives Riemann Sums for a definite integral called a double integral:

$$\sum_{i,j=1}^{N,M} f(x_i, y_j) A_{i,j} \approx \sum_{i,j=1}^{N,M} f(x_i, y_j) \Delta x_i \Delta y_j \approx \int\int_R f(x,y) \, dxdy.$$

This double integral can be interpreted as the net volume under the graph of $f$ over the region $R$. If one sums first over $j$ holding $i$ fixed, and then sums over $i$, one obtains

$$\sum_{i,j=1}^{N,M} f(x_i, y_j) \Delta x_i \Delta y_j = \sum_{i=1}^{N} \left( \sum_{j=1}^{M} f(x_i, y_j) \Delta y_j \right) \Delta x_i$$

$$\approx \sum_{i=1}^{N} \left( \int_{f_1(x_i)}^{f_2(x_i)} f(x_i, y) \, dy \right) \Delta x_i \approx \int_a^b \left( \int_{f_1(x)}^{f_2(x)} f(x,y) \, dy \right) dx.$$

The conclusion is that the double integral can be evaluated by an iterated integral:

$$\int\int_R f(x,y) \, dxdy = \int_a^b \left( \int_{f_1(x)}^{f_2(x)} f(x,y) \, dy \right) dx.$$

This is one of three main techniques available to us for computing a double integral. That is, one technique is to perform the appropriate iterated integral. The second technique available to us later is to change variables by reparameterizing the region $R$, which will result in a simpler integrand, or a simpler base region over which we integrate, or both. A third technique will be to deploy a generalization of the Fundamental Theorem of Calculus when appropriate conditions hold.

If the region $R$ could have also been expressed as the region between two curves $x = g_1(y), x = g_2(y)$ over the interval $[c, d]$ along the $y$-axis, then summing over the $i$ first while holding $j$ fixed, and then summing over the $j$, gives a completely analogous result that

$$\int\int_R f(x,y) \, dxdy = \int_c^d \left( \int_{g_1(y)}^{g_2(y)} f(x,y) \, dx \right) dy.$$

Since either of these two iterated integrals gives the same net volume under $f$ over $R$, one obtains *Fubini's theorem*:

$$\int\int_R f(x,y)\,dxdy = \int_a^b \left(\int_{f_1(x)}^{f_2(x)} f(x,y)\,dy\right)dx = \int_c^d \left(\int_{g_1(y)}^{g_2(y)} f(x,y)\,dx\right)dy.$$

Sometimes exchanging the order of integration in the iterated integral yields a simpler evaluation.

### EXAMPLE 6.6

Evaluate the double integral $\int\int_R e^{x^2}\,dxdy$ on the region over the interval $[0,1]$ above the $x$-axis and underneath the line $y = x$.

If one integrates over $x$ first, one obtains

$$\int_0^1 \left(\int_y^1 e^{x^2}\,dx\right)dy,$$

where the inner integral is difficult to evaluate without power series techniques.

On the other hand, if one integrates over $y$ first, one obtains

$$\int_0^1 \left(\int_0^x e^{x^2}\,dy\right)dx = \int_0^1 \left(e^{x^2}\int_0^x dy\right)dx = \int_0^1 \left(e^{x^2}\,y\big|_0^x\right)dx$$

$$= \int_0^1 \left(e^{x^2}(x-0)\right)dx = e^{x^2}/2\big|_0^1 = e^1/2 - 1/2.$$

### ■ Exercises

1. Evaluate $\int\int_R x^2 y^3\,dxdy$ for the rectangle $R = \{(x,y)^T | 0 \leq x \leq 1; 0 \leq y \leq 1\}$.
2. Evaluate $\int\int_R xy\,dxdy$ for the region $R = \{(x,y)^T | 0 \leq x \leq 1; 0 \leq y \leq x^2\}$.
3. Evaluate $\int_0^1 \left(\int_0^1 x^2 y^3\,dx\right)dy$.

**4.** Evaluate $\int_0^1 \left( \int_0^1 x^2 y^3 dy \right) dx$.

**5.** Evaluate $\int_0^1 \left( \int_y^1 \cos(x^2) dx \right) dy$ by reversing the order of integration.

### ■ 6.8.2 Triple Integrals

For a solid region $R$ in $\mathbb{R}^3$ one can analogously find a *triple integral*

$$\int\int\int_R f(x,y,z)\, dV$$

with similar techniques as applied in the double integral setting, accounting for one more dimension.

Partition the whole region $R$ by slicing it with an $N \times M \times K$ grid of boxes of widths $\Delta x_i$, lengths $\Delta y_j$, and heights $\Delta z_k$. Letting $1 \leq i \leq N$, $1 \leq j \leq M$, and $1 \leq k \leq K$ index the grid, we index each box as the $(i,j,k)$ box. Intersecting the $(i,j,k)$ box with $R$ will form subregions $R_{i,j,k}$ (most of which are boxes, except near the edges) of volume $V_{i,j,k}$. We approximate $V_{i,j,k}$ by the volume $\Delta x_i \cdot \Delta y_j \cdot \Delta z_k$ of the $(i,j,k)$ rectangle containing $R_{i,j,k}$, noting that in most cases $V_{i,j,k} = \Delta x_i \cdot \Delta y_j \cdot \Delta z_k$. One chooses $(x_i, y_j, z_k)^T$ to be a point in the $(i,j,k)$ box and computes $f(x_i, y_j, z_k)$. The triple integral $\int\int\int_R f(x,y,z)dV$ is then approximated by the Riemann Sum $\sum_{i=1}^N \sum_{j=1}^M \sum_{k=1}^K f(x_i, y_j, z_k)\Delta x_i \Delta y_j \Delta z_k$ (where we agree not to count the volume of any boxes completely missing $R$). These Riemann Sums for the triple integral converge under a wide array of conditions, including when $f$ is continuous and the boundary $\partial R$ is well behaved.

$$\sum_{i=1}^N \sum_{j=1}^M \sum_{k=1}^K f(x_i, y_j, z_k)\Delta x_i \Delta y_j \Delta z_k \approx \int\int\int_R f(x,y,z)\, dV. \tag{6.44}$$

One could have gridded $R$ with any geometric shapes and summed their areas as another type of Riemann Sum for the triple integral.

If the region $R$ can be described as

$$R = \{(x,y,z)^T | a \leq x \leq b; f_1(x) \leq y \leq f_2(x), g_1(x,y) \leq z \leq g_2(x,y)\}$$

one can sum over $k$ first holding $i, j$ fixed, then sum over $j$ holding $i$ fixed, then sum over $i$ to obtain an iterated sum:

$$\sum_{i=1}^{N}\sum_{j=1}^{M}\sum_{k=1}^{K} f(x_i, y_j, z_k) \Delta x_i \Delta y_j \Delta z_k$$

$$= \sum_{i=1}^{N} \left\{ \sum_{j=1}^{M} \left[ \sum_{k=1}^{K} f(x_i, y_j, z_k) \Delta z_k \right] \Delta y_j \right\} \Delta x_i$$

$$\approx \sum_{i=1}^{N} \left\{ \sum_{j=1}^{M} \left[ \int_{g_1(x_i,y_k)}^{g_2(x_i,y_k)} f(x_i, y_j, z) \, dz \right] \Delta y_j \right\} \Delta x_i$$

$$\approx \sum_{i=1}^{N} \left\{ \int_{f_1(x_i)}^{f_2(x_i)} \left[ \int_{g_1(x_i,y)}^{g_2(x_i,y)} f(x_i, y, z) \, dz \right] dy \right\} \Delta x_i$$

$$\approx \int_{a}^{b} \left\{ \int_{f_1(x)}^{f_2(x)} \left[ \int_{g_1(x,y)}^{g_2(x,y)} f(x, y, z) \, dz \right] dy \right\} dx.$$

One concludes in this case that:

$$\iiint_R f(x,y,z) \, dV = \iiint_R f(x,y,z) \, dxdydz = \int_{a}^{b} \left\{ \int_{f_1(x)}^{f_2(x)} \left[ \int_{g_1(x,y)}^{g_2(x,y)} f(x,y,z) \, dz \right] dy \right\} dx. \tag{6.45}$$

**EXAMPLE 6.7**

Find the volume of the region $R$ above the triangle $0 \le x \le 1, 0 \le y \le x$ in the $x$-$y$ plane and below the surface $z = x^2 + y^2$.

In the preceding notation $R = \{(x,y,z)^T | a = 0 \le x \le 1 = b, f_1(x) = 0 \le y \le x = f_2(x),$ and $g_1(x,y) = 0 \le z \le x^2 + y^2 = g_2(x,y)\}$. Then the volume is computed as

$$\iiint_R 1 \, dxdydz = \int_0^1 \left\{ \int_0^x \left[ \int_0^{x^2+y^2} 1 \, dz \right] dy \right\} dx$$

$$= \int_0^1 \left\{ \int_0^x \left[ z \big|_0^{x^2+y^2} \right] dy \right\} dx = \int_0^1 \left\{ \int_0^x [x^2 + y^2 - 0] \, dy \right\} dx$$

$$= \int_0^1 \left\{ x^2 y + y^3/3 \big|_0^x \right\} dx = \int_0^1 \{x^3 + x^3/3\} \, dx$$

$$= \int_0^1 \{4x^3/3\} \, dx = x^4/3 \big|_0^1 = 1/3.$$

One could also describe $R$ as $R = \{(x,y,z)^T | a = 0 \le y \le 1 = b, f_1(y) = y \le x \le 1 = f_2(y)$, and $g_1(x,y) = 0 \le z \le x^2 + y^2 = g_2(x,y)\}$ obtaining the reordered iterated integral

$$\iiint_R 1\, dxdydz = \int_0^1 \left\{ \int_y^1 \left[ \int_0^{x^2+y^2} 1\, dz \right] dx \right\} dy$$

$$= \int_0^1 \left\{ \int_y^1 \left[ z\big|_0^{x^2+y^2} \right] dx \right\} dy = \int_0^1 \left\{ \int_y^1 [x^2 + y^2 - 0]\, dx \right\} dy$$

$$= \int_0^1 \left\{ x^3/3 + xy^2 \big|_y^1 \right\} dy = \int_0^1 \{1/3 + y^2 - y^3/3 - y^3\}\, dy$$

$$= \int_0^1 \{1/3 + y^2 - 4y^3/3\}\, dy = y/3 + y^3/3 - y^4/3 \big|_0^1 = 1/3 + 1/3 - 1/3 - 0 = 1/3.$$

An important application of triple integrals is the computation of the center of mass of a region $R$. The region $R$ will act as if all of its mass is centered at the point $(\bar{x}, \bar{y}, \bar{z})^T$, and this is critical in the placement of support structures in engineering and in architectural design. Let $\delta(x,y,z)$ be the density of the region at the point $(x,y,z)^T$. The total mass is obtained by summing $\delta(x,y,z)dxdydz$, the density times the volume element, over the region $R$ to obtain the mass $M$ as

$$M = \iiint_R \delta(x,y,z)\, dxdydz.$$

One deduces that $R$ should balance if supported at its center of mass, and so the total torque from $\bar{x}$ should be the sum of the $x$ distance from $\bar{x}$ times mass, which should vanish by the fact that balance is achieved. That is,

$$0 = \iiint_R (x - \bar{x})\delta(x,y,z)\, dxdydz$$

$$= \iiint_R x\, \delta(x,y,z)\, dxdydz - \bar{x} \iiint_R \delta(x,y,z)\, dxdydz.$$

Solving for $\bar{x}$ yields

$$\bar{x} = \frac{\iiint_R x\, \delta(x,y,z)\, dxdydz}{\iiint_R \delta(x,y,z)\, dxdydz}. \tag{6.46}$$

Similarly

$$\bar{y} = \frac{\iiint_R y\, \delta(x,y,z)\, dxdydz}{\iiint_R \delta(x,y,z)\, dxdydz} \qquad \bar{z} = \frac{\iiint_R z\, \delta(x,y,z)\, dxdydz}{\iiint_R \delta(x,y,z)\, dxdydz}. \tag{6.47}$$

One can think of each of these as weighted averages of the coordinate values.

If the density is a constant over a homogeneous region, that is $\delta(x,y,z) = \delta$, then this constant $\delta$ pulls out of all the integrals in Eq. (6.46) and Eq. (6.47) and cancels out in ratio to give the coordinates of the *centroid* as:

$$\bar{x} = \frac{\int\int\int_R x\,dxdydz}{\int\int\int_R dxdydz} \quad \bar{y} = \frac{\int\int\int_R y\,dxdydz}{\int\int\int_R dxdydz} \quad \bar{z} = \frac{\int\int\int_R z\,dxdydz}{\int\int\int_R dxdydz}. \quad (6.48)$$

### ■ Exercises

1. Evaluate $\int\int\int_R 1\,dxdydz$ for the region $R = \{(x,y,z)^T | 0 \leq x \leq 1; 0 \leq y \leq 1-x; 0 \leq z \leq 1-x-y\}$.
2. Evaluate $\int\int\int_R x\,dxdydz$ for the region $R = \{(x,y,z)^T | 0 \leq x \leq 1; 0 \leq y \leq 1-x; 0 \leq z \leq 1-x-y\}$.
3. Use Exercises (1) and (2) to find $\bar{x}$, the $x$-coordinate of the center of mass of the region $R = \{(x,y,z)^T | 0 \leq x \leq 1; 0 \leq y \leq 1-x; 0 \leq z \leq 1-x-y\}$ if the density function is constant. Predict $\bar{y}, \bar{z}$ by symmetry, or compute them directly.

### ■ 6.8.3 Change of Variables; Reparameterization

If the region $R$ in $\mathbb{R}^2$ is *reparameterized* by the one-to-one, onto, continuously differentiable mapping $\vec{f}(u,v) = (x(u,v), y(u,v))^T : W \to R = \vec{f}(W)$, one can rely on linearization to see the effect of $\vec{f}$ on a small rectangle of dimension $\Delta u \times \Delta v$ at the lower left corner point $(u_0, v_0)^T$. This rectangle can be seen as

$$R \equiv \{(u_0 + t, v_0 + s)^T \mid 0 \leq t \leq \Delta u, \ 0 \leq s \leq \Delta v\}.$$

Then ignoring higher-order terms and relying on linearization, one sees

$$\begin{aligned}
\vec{f}(u_0 + t, v_0 + s) &\approx \vec{f}(u_0, v_0) + D\vec{f}(u_0, v_0)[u_0 + t - u_0, v_0 + s - v_0]^T \\
&= \vec{f}(u_0, v_0) + \begin{pmatrix} \partial x/\partial u & \partial x/\partial v \\ \partial y/\partial u & \partial y/\partial v \end{pmatrix} \begin{bmatrix} t \\ s \end{bmatrix} \\
&= \vec{f}(u_0, v_0) + \begin{pmatrix} \partial x/\partial u \\ \partial y/\partial u \end{pmatrix} \cdot t + \begin{pmatrix} \partial x/\partial v \\ \partial y/\partial v \end{pmatrix} \cdot s \\
&= \vec{f}(u_0, v_0) + \frac{\partial \vec{f}}{\partial u} \cdot t + \frac{\partial \vec{f}}{\partial v} \cdot s
\end{aligned}$$

for $0 \leq t \leq \Delta u, 0 \leq s \leq \Delta v$. This tells us that the image under $\vec{f}$ is very close to the parallelogram $P$ spanned by $\frac{\partial \vec{f}}{\partial u}\Delta u = [(\partial x/\partial u)\Delta u, (\partial y/\partial u)\Delta u]^T$ and $\frac{\partial \vec{f}}{\partial v}\Delta v = [(\partial x/\partial v)\Delta v, (\partial y/\partial v)\Delta v]^T$ emanating from the corner point $\vec{f}(u_0, v_0)$. The area of this parallelogram $A(P)$ is now computed from the $2 \times 2$ determinant

$$\left|\det\begin{pmatrix} \partial x/\partial u \Delta u & \partial y/\partial u \Delta u \\ \partial x/\partial v \Delta v & \partial y/\partial v \Delta v \end{pmatrix}\right| = \left|\det\begin{pmatrix} \partial x/\partial u & \partial y/\partial u \\ \partial x/\partial v & \partial y/\partial v \end{pmatrix}\right|\Delta u \Delta v$$

$$= \left|\det\begin{pmatrix} \partial x/\partial u & \partial x/\partial v \\ \partial y/\partial u & \partial y/\partial v \end{pmatrix}\right|\Delta u \Delta v = \left|\det(D\vec{f})\right|\Delta u \Delta v = A(P).$$

The rectangle of dimensions $\Delta u$ by $\Delta v$ very nearly transforms to the parallelogram spanned by $(\partial \vec{f}/\partial u)\Delta u$ and $(\partial \vec{f}/\partial v)\Delta v$; and the area element $\Delta u \Delta v$ transforms to the area element $|\det(D\vec{f})|\Delta u \Delta v$.

In a similar manner, if one were to reparameterize in $\mathbb{R}^3$ by the mapping $\vec{f}(u,v,w) = (x(u,v,w), y(u,v,w), z(u,v,w))^T : W \to R = \vec{f}(W)$ one would see by analyzing the corresponding linearization that the small rectangle of dimensions $\Delta u \Delta v \Delta w$ is transformed nearly exactly to the parallelepiped spanned by the vectors $(\partial \vec{f}/\partial u)\Delta u$, $(\partial \vec{f}/\partial v)\Delta v$, and $(\partial \vec{f}/\partial w)\Delta w$. The volume element $\Delta u \Delta v \Delta w$ is transformed to the volume element of the parallelepiped, which is given by

$$|\{\partial \vec{f}/\partial u \Delta u\} \cdot \{(\partial \vec{f}/\partial v \Delta v) \times (\partial \vec{f}/\partial w \Delta w)\}|$$
$$= |\{\partial \vec{f}/\partial u\} \cdot \{\partial \vec{f}/\partial v \times \partial \vec{f}/\partial w\}|\Delta u \Delta v \Delta w$$
$$= |det D\vec{f}|\Delta u \Delta v \Delta w.$$

In each case one sees that the determinant of the Jacobian matrix in absolute value $|\det D\vec{f}|$ evaluated at $(u_0, v_0)^T$ or $(u_0, v_0, w_0)^T$ determines the magnification factor by which area or volume element transforms near the point $(u_0, v_0)^T$ or $(u_0, v_0, w_0)^T$ under the reparameterization $\vec{f}$.

These procedures give us a strategy for computing integrals under reparameterization. We illustrate this in $\mathbb{R}^2$. First grid $W$ with the $N \times M$ grid of rectangles with corner points $(u_i, v_j)^T$ and widths $\Delta u_i$ and lengths $\Delta v_j$ for $0 \leq i \leq N, 0 \leq j \leq M$. Take the area element corresponding to the $(i,j)$ rectangle $\Delta u_i \Delta v_j$ and transform it under the reparameterization $\vec{f}(u,v) = (x(u,v), y(u,v))^T$ to the area element $\left|\det(D\vec{f})(u_i, v_j)\right|\Delta u_i \Delta v_j = A(P_{i,j})$ corresponding to the parallelogram $P_{i,j}$ spanned by the vectors $(\partial \vec{f}/\partial u)(u_i, v_j)\Delta u$ and $(\partial \vec{f}/\partial v)(u_i, v_j)\Delta v$. Use the parallelograms $P_{i,j}$ as an approximate grid for $R = \vec{f}(W)$. The corresponding Riemann

Sum for the double integral $\int\int_R g(x,y)dxdy$ would be

$$\sum_{i=1}^{N}\sum_{j=1}^{M} g(x(u_i,v_j), y(u_i,v_j))\, A(P_{i,j}) \tag{6.49}$$

$$= \sum_{i=1}^{N}\sum_{j=1}^{M} g(x(u_i,v_j), y(u_i,v_j))\, \left|\det(D\vec{f})(u_i,v_j)\right| \Delta u_i \Delta v_j. \tag{6.50}$$

Since Eq. (6.49) is a Riemann Sum for the double integral of $g(x,y)$ over $R$, one has

$$\sum_{i=1}^{N}\sum_{j=1}^{M} g(x(u_i,v_j), y(u_i,v_j))\, A(P_{i,j}) \approx \int\int_R g(x,y)dxdy. \tag{6.51}$$

And since Eq. (6.50) is a Riemann Sum for the double integral of $g(\vec{f}(u,v))\left|\det(D\vec{f})(u,v)\right|$ over $W$, one also sees

$$\sum_{i=1}^{N}\sum_{j=1}^{M} g(x(u_i,v_j), y(u_i,v_j))\, \left|\det(D\vec{f})(u_i,v_j)\right| \Delta u_i \Delta v_j$$

$$\approx \int\int_W g(\vec{f}(u,v))\left|\det(D\vec{f})(u,v)\right| du\,dv. \tag{6.52}$$

Because the Riemann sums in Eq. (6.49) and Eq. (6.50) are equal, their limiting values, the two definite integrals in Eq. (6.51) and Eq. (6.52), must be the same. The conclusion is known as the Change of Variables formula [68].

### ■ Theorem 6.8
### Change of Variables in $\mathbb{R}^2$

Let $\vec{f}: W \to R = \vec{f}(W)$ be a one-to-one, onto, continuously differentiable, regular function from $W \subset \mathbb{R}^2$ to $R \subset \mathbb{R}^2$. Let $g: R \to \mathbb{R}$ be continuous. Then

$$\int\int_R g(x,y)dxdy = \int\int_W g(\vec{f}(u,v))\left|\det(D\vec{f})(u,v)\right| du\,dv. \tag{6.53}$$

The Change of Variables formula in $\mathbb{R}^3$ is seen in an entirely analogous manner. Transform the volume element $\Delta u \Delta v \Delta w$ to the corresponding volume element

$\left|\det(D\vec{f})(u,v,w)\right|\Delta u\Delta v\Delta w$ for the transformed parallelepiped element. Multiplying these volume elements by $g$ at the appropriate corner points and adding provides two equal Riemann Sums for two equivalent expressions of the same integral.

■ **Theorem 6.9**

**Change of Variables in $\mathbb{R}^3$**

Let $\vec{f}: W \to R = \vec{f}(W)$ be a one-to-one, onto, continuously differentiable, regular function from $W \subset \mathbb{R}^3$ to $R \subset \mathbb{R}^3$. Let $g: R \to \mathbb{R}$ be continuous. Then

$$\int\int\int_R g(x,y,z)dxdydz = \int\int\int_W g(\vec{f}(u,v,w))\left|\det(D\vec{f})(u,v,w)\right|dudvdw.$$
(6.54)

The Change of Variables formula is seen to hold in all dimensions $n$ [68], after verifying that $n \times n$ determinants measure $n$-volume.

In practice, it may be difficult to find the one-to-one, onto, regular map from $W$ to *all* of $R$. If there is overlap along a thin region of 0 area or volume it will have no impact on the multiple integral, so one-to-oneness needs only hold everywhere but on a thin set. Or $f(W)$ may miss only a thin set in $R$ with a similar lack of impact, and in this circumstance the Change of Variable formula is still true. A final relaxation of the hypotheses can occur if we let regularity lapse on a thin set, as long as the one-to-one property is not violated. Thus if $\det D\vec{f}$ vanishes on a thin set under these circumstances, one can still use the Change of Variables formula.

### EXAMPLE 6.8

Find the area $A$ of the circle of radius $\rho$.

The area $A$ is computable as the integral $\int\int_R 1\ dxdy$ where $R = \{(x,y)^T | x^2 + y^2 \leq \rho^2\}$. In Euclidean coordinates this integral is computable as

$$\int_{-\rho}^{\rho}\int_{-\sqrt{\rho^2-x^2}}^{\sqrt{\rho^2-x^2}} 1\ dydx = \int_{-\rho}^{\rho} 2\sqrt{\rho^2-x^2}\ dx.$$

This last integral requires a trigonometric substitution $x = \rho\sin(\theta)$ and then a trigonometric reduction formula to finish the problem and obtain $\pi\rho^2$.

An alternate approach is to change variables to polar coordinates. Let $x = r\cos(\theta), y = r\sin(\theta)$ where $r$ corresponds to $u$ and $\theta$ corresponds to $v$ in the Change of Variables formula. Let $W$ be the rectangle $0 \leq r \leq \rho, 0 \leq \theta \leq 2\pi$. The reparameterization mapping $\vec{f}(r,\theta) = (r\cos(\theta), r\sin(\theta))^T$

is seen to be one-to-one off the thin set $r=0$ in $W$, and off the thin set $\theta=0, \theta=2\pi$. $\vec{f}$ is also seen to be regular off the thin set $r=0$ as

$$\det D\vec{f} = \det\begin{pmatrix} \partial x/\partial r & \partial x/\partial \theta \\ \partial y/\partial r & \partial y/\partial \theta \end{pmatrix} = \det\begin{pmatrix} \cos(\theta) & -r\sin(\theta) \\ \sin(\theta) & r\cos(\theta) \end{pmatrix} = r(\cos^2(\theta) + \sin^2(\theta)) = r.$$

Now the Change of Variables formula gives

$$\iint_R 1\, dx\, dy = \iint_W (1 \circ \vec{f}(r,\theta))|\det D\vec{f}(r,\theta)|\, dr d\theta = \int_0^{2\pi}\int_0^{\rho} 1 \cdot r\, dr\, d\theta = \int_0^{2\pi} r^2/2\Big|_0^{\rho}\, d\theta$$

$$= \int_0^{2\pi} \rho^2/2\, d\theta = (\rho^2/2)\, \theta\Big|_0^{2\pi} = \pi\rho^2.$$

This illustrates that the Change of Variables formula can drastically simplify an integral computation if $W$ and $\vec{f}$ are chosen judiciously.

### EXAMPLE 6.9

Find the volume of a sphere of radius $\rho$.

We have previously seen the requisite parameterization, spherical coordinates. Let $\vec{f}(\theta,\phi,w) = (x(\theta,\phi,w), y(\theta,\phi,w), z(\theta,\phi,w))^T = (w\cos(\theta)\sin(\phi), w\sin(\theta)\sin(\phi), w\cos(\phi))^T$ with $\det D\vec{f} = -w^2\sin(\phi)$. Let $W$ be the box given by $0 \le \theta \le 2\pi; 0 \le \phi \le \pi; 0 \le w \le \rho$. Then $R = \vec{f}(W)$ is the sphere of radius $\rho$ centered at $\vec{0}$. The hypotheses of the Change of Variables theorem are satisfied off of a thin set. The volume integral becomes

$$\iiint_R 1\, dx\, dy\, dz = \int_0^{2\pi}\int_0^{\pi}\int_0^{\rho} 1 \cdot |-w^2 \sin(\phi)|\, dw\, d\phi d\theta = \int_0^{2\pi}\int_0^{\pi} (w^3/3)\sin(\phi)\Big|_0^{\rho}\, d\phi\, d\theta$$

$$= \int_0^{2\pi}\int_0^{\pi} (\rho^3/3)\sin(\phi)\, d\phi\, d\theta = \int_0^{2\pi} (\rho^3/3)(-\cos(\phi))\Big|_0^{\pi}\, d\theta = \int_0^{2\pi} (\rho^3/3)\, 2\, d\theta$$

$$= 2(\rho^3/3)\, \theta\Big|_0^{2\pi} = 2(\rho^3/3)\, 2\pi = (4\pi/3)\rho^3.$$

In statistics it is important to know the area under the bell curve $e^{-x^2/2}$. To find the value of $c_1 = \int_{-\infty}^{\infty} e^{-x^2/2}\, dx$ is then the goal. This can be done with double integrals by turning to the bivariate function $f(x,y) = e^{-x^2/2}e^{-y^2/2} = e^{-(x^2+y^2)/2}$. First notice that

$$\int\int_{\mathbb{R}^2} f(x,y)\,dx\,dy = \int_{-\infty}^{\infty}\left(\int_{-\infty}^{\infty} e^{-x^2/2}e^{-y^2/2}\,dy\right)dx$$
$$= \int_{-\infty}^{\infty} e^{-x^2/2}\left(\int_{-\infty}^{\infty} e^{-y^2/2}\,dy\right)dx = \int_{-\infty}^{\infty} e^{-x^2/2}(c_1)\,dx$$
$$= (c_1)\int_{-\infty}^{\infty} e^{-x^2/2}\,dx = (c_1)(c_1) = c_1^2. \tag{6.55}$$

On the other hand, one can compute the double integral of $f$ by converting to polar coordinates under the transformation $\vec{f}(r,\theta) = (r\cos(\theta), r\sin(\theta))^T = (x,y)^T$ with Jacobian matrix having determinant $\det D\vec{f} = r$, as in Example (6.8). In this case, to fill out the plane, we have $0 \leq r < \infty; 0 \leq \theta \leq 2\pi$. Thus

$$\int\int_{\mathbb{R}^2} f(x,y)\,dx\,dy = \int_0^{2\pi}\int_0^{\infty} f(r\cos(\theta), r\sin(\theta))\,r\,dr\,d\theta$$
$$= \int_0^{2\pi}\int_0^{\infty} e^{-\{(r\cos(\theta))^2+(r\sin(\theta))^2\}/2}\,r\,dr\,d\theta = \int_0^{2\pi}\left(\int_0^{\infty} e^{-r^2/2}\,r\,dr\right)d\theta$$
$$= \int_0^{2\pi} -e^{-r^2/2}\Big|_0^{\infty} d\theta = \int_0^{2\pi}(0+e^0)d\theta = \int_0^{2\pi} 1\,d\theta = \theta\big|_0^{2\pi} = 2\pi. \tag{6.56}$$

Equating Eq. (6.55) and Eq. (6.56), one sees that $c_1^2 = 2\pi$, or equivalently

$$c_1 = \sqrt{2\pi} = \int_{-\infty}^{\infty} e^{-x^2/2}\,dx. \tag{6.57}$$

Equation (6.57) explains the fact that the standard normal density function has the form

$$\rho(0,1;x) = \frac{1}{\sqrt{2\pi}}e^{-x^2/2}$$

with the $\sqrt{2\pi}$ in the denominator. If one divides all sides of Eq. (6.57) by $\sqrt{2\pi}$ the result is

$$1 = \int_{-\infty}^{\infty} \frac{1}{\sqrt{2\pi}}e^{-x^2/2}\,dx, \tag{6.58}$$

corresponding to the fact that the total area must be 1 for a probability density function.

■ **Exercises**

1. Find the integral $\int\int_R \sqrt{x^2+y^2}\,dxdy$ where $R$ is the unit circle centered at $\vec{0}$ by changing to polar coordinates.

2. Find the integral $\int\int\int_R (x^2 + y^2 + z^2)^{3/2}\, dx\,dy\,dz$ where $R$ is the sphere of radius 2 centered at $\vec{0}$ by changing to spherical coordinates.
3. Let $\vec{f}(u,v) = (u - v, u + v)^T$.
   (a) Show $\vec{f}$ takes the unit square $W = \{(u,v)^T \mid 0 \leq u \leq 1; 0 \leq v \leq 1\}$ to the parallelogram $R$ with vertices $(0,0)^T, (1,1)^T, (-1,1)^T, (0,2)^T$.
   (b) Find the Jacobian matrix $D\vec{f}$ and $\det D\vec{f}$.
   (c) Find $\int\int_R 1\, dx\,dy$ by making the change of variables $x = u - v; y = u + v$.
   (d) How does the area of $R$ relate to the area of $W$? Compare this to the magnification scale $|\det D\vec{f}|$ under change of variables corresponding to $\vec{f}$.

### ■ 6.8.4 Surface Integrals

One will need to be able to perform an integral on a regular parameterized surface in order to understand some of the main physical applications to come. Given a function $g$ on the surface $S$, the idea behind a surface integral is to grid $S$ by small surface elements $S_{i,j}$ of area $A(S_{i,j})$, then take the product $g(p_{i,j}) \cdot A(S_{i,j})$ of $g$ at a point $p_{i,j}$ in the $(i,j)$ surface element $S_{i,j}$. The sum $\sum_{i,j} g(p_{i,j}) \cdot A(S_{i,j})$ is the Riemann Sum for a surface integral $\int\int_S g\, dA = \int\int_S g\, d\sigma$, where $dA = d\sigma$ both denote the surface area element.

One can change variables via a parameterization to convert a surface integral to a regular double integral on a region $W$ in $\mathbb{R}^2$ with the following methods. Parameterize each coordinate in $(x,y,z)^T$ by parameters $(u,v)^T$ to obtain a parameterization for the surface $\vec{f}(u,v) = (x(u,v), y(u,v), z(u,v))^T : W \to S$ where each coordinate is a continuously differentiable function of $(u,v)^T$, and where $\vec{f}$ is one-to-one, onto $S$, and regular (that is, the Jacobian matrix $D\vec{f}$ is of maximal rank). Grid the region $W$ and examine the $(i,j)$ rectangle at corner point $(u_i, v_j)$ with dimensions $\Delta u_i \Delta v_j$. Under $\vec{f}$ this rectangle is transformed to be nearly its linearized transformation at the point $(u_i, v_j)^T$. That is, for $0 \leq t \leq \Delta u_i; 0 \leq s \leq \Delta v_j$ one has

$$\begin{aligned}
\vec{f}(u_i + t, v_j + s) &\approx \vec{f}(u_i, v_j) + D\vec{f}(u_i, v_j)[u_i + t - u_i, v_j + s - v_j]^T \\
&= \vec{f}(u_i, v_j) + \begin{pmatrix} \partial x/\partial u & \partial x/\partial v \\ \partial y/\partial u & \partial y/\partial v \\ \partial z/\partial u & \partial z/\partial v \end{pmatrix} \begin{bmatrix} t \\ s \end{bmatrix} \\
&= \vec{f}(u_i, v_j) + \begin{pmatrix} \partial x/\partial u \\ \partial y/\partial u \\ \partial z/\partial u \end{pmatrix} t + \begin{pmatrix} \partial x/\partial v \\ \partial y/\partial v \\ \partial z/\partial v \end{pmatrix} s \\
&= \vec{f}(u_i, v_j) + \frac{\partial \vec{f}}{\partial u} t + \frac{\partial \vec{f}}{\partial v} s.
\end{aligned}$$

Thus the $(i,j)$ rectangle is transformed to a region $S_{i,j}$ on the surface that is very close to the parallelogram $P_{i,j}$ spanned by $\partial \vec{f}/\partial u \Delta u_i$ and $\partial \vec{f}/\partial v \Delta v_j$

emanating from the point $\vec{f}(u_i, v_j)$. The resulting parallelogram then has area given by $A(P_{i,j}) = |(\partial \vec{f}/\partial u) \times (\partial \vec{f}/\partial v)|(u_i, v_j)\Delta u_i \Delta v_j$, which is exceedingly close to the area $A(S_{i,j})$ of the image $\vec{f}$ of the $(i,j)$ rectangle. The Riemann Sums then satisfy

$$\sum_{i=1}^{N}\sum_{j=1}^{M} g(x(u_i,v_j), y(u_i,v_j), z(u_i,v_j))A(P_{i,j}) = \qquad (6.59)$$

$$\sum_{i=1}^{N}\sum_{j=1}^{M} g(x(u_i,v_j), y(u_i,v_j), z(u_i,v_j)) \left|\frac{\partial \vec{f}}{\partial u} \times \frac{\partial \vec{f}}{\partial v}\right|(u_i,v_j)\Delta u_i \Delta v_j. \quad (6.60)$$

Since Eq. (6.59) is a Riemann Sum for the surface integral of $g(x,y,z)$ over $S$, one has

$$\sum_{i=1}^{N}\sum_{j=1}^{M} g(x(u_i,v_j), y(u_i,v_j), z(u_i,v_j))A(P_{i,j}) \approx \int\!\!\int_S g(x,y,z)\, dA. \qquad (6.61)$$

And since Eq. (6.60) is a Riemann Sum for the double integral of the function $g(\vec{f}(u,v))|(\partial \vec{f}/\partial u) \times (\partial \vec{f}/\partial v)|(u,v)$ over $W$, one also sees

$$\sum_{i=1}^{N}\sum_{j=1}^{M} g(x(u_i,v_j), y(u_i,v_j), z(u_i,v_j)) \left|\frac{\partial \vec{f}}{\partial u} \times \frac{\partial \vec{f}}{\partial v}\right|(u_i,v_j)\Delta u_i \Delta v_j$$

$$\approx \int\!\!\int_W g(\vec{f}(u,v))\left|\frac{\partial \vec{f}}{\partial u} \times \frac{\partial \vec{f}}{\partial v}\right|(u,v)\,du\,dv. \qquad (6.62)$$

Because the Riemann Sums in Eq. (6.59) and Eq. (6.60) are equal, their limiting values, the two definite integrals of Eq. (6.61) and Eq. (6.62), must be the same. The conclusion is then:

■ **Theorem 6.10**

Let $\vec{f}: W \to S = \vec{f}(W)$ be a one-to-one, onto, continuously differentiable, regular function from $W \subset \mathbb{R}^2$ to a surface $S \subset \mathbb{R}^3$. Let $g: S \to \mathbb{R}$ be continuous. Then

$$\int\!\!\int_S g(x,y,z)\, dA = \int\!\!\int_W g(\vec{f}(u,v))\left|\frac{\partial \vec{f}}{\partial u} \times \frac{\partial \vec{f}}{\partial v}\right|(u,v)\, du\,dv. \qquad (6.63)$$

## 6.8 Integration, Green's Theorem, and the Divergence Theorem

The preceding procedure is very similar to the two-dimensional change of variables process in the plane, except that now the range falls in $\mathbb{R}^3$.

---
**EXAMPLE 6.10**

Find the surface area of the sphere of radius $\rho$.

Let $S$ be the sphere of radius $\rho$, and parameterize $S$ via spherical coordinates: $\vec{f}(\theta, \phi) = (x(\theta, \phi), y(\theta, \phi), z(\theta, \phi))^T = (\rho \cos(\theta) \sin(\phi), \rho \sin(\theta) \sin(\phi), \rho \cos(\phi))^T$, where $W$ is the rectangle given by $0 \leq \theta \leq 2\pi; 0 \leq \phi \leq \pi$. Then $\vec{f}(W) = S$. From this one has

$$\partial \vec{f}/\partial \theta = (-\rho \sin(\theta) \sin(\phi), \rho \cos(\theta) \sin(\phi), 0)^T,$$
$$\partial \vec{f}/\partial \phi = (\rho \cos(\theta) \cos(\phi), \rho \sin(\theta) \cos(\phi), -\rho \sin(\phi))^T,$$
$$|(\partial \vec{f}/\partial \theta) \times (\partial \vec{f}/\partial \phi)| = |(-\rho^2 \cos(\theta) \sin^2(\phi), -\rho^2 \sin(\theta) \sin^2(\phi), -\rho^2 \sin(\phi) \cos(\phi))^T| = \rho^2 \sin(\phi).$$

Thus, integration of the surface $S$ is performed as follows:

$$\iint_S 1 \, dA = \int_0^{2\pi} \int_0^{\pi} 1 \cdot |(\partial \vec{f}/\partial \theta) \times (\partial \vec{f}/\partial \phi)| \, d\phi d\theta = \int_0^{2\pi} \int_0^{\pi} 1 \cdot \rho^2 \sin(\phi) \, d\phi d\theta$$
$$= \int_0^{2\pi} \rho^2 (-\cos(\phi))\big|_0^{\pi} \, d\theta = \int_0^{2\pi} \rho^2 2 d\theta = \rho^2 2\theta \big|_0^{2\pi} = 4\pi \rho^2.$$

---

### ■ Exercises

1. Given the parameterization $\vec{f}(x, y) = (x, y, f_3(x, y))^T$ of the graph of $z = f_3(x, y)$, use Eq. (6.63) to verify that this parameterization computes the surface area $\mathcal{A}$ of the graph over $R$ with the following formula:
$\mathcal{A} = \int \int_R \sqrt{1 + (\partial f_3/\partial x)^2 + (\partial f_3/\partial y)^2} \, dx dy$.

2. Use the result of the previous problem to show that the graph of a paraboloid $z = f_3(x, y) \equiv x^2 + y^2$, over the region $R$, which is the unit circle centered at $(0, 0)^T$, has surface area $\mathcal{A} = \int \int_R \sqrt{1 + 4x^2 + 4y^2} \, dx dy$. Then use a change of variables from Cartesian to polar coordinates and evaluate the surface area of the paraboloid $\mathcal{A}$.

### ■ 6.8.5 Line Integrals

Let $\vec{r}(t) = (x_1(t), x_2(t), \ldots, x_n(t))^T : [a, b] \to \mathbb{R}^n$ be any differentiable function tracing out a given curve $\mathcal{C} = \vec{r}([a, b]) \subset \mathbb{R}^n$, and let $f_i(\vec{x}) : \Omega \to \mathbb{R}$ for $i = 1, \ldots, n$ be continuously differentiable functions from an open set $\Omega \subset \mathbb{R}^n$ with $\mathcal{C} \subset \Omega$, with

$\vec{f} = (f_1, f_2, \ldots, f_n)$. The *line integral* $\int_C f_1 dx_1 + f_2 dx_2 + \cdots + f_n dx_n$ is defined to be

$$\int_C f_1 dx_1 + f_2 dx_2 + \cdots + f_n dx_n = \int_C \vec{f} \cdot d\vec{x}$$
$$\equiv \int_a^b \left( f_1(\vec{r}(t)) \frac{dx_1}{dt} + f_2(\vec{r}(t)) \frac{dx_2}{dt} + \cdots + f_n(\vec{r}(t)) \frac{dx_n}{dt} \right) dt \quad (6.64)$$
$$= \int_a^b \vec{f}(\vec{r}(t)) \cdot \vec{r}\,'(t)\, dt.$$

The line integral is independent of parameterization up to a sign depending on the forward orientation (from $\vec{r}(a)$ toward $\vec{r}(b)$) or backward orientation (from $\vec{r}(b)$ toward $\vec{r}(a)$) of travel along the path, which can verified by a change of variable to arclength $s = s(t) = \int_a^t |\vec{r}\,'(u)|\, du$.

### EXAMPLE 6.11

Evaluate the line integral $\int_C (-y/2) dx + (x/2) dy$, for $C$ the circle of radius $r$ oriented positively, that is, counterclockwise.

Parameterize the circle by $\vec{r}(t) = (x(t), y(t))^T = (r\cos(t), r\sin(t))^T$ for $t \in [0, 2\pi]$. Then $d\vec{r}\,'(t) = (dx/dt, dy/dt)^T = (-r\sin(t), r\cos(t))^T$, and

$$\int_C (-y/2)\, dx + (x/2)\, dy = \int_0^{2\pi} [(-y(t)/2)\, dx/dt + (x(t)/2)\, dy/dt]\, dt$$
$$= \int_0^{2\pi} [(-r\sin(t)/2)(-r\sin(t)) + (r\cos(t)/2)(r\cos(t))]\, dt$$
$$= \int_0^{2\pi} r^2/2\, dt = (r^2/2) \cdot 2\pi = \pi r^2.$$

If one chose a different parameterization, say at twice the previous speed for half the time, then one has $\vec{r}(t) = (x(t), y(t))^T = (r\cos(2t), r\sin(2t))^T$ for $t \in [0, \pi]$. Then $d\vec{r}\,'(t) = (dx/dt, dy/dt)^T = (-2r\sin(2t), 2r\cos(2t))^T$, and

$$\int_C (-y/2)\, dx + (x/2)\, dy = \int_0^{\pi} [(-y(t)/2)\, dx/dt + (x(t)/2)\, dy/dt]\, dt$$
$$= \int_0^{\pi} [(-r\sin(2t)/2)(-2r\sin(t)) + (r\cos(2t)/2)(2r\cos(t))]\, dt$$
$$= \int_0^{\pi} r^2\, dt = r^2 \cdot \pi = \pi r^2,$$

as before. All other counterclockwise orientations of $C$ give this same result $\pi r^2$.

If one chose the negative orientation (in the clockwise direction) one could choose the parameterization $\vec{r}(t) = (x(t), y(t))^T = (r\cos(-t), r\sin(-t))^T$ for $t \in [0, 2\pi]$. Then $\vec{r}\,'(t) = (dx/dt, dy/dt)^T = (-r\sin(-t)(-1), r\cos(-t)(-1))^T$, and

$$\int_C (-y/2)\,dx + (x/2)\,dy = \int_0^{2\pi} [(-y(t)/2)\,dx/dt + (x(t)/2)\,dy/dt]\,dt$$

$$= \int_0^{2\pi} [(-r\sin(-t)/2)(r\sin(-t)) + (r\cos(-t)/2)(-r\cos(-t))]\,dt$$

$$= \int_0^{2\pi} -r^2/2\,dt = -(r^2/2) \cdot 2\pi = -\pi r^2,$$

giving the opposite value to that computed with the positive orientation. All other negatively oriented (clockwise) parameterizations of $C$ give the same result of $-\pi r^2$.

It is no accident that the preceding example computes the area of the circle. If we let $R$ be the region given by the disc of radius $r$, the positively oriented boundary $\partial R$ is just the circle oriented positively (counterclockwise). Then

$$\int_{\partial R} (-y/2)\,dx + (x/2)\,dy = \int\int_R 1\,dxdy = \pi r^2$$

is a first two-dimensional example of an integral computation along the boundary $\partial R$ equalling a related integral computation along the region $R$. We had seen this earlier when mentioning that the Fundamental Theorem of Calculus showed that an antiderivative $F$ evaluated along the boundary end points $a, b$ gave the integral of the derivative $F' = f$ over the interval $[a, b]$. The two-dimensional example is also an example of the Fundamental Theorem of Calculus deployed in higher dimensions. It is also true that

$$\int_{\partial R} (-y/2)\,dx + (x/2)\,dy = \int\int_R 1\,dxdy$$

holds for any region $R$ in $\mathbb{R}^2$ with well-behaved boundary a differentiable curve.

### ■ Exercises

1. Given the line integral $\int_C x\,dx + y\,dy$ for $C$ the portion of the parabola $y = x^2$ over the interval $[0, 1]$,

(a) parameterize $\mathcal{C}$ by $\gamma(t) = (t, t^2)^T$ for $t \in [0,1]$ and evaluate the line integral;

(b) parameterize $\mathcal{C}$ by $\gamma(t) = (2t, 4t^2)^T$ for $t \in [0, 1/2]$, evaluate the line integral, and compare with your previous answer; and

(c) parameterize $\mathcal{C}$ by $\gamma(t) = (1-t, (1-t)^2)^T$ for $t \in [0,1]$, evaluate the line integral, and compare with your previous answers; examine the orientation of this curve when comparing.

2. Assume that $F(x,y)$ satisfies $\vec{\nabla} F = (f_1, f_2)^T$ on an open set $\Omega \subset \mathbb{R}^2$. Show that for any path in $\Omega$ parameterized by $\gamma(t)$ for $t \in [a,b]$, that $F(\gamma(t))' = (f_1(\gamma(t)), f_2(\gamma(t)))^T \cdot \gamma'(t)$, and hence the line integral $\int_a^b (f_1(\gamma(t)), f_2(\gamma(t)))^T \cdot \gamma'(t) dt$ depends only on the end points $\gamma(b)$ and $\gamma(a)$, and conclude that the line integral $\int f_1 dx + f_2 dy$ is independent of path in $\Omega$.

### ■ 6.8.6 Green's Theorem; the Divergence Theorem

Let $R$ be a region in the plane $\mathbb{R}^2$ with boundary $\partial R$ parameterizable with positive (counterclockwise) orientation as a closed nonself-intersecting continuously differentiable curve. Let $P(x,y)$ and $Q(x,y)$ have continuous partial derivatives in an open set $\Omega$ containing $R$. Under these hypotheses one has a relation between line integrals on the boundary and double integrals over the region described as:

$$\int_{\partial R} P\, dx + Q\, dy = \int\int_R \left(\frac{\partial Q}{\partial x} - \frac{\partial P}{\partial y}\right) dx dy, \qquad (6.65)$$

and known as Green's theorem. The assumptions on the boundary $\partial R$ can be relaxed so that it is the union of finitely many continuously differentiable curves (a *piecewise continuously differentiable* curve).

We show Green's theorem first on the unit square in $\mathbb{R}^2$. Let $R$ be given by $0 \le x \le 1; 0 \le y \le 1$. First observe that the boundary of this square can be parameterized over its four edges by $\gamma = \gamma_1 + \gamma_2 - \gamma_3 - \gamma_4$ where $\gamma_1(t) = (t,0)^T$, $\gamma_2(t) = (1,t)^T$, $\gamma_3(t) = (t,1)^T$, $\gamma_4(t) = (0,t)^T$, for $t \in [0,1]$. Note the minus sign on, for instance $\gamma_3$, says to move in the opposite orientation than the direction moved along $\gamma_3$ when $t$ moves from 0 to 1. Also note that if one moves along $\gamma$, that is first along $\gamma_1$ then $\gamma_2$ then $-\gamma_3$ then $-\gamma_4$ in succession, one has traversed the boundary $\partial R$ once in the positive counterclockwise direction. We list the derivative of each parameterized component curve: $\gamma_1'(t) = (dx/dt, dy/dt)^T = (t', 0')^T = (1,0)^T$, $\gamma_2'(t) = (1', t')^T = (0,1)^T$, $\gamma_3'(t) = (t', 1')^T = (1,0)^T$, $\gamma_4'(t) = (0', t')^T = (0,1)^T$.

## 6.8 Integration, Green's Theorem, and the Divergence Theorem

First the line integral along the boundary in Eq. (6.65) is explicitly computed via the preceding parameterization:

$$\int_{\partial R} P\, dx + Q\, dy = \int_{\gamma(t)} (P\, dx/dt + Q\, dy/dt)\, dt$$

$$= \int_{\gamma_1(t)} (P\, dx/dt + Q\, dy/dt)\, dt + \int_{\gamma_2(t)} (P\, dx/dt + Q\, dy/dt)\, dt$$

$$- \int_{\gamma_3(t)} (P\, dx/dt + Q\, dy/dt)\, dt - \int_{\gamma_4(t)} (P\, dx/dt + Q\, dy/dt)\, dt$$

$$= \int_0^1 [P(t,0)\cdot 1 + Q(t,0)\cdot 0]\, dt + \int_0^1 [P(1,t)\cdot 0 + Q(1,t)\cdot 1]\, dt$$

$$- \int_0^1 [P(t,1)\cdot 1 + Q(t,1)\cdot 0]\, dt - \int_0^1 [P(0,t)\cdot 0 + Q(0,t)\cdot 1]\, dt$$

$$= \int_0^1 [P(t,0) - P(t,1)]\, dt + \int_0^1 [Q(1,t) - Q(0,t)]\, dt. \qquad (6.66)$$

Next one works on each summand in the double integral of Eq. (6.65), seen as an appropriate iterated integral,

$$\iint_R \frac{\partial Q}{\partial x}\, dx\, dy = \int_0^1 \left( \int_0^1 \frac{\partial Q}{\partial x}\, dx \right) dy \qquad (6.67)$$

$$= \int_0^1 Q(x,y)\big|_0^1\, dy = \int_0^1 (Q(1,y) - Q(0,y))\, dy. \qquad (6.68)$$

Note that the Fundamental Theorem of Calculus has been used to antidifferentiate $\partial Q/\partial x$ with respect to $x$ to obtain $Q$ in evaluating the inner integral $\int_0^1 \partial Q/\partial x\, dx$ in Eq. (6.67) as $Q(x,y)|_0^1$ in Eq. (6.68).

The second term in the double integral of Eq. (6.65) is similarly reduced to a line integral.

$$\iint_R -\frac{\partial P}{\partial y}\, dx\, dy = \int_0^1 \left( \int_0^1 -\frac{\partial P}{\partial y}\, dy \right) dx \qquad (6.69)$$

$$= \int_0^1 -P(x,y)\big|_0^1\, dx = \int_0^1 (P(x,0) - P(x,1))\, dx, \qquad (6.70)$$

where the Fundamental Theorem of Calculus has been used to antidifferentiate $\partial P/\partial y$ with respect to $y$ to obtain $P$ in evaluating the inner integral $\int_0^1 -\partial P/\partial y\, dy$ in Eq. (6.69) as $-P(x,y)|_0^1$ in Eq. (6.70).

Finally one observes that adding Eq. (6.68) and Eq. (6.70) gives Eq. (6.66)

$$\iint_R \frac{\partial Q}{\partial x} \, dx \, dy + \iint_R -\frac{\partial P}{\partial y} \, dx \, dy = \iint_R \left( \frac{\partial Q}{\partial x} - \frac{\partial P}{\partial y} \right) dx \, dy$$

$$= \int_0^1 [P(t,0) - P(t,1)] \, dt + \int_0^1 [Q(1,t) - Q(0,t)] \, dt$$

$$= \int_{\partial R} P \, dx + Q \, dy$$

and Green's theorem is now known on the unit square.

The next objective is to extend to more general regions by making use of a change of variables parameterization taking the unit square in $\mathbb{R}^2$, now denoted $W$, to a more general region $R$ in $\mathbb{R}^2$.

### ■ Theorem 6.11
#### Green's Theorem

Let $\vec{f}(u,v) = (x(u,v), y(u,v))^T : W \to R = \vec{f}(W)$ be a continuously differentiable, regular, one-to-one, onto transformation with $\det D\vec{f} > 0$ and with all second-order partials of $x$ and $y$ continuous, taking the unit square $W$ to a region $R$ in $\mathbb{R}^2$. Let $P(x,y)$, and $Q(x,y)$ have continuous partial derivatives on $R$. Then for $\partial R$ the positively oriented boundary of $R$, one has

$$\int_{\partial R} P \, dx + Q \, dy = \iint_R \left( \frac{\partial Q}{\partial x} - \frac{\partial P}{\partial y} \right) dx dy. \tag{6.71}$$

**Proof** Since

$$0 < |\det D\vec{f}| = \det D\vec{f} = \det \begin{pmatrix} \partial x/\partial u & \partial x/\partial v \\ \partial y/\partial u & \partial y/\partial v \end{pmatrix} = \frac{\partial x}{\partial u}\frac{\partial y}{\partial v} - \frac{\partial x}{\partial v}\frac{\partial y}{\partial u},$$

one has, by the Change of Variables formula, that

$$\iint_R \left( \frac{\partial Q}{\partial x} - \frac{\partial P}{\partial y} \right) dx \, dy$$
$$= \iint_W \left( \frac{\partial Q}{\partial x} - \frac{\partial P}{\partial y} \right) \left( \frac{\partial x}{\partial u}\frac{\partial y}{\partial v} - \frac{\partial x}{\partial v}\frac{\partial y}{\partial u} \right) du \, dv. \tag{6.72}$$

## 6.8 Integration, Green's Theorem, and the Divergence Theorem

Let $\gamma(t) = (u(t), v(t))^T$ be a positively oriented parameterization of the boundary $\partial W$. Under composition with $\vec{f} = (x(u,v), y(u,v))^T$ one has $\vec{f}(\gamma(t)) = (x(u(t), v(t)), y(u(t), v(t)))^T$, from which one sees

$$\begin{aligned}
dx &= \frac{dx}{dt} dt = \frac{d(x(u(t), v(t))^T}{dt} dt = \vec{\nabla} x \cdot (u'(t), v'(t))^T dt \\
&= \left(\frac{\partial x}{\partial u}, \frac{\partial x}{\partial v}\right)^T \cdot \left(\frac{du}{dt}, \frac{dv}{dt}\right)^T dt = \left(\frac{\partial x}{\partial u}\frac{du}{dt} + \frac{\partial x}{\partial v}\frac{dv}{dt}\right) dt = \frac{\partial x}{\partial u} du + \frac{\partial x}{\partial v} dv,
\end{aligned}$$

for $du = du/dt\, dt$ and $dv = dv/dt\, dt$. Similarly under composition

$$dy = \frac{\partial y}{\partial u} du + \frac{\partial y}{\partial v} dv.$$

Thus the line integral $\int_{\partial R} P\, dx + Q\, dy$ on the boundary $\partial R$ *pulls back* to the line integral on the boundary $\partial W$ given by

$$\begin{aligned}
&\int_{\partial R} P\, dx + Q\, dy \\
&= \int_{\partial W} P \left(\frac{\partial x}{\partial u} du + \frac{\partial x}{\partial v} dv\right) + Q \left(\frac{\partial y}{\partial u} du + \frac{\partial y}{\partial v} dv\right) \\
&= \int_{\partial W} \left(P\frac{\partial x}{\partial u} + Q\frac{\partial y}{\partial u}\right) du + \left(P\frac{\partial x}{\partial v} + Q\frac{\partial y}{\partial v}\right) dv \\
&= \int\int_W \frac{\partial}{\partial u}\left(P\frac{\partial x}{\partial v} + Q\frac{\partial y}{\partial v}\right) - \frac{\partial}{\partial v}\left(P\frac{\partial x}{\partial u} + Q\frac{\partial y}{\partial u}\right) du\, dv, \quad (6.73)
\end{aligned}$$

where the last equality converting to a double integral was obtained by application of Green's theorem on $W$.

The remainder of the proof consists of showing that the integrand in Eq. (6.73) equals the integrand in Eq. (6.72), which gives the two corresponding integrals are equal and results in Green's theorem on $R$.

Now

$$\frac{\partial P}{\partial u} = \frac{\partial P(x(u,v), y(u,v))}{\partial u} = \frac{\partial P}{\partial x}\frac{\partial x}{\partial u} + \frac{\partial P}{\partial y}\frac{\partial y}{\partial u}$$

by an application of the Chain Rule. Similarly

$$\frac{\partial P}{\partial v} = \frac{\partial P(x(u,v), y(u,v))}{\partial v} = \frac{\partial P}{\partial x}\frac{\partial x}{\partial v} + \frac{\partial P}{\partial y}\frac{\partial y}{\partial v}$$
$$\frac{\partial Q}{\partial u} = \frac{\partial Q(x(u,v), y(u,v))}{\partial u} = \frac{\partial Q}{\partial x}\frac{\partial x}{\partial u} + \frac{\partial Q}{\partial y}\frac{\partial y}{\partial u}$$
$$\frac{\partial Q}{\partial v} = \frac{\partial Q(x(u,v), y(u,v))}{\partial v} = \frac{\partial Q}{\partial x}\frac{\partial x}{\partial v} + \frac{\partial Q}{\partial y}\frac{\partial y}{\partial v}.$$

Thus the integrand in Eq. (6.73) becomes

$$\frac{\partial}{\partial u}\left(P\frac{\partial x}{\partial v} + Q\frac{\partial y}{\partial v}\right) - \frac{\partial}{\partial v}\left(P\frac{\partial x}{\partial u} + Q\frac{\partial y}{\partial u}\right)$$
$$= \frac{\partial P}{\partial u}\frac{\partial x}{\partial v} + P\frac{\partial^2 x}{\partial u \partial v} + \frac{\partial Q}{\partial u}\frac{\partial y}{\partial v} + Q\frac{\partial^2 y}{\partial u \partial v}$$
$$\quad - \frac{\partial P}{\partial v}\frac{\partial x}{\partial u} - P\frac{\partial^2 x}{\partial v \partial u} - \frac{\partial Q}{\partial v}\frac{\partial y}{\partial u} - Q\frac{\partial^2 y}{\partial v \partial u}$$
$$= \frac{\partial P}{\partial u}\frac{\partial x}{\partial v} + \frac{\partial Q}{\partial u}\frac{\partial y}{\partial v} - \frac{\partial P}{\partial v}\frac{\partial x}{\partial u} - \frac{\partial Q}{\partial v}\frac{\partial y}{\partial u}$$
$$= \left(\frac{\partial P}{\partial x}\frac{\partial x}{\partial u} + \frac{\partial P}{\partial y}\frac{\partial y}{\partial u}\right)\frac{\partial x}{\partial v} + \left(\frac{\partial Q}{\partial x}\frac{\partial x}{\partial u} + \frac{\partial Q}{\partial y}\frac{\partial y}{\partial u}\right)\frac{\partial y}{\partial v}$$
$$\quad - \left(\frac{\partial P}{\partial x}\frac{\partial x}{\partial v} + \frac{\partial P}{\partial y}\frac{\partial y}{\partial v}\right)\frac{\partial x}{\partial u} - \left(\frac{\partial Q}{\partial x}\frac{\partial x}{\partial v} + \frac{\partial Q}{\partial y}\frac{\partial y}{\partial v}\right)\frac{\partial y}{\partial u}$$
$$= \frac{\partial P}{\partial y}\left(\frac{\partial y}{\partial u}\frac{\partial x}{\partial v} - \frac{\partial y}{\partial v}\frac{\partial x}{\partial u}\right) + \frac{\partial Q}{\partial x}\left(\frac{\partial x}{\partial u}\frac{\partial y}{\partial v} - \frac{\partial x}{\partial v}\frac{\partial y}{\partial u}\right)$$
$$= \left(\frac{\partial Q}{\partial x} - \frac{\partial P}{\partial y}\right)\left(\frac{\partial x}{\partial u}\frac{\partial y}{\partial v} - \frac{\partial x}{\partial v}\frac{\partial y}{\partial u}\right),$$

which is the integrand in Eq. (6.72). Thus the integrands and the integrals in Eq. (6.73) and Eq. (6.72) are equal, and one sees that

$$\int_{\partial R} P\,dx + Q\,dy = \int\int_R \left(\frac{\partial Q}{\partial x} - \frac{\partial P}{\partial y}\right) dx dy$$

via change of variables. □

By piecing together adjacent regions $R$ on which Green's theorem holds, one sees that the orientations on the adjacent overlapping edges are opposite, and they cancel out. The result is that Green's theorem can be seen to hold on very general regions $\tilde{R}$ that can be obtained from a grid of adjacent smaller regions $R$ on which the theorem is known to hold. By piecing together such regions, one can also see that if the resulting region $\tilde{R}$ contains any holes in the interior, then the positive orientation induced from $\partial \tilde{R}$ is clockwise on the edges of the holes.

## Corollary 6.9

Let $R$ be a region satisfying the hypotheses in Green's theorem. Then the area $A(R)$ of $R$ is given by

$$A(R) = \frac{1}{2} \int_{\partial R} -y\,dx + x\,dy = \int_{\partial R} -y\,dx = \int_{\partial R} x\,dy.$$

**Proof** By Green's Theorem

$$\int_{\partial R} -y\,dx + x\,dy = \int\int_R \frac{\partial x}{\partial x} - \frac{\partial(-y)}{\partial y}\,dxdy = \int\int_R 2\,dxdy = 2A(R),$$

and dividing by 2 gives the first equality in the theorem. Similarly

$$\int_{\partial R} -y\,dx = \int\int_R -\frac{\partial(-y)}{\partial y}\,dxdy = \int\int_R 1\,dxdy$$
$$= A(R) = \int\int_R \frac{\partial x}{\partial x}\,dxdy = \int_{\partial R} x\,dy.$$

□

Another important example will be that of the *flux* of a vector field across the boundary of a region $R$. Let $\vec{V} = (V_1, V_2)^T$ be a continuously differentiable vector-valued function on $\mathbb{R}^2$, that is each $V_i = V_i(x_1, x_2)$ is continuously differentiable. Such a $\vec{V}$ is called a vector field on $\mathbb{R}^2$. Then the flux integral of $\vec{V}$ along a positively oriented curve $\partial R$ bounding a region $R$ in $\mathbb{R}^2$ is given by the line integral:

$$\int_{\partial R} -V_2 dx_1 + V_1 dx_2.$$

This can be represented as a two-dimensional integral of the *divergence* of $\vec{V}$ that is defined by $\text{div}(\vec{V}) = \partial V_1/\partial x_1 + \partial V_2/\partial x_2$.

### ■ Theorem 6.12
**Divergence Theorem in $\mathbb{R}^2$**
Let $R$ be a region with parameterizable oriented boundary $\partial R$ in $\mathbb{R}^2$ to which Green's theorem is applicable. Let $\vec{V} = (V_1, V_2)^T$ be a continuously differentiable

vector-valued function from $\mathbb{R}^2$ to $\mathbb{R}^2$. Then the flux integral of $\vec{V}$ on $\partial R$ equals the integral of the divergence of $\vec{V}$ on $R$. That is

$$\int_{\partial R} -V_2 dx_1 + V_1 dx_2 = \int\int_R \frac{\partial V_1}{\partial x_1} + \frac{\partial V_2}{\partial x_2} \, dx_1 dx_2. \tag{6.74}$$

**Proof** Apply Green's theorem to the flux line integral to obtain

$$\int_{\partial R} -V_2 \, dx_1 + V_1 \, dx_2 = \int\int_R \frac{\partial V_1}{\partial x_1} - \frac{\partial(-V_2)}{\partial x_2} \, dx_1 dx_2$$
$$= \int\int_R \frac{\partial V_1}{\partial x_1} + \frac{\partial V_2}{\partial x_2} \, dx_1 dx_2 = \int\int_R \text{div}(\vec{V}) \, dx_1 dx_2.$$

$\square$

There is an important interpretation of the flux line integral $\int_{\partial R} -V_2 dx_1 + V_1 dx_2$ associated to the *vector field* given by $\vec{V} = (V_1, V_2)^T$. This can be seen as follows. Let $\partial R$ be parameterized by $\gamma(t) = (x_1(t), x_2(t))^T$ for $t \in [a,b]$, and compute the flux line integral with this parameterization:

$$\int_{\partial R} -V_2 \, dx_1 + V_1 \, dx_2$$
$$= \int_a^b \left(-V_2 \frac{dx_1}{dt} + V_1 \frac{dx_2}{dt}\right) dt = \int_a^b (V_1, V_2)^T \cdot \left(\frac{dx_2}{dt}, -\frac{dx_1}{dt}\right)^T dt$$
$$= \int_a^b |\vec{V}| \cos(\theta(t)) \left|\left(\frac{dx_2}{dt}, -\frac{dx_1}{dt}\right)^T\right| dt \tag{6.75}$$
$$= \int_a^b |\vec{V}| \cos(\theta(t)) \left|\left(\frac{dx_1}{dt}, \frac{dx_2}{dt}\right)^T\right| dt = \int_a^b |\vec{V}| \cos(\theta(t)) |\gamma'(t)| \, dt. \tag{6.76}$$

Here $(dx_2/dt, -dx_1/dt)^T = \vec{N}$ forms an outer normal vector $\vec{N}$ to the boundary $\partial R$ because: it is perpendicular to the tangent vector $(dx_1/dt, dx_2/dt)^T$ by the dot product vanishing; and, by Exercise (3) in Section 6.1.2, $\vec{N}$ is a rotation clockwise by $\pi/2$ of the tangent vector. Also, $\theta(t)$ in Eq. (6.75) is taken to be the angle between $\vec{V}$ and $\vec{N}$ at time $t$ at position $\gamma(t)$. The element $|\gamma'(t)| dt$ is seen as an arclength element because it is the speed of travel $|\gamma'(t)|$ times the time increment $dt$, giving the distance traveled on the boundary for a small time increment. If $\vec{V}$ is thought of as the velocity field to a flow and $|\gamma'(t)| dt$ is thought of as the length of a small piece of the boundary curve near $\gamma(t)$ then this small piece of the boundary would flow out along $\vec{V}$ in unit time, sweeping out a parallelogram of area given by a product of the length $|\gamma'(t)|dt$ times the altitude $|\vec{V}|\cos(\theta(t))$ in the outer normal direction. This represents a rate of flow over unit time of the velocity $\vec{V}$ across the piece of the boundary segment of length $|\gamma'(t)|dt$. Summing these products $|\vec{V}|\cos(\theta(t))\,|\gamma'(t)|\,dt$ all up around $\partial R$ gives the net rate of flow of $\vec{V}$ across $\partial R$ for a unit time. This is called the *flux* of $\vec{V}$ across $\partial R$. When $0 \leq \theta(t) < \pi/2$ then $\vec{V}$ flows out of $R$; when $\pi/2 < \theta(t) \leq \pi$ then $\vec{V}$ flows into $R$; the overall integral is a net rate of flow across the boundary per unit time, or flux as seen in Figure 6.2.

In some references the flux elements $|\vec{V}|\cos(\theta(t))\,|\gamma'(t)|\,dt$ can be represented as $\vec{V}\cdot\vec{n}\,|\gamma'(t)|\,dt$ where $\vec{n} = \vec{N}/|\vec{N}|$ is the unit outer normal. This follows by virtue of the fact that $\vec{V}\cdot\vec{n} = |\vec{V}||\vec{n}|\cos(\theta) = |\vec{V}|\,1\,\cos(\theta)$. This, together with the fact that $ds = |\gamma'(t)|dt$ denotes the arclength element, allows for the flux integral to be often represented via Eq. (6.76) as

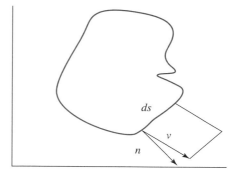

**FIGURE 6.2** ■ Region with boundary flow.

$$\int_{\partial R} -V_2\,dx_1 + V_1\,dx_2 = \int_{\partial R} \vec{V}\cdot\vec{n}\,ds.$$

In this notation the Divergence theorem says that the flux across the boundary $\partial R$ is the integral of the divergence over $R$ [68]:

$$\int_{\partial R} \vec{V}\cdot\vec{n}\,ds = \int\int_R \text{div}(\vec{V})\,dx_1 dx_2. \tag{6.77}$$

### ■ Exercises

1. Use Green's theorem to evaluate $\int_{\partial R} xy^2\,dx + x^2 y\,dy$ for a closed differentiable curve bounding *any* region $R$.

2. Express the flux integral $\int_{\partial R} -V_2 dx + V_1 dy$ for each vector field $\vec{V} = (V_1, V_2)^T$ that follows. Then use the Divergence theorem to compute the flux integral for $R$, which is any region of finite area $A(R)$ with boundary $\partial R$ a closed differentiable curve,

   (a) for $\vec{V} = (x, y)^T$;
   (b) for $\vec{V} = (-x, -y)^T$;

(c) for $\vec{V} = (-y, x)^T$; and
(d) for $\vec{V} = (1, 5)^T$.

## ■ 6.9 Vector Fields, Divergence, and Curl

A *vector field* is a continuously differentiable vector-valued function $\vec{V}$ from an open subset $\Omega \subset \mathbb{R}^n$ to $\mathbb{R}^n$. One visualizes a vector field as a collection of vectors with each $\vec{V}(\vec{x})$ emanating from the base point $\vec{x}$. A vector field can be thought of as the velocity vector at position $\vec{x}$ of a flow $\vec{r}(t)$ through $\vec{x}$ satisfying the equation $\vec{r}\,'(t) = \vec{V}(\vec{r}(t))$. Here $\vec{V}$ is given by an $n$-tuple of continuously differentiable coordinate functions $V_i(x_1, x_2, \ldots, x_n)$, that is $\vec{V}(\vec{x}) = (V_1(\vec{x}), V_2(\vec{x}), \ldots, V_n(\vec{x}))^T$.

**Definition 6.4** The *divergence* of a vector field $\vec{V} : \Omega \to \mathbb{R}^n$ for open $\Omega \subset \mathbb{R}^n$ is given by

$$\begin{aligned}
\operatorname{div}(\vec{V}) &= \vec{\nabla} \cdot \vec{V} = (\partial_{x_1}, \partial_{x_2}, \ldots, \partial_{x_n})^T \cdot (V_1, V_2, \ldots, V_n)^T \\
&= \partial_{x_1} V_1 + \partial_{x_2} V_2 + \cdots + \partial_{x_n} V_n = \sum_{i=1}^n \partial_{x_i} V_i = \sum_{i=1}^n \frac{\partial V_i}{\partial x_i}.
\end{aligned}$$

The physical significance of the divergence of $\vec{V}$ is that it measures the rate of the vector field's expansion (divergence) or contraction (convergence) per volume per unit time if one flowed along the vector field.

As an example let $\vec{V}(x, y, z) = (x, y, z)^T$. Then $\operatorname{div}(\vec{V}) = \partial x/\partial x + \partial y/\partial y + \partial z/\partial z = 3$. One could flow along the path $\vec{r}(t) = (xe^t, ye^t, ze^t)^T$ through each point $(x, y, z)^T$ with $\vec{r}\,'(t) = (xe^t, ye^t, ze^t)^T = \vec{V}(\vec{r}(t))$. This flow would take a small box of dimensions $\Delta x \times \Delta y \times \Delta z$ at corner point $(x, y, z)^T$ to a larger box with corner point $(xe^t, ye^t, ze^t)^T$ and dimensions $e^t \Delta x \times e^t \Delta y \times e^t \Delta z$. This can be seen in the following way. Let the original small box be parameterized by $(x + a, y + b, z + c)^T$ for $0 \le a \le \Delta x, 0 \le b \le \Delta y$, and $0 \le c \le \Delta z$. Then $(x + a, y + b, z + c)^T$ flows to $((x+a)e^t, (y+b)e^t, (z+c)e^t)^T$ with corner $(xe^t, ye^t, ze^t)^T$ and $0 \le e^t a \le e^t \Delta x$, $0 \le e^t b \le e^t \Delta y$, and $0 \le e^t c \le e^t \Delta z$ so the small box has flowed to a new box of dimensions $e^t \Delta x \times e^t \Delta y \times e^t \Delta z$ as claimed. The new volume per the original volume is $e^{3t}(\Delta x \Delta y \Delta z)/(\Delta x \Delta y \Delta z) = e^{3t}$ and the rate of change of this ratio is $e^{3t} 3$, which at time 0 is $3 = \operatorname{div}(\vec{V})$.

Because the divergence $\operatorname{div}(\vec{V})$ is the rate of expansion per unit $n$-volume per unit time of the vector field, when multiplied by an $n$-volume element $dx_1 dx_2 \ldots dx_n$ and summed over a region $R$ one obtains the integral of the divergence $\int \cdots \int_R \operatorname{div}(\vec{V}) dx_1, \ldots, dx_n$ is the aggregate rate of expansion of the fluid flowing along $\vec{V}$ in $R$. One might expect that this would be related to a rate of flow across the boundary $\partial R$, and this intuition is confirmed by the Divergence theorem. The aggregate rate of expansion of the fluid over $R$ per unit time, that is the integral of the divergence over $R$, is indeed the rate of flow across the boundary $\partial R$ per unit time, or the flux integral. With this hindsight, the Divergence theorem for $\mathbb{R}^2$ seems more intuitive. This same intuition tells us that the Divergence theorem should hold for all dimensions. That is indeed the case. As an example in $\mathbb{R}^3$, take a region $R$ with boundary $\partial R$ a closed regular surface with $\vec{n}$ a unit outer normal to the surface. The vector field $\vec{V}$ should take a small section (an area element $dA$) from the surface and flow it in the direction of $\vec{V}$ after unit time. The volume of fluid passing through $dA$ should be the area of the base $dA$ times the altitude $|\vec{V}|\cos(\theta) = \vec{V}\cdot\vec{n}$ where $\theta$ is the angle between $\vec{V}$ and $\vec{n}$. Summing these terms $\vec{V}\cdot\vec{n}\,dA$ gives the net rate of flow across $\partial R$ per unit time as the integral $\int\int_R \vec{V}\cdot\vec{n}\,dA$. This net rate of flow, or flux, should also be equal to the integral of the divergence [68], as just discussed.

### ■ Theorem 6.13
**Divergence Theorem in $\mathbb{R}^3$**

$$\int\int_R \vec{V}\cdot\vec{n}\,dA = \int\int\int_R \text{div}(\vec{V})\,dxdydz \qquad (6.78)$$

An important application of the Divergence theorem is to the area of differential equations. These equations govern the flow of heat, fluid motion, electromagnetic behavior, and the behavior of functions over the complex numbers, among many in a long list of applications.

As an example, we will model the behavior of heat flow using a familiar observation made in day-to-day life: heat flows from hot to cold. This simple observation is then paired with the knowledge that the gradient is the direction of maximal ascent, and minus the gradient is the direction of maximal descent. Suppose $T(x, y, z, t)$ denotes the temperature at point $(x, y, z)^T \in \mathbb{R}^3$ at time $t$. The spatial direction of maximal descent in $T$ namely $-(\partial T/\partial x, \partial T/\partial y, \partial T/\partial z)^T$ at a given time $t$ should then give the direction of the flow of heat at that instant of time. We let the vector field $\vec{V}$ at time $t$ then be $\vec{V} = -c_1(\partial T/\partial x, \partial T/\partial y, \partial T/\partial z)^T$ where $c_1 > 0$ depends on the material through which the heat is flowing. Given this $\vec{V}$ as the velocity field of heat flow, one examines an arbitrary small region $R$, computing the flux of $\vec{V}$ across the boundary as the net rate of flow of heat across the boundary $\partial R$ per unit time. This net rate of flow of heat across the boundary $\partial R$ is clearly proportional to the rate of change of temperature and energy of the region $R$. One deduces that:

$$\iint_{\partial R} \vec{V} \cdot \vec{n}\, dA = c_2 \frac{\partial}{\partial t} \iiint_R T\, dxdydz.$$

Applying the Divergence theorem to re-express the left-hand side, and differentiating under the integral sign on the right-hand side, one obtains:

$$\iiint_R \operatorname{div}[-c_1(\partial T/\partial x, \partial T/\partial y, \partial T/\partial z)^T]\, dxdydz = \iiint_R c_2 \frac{\partial T}{\partial t}\, dxdydz,$$

which becomes

$$\iiint_R -c_1 \left( \frac{\partial^2 T}{\partial x^2} + \frac{\partial^2 T}{\partial y^2} + \frac{\partial T^2}{\partial z^2} \right) dxdydz = \iiint_R c_2 \frac{\partial T}{\partial t}\, dxdydz.$$

A quick reflection says that when the flux is positive the net rate of heat flow per unit time is outward, causing the temperature and energy in $R$ to drop, forcing negativity of the constant $c_2 < 0$. Next one observes that since the integrals are the same over *any* region $R$, the integrands must be the same, from which one concludes, upon dividing both sides by $c_2$ and setting $c^2 = -c_1/c_2$, that:

$$c^2 \left( \frac{\partial^2 T}{\partial x^2} + \frac{\partial^2 T}{\partial y^2} + \frac{\partial T^2}{\partial z^2} \right) = \frac{\partial T}{\partial t}. \tag{6.79}$$

Equation (6.79) is the *Heat Equation* in $\mathbb{R}^3$ and is studied in the area of partial differential equations. The sum of second partials in Eq. (6.79) is called the *Laplacian* of $T$.

Another important notion is the *curl* of a vector field $\vec{V} = (V_1, V_2, V_3)^T$ in $\mathbb{R}^3$.

---

**Definition 6.5** The *curl* of a vector field $\vec{V} : \Omega \to \mathbb{R}^3$ for open $\Omega \subset \mathbb{R}^3$ is given by

$$\begin{aligned} \operatorname{curl}(\vec{V}) &= \vec{\nabla} \times \vec{V} = (\partial_{x_1}, \partial_{x_2}, \partial_{x_3})^T \times (V_1, V_2, V_3)^T \\ &= (\partial_{x_2} V_3 - \partial_{x_3} V_2,\, -\partial_{x_1} V_3 + \partial_{x_3} V_1,\, \partial_{x_1} V_2 - \partial_{x_2} V_1)^T. \end{aligned}$$

For vector fields in the plane $\vec{U} : \Lambda \to \mathbb{R}^2$ for open $\Lambda \subset \mathbb{R}^2$, define the scalar function $\operatorname{curl}(\vec{U}) = \partial_{x_1} U_2 - \partial_{x_2} U_1$.

---

The significance of the *curl* of a vector field $\vec{V}$ is that it measures the tendency toward rotation of a small rigid body in the flow of $\vec{V}$. To see this consider the example of rotation about the $z$-axis given by the flow $\vec{r}(t) = (r\cos(\omega t), r\sin(\omega t), z)^T$

with angular velocity $\omega$. Differentiating with respect to time $t$ one obtains $\vec{r}\,'(t) = (-r\omega\sin(\omega t), r\omega\cos(\omega t), 0)^T$. Letting $x = r\cos(\omega t)$, and $y = r\sin(\omega t)$, we see the associated vector field corresponding to the rotational flow $\vec{r}$ is $\vec{V}(x,y,z) = (-\omega y, \omega x, 0)^T$ because $\vec{r}\,'(t) = \vec{V}(\vec{r}(t))$. The curl of $\vec{V}$ is now computed as $\operatorname{curl}(\vec{V}) = (\partial_x, \partial_y, \partial_z)^T \times (-\omega y, \omega x, 0)^T = (\partial_y 0 - \partial_z(\omega x), -\partial_x 0 + \partial_z(-\omega y), \partial_x(\omega x) - \partial_y(-\omega y))^T = (0, 0, 2\omega)^T$, which is a vector of length twice the angular velocity pointing in the same direction as the axis of rotation, the $z$-axis.

There is a type of Green's theorem involving the curl of a vector, which is recorded here without proof.

### ■ Theorem 6.14

**Stokes Theorem I** [68]

Let $R$ be a regular positively oriented surface equipped with unit normal vector $\vec{n}$ in $\mathbb{R}^3$ and bounded by an oriented continuously differentiable curve $\partial R$ parameterized by $\gamma(t)$. Let $\vec{V}$ be a continuously differentiable vector field on an open set containing $R$. Then:

$$\int_{\partial R} \vec{V} \cdot \gamma\,'(t)\, dt = \int\!\!\int_R \operatorname{curl}(\vec{V}) \cdot \vec{n}\, dA \tag{6.80}$$

All the integration theorems studied so far, Greens theorem, the Divergence theorem, and Stokes theorem are generalizations of the Fundamental Theorem of Calculus. They are all subsumed as special cases of one overall formula also known as Stokes theorem [68], studied in the area of differential forms [68].

### ■ Exercises

1. Find $\operatorname{div}(\vec{\nabla} f)$ where $f(x_1, x_2, \ldots, x_n)$ is twice differentiable in all variables.
2. Find and simplify $\operatorname{div}(\vec{V})$ for $\vec{V} = (x^2, y)^T$.
3. Consider a vector field in the plane $\vec{U}(x,y) = (-y, x)^T$ and the flow curves $\vec{r}(t) = (x(t), y(t))^T$:
   (a) Find $\operatorname{div}(\vec{U})$ and $\operatorname{curl}(\vec{U})$.
   (b) Let $x(t) = r(t) \cdot \cos(\theta(t))$ and compute $x'(t)$ in terms of $r$ and $\theta$, which has two terms. Also compute the five terms of $x''(t)$.
   (c) Let $y(t) = r(t) \cdot \sin(\theta(t))$ and compute $y'(t)$ in terms of $r$ and $\theta$, which has two terms. Also compute $y''(t)$, which will have five terms.

(d) Consider the special case $\vec{r}(t) = (r_0 \cos(t+t_0), r_0 \sin(t+t_0))^T$, for each fixed $r_0 \geq 0$ and $t_0 \in \mathbb{R}$. Find the polar coordinates $(r(t), \theta(t))^T$ and compute $r'(t)$, $r''(t)$, $\theta'(t)$, and $\theta''(t)$.

(e) Using the simple expressions for $r'$, $r''$, $\theta'$, and $\theta''$ show that $x'(t)$, $x''(t)$, $y'(t)$, and $y''(t)$ reduce to single terms.

(f) Show the flow curves $\vec{r}(t)$ satisfy $\vec{r}\,'(t) = \vec{U}(\vec{r}(t))$.

(g) Compute $\|\vec{r}(t)\|$ and $\|\vec{U}(\vec{r}(t))\|$.

(h) Draw the flow curves for $r = 1$ and $2$, and observe that this is a rotational flow. Compute $\vec{r}(t) \cdot \vec{r}\,'(t)$ and use geometry to explain your answer in (a).

## 6.10 Optimization and Linear Regression

To optimize a real-valued differentiable function $f(x_1, x_2, \ldots, x_n)$ of several variables defined on an open set, one searches for critical points and solves the equation $\vec{\nabla} f = \vec{0}$. That is, at a local extremum, all the partial derivatives $\partial f / \partial x_i = 0$ for $i = 1, \ldots, n$. To see this, consider $x_i$ to be varying at the extremum, while the other variables are held fixed; then the restricted one variable function has 0 derivative, or equivalently $\partial f / \partial x_i = 0$ at the optimal point. This must be true for all $i$, so the gradient vanishes at an extremum value of the function $f$.

Since the linearization of $f$ at the critical point is then a constant, the second-order terms come into play. These are related to second partial derivatives of $f$ at the critical point. The *Hessian* matrix $H$ of second-order partials given by

$$H = \begin{pmatrix} \frac{\partial^2 f}{\partial x_1 \partial x_1} & \frac{\partial^2 f}{\partial x_1 \partial x_2} & \cdots & \frac{\partial^2 f}{\partial x_1 \partial x_n} \\ \frac{\partial^2 f}{\partial x_2 \partial x_1} & \frac{\partial^2 f}{\partial x_2 \partial x_2} & \cdots & \frac{\partial^2 f}{\partial x_2 \partial x_n} \\ \vdots & \vdots & \frac{\partial^2 f}{\partial x_i \partial x_j} & \vdots \\ \frac{\partial^2 f}{\partial x_n \partial x_1} & \frac{\partial^2 f}{\partial x_n \partial x_2} & \cdots & \frac{\partial^2 f}{\partial x_n \partial x_n} \end{pmatrix} \tag{6.81}$$

can then often be utilized in assessing the type of critical point.

For a function of two variables one can solve $\partial f / \partial x_1 = 0 = \partial f / \partial x_2$ to find a critical point $p$, and then rely on the following extra information obtainable from the Hessian evaluated at $p$ to determine the nature of the critical point:

1. If $\det H < 0$ there is neither a maximum nor a minimum at $p$.
2. If $\det H > 0$ and trace $H > 0$ there is a local minimum at $p$.
3. If $\det H > 0$ and trace $H < 0$ there is a local maximum at $p$.

As a set of examples where the answer is clear geometrically, examine the functions $f(x,y) = x^2 + y^2$, $g(x,y) = -x^2 - y^2$ and $h(x,y) = x^2 - y^2$. The only critical point for each function is $(0,0)^T$, since for instance in the case of $f$ $\vec{\nabla} f = (2x, 2y)^T = (0,0)^T$ is solved only when $x = 0, y = 0$. The respective Hessians are then computed as

$$H_f = \begin{pmatrix} 2 & 0 \\ 0 & 2 \end{pmatrix} \quad H_g = \begin{pmatrix} -2 & 0 \\ 0 & -2 \end{pmatrix} \quad H_h = \begin{pmatrix} 2 & 0 \\ 0 & -2 \end{pmatrix}.$$

In the case of $f$, $\det H_f = 4 > 0$ and trace $H_f = 4 > 0$ so there is a local (and global) minimum at $(0,0)$. For $g$, $\det H_g = 4 > 0$ and trace $H_f = -4 < 0$ so there is a local (and global) maximum at $(0,0)^T$. For $h$, $\det H_h = -4 < 0$ so there is neither a local maximum nor a local minimum at $(0,0)$.

Optimization of functions of several variables can have substantial consequences. As an example of this we turn our attention to the line of least squares. Suppose a set of paired data points of form $(x_i, y_i)^T$ for $i = 1, \ldots, n$ has been obtained experimentally. One would like to model these by a function $f(x)$, where $f$ is often a line, a polynomial, or other well-understood function. One measures the residuals $r_i = y_i - f(x_i)$ to measure the fit of the model $f(x)$ to the data points by squaring the residuals and then summing to obtain the sum of squares function:

$$SS(f) = \sum_{i=1}^n r_i^2 = \sum_{i=1}^n [y_i - f(x_i)]^2.$$

One then varies the family of parameters describing the category of $f$, minimizing the sum of squares function, to obtain a function of least squares.

Here is the process for a line of least squares $f(x) = mx + b$, where $m, b$ are the two real parameters necessary to describe a linear function. For succinctness, the sum of squares function is denoted as a function $g$ of $m$ and $b$:

$$g(m,b) = SS(f) = \sum_{i=1}^n [y_i - (mx_i + b)]^2. \tag{6.82}$$

The process is to minimize $g$ as a function of the two variables $m, b$. Critical points are found by finding the points where the gradient vanishes, $(\partial g/\partial m, \partial g/\partial b)^T = (0,0)^T$. This becomes

$$\partial g/\partial m = \sum_{i=1}^{n} 2[y_i - (mx_i + b)](-x_i) = -2\sum_{i=1}^{n} y_i x_i + 2m\sum_{i=1}^{n} x_i^2 + 2b\sum_{i=1}^{n} x_i = 0$$

$$\partial g/\partial b = \sum_{i=1}^{n} 2[y_i - (mx_i + b)](-1) = -2\sum_{i=1}^{n} y_i + 2m\sum_{i=1}^{n} x_i + 2b\sum_{i=1}^{n} 1 = 0.$$

Dividing both equations by 2 and putting them in matrix form gives

$$\begin{pmatrix} \sum_{i=1}^{n} x_i^2 & \sum_{i=1}^{n} x_i \\ \sum_{i=1}^{n} x_i & n \end{pmatrix} \begin{pmatrix} m \\ b \end{pmatrix} = \begin{pmatrix} \sum_{i=1}^{n} y_i x_i \\ \sum_{i=1}^{n} y_i \end{pmatrix},$$

from which one sees that

$$\begin{pmatrix} m \\ b \end{pmatrix} = \frac{1}{n\sum_{i=1}^{n} x_i^2 - (\sum_{i=1}^{n} x_i)^2} \begin{pmatrix} n & -\sum_{i=1}^{n} x_i \\ -\sum_{i=1}^{n} x_i & \sum_{i=1}^{n} x_i^2 \end{pmatrix} \begin{pmatrix} \sum_{i=1}^{n} y_i x_i \\ \sum_{i=1}^{n} y_i \end{pmatrix}.$$

The coefficients for the line of least squares, or line of regression, are seen to be:

$$m = \frac{n\sum_{i=1}^{n} y_i x_i - \sum_{i=1}^{n} x_i \sum_{i=1}^{n} y_i}{n\sum_{i=1}^{n} x_i^2 - (\sum_{i=1}^{n} x_i)^2} = \frac{\overline{yx} - (\overline{x})(\overline{y})}{\overline{x^2} - (\overline{x})^2}, \tag{6.83}$$

and

$$b = \frac{-\sum_{i=1}^{n} x_i \sum_{i=1}^{n} y_i x_i + \sum_{i=1}^{n} x_i^2 \sum_{i=1}^{n} y_i}{n\sum_{i=1}^{n} x_i^2 - (\sum_{i=1}^{n} x_i)^2} = \frac{-(\overline{x})(\overline{yx}) + (\overline{x^2})(\overline{y})}{\overline{x^2} - (\overline{x})^2}, \tag{6.84}$$

where $\overline{E} = (\sum_{i=1}^{n} E_i)/n$ for each expression $E$. One could utilize the Hessian to confirm an absolute minimum. However, since there is only one critical point, and since $g(m, b)$ must grow large as $|(m, b)^T|$ approaches infinity, the preceding critical values give the absolute minimum of $g$.

■ **Exercises**

1. Find critical points and the absolute minimum value assumed by the function $f(x, y) = x + x^2 + 3y + y^2$ over $\mathbb{R}^2$.
2. Find the line of least squares for the three data points: $(1, 1)^T$, $(2, 3)^T$, and $(3, 3)^T$.

3. Compute the Hessian of $g(m,b)$ in Eq. (6.82) and explain why the only critical point for $g$ is a minimum.

## 6.11 A Revisit to Gradient Flow

Suppose $f(\vec{x})$ represents some scalar physical quantity, like temperature, or pressure, or concentration, or density, etc. If the distribution of this physical quantity is not uniform, also called *nonhomogeneous*, then at each point $\vec{x}$, the directions in which the quantity will increase the fastest is,

$$\hat{\nabla} f(\vec{x}) = \left(1/\|\vec{\nabla} f\|\right) \cdot \vec{\nabla} f. \tag{6.85}$$

Orthogonal to $\hat{\nabla} f(\vec{x})$ is a surface of constant quantities, also called *isoclines*. These have special terminologies:

**Temperature** $T$: Isotherm $\perp$ Convection;
**Pressure** $P$: Isobar $\perp$ Pressure gradient;
**Concentration** $C$: Iso-segregation $\perp$ Diffusion; and
**Density** $\rho$: Isopycnal $\perp$ Advection;

The reason that particles move or temperatures change is that systems tend toward equilibria. Thus, one may suspect that the *flow* of a physical quantity is correlated with its *gradient* [39]. This is indeed observed to be the case, and this is expressed as different laws of physics:

**Temperature:** Fourier's Law of Thermal Conductivity states that:
$$\textit{Heat-Energy Flow} \propto \|\vec{\nabla} T\|;$$
**Pressure:** Newton's $2^{nd}$ Law for Gases states that: *Pressure Force* $\propto \|\vec{\nabla} P\|$;
**Concentration:** Fick's Law of Diffusion states that: *Flow of Particles* $\propto \|\vec{\nabla} C\|$;
**Density:** Euler's Continuity Equation states that: *Velocity Divergence* $\propto \|\vec{\nabla} \rho\|$.

The presence of a gradient suggests the existence of a force field and this sets up the dynamics by which a system can tend toward an equilibrium. To maintain the force, leading to an eventual *steady state*, requires the flow of some quantity into a *sink* and out of a *source*. For example, if there is no source for a river, then it will dry up after it empties into a stationary and still body of water, like a pond.

## Exercises

1. For the functions $f$ compute the gradient and describe the isoclines $f = 1$:
   (a) $f(x, y) = (x^2 + y^2)/2$.
   (b) $f(x, y, z) = x^2/2 + 2y^2 + z^2/18$.
   (c) $f(x, y) = 2x - y$.
   (d) $f(x, y, z) = 3x + 2y - z$.

2. Standing at the reference point $P_0 = (1/2, 1, -1)^T$ the temperature is $T_0 > 0$. Moving 1 unit in the $x$-direction from $P_0$ gives a temperature of $T_0 \cdot 17/9$. Moving 1 unit in the $y$-direction from $P_0$ gives a temperature of $T_0 \cdot 7/3$. Moving 1 unit in the $z$-direction from $P_0$ gives a temperature of $T_0 \cdot 5/9$.
   (a) Compute $\Delta T_\alpha$ for the three movements $\alpha \in \{x, y, z\}$.
   (b) Construct the function $f(a, b, c) = (a\hat{i} + b\hat{j} + c\hat{k}) \cdot \vec{\nabla} T$ using the approximations $\partial_x T(P_0) \simeq \Delta T_x$, $\partial_y T(P_0) \simeq \Delta T_y$, and $\partial_z T(P_0) \simeq \Delta T_z$.
   (c) Find an algebraic equation ensuring that the vector $\hat{u} \equiv (a\hat{i} + b\hat{j} + c\hat{k})$ is a unit vector and express as $g(a, b, c) = 1$.
   (d) Introduce a new parameter $\lambda \in \mathbb{R}$, called a *Lagrange multiplier*, so that $\vec{\nabla} f(a, b, c) = \lambda \vec{\nabla} g(a, b, c)$ and obtain three more algebraic equations.
   (e) Using division, solve the four equations in the four unknowns $a, b, c,$ and $\lambda$.
   (f) Find $\hat{u}$, which is a reasonable approximation to $\hat{\nabla} T(P_0)$.
   (g) Find the angle between $P_0$ and $\hat{u} \simeq \hat{\nabla} T(P_0)$ using the cosine law.

## ■ 6.12 The Multivariable Newton's Method for Finding Zeros of a Function

Consider $\vec{g} : \mathbb{R}^m \to \mathbb{R}^m$ to be continuously differentiable in all components. When $\vec{g}(\vec{x}_*) = \vec{0}$ does not have an obvious solution, then if $\vec{x}_n$ is close to $\vec{x}_*$ and $\vec{x}_{n+1}$ is even closer, one can employ linearization of $\vec{g}$ at the point of expansion $\vec{x}_n$ to obtain

$$\vec{0} \approx \vec{g}(\vec{x}_{n+1}) \approx \vec{g}(\vec{x}_n) + (D\vec{g}(\vec{x}_n))(\vec{x}_{n+1} - \vec{x}_n). \tag{6.86}$$

Thus, if the $m \times m$ Jacobian matrix $D\vec{g}(x_*)$ is invertible then one can solve for $\vec{x}_{n+1}$ via the iterative process

$$\vec{x}_{n+1} \approx \vec{x}_n - (D\vec{g}(\vec{x}_n))^{-1} \vec{g}(\vec{x}_n). \tag{6.87}$$

## 6.12 The Multivariable Newton's Method for Finding Zeros of a Function

Experience shows that choosing a *relaxation parameter* $\lambda \in (0,1]$ and iterating under the scheme

$$\vec{x}_{n+1} \approx \vec{x}_n - \lambda (D\vec{g}(\vec{x}_n))^{-1} \vec{g}(\vec{x}_n) \tag{6.88}$$

may actually prevent *overshooting* the solution, and can result in faster convergence.

When the objective is to find an extreme point for $f \in C^2(\mathbb{R}^m \to \mathbb{R})$, then one needs to find a critical point $\vec{x}_*$ solving $\vec{g}(\vec{x}_*) = \vec{\nabla} f(\vec{x}_*) = \vec{0}$. In this case the Jacobian matrix $D\vec{g} = D(\vec{\nabla} f) = H[f]$ where $H[f]$ is the Hessian matrix of $f$. Then the iteration scheme becomes

$$\vec{x}_{n+1} \approx \vec{x}_n - \lambda \cdot (H[f](\vec{x}_n))^{-1} \cdot \vec{\nabla} f(\vec{x}_n), \tag{6.89}$$

assuming that the Hessian of $f(\vec{x})$ is invertible at $\vec{x}_*$, an extreme point of $f$.

### ■ Exercises

1. Consider the function $f(x,y) = x^2 + y^4$ with initial guess of $\vec{x}_0 = (1,2)^T$.
   (a) Compute $\vec{\nabla} f(\vec{x}_0)$, $H[f](\vec{x}_0)$, and $H[f](\vec{x}_0)^{-1}$.
   (b) Compute $\vec{x}_1$ using Eq. (6.89) with $\lambda = 1$.
   (c) Compute $\vec{x}_2$ with $\lambda = 1$ along with $\|\vec{x}_2 - \vec{x}_1\|$, $\|\vec{x}_1\|$, and $\|\vec{x}_2\|$.
   (d) Compute $\vec{x}_1$ using Eq. (6.89) with $\lambda = 1.5$.
   (e) Compute $\vec{x}_2$ with $\lambda = 1.5$ along with $\|\vec{x}_2 - \vec{x}_1\|$, $\|\vec{x}_1\|$, and $\|\vec{x}_2\|$.

# III Applications

In Part III mathematical techniques are used to understand scientific issues. Typical results from difference and differential equations, which connect mathematics to science, will be presented. Both quantitative and qualitative issues are considered.

Behind all modeling is the uncertainty associated with measurement and comprehension. Be that as it may, a meaningful philosophy, to understand the natural world, is developed as one extends themselves to new fields of interest.

Explaining new discoveries requires a framework of mathematics. Giving significance to new knowledge needs a foundation of science.

# 7 Mathematical Modeling

The process of mathematical modeling requires a consideration of the needs of those who will use a model. The specific behavior of a solution may be necessary, in which case using an algorithm to approximate the response of a system may provide the appropriate insight. However, there are instances where a more qualitative analysis in required. Does a system have stable equilibria? Are solutions bounded and do they oscillate? If a solution is unbounded, does it become infinite in finite time? Some of these are questions that can be answered, in part, with inequalities or probability measures, rather than a conclusive statement.

## Introduction

Mathematical concepts are a natural part of experience. Fundamental to that experience is counting, collecting, computing ratios, estimating likelihoods, making predictions, coming to conclusions, setting policies, and making sense of past errors. To even begin a process of analysis, relevant items need to be identified and named. In this way, the elements are *abstracted* and a lucid interaction between objects can take place in the unobstructed world of the mind. The problem with this world is that it often lacks rigor. This is the tradeoff one must deal with in order to be creative. The following is one way of understanding the process of mathematical modeling:

$$Variables \to$$
$$Hypothesize\ relationship \to Parameters \to Access\ model$$
$$\to Hypothesize\ additional\ relationships$$
$$\to Introduce\ more\ parameters \to Reevaluate\ new\ model\ldots$$

One may question the need to keep improving a mathematical model, however, this is a relative issue. Who should decide when a model is sufficient: a mathematician, a scientist, an engineer, an economist, a politician?

The theory of difference and differential equations and their applications is presented. The use of functions to model behavior is natural. One may consider the process of constructing a *model function* in terms of its form and parameters.

## ■ 7.1 Objectives of Mathematical Modeling

There are three basic goals of mathematical modeling:

(1) *Prediction (future)*  (2) *Cataloging (present)*  (3) *Postdiction (past)*

Different scientists will concentrate on different aspects related to the issue of time.

### ■ 7.1.1 Prediction

In a completely *deterministic* universe, the future behavior of a system depends on the present state of the system, if it is in equilibrium, and its evolution equations, if it is changing. A simple example is given by cell division:

- Let $c_n \in \mathbb{N}_0$ be the number of cells after the $n$th division.
- When a cell divides there is a mother cell and a daughter cell.
- A mother cell divides at regular intervals.
- A daughter cell needs one interval to mature into a mother cell before dividing.

Then if $c_{n+1}$ is the new number of cells, the *Fibonacci* evolution equation is

$$\text{new cell total} = c_{n+1} = c_n + c_{n-1} = \text{daughters and mothers}.$$

To analyze this equation, writing as a matrix system $\vec{c}_{n+1} = M \cdot \vec{c}_n$, gives

$$\begin{pmatrix} c_{n+1} \\ c_n \end{pmatrix} = \begin{pmatrix} 1 & 1 \\ 1 & 0 \end{pmatrix} \cdot \begin{pmatrix} c_n \\ c_{n-1} \end{pmatrix}, \ \vec{c}_0 \equiv \begin{pmatrix} c_1 \\ c_0 \end{pmatrix} = \begin{pmatrix} 0 \\ 1 \end{pmatrix}, \ M \equiv \begin{pmatrix} 1 & 1 \\ 1 & 0 \end{pmatrix}, \quad (7.1)$$

where $\vec{c}_0$ indicates that initially there were no mother cells and one daughter cell. In this special situation, the equation simplifies to $\vec{c}_n = M^n \cdot \vec{c}_0$. Consequently, $M$ is called the *propagation matrix* and each application represents a new generation.

### ■ 7.1.2 Cataloging

Before moving in time with our understanding of variables, the process of collecting and classifying data must be performed thoroughly. In biology, this involves creativity when a new species is found and must be named appropriately. Also, sufficient expertise is needed to know when a species has already been *identified* by others. Before any analysis can proceed, there has to be a process of labeling and naming.

To make any decisions on a system, one first needs to understand the state of the system. In the case of a patient complaining about a symptom(s), a primary care physician (PCP) needs to identify various possible medical conditions that would lead to the symptoms being reported, and then determine the most likely condition or medical state of the patient. This involves a *qualitative probability analysis* by the PCP, where the symptoms are responses of a patient's system. Once a diagnosis is made, a prognosis or prediction can be made, based on experience. A treatment, when available, involves changing the parameters of the patient's physiological and/or psychological systems. Only then is it possible to effect a change in the prognosis.

### ■ 7.1.3 Postdiction

At the present, our senses are telling us about the past. This is based on the assumption that the universe is completely *causal*, i.e., the past determines the present and future states. To understand what caused that which is being observed, we must be capable of evolving backward. In this way we can endeavor to *postdict the past* and know what made the dinosaurs become extinct, when life began, and how the universe was created. As an example, suppose that a large number of cells are counted $b_*$ and an estimate is to be made about how long ago the first cell began dividing. The corresponding equation to Eq. (7.1), as a matrix system, is

$$\begin{pmatrix} b_{-n+1} \\ b_{-n} \end{pmatrix} = \begin{pmatrix} 1 & 1 \\ 1 & 0 \end{pmatrix} \cdot \begin{pmatrix} b_{-n} \\ b_{-n-1} \end{pmatrix} \implies \begin{pmatrix} b_{-n} \\ b_{-n-1} \end{pmatrix} = \begin{pmatrix} 0 & 1 \\ 1 & -1 \end{pmatrix} \cdot \begin{pmatrix} b_{-n+1} \\ b_{-n} \end{pmatrix}$$

$$\implies \begin{pmatrix} b_0 \\ b_{-1} \end{pmatrix} = N \cdot \begin{pmatrix} b_1 \\ b_0 \end{pmatrix} \implies \begin{pmatrix} b_{-n-1} \\ b_{-n} \end{pmatrix} = N^n \cdot \begin{pmatrix} b_0 \\ b_{-1} \end{pmatrix}. \tag{7.2}$$

The goal is to find $n$ so that the components $b_i$ are as close to 0 as possible, without being negative. Furthermore, one expects an initial state with mostly daughters, expressed as $b_{-1} \geq b_0 \geq 0$. In this case, $N = M^{-1}$ is called the *back-propagation matrix* and each application postdicts the state of the previous generation.

### ■ Exercises

1. Use postdiction with matrix $N$ in Eq. (7.2) to find the number of steps $n$ that lead to the population vector $\vec{p} = (8\ 5)^T$.

2. Use postdiction as defined in Eq. (7.2) to find the number of steps $n$ that lead to the population vector $\vec{q} = (80\ 50)^T$ by first computing $N^2$, $(N^2)^2 = N^4$, and $(N^4)^2 = N^8$. It may also help to note that $M \cdot N = I$.

## 7.2 Difference Equations

The basic theme presented here is that from past knowledge, and present information, one obtains some ability to make predictions. These are limited by the accuracy of the parameters and the information gathered on the present and past states of the system. Thus there is a disintegration in the confidence of future estimates, which is reflected in an increased size in the standard deviations of the dependent variables. This can happen so fast that exact determinability becomes less important than knowing all the likely long-term behaviors of a system.

### 7.2.1 Change Versus Equilibrium

Most of this chapter will consider differential equations. The reason is that the derivative of a quantity $\vec{f}$ is a good model for its rate of change, denoted $\vec{f'}$. Choosing

| Differential Equations | Equations of Differentials | Difference Equations |
|---|---|---|
| $d\vec{f}/dt = \vec{F}(t,\vec{f})$ | $d\vec{f} = \vec{F}(t,\vec{f}) \cdot dt$ | $\Delta\vec{f} = \vec{F}(t,\vec{f}) \cdot \Delta t$ |

a small step-size $\Delta t > 0$, with $\Delta \vec{f} = \vec{f}_{i+1} - \vec{f}_i$, one obtains Euler's Iteration Method

$$\begin{aligned} IC \quad &: \quad (t_0, \vec{f}_0) = (t_0, \vec{f}(t_0)), \\ t_{i+1} &= t_i + \Delta t, \\ \vec{f}_{i+1} &= \vec{f}_i + \vec{F}(t_i, \vec{f}_i) \cdot \Delta t, \end{aligned} \tag{7.3}$$

where $(t_0, \vec{f}_0)$ is the *Initial Condition* (IC). For postdiction $\Delta \vec{f} = \vec{f}_{-i} - \vec{f}_{-i-1}$, so

$$\begin{aligned} t_{-i-1} &= t_{-i} - \Delta t, \\ \vec{f}_{-i-1} &= \vec{f}_{-i} - \vec{F}(t_{-i}, \vec{f}_{-i}) \cdot \Delta t. \end{aligned} \tag{7.4}$$

These systems can have many types of behaviors. In particular, one looks for stationary points, or *equilibria*, where $\vec{F}(t_*, \vec{f}_*) = \vec{0}$. Typically a function $\vec{f}$ will move away from $\vec{f}_*$ slowly as $t$ moves from $t_*$.[1] By fixing time $t = t_*$, one can use Newton's Iteration Method

$$\vec{f}_{i+1} = \vec{f}_i - \left[ D\vec{F}(t_*, \vec{f}_{-i}) \right]^{-1} \cdot \vec{f}_i, \text{ seed:} \vec{f}_0 \simeq \vec{f}_*, \tag{7.5}$$

---

[1] At equilibrium there is no change in $\vec{f}$, but natural variability is always present in real systems so $\vec{f}(t_*)$ actually expresses $\vec{f}(t)$ near $t_*$.

which, if this sequence converges, leads to a precise value for $\vec{f}_*$. Identifying and cataloging equilibria with regard to their stability properties allows an assessment of those points where the system will spend most if its time.

An understanding of the behavior of a system may require finding numerical solutions to $\vec{f}(t)$ near the equilibria $\{(t_*, \vec{f}_*)\}$. A system becomes interesting to study when we are able to observe dramatic changes as part of transitions from one equilibrium state to another.

## ■ Exercises

1. Consider the iterative pair of equations

$$x_{n+1} = 0.5 \cdot x_n + 0.8 \cdot y_n, \quad y_{n+1} = 0.4 \cdot x_n + 0.1 \cdot y_n.$$

   (a) Rewrite as a vector equation, and then as a matrix equation.
   (b) Find the eigenvalues for the propagating matrix.
   (c) For the negative eigenvalue, find the eigenvector of the form $\vec{v} = (1 \ b)^T$.
   (d) Using $\vec{v}$ as an initial condition obtain two iterations and compute the percentage decrease in the first component.

## ■ 7.3 First-Order Single-Variable Differential Equations

The simplest differential equation is the Initial Value Problem

$$\frac{df}{dt} = g(t), \quad f(t_0) = f_0, \tag{7.6}$$

or simply $f' = g$, which says that the rate of change of $f(t)$ equals $g(t)$. This equation has the solution, using the Fundamental Theorem of Calculus,

$$f(t) = f_0 + \int_{s=t_0}^{t} g(s)ds. \tag{7.7}$$

In order for Eq. (7.6) and its solution in Eq. (7.7) to be useful in an application one must have a method for measuring the rate of change of $f(t)$, which is $g(t)$. Suppose $g(t) = r_0$ corresponding to a constant rate of change. Then the solution is

$$f(t) = f_0 + r_0 \cdot (t - t_0).$$

This results in an *explicit* linear relationship between the dependent variable $f$ and the independent variable $t$. The quantities $f_0, r_0$, and $t_0$ are parameters of the model, from Eq. (7.6).

Next, suppose that the rate of change depends only on the dependent variable,

$$\frac{df}{dt} = F[f(t)], \quad f(t_0) = f_0. \tag{7.8}$$

Then the Fundamental Theorem of Calculus cannot be used directly. However, if the equation is flipped so that $t$ becomes the dependent variable and $f$ the independent variable, then

$$\frac{dt}{df} = \frac{1}{F[f(t)]} \quad \Longleftrightarrow \quad dt = \frac{df}{F(f)}, \tag{7.9}$$

with initial condition $t(f_0) = t_0$. The solution is

$$t(f) = t_0 + \int_{s=f_0}^{f} \frac{1}{F(s)} ds. \tag{7.10}$$

In this way $f$ is only defined implicitly as a function of $t$. Using an inverse function of the right-hand side of Eq. (7.10), $f(t)$ may be obtained explicitly. Theorem 6.6 gives conditions under which an inverse of $f$ can be found. This also allows approximations of the explicit solution.

### EXAMPLE 7.1

Let $P(t)$ represent the size of a population, and let $t$ be time. The Malthus model assumes

$$\frac{dP(t)}{dt} \propto P(t) \implies P'(t) = k \cdot P(t), \tag{7.11}$$

where $k$ is a parameter that represents the rate of population growth. Typically one finds that an environment, with only one main species, can be classified as

$$k < 0 \text{ for predators}, \quad k = 0 \text{ in equilibrium}, \quad k > 0 \text{ for prey}. \tag{7.12}$$

The behaviors of the solutions to these three models are quite different.

### ■ Exercises

1. Use the flipping technique to solve $P' = 0.2 \cdot P$, with $P(2000) = 6$. Be sure to express $P(t)$ explicitly as a function of $t$.

2. Use the flipping technique to solve $P' = 0.2 \cdot P \cdot (1 - P/10)$, with $P(2000) = 6$. (*Hint:* It will help to know that $[P \cdot (1 - P/10)]^{-1} = [P]^{-1} + [10 - P]^{-1}$.)

## 7.4 First-Order Linear Differential Equations with Forcing

The growth rate constant $k$ of a species is an important characteristic that is often determined by measurement. The Malthus population model can then be expressed in its homogeneous form,

$$\frac{dP}{dt} - k \cdot P(t) = 0 \implies P(t) = c \cdot e^{k \cdot t}, \tag{7.13}$$

where $c \in \mathbb{R}$ is a parameter that represents the population at $t = 0$. The population grows if $k > 0$, disappears if $k < 0$, and stays constant if $k = 0$. Now suppose that a stable constant population is desired but $k \neq 0$. If $k > 0$, a *harvest* rate $h > 0$ can be imposed, but if $k < 0$, a *restocking* rate $-h > 0$ is required. Both cases are expressed as

$$\frac{dP}{dt} - k \cdot P(t) = -h \implies P(t) = \frac{h}{k} + c \cdot e^{k \cdot t}. \tag{7.14}$$

When $k, h < 0$, then an asymptotic population of $P(\infty) = (h/k) > 0$ is reached, theoretically. However, for exponentially growing populations $k > 0$, with $c > 0$, a constant harvest or *extermination* rate is clearly not sufficient to control the population. One approach is to choose a harvest rate that depends on the population size $h = h(P) > 0$. In this case, conditions on $h(P)$ can be computed.

(1) Find $h(P)$ so that $P(\infty) = M > 0$, where $M$ is the target population.
(2) Find $h(P)$ so that $P = M$ is a stable equilibrium point.

These lead to the two mathematical conditions:

(1) $0 = k \cdot M - h(M)$; and
(2) $0 > k \cdot 1 - h'(M)$.

A family of acceptable models are given by $\{h(P) \equiv \beta \cdot P^m : \beta > 0, m > 0\}$, where $\beta$ and $m$ are new parameters that must be determined by the two conditions:

(1) $h(M) \equiv \beta \cdot M^m = k \cdot M \implies \ln(\beta) + (m-1) \cdot \ln(M) = \ln(k)$; and
(2) $k < \beta \cdot m \cdot M^{m-1} \equiv h'(M) \implies \ln(\beta) + \ln(m) + (m-1) \cdot \ln(M) > \ln(k)$.

Combining these two equations gives $\ln(m) > 0$, which implies $m > 1$. The family of rate of change models

$$\frac{dP(t)}{dt} - k \cdot P(t) = -\beta \cdot P^m(t), \tag{7.15}$$

are called *Bernoulli Equations*. More general cases are discussed in [23].

### ■ 7.4.1 Forced Linear Equations

A *monic* differential equation is one where the highest derivative has a 1 as a coefficient. For example, consider the monic, forced linear equation,

$$\frac{dy}{dt} + a \cdot y(t) = f(t), \ y(t_0) = y_0 \in \mathbb{R}, \tag{7.16}$$

where $f(t)$ is the external forcing, $a$ is the rate of change parameter for the system, and $y(t)$ is the solution, which represents the changing output of the system. The output $y(t)$ is desired for time $t > t_0$ where the initial value is $y_0$. To obtain a solution of Eq. (7.16) suppose that the left-hand side originated from a Product Rule. This is a mathematical consideration, and may have nothing to do with the scientific origin of the equation. To succeed with this idea, consider multiplying both sides of Eq. (7.16) by an *integrating factor* $\mu(t) = \exp(a \cdot t)$, which results in

$$\mu \cdot y' + a \cdot \mu \cdot y \ = \ \mu \cdot y' + \mu' \cdot y \ = \ (\mu \cdot y)' \ = \ \mu \cdot f. \tag{7.17}$$

Now both sides can be integrated, using the initial conditions $y(t_0) = y_0$, to obtain

$$\int_{s=t_0}^{t} ((\mu \cdot y)' \ = \ \mu \cdot f) \, ds \ \implies \ \mu(t) \cdot y(t) - \mu(t_0) \cdot y_0 \ = \ \int_{s=t_0}^{t} \mu(s) \cdot f(s) ds.$$

Multiplying both sides by the reciprocal of the integrating factor $\mu^{-1} = \exp(-a \cdot t)$ and rearranging, gives

$$y(t) \ = \ e^{-a \cdot t} \cdot e^{a \cdot t_0} \cdot y_0 \ + \ e^{-a \cdot t} \cdot \int_{s=t_0}^{t} e^{a \cdot s} \cdot f(s) ds. \tag{7.18}$$

Then using properties of the exponential function,

$$y(t) \ = \ e^{-a \cdot (t-t_0)} \cdot y_0 \ + \ \int_{s=t_0}^{t} e^{-a \cdot (t-s)} \cdot f(s) ds. \tag{7.19}$$

If $a > 0$ then the effect of the initial condition $y_0$ diminishes as $t > t_0$ increases, and eventually only the recent forcing $f(s \lesssim t)$ will effect the behavior of $y(t)$.

## 7.4.2 Variable Coefficients

When coefficients of differential equations vary as functions of the independent variable, solutions are more difficult to find. However, in the case of first-order linear equations, an integrating factor can be found systematically. Consider the equation

$$\frac{dy}{dt} + p(t) \cdot y(t) = q(t), \quad y(t_0) = y_0. \tag{7.20}$$

Then, inserting $p(t)$ and $q(t)$ into Table 7.1 results in the function $I(t)$.

| Parameter Functions | $p = p(t)$ | $q = q(t)$ | New Integrand |
|---|---|---|---|
| New Parameters | $H(t) = \int p(t)\,dt$ | $I(t) = \exp[H(t)]$ | $I(t) \cdot q(t)$ |

**TABLE 7.1** ■ The Integrating Factor Method for $y' + p(t) \cdot y = q(t)$.

The solution can be directly written, similar to Eq. (7.19), as

$$y(t) = \frac{I(t_0)}{I(t)} \cdot y_0 + \frac{1}{I(t)} \cdot \int_{s=t_0}^{t} I(s) \cdot q(s)\,ds. \tag{7.21}$$

To be solvable, the integrating factor $I(t)$ must not vanish. This requires that $H(t)$, which is an antiderivative of $p(t)$, remain finite.

### EXAMPLE 7.2

For a parameter $a > 0$, the equation $y' + a \cdot y = 0$, with $y(t_0) = y_0$ has the solution $y(t) = e^{-a(t-t_0)}y_0$, which vanishes at $t \to \infty$. For a parameter $b > 0$, the equation

$$y' + a \cdot y = b, \quad y(t_0) = y_0, \tag{7.22}$$

can be solved using the substitution $y = u + b/a$, which leads to the equation $u' + a \cdot u = 0$, with $u(t_0) = y_0 - b/a$ with solution $u(t) = e^{-a(t-t_0)}(y_0 - b/a)$. Converting back to the $y$-variable gives the solution to Eq. (7.22) to be

$$y(t) = e^{-a(t-t_0)}(y_0 - b/a) + b/a = (b/a) \cdot \left(1 - e^{-a(t-t_0)}\right) + e^{-a(t-t_0)} \cdot y_0. \tag{7.23}$$

Thus $y(t)$ converges to $b/a$ as $t \to \infty$. This can be obtained from Eq. (7.22) by setting $y' = 0$, corresponding to an equilibrium, or stationary point for the system in Eq. (7.22). If $a > 0$, the system will tend toward its equilibrium.

## Exercises

1. Solve $P' + 2 \cdot P = 6 \cdot P^3$ with $P(0) = 1$, using the change of variables $u = P^{-2}$. Does the solution approach an equilibrium as $t \to \infty$?

2. Use Eq. (7.19) to obtain solution (7.23) from Eq. (7.22).

3. Solve the following separable initial value problems (IVPs):

    (a) $dy/dt = 3 \cdot t^2 - 4 \cdot t + 5$, where $y(0) = 1$ (explicit).
    (b) $dy/dt = 7 \cdot y$, where $y(0) = 8$ (autonomous).
    (c) $dP/dt = 5 \cdot t^4 \cdot P^2$, where $P(0) = 2$ (separable).

4. Solve the following linear nonhomogeneous IVP using an integrating factor: $dy/dt - 3 \cdot y = e^{3 \cdot t} + 6$, where $y(0) = 2$.

5. Suppose a forcing $f(t) = \beta \cdot e^{b \cdot t}$ is applied in Eq. (7.16) with $b > 0$, $\beta > 0$, and $a > -b$.

    (a) For $y_0 \neq 0$ and $t_0 = 0$, solve Eq. (7.16) explicitly to obtain $y(t)$ by computing the integral in Eq. (7.18) and then simplifying.
    (b) Compute the limit: $\lim_{t \to \infty} y(t)/f(t)$ and comment on when the limit $\nexists$.
    (c) When $a = -b$, solve Eq. (7.19) and explain why the forcing $f(t)$ still dominates over any initial condition $y_0$.
    (d) For $y_0 = 1$, $a = 1$, $b = 1$, and $\beta = 1$ plot the solution $y(t)$ for $t \in [0, 5]$.

6. Suppose a forcing $f(t) = C \cdot \sin(\omega \cdot t)$ is applied in Eq. (7.16) with $\omega > 0$, $C > 0$, and $a > 0$.

    (a) For $y_0 \neq 0$ and $t_0 = 0$, solve Eq. (7.16) explicitly to obtain $y(t)$ by computing the integral in Eq. (7.18) using integration by parts and simplifying.
    (b) If there is no forcing, what does the solution reduce to?
    (c) For large $\omega > 0$ find an approximate value $t_2$ corresponding to when the effect of the forcing $f(t)$ in the solution $y(t)$ is twice the effect of the initial condition $y_0$ in $y(t)$. (*Hint:* Set $|\sin(\omega \cdot t)| \simeq 1$ and $|\cos(\omega \cdot t)| \simeq 1$.)
    (d) As $a > 0$ increases, does $t_2$ increase or decrease? What does this say about a large $a > 0$ system's responsiveness to external forcing?
    (e) For $y_0 = 0$, $a = 1$, $C = 2$, and $\omega = \pi$, solve Eq. (7.19) and plot $y(t)$ for $t \in [0, 10]$.

## 7.5 Constant-Coefficient Systems of Linear Differential Equations

When the rate of change of several dependent variables depend on each other, a straightforward solution is typically elusive. If the rates of change are linear in the

## 7.5 Constant-Coefficient Systems of Linear Differential Equations

dependent variables, then we obtain a linear system of equations

$$\frac{d\vec{P}(t)}{dt} = K \cdot \vec{P}(t), \ \text{IC}: \vec{P}(t_0) = \vec{P}_0 \in \mathbb{R}^n. \tag{7.24}$$

The components of the Initial Condition (IC) vector $\vec{P}_0$, and the constant components of the square-matrix $K$ are obtained by measurement. Then, similar to Eq. (7.13), the solution can be expressed as

$$\vec{P}(t) = e^{K \cdot t} \cdot \vec{P}_0, \tag{7.25}$$

where the vector $\vec{P}_0$ replaces $c$ as the initial condition, and the *generator matrix* $K$ replaces the growth rate constant $k$. Also, $\vec{P}_0$ is written after the exponential since $e^{K \cdot t}$ acts as an *evolution operation* pushing $\vec{P}_0$ to $P(t)$, and satisfying the matrix Initial Value Problem (IVP)

$$\frac{d}{dt} e^{K \cdot t} = K \cdot e^{K \cdot t}, \ e^{K \cdot 0} = e^{0_{n \times n}} = I_{n \times n}. \tag{7.26}$$

The solution is obvious from the formal series expansion

$$e^{K \cdot t} = \sum_{n=0}^{\infty} \frac{1}{n!}(K \cdot t)^n = I + t \cdot K + \frac{t^2}{2} \cdot K^2 + \frac{t^3}{6} \cdot K^3 + \mathcal{O}(t^4). \tag{7.27}$$

This expression is useful for small values of $t$, but to determine the long-term behavior of a system, it is far more insightful to diagonalize $K$, if possible. Suppose there exists an $n \times n$ invertible matrix $S$ so that

$$S \cdot K \cdot S^{-1} = \text{diag}\{\lambda_1, \lambda_2, \ldots, \lambda_n\} = D, \ \hat{e}_j = S \cdot \vec{v}_j,$$

where $\{\lambda_j\}_{j=1}^n$ are the eigenvalues of $K$ with eigenvectors $\{\vec{v}_j\}_{j=1}^n$. If $K$ is not defective, then the construction $S^{-1} \equiv (\vec{v}_1, \vec{v}_2, \ldots, \vec{v}_n)$ gives a diagonalization for $K$. The evolution operator can then be expressed exactly as

$$\begin{aligned}
e^{K \cdot t} &= S^{-1} \sum_{n=0}^{\infty} \frac{t^n}{n!} S \cdot K^n \cdot S^{-1} \cdot S = S^{-1} \sum_{n=0}^{\infty} \frac{t^n}{n!} (S \cdot K \cdot S^{-1})^n \cdot S \\
&= S^{-1} \sum_{n=0}^{\infty} \frac{1}{n!} D^n \cdot t^n \cdot S = S^{-1} \cdot e^{D \cdot t} \cdot S \\
&= S^{-1} \cdot \text{diag}\{e^{\lambda_1 \cdot t}, e^{\lambda_2 \cdot t}, \ldots e^{\lambda_n \cdot t}\} \cdot S. \tag{7.28}
\end{aligned}$$

The advantage of this form is that only three basic computations are needed to solve the IVP:

(1) multiplication of $\vec{P}_0$ by $S$;
(2) multiplication by the diagonal evolution operator; and
(3) multiplication by $S^{-1}$.

To obtain a general solution one can form the fundamental set $\mathcal{F}$ and corresponding fundamental matrix $M(t)$ of vector eigenstates of Eq. (7.24),

$$\mathcal{F} \equiv \{e^{\lambda_j \cdot t} \cdot \vec{v}_j\}_{j=1}^{n}, \quad M(t) \equiv \left(e^{\lambda_1 \cdot t} \cdot \vec{v}_1,\ e^{\lambda_2 \cdot t} \cdot \vec{v}_2,\ \ldots,\ e^{\lambda_n \cdot t} \cdot \vec{v}_n\right).$$

Then the evolution operator, or *transition matrix* [23], is constructed from the fundamental matrix using $e^{K \cdot t} = M(t) \cdot M(0)^{-1}$. The long-term behavior of solutions to Eq. (7.24) depend on the eigenvalues of $K$. In particular, let

$$\Lambda \equiv \max\{\Re(\lambda_j) : j = 1\ldots n\}, \quad \Omega \equiv \max\{\Im(\lambda_j) : j = 1\ldots n\} \tag{7.29}$$

where $\lambda_j = \Re(\lambda_j) + i \cdot \Im(\lambda_j) \in \mathbb{C}$. If $\Lambda > 0$, the system has an *instability*, so that, depending on the IC, a solution will likely have a component that grows exponentially. Otherwise, if $\Lambda \leq 0$, all components remain bounded. In particular, for the vector function in Eq. (7.25),

$$\Lambda < 0 \implies \lim_{t \to +\infty} \vec{P}(t) = \vec{0},$$

in which case $\vec{0}$ is considered to be a stable equilibrium point. In addition, if $\Omega > 0$ then the solutions in Eq. (7.25) of Eq. (7.24) may circle the equilibrium. The actual path may spiral toward ($\Lambda < 0$), or away ($\Lambda > 0$), from the origin. The eigenvalues, and in particular $\Lambda$ and $\Omega$, are characteristics of the system that depend on its parameters, in particular $K$ and $\vec{P}_0$.

### EXAMPLE 7.3

Let $M = \begin{pmatrix} a & b \\ c & d \end{pmatrix}$. The trace is $\mathcal{T} = a + d$ and the determinant is $\mathcal{D} = ad - bc$. The eigenvalues $\lambda$ satisfy $\lambda^2 - \mathcal{T}\lambda + \mathcal{D} = 0$, which, by the quadratic formula, implies

$$\lambda = \frac{\mathcal{T} \pm \sqrt{\mathcal{T}^2 - 4 \cdot \mathcal{D}}}{2} = \frac{a + d \pm \sqrt{(a-d)^2 + 2 \cdot b \cdot c}}{2}. \tag{7.30}$$

If $\mathcal{D} > 0$ and $\mathcal{T} < 0$, then the system satisfies the equivalent Routh-Hurwitz stability conditions [35]

$$a \cdot d - b \cdot c > 0, \ a + d < 0. \tag{7.31}$$

In this case, $\Lambda = \mathcal{T}/2 < 0$ if $\mathcal{T}^2 \leq 4\mathcal{D}$ and $2 \cdot \Lambda = \mathcal{T} + \sqrt{\mathcal{T}^2 - 4\mathcal{D}} < 0$ if $\mathcal{T}^2 \geq 4\mathcal{D}$. Conversely, $\mathcal{D} < 0 \implies \Lambda \geq -\mathcal{D} > 0$, whereas $\mathcal{T} > 0 \implies \Lambda \geq \mathcal{T}/2 > 0$. In both these cases the system is unstable. Furthermore, the axes $\mathcal{D} = 0$ or $\mathcal{T} = 0$ represent potential instabilities for the system.

The two-dimensional systems in Eq. (7.24) are completely understood, with the help of the characteristics of $K$ in Eq. (7.29) and conditions like Eq. (7.31) or as in Table 7.2. However, there are subtleties even for these systems.

| Parameter Conditions | Characteristics $(\Lambda, \Omega)$ | Classification |
|---|---|---|
| $\mathcal{T} = 0 \wedge \mathcal{D} > 0$ | $(0, +/-)$ | Circle (bounded) |
| $\mathcal{D} < 0$ | $(+, 0)$ | Saddle (unstable) |
| $\mathcal{T} > 0 \wedge \mathcal{T}^2 > 4 \cdot \mathcal{D}$ | $(+, 0)$ | Source (unstable) |
| $\mathcal{T} < 0 \wedge \mathcal{T}^2 > 4 \cdot \mathcal{D}$ | $(-, 0)$ | Sink (stable) |
| $\mathcal{T} > 0 \wedge \mathcal{T}^2 < 4 \cdot \mathcal{D}$ | $(+, +/-)$ | Source (unstable) |
| $\mathcal{T} < 0 \wedge \mathcal{T}^2 < 4 \cdot \mathcal{D}$ | $(-, +/-)$ | Sink (stable) |

**TABLE 7.2** ■ Table of different $2 \times 2$ matrix generators.

### EXAMPLE 7.4

Consider the borderline case $\mathcal{T}^2 = 4\mathcal{D}$, or in terms of the parameters, $(a-d)^2 = 4 \cdot b \cdot c$. Then the only eigenvalue is $\lambda = \Lambda = \mathcal{T}/2$, which has an algebraic multiplicity of 2. To find the geometric multiplicity, the eigenvector equation $K \cdot \vec{v} = \lambda \cdot \vec{v}$ for $\vec{v} = (v_1 \ v_2)^T$ must be solved:

$$\begin{pmatrix} a & b \\ c & d \end{pmatrix} \cdot \begin{pmatrix} v_1 \\ v_2 \end{pmatrix} = \frac{a+d}{2} \cdot \begin{pmatrix} v_1 \\ v_2 \end{pmatrix}, \tag{7.32}$$

which implies the $2 \times 2$ linear system of rank 1,

$$\begin{cases} a \cdot v_1 + b \cdot v_2 = (a+d)/2 \cdot v_1 \\ c \cdot v_1 + d \cdot v_2 = (a+d)/2 \cdot v_2 \end{cases} \implies \begin{cases} (a-d) \cdot v_1 = -2 \cdot b \cdot v_2, \\ 2 \cdot c \cdot v_1 = (a-d) \cdot v_2. \end{cases}$$

If $a = d$ then $b \cdot c = 0$ so $\lambda = a$. Furthermore, if $b = 0$ but $c \neq 0$, then $v_1 = 0$ must hold, which implies that the set of all eigenvectors is $E_a^{K_{a,c}}$, a one-dimensional subspace of $\mathbb{R}^2$,

$$K_{a,c} = \begin{pmatrix} a & 0 \\ c & a \end{pmatrix} \implies \sigma(K_{a,c}) = \{a\} \implies E_a^{K_{a,c}} = \left\{ \begin{pmatrix} 0 \\ v_2 \end{pmatrix} : v_2 \in \mathbb{R} \right\}. \quad (7.33)$$

Since the algebraic multiplicity of $\lambda = T/2 = a$ is 2, and the geometric multiplicity is 1, $K_{a,c}$ is defective and so cannot be diagonalized to obtain a simple form for the evolution operator $e^{K_{a,c} \cdot t}$.

### ■ 7.5.1 Forced Constant-Coefficient Linear-Vector Equations

Recall, a *monic* differential equation is one where the highest derivative has a 1 as its coefficient. In general, let $A$ be an $n \times n$ parameter matrix with constant components. An associated $n$-dimensional, monic, forced linear-vector equation, is

$$\frac{d\vec{Y}}{dt} - A \cdot \vec{Y}(t) = \vec{f}(t), \quad \text{IC}: \vec{Y}(t_0) = \vec{Y}_0 \in \mathbb{R}^n \quad (7.34)$$

where $\vec{f}(t) = (f_1(t), f_2(t), \ldots, f_n(t))^T$ is the external forcing. The vector solution $\vec{Y}(t) = (y_1(t), y_2(t), \ldots, y_n(t))^T$ represents the evolving *state of the system* for time $t > t_0$, where the initial state is $\vec{Y}_0$. To move toward a solution of Eq. (7.34) it is useful to consider the left-hand side as having been derived from a Product Rule, motivating the search for an $n \times n$-*integrating factor*. Consider multiplying both sides of Eq. (7.34) on the left by the factor $\exp(-A \cdot t)$. Then the system becomes

$$\frac{d}{dt}\left[e^{-A \cdot t} \cdot \vec{Y}(t)\right] = e^{-A \cdot t} \cdot \vec{f}(t), \quad (7.35)$$

which can be integrated from $t_0 \to t$, using the IC: $\vec{Y}(t_0) = \vec{Y}_0$, to obtain

$$e^{-A \cdot t} \cdot \vec{Y}(t) - e^{-A \cdot t_0} \cdot \vec{Y}_0 = \int_{s=t_0}^{t} e^{-A \cdot s} \cdot \vec{f}(s) ds. \quad (7.36)$$

In this context, the matrix *evolution operator* $\exp(-A \cdot t)$ is called an *integrating factor* since it allowed an integration of the differential equation (7.34). Using matrix algebra, which states that

$$\left[e^{-A \cdot t}\right]^{-1} = e^{A \cdot t} \implies e^{-A \cdot t} \cdot e^{A \cdot t} = e^{0_{n \times n}} = I_{n \times n}, \quad (7.37)$$

one can multiply Eq. (7.35) by the inverse of the integrating factor to obtain

$$\vec{Y}(t) = e^{A \cdot (t-t_0)} \cdot \vec{Y_0} + \int_{s=t_0}^{t} e^{A \cdot (t-s)} \cdot \vec{f}(s) ds. \qquad (7.38)$$

Such formulae represent a mixture of matrix algebra and integral calculus.

### ■ 7.5.2 Linear Electrical Circuits

A closed electrical circuit consists of a forcing voltage $E(t)$, also called an *electromotive force* (EMF), which drives a current $I(t)$ that results in a corresponding buildup of charges $Q(t)$ at various fixed points, or components, in the circuit. The rate of charge buildup across a component is $dQ/dt \equiv Q'(t) = I(t)$, and the response voltage is $V(t)$. There are three electric-circuit components that respond linearly to the time-dependent variables $Q$, $I$, and $V$:

$$Q(t) = C \cdot V(t), \quad I(t) = \frac{1}{R} \cdot V(t), \quad I'(t) = \frac{1}{L} \cdot V(t). \qquad (7.39)$$

The parameters $C$, $R$, and $L$ are constants of proportionality between charge, current, and voltage. These are linear idealizations of real components in a circuit:

**Resistor:** $(1/R)$ is conductance, where $R \geq 0$ is the resistance;
**Capacitor:** $C \geq 0$ is the capacitance; and
**Inductor:** $L \geq 0$ is the inductance.

Measuring the charge buildups due to Eq. (7.39), and using Kirchhoff's law of *conservation of charge*, gives the equation for an $LRC$ circuit,

$$V_C + V_R + V_L + EMF \equiv \frac{1}{C} \cdot Q + R \cdot Q' + L \cdot Q'' + E(t) = 0. \qquad (7.40)$$

Rearranging Eq. (7.40) gives the monic second-order linear differential equation

$$Q'' + a \cdot Q' + b \cdot Q = f(t), \qquad (7.41)$$

where $a \equiv R/L$, $b \equiv 1/(L \cdot C)$, and $f(t) \equiv -E(t)/L$.

If an EMF is applied to the system in Eq. (7.40), and then removed, the charges will disperse until equilibrium $Q = 0$ is attained. This is a certainty because $R$, $C$, and $L$ are always at least slightly positive for any electrical circuit. However, in the ideal resistance-free case that $R = 0$, corresponding to $a = 0$ in Eq. (7.41), one can suppose that $b \equiv 1/(L \cdot C) = \omega^2 > 0$. Then a general solution to Eq. (7.41) for $f = 0$ is given by a harmonic oscillator model function

$$Q(t) = \alpha \cdot \cos(\omega \cdot t) + \beta \cdot \sin(\omega \cdot t), \qquad (7.42)$$

where the amplitude of oscillation is $\mathcal{A} \equiv \sqrt{\alpha^2 + \beta^2}$, by Pythagorean's theorem.[2] Also, since $\sin(t)$ and $\cos(t)$ have period $2\pi$, the period for $Q(t)$ is $T = 2\pi/\omega = 1/\nu$. Thus the system in Eq. (7.41), without any vibratory forcing, will have a natural tendency to oscillate.

| Mathematics | Physics |
|---|---|
| Angular frequency | Natural frequency |
| $\omega = 1/\sqrt{L \cdot C} = 2\pi \cdot \nu$ | $\nu = (2\pi \cdot \sqrt{L \cdot C})^{-1} = \omega/(2\pi)$ |

The point here is that $\nu$ is a measurable quantity, whereas $\omega$ is derivable from an idealized model. If a forcing is imposed on Eq. (7.41) with the same *natural resonant frequency* $\nu = \omega/(2\pi)$ as in Eq. (7.42), then $Q(t)$ will absorb the input, and its amplitude $\mathcal{A}$ will grow. This can clearly be seen in experiments.

### EXAMPLE 7.5

Consider the equation, for parameter $\omega > 0$,

$$Q'' + 2 \cdot \omega \cdot Q' + \omega^2 \cdot Q = 0, \text{ IC}: Q(0) = -1, Q'(0) = 2.$$

Then define the new variable $P \equiv Q'$. This implies that $P' \equiv Q''$ giving the equivalent equation

$$P' + 2 \cdot \omega \cdot P + \omega^2 \cdot Q = 0.$$

This can now be expressed as a $2 \times 2$ system of equations,

$$\begin{pmatrix} Q \\ P \end{pmatrix}' = \begin{pmatrix} 0 & 1 \\ -\omega^2 & -2 \cdot \omega \end{pmatrix} \cdot \begin{pmatrix} Q \\ P \end{pmatrix}, \text{ IC}: \begin{pmatrix} Q(0) \\ P(0) \end{pmatrix} = \begin{pmatrix} -1 \\ 2 \end{pmatrix}. \qquad (7.43)$$

Since $\mathcal{T}^2 = (-2 \cdot \omega)^2 = 4 \cdot \omega^2 = 4\mathcal{D}$, then Example 7.4 and Eq. (7.30) give only one eigenvalue $\lambda = -\omega$. A corresponding eigenvector is $\vec{v}_\omega = (1, -\omega)^T$ and a solution to Eq. (7.43) is $\vec{Q}_1(t) = \mathcal{A}e^{-\omega t}(1, -\omega)^T$. However, this solution cannot satisfy the initial conditions for any value of $\mathcal{A}$ unless $\omega = 2$.

---

[2]General properties of combinations of oscillatory functions are presented in [17].

## Exercises

1. Consider the IVP: $\vec{Y}' = K_{2,3} \cdot \vec{Y}$, with IC: $\vec{Y}(0) = (-1\ 1)^T$, for $K_{2,3}$ from Example 7.4 where in Eq. (7.33) choose $a = 2$ and $c = 3$.

   (a) Show that $\vec{Y}_1(t) = (0\ e^{2 \cdot t})^T$ solves the differential equation.
   (b) Show that $\vec{0} = (K_{2,3} - 2 \cdot I)^2 \cdot \vec{v}$ for all $\vec{v} \in \mathbb{R}^2$.
   (c) Show that $\vec{Y}_2(t) = (1\ 3 \cdot t)^T \cdot e^{2 \cdot t}$ solves the differential equation.
   (d) Use the linear combination $\vec{Y} = \alpha \cdot \vec{Y}_1 + \beta \cdot \vec{Y}_2$ to solve the IVP for $x(t)$ and $y(t)$ by using the IC to solve for $\alpha$ and $\beta$.

2. Rewrite the system of equations $x' = 4 \cdot y - 2 \cdot x + 8 \cdot t$ and $y' = -x + 3 \cdot y - 4 \cdot t$ as a first-order system Eq. (7.34), with initial conditions that $x(0) = -2$ and $y(0) = 1$. For the parameter matrix $A$ find the eigenvalues, eigenvectors, the diagonalizing matrix $S$, and the evolution operator $\exp(A \cdot t)$. Then express the solution in the form of an integral, as in Eq. (7.38), integrate, simplify, and find $x(t)$ and $y(t)$.

3. Show that the maximum and minimum of $Q(t)$ in Eq. (7.42) occur when $\tan(\omega \cdot t) = \beta/\alpha$. Then use the alternate form of Pythagorean's theorem $\tan^2(\theta) + 1 = \sec^2(\theta)$ to show that $max|Q(t)| = \sqrt{\alpha^2 + \beta^2} = \mathcal{A}$.

4. Consider the system in Eq. (7.41) with $a = 0$ and $b = 4$ where a forcing $f(t) = \cos(2 \cdot t)$ is applied to Eq. (7.41).

   (a) Show that a solution of Eq. (7.42) cannot work, even with $\omega = 2$.
   (b) Assume a solution of the form $Q(t) = \mathcal{A} \cdot t \cdot \sin(2 \cdot t)$. Show that this satisfies the IC: $Q(0) = 0$, $Q'(0) = 0$.
   (c) Find the value of $\mathcal{A}$ that will solve the forcing problem.
   (d) Draw the functions $\pm |t/4|$ on the interval $[0, 2\pi]$. Draw, on the same diagram, the solution $Q(t)$.

5. Complete Example 7.5 using the following steps:

   (a) Verify that $\vec{v} = (1\ b)^T$ is an eigenvector of $\begin{pmatrix} 0 & 1 \\ -\omega^2 & -2 \cdot \omega \end{pmatrix}$ only if $b = -\omega$.
   (b) Show that $\vec{Q}_2(t) = \mathcal{B} \cdot e^{-\omega t} \cdot \left( (1 - \omega t)\ (-2\omega + \omega^2 t) \right)^T$ also solves Eq. (7.43).
   (c) Find $\mathcal{A}$ and $\mathcal{B}$ so that $\vec{Q}_1(t) + \vec{Q}_2(t)$ solves the IC of Eq. (7.43). Write the vector solution as simply as possible.
   (d) For $\omega = 1$, draw the functions $Q(t)$ and $P(t)$, which are the first and second components of $\vec{Q}_1(t) + \vec{Q}_2(t)$, on the same diagram, for $t \in [0, 3]$.

6. Consider the following $2 \times 2$ matrices,

$$K_1 = \begin{pmatrix} -2 & -3 \\ 3 & -2 \end{pmatrix}, \quad K_2 = \begin{pmatrix} -2 & -1 \\ 1 & -4 \end{pmatrix}, \quad K_3 = \begin{pmatrix} 0 & 1 \\ -4 & 5 \end{pmatrix},$$

$$K_4 = \begin{pmatrix} 0 & 1 \\ -10 & -7 \end{pmatrix}, \quad K_5 = \begin{pmatrix} 3 & 2 \\ 0 & -2 \end{pmatrix}, \quad K_6 = \begin{pmatrix} 1 & -6 \\ 2 & 1 \end{pmatrix}.$$

   (a) Find the pair $(\mathcal{T}_j, \mathcal{D}_j)$ for each matrix $K_j$.
   (b) Find the set of eigenvalues $\sigma(K_j)$ for each $j \in \mathbb{Z}_6$.
   (c) Find the pair $(\Lambda_j, \Omega_j)$ for each matrix $K_j$ from the set of eigenvalues.
   (d) The linear systems $\vec{P}\,' = K_j \cdot \vec{P}$ have an equilibrium point at $(0\ 0)^T$. Use the terminologies, defined in Eq. (7.2), to qualitatively classify this point for each $K_j$ based on $\sigma(K_j)$.

7. Solve the system $\vec{P}\,' = \begin{pmatrix} -1 & -1 \\ 1 & -1 \end{pmatrix} \vec{P}$ with IC: $\vec{P}_0 = \begin{pmatrix} 2 \\ 1 \end{pmatrix}$ using the following steps:

   (a) For $K \equiv \begin{pmatrix} -1 & -1 \\ 1 & -1 \end{pmatrix}$ find the trace $\mathcal{T}$ and determinant $\mathcal{D}$.
   (b) Find the eigenvalues $\lambda_1$ and $\lambda_2$ of $K$.
   (c) Find the eigenvectors $\vec{v}_1$ and $\vec{v}_2$.
   (d) Create the fundamental set $\{e^{\lambda_1 t} \cdot \vec{v}_1, e^{\lambda_2 t} \cdot \vec{v}_2\}$.
   (e) Create the fundamental matrix $M(t) \equiv (e^{\lambda_1 t} \vec{v}_1\ e^{\lambda_2 t} \vec{v}_2)$.
   (f) Compute $M(0)$ and its inverse matrix $M(0)^{-1}$.
   (g) Construct and simplify the transition matrix $M(t) \cdot M(0)^{-1}$. Using Euler's formula, the final form should have no imaginary part.
   (h) Check that the transition matrix satisfies the same properties as the evolution operator $\exp(K \cdot t)$ in Eq. (7.26).
   (i) Now construct and simplify the solution $\vec{P}(t) = e^{K \cdot t} \vec{P}_0$.

8. Consider the system of linear differential equations:

$$x' = -2y, \quad y' = x + 3y. \tag{7.44}$$

   (a) Write Eq. (7.44) as a first-order $2 \times 2$ matrix system: $\begin{pmatrix} x \\ y \end{pmatrix}' = K \begin{pmatrix} x \\ y \end{pmatrix}$.
   (b) Find the trace $\mathcal{T}$ and determinant $\mathcal{D}$ of the matrix $K$.
   (c) Write and solve the characteristic equation for $K$: $\lambda^2 - \mathcal{T}\lambda + \mathcal{D} = 0$.
   (d) Using the eigenvalues, find the two eigenvectors of the matrix $K$.
   (e) Write the general solution of the linear system $(x(t), y(t))^T$.

## 7.6 Systems of Nonlinear Differential Equations

When there is only one independent variable $t \in \mathbb{R}$ and $n \in \mathbb{N}$ dependent unknown variables $\{x_j(t)\}_{j=1}^n$, a vector state of the system is often defined as $\vec{x}(t) = (x_1(t), x_2(t), \ldots, x_n(t))^T$. The governing equations for such a system must associate the rate of change of the state to $n$-continuous functions, of $n+1$ variables,

$$F_j \in C^0\left(\mathbb{R}^{1+n} \to \mathbb{R}\right), \forall j \in \mathbb{Z}_n.$$

The system of equations are generally written as the IVP

$$\frac{d\vec{x}(t)}{dt} = \vec{F}(t, \vec{x}(t)), \quad \text{IC}: \vec{x}(t_0) = \vec{x}_0 \in \mathbb{R}^n. \tag{7.45}$$

Solving these equations for nonlinear functions $F_j$ requires special mathematical transformations or numerical techniques. Before an analysis begins, however, it is important to be sure that at least a local solution, for $t$ around $t_0$, is possible.

### 7.6.1 Existence and Uniqueness

A sequence of approximating solutions $\{\vec{x}_k(t)\}_{k=0}^\infty$ is obtained by setting the initial function to be the constant $\vec{x}_0(t) = \vec{x}_0$, and iterating the expression, using the Fundamental Theorem of Calculus,

$$\vec{x}_{k+1}(t) = \vec{x}_0 + \int_{s=t_0}^{t} \vec{F}\left(s, \vec{x}_k(s)\right) ds, \tag{7.46}$$

where truncation gives a partial-sum approximation to $\vec{x}(t)$ in Eq. (7.45). The solution can be written as the following series, assuming convergence,

$$\vec{x}(t) = \vec{x}_0 + \sum_{k=0}^{\infty} \left(\vec{x}_{k+1}(t) - \vec{x}_k(t)\right). \tag{7.47}$$

Suppose that for some $\Delta t > 0$, and Lipschitz constant $L \in (0, 1)$, that

$$\begin{aligned}\|\vec{F}(t, \vec{v}) - \vec{F}(t, \vec{w})\| &\leq L \cdot \|\vec{v} - \vec{w}\|, \\ \forall t \in [t_0 - \Delta t, t_0 + \Delta t], \; \forall \vec{v}, \vec{w} &\in \mathbb{R}^n.\end{aligned} \tag{7.48}$$

This condition is restrictive enough to imply that all the components of $\vec{F}$ are finite. In particular, for the given IC: $\vec{x}_0 \in \mathbb{R}^n$ and $\Delta t > 0$, $\exists M > 0$ so that

$$\|\vec{F}(t, \vec{x}_0)\| \leq M, \ \forall t \in [t_0 - \Delta t, t_0 + \Delta t]. \tag{7.49}$$

Then the first approximation $\vec{x}_1(t)$ can be made arbitrarily close to the initial condition $\vec{x}_0$, by choosing $|t - t_0| \leq \Delta t$ sufficiently small. In particular, let $\Delta t = L/(2 \cdot M)$. Then the difference between $\vec{x}_1$ and $\vec{x}_0$ is estimated to be

$$\sup_{|t-t_0| \leq \Delta t} \|\vec{x}_1(t) - \vec{x}_0\| \leq \int_{s=t_0-\Delta t}^{t+\Delta t} \|\vec{F}(s, \vec{x}_0)\| ds \leq M \cdot 2\Delta t = L. \tag{7.50}$$

Consequently, the partial sums of Eq. (7.47) are precisely $\vec{x}_{k+1}(t)$ and condition (7.48) guarantees that the sequence $\{\vec{x}_k(t)\}_{k=0}^{\infty}$ is Cauchy on the space $\mathcal{C}^0([t_0 - \Delta t, t_0 + \Delta t] \to \mathbb{R}^n)$. Indeed, for each $\epsilon > 0$ and $\Delta t > 0$, define

$$N_\epsilon \equiv \max\left\{1, \lceil \ln\left[(1-L) \cdot \epsilon\right] / \ln(L) \rceil \right\}. \tag{7.51}$$

Then, $(\forall \epsilon > 0) \implies (\exists N_\epsilon \in \mathbb{N})$ so that $(\forall m > k \geq N_\epsilon)$ and $t \in [t_0 - \Delta t, t_0 + \Delta t]$,

$$\|\vec{x}_m(t) - \vec{x}_k(t)\| \leq \sum_{j=0}^{m-k-1} \|\vec{x}_{k+j+1}(t) - \vec{x}_{k+j}(t)\| \tag{7.52}$$

$$\leq \sum_{j=0}^{m-k-1} L^j \cdot \sup_{|s-t_0| \leq \Delta t} \|\vec{x}_{k+1}(s) - \vec{x}_k(s)\| \tag{7.53}$$

$$= \frac{1 - L^{m-k}}{1 - L} \cdot \sup_{|s-t_0| \leq \Delta t} \|\vec{x}_{k+1}(s) - \vec{x}_k(s)\| \tag{7.54}$$

$$\leq \frac{1 - 0}{1 - L} \cdot L^k \cdot \sup_{|s-t_0| \leq \Delta t} \|\vec{x}_1(s) - \vec{x}_0\|$$

$$\leq \frac{L^{k+1}}{1 - L} < \epsilon.$$

This shows, under rather restrictive conditions on $\vec{F}$, that a solution to the nonlinear Eq. (7.45) exists for $t$ before and after $t_0$. However, this study requires many nontrivial facts from mathematics:

- Linearity of the vector addition in Eq. (7.47);
- The Fundamental Theorem of Calculus on Eq. (7.45) suggesting Eq. (7.46);
- The triangle inequality for vectors used in Eq. (7.52);
- The Lipschitz condition in Eq. (7.48) gives Eq. (7.53);

- The triangle inequality for integrals, as used in Eq. (7.50); and
- The Geometric series identity for Eq. (7.54).

The study of nonlinear systems of differential equations is on a solid foundation due to arguments as just presented. A more general and detailed treatment can be found in [8].

### ■ 7.6.2 Linear Approximations of Nonlinear Systems and Quasilinearity

Although there are many methods for finding iterative solutions to nonlinear systems [47]useful to have analytic approximations of the true solution. Thus, if the behavior of a system is needed only near a point $\vec{x}_0$ around time $t_0$, then one should define a new state vector $\vec{y}$, and rate of change vector $\vec{G}$, as

$$\vec{y}(t) \equiv \vec{x}(t+t_0) - \vec{x}_0, \quad \vec{G}(t, \vec{y}(t)) \equiv \vec{F}(t+t_0, \vec{y}(t) + \vec{x}_0),$$

which transforms Eq. (7.45) into the nonlinear system

$$\frac{d\vec{y}(t)}{dt} = \vec{G}(t, \vec{y}(t)), \quad \vec{y}(0) = \vec{0}. \tag{7.55}$$

A linear approximation to the new rate of change vector $\vec{G}$ is

$$\vec{G}(t, \vec{y}(t)) \simeq \vec{G}(t, \vec{0}) + \left(\frac{\partial \vec{G}(t, \vec{u})}{\partial \vec{u}}\right)_{\vec{u}=\vec{0}} \cdot \vec{y}(t) = \vec{G}(t, \vec{0}) + D\vec{G}(t, \vec{0}) \cdot \vec{y}(t).$$

Now, the system in Eq. (7.55) can be approximated as the linear equation,

$$\frac{d\vec{y}(t)}{dt} - D\vec{G}(t, \vec{0}) \cdot \vec{y}(t) = \vec{G}(t, \vec{0}), \quad \vec{y}(0) = \vec{0}, \tag{7.56}$$

which can be studied as in Eq. (7.34). However, in this situation the square matrix $A(t) = D\vec{G}(t, \vec{0})$ may have time-varying components. In this case, the expression in Eq. (7.35) can only be used if the coefficients vary slowly in time. To obtain some level of approximation, the $n \times n$ fundamental matrix, or transition matrix $\Phi(t)$, needs to be constructed [23], which solves

$$\frac{d}{dt}\Phi(t) = A(t) \cdot \Phi(t), \quad \Phi(0) = I_{n \times n}. \tag{7.57}$$

The Wilhelm Magnus Method [4] involves commutation of $A(t)$ with itself at different times,

$$[A(t_1), A(t_2)] \equiv A(t_1) \cdot A(t_2) - A(t_2) \cdot A(t_1), \quad (7.58)$$

which, in general, is nonzero. When $t_1 = t_2$ then $[A(t_1), A(t_1)] = 0_{n \times n}$. An alternative uses the Fundamental Theorem of Calculus, assuming that the components of $A(t)$ are continuous in $t$,

$$\Phi(t) = I + \int_{s_1=0}^{t} A(s_1) ds_1 + \int_{s_1=0}^{t} A(s_1) \cdot \int_{s_2=0}^{s_1} A(s_2) ds_2 ds_1 \quad (7.59)$$
$$+ \int_{s_1=0}^{t} A(s_1) \cdot \int_{s_2=0}^{s_1} A(s_2) \cdot \int_{s_3=0}^{s_2} A(s_3) ds_3 ds_2 ds_1 + \ldots,$$

where the order that the matrices appear is important. For $|t| \sim 0$ the fundamental matrix has an inverse $\Phi^{-1}(t)$ that now can be used as an integrating factor in Eq. (7.56) with $A(t) = D\vec{G}(t, \vec{0})$. This leads to the formula

$$\vec{y}(t) = \Phi(t) \cdot \int_0^t \Phi^{-1}(s) \cdot \vec{G}(s, \vec{0}) ds. \quad (7.60)$$

An application of this procedure is found in another approximation method that one should consider. Suppose it can be observed that

$$\vec{G}(t, \vec{y}(t)) = A(t) \cdot \vec{y}(t) + \vec{G}_p(t, \vec{y}(t)), \quad (7.61)$$

where the perturbation vector $\vec{G}_p(t, \vec{y})$ may be linear in $\vec{y}$ but have a $t$ dependence that is small, or even oscillatory but averaging out to 0. The equation

$$\frac{d\vec{y}(t)}{dt} - A(t) \cdot \vec{y}(t) = \vec{G}_p(t, \vec{y}(t)), \quad \vec{y}(0) = \vec{0}, \quad (7.62)$$

is a quasilinear form of Eq. (7.55). There are several possible choices for $A(t)$ and an optimal choice depends on the region of interest or applicability of Eq. (7.55). Understanding the behavior of solutions to Eq. (7.62) requires quite a bit of investigation. One procedure involves creating a sequence of solutions $\{\vec{y}_n(t)\}$ where $\vec{y}_0 = \vec{0}$ is the IC. Then Eq. (7.60) applies to Eq. (7.55) using Eq. (7.61) to give the iteration scheme,

$$\vec{y}_{n+1}(t) = \Phi(t) \cdot \int_0^t \Phi^{-1}(s) \cdot \vec{G}_p(s, \vec{y}_n(s)) ds. \quad (7.63)$$

## 7.6 Systems of Nonlinear Differential Equations

These various methods can simplify a mathematical model through linearization and a change of variables. This aids in the analysis of the original nonlinear system. However, if a system is overanalyzed and transformed, the new system can hide phenomena, like the transition process between equilibrium points.

### ■ Exercises

1. Consider Eq. (7.57) with time-dependent matrix $A(t) = \begin{pmatrix} 1 & t \\ 0 & 2 \end{pmatrix}$.

   (a) Compute $[A(t), A(0)]$ and thus show that it is nonzero in general.
   (b) Use the formula in Eq. (7.59) to obtain a cubic in $t$ approximation to $\Phi(t)$.
   (c) Find $A(t)^{-1}$, compute $\Phi'(t)$, and verify that $A(t)^{-1} \cdot \Phi'(t) = \Phi(t) + \mathcal{O}(t^3)$.
   (d) Find the quadratic in $t$ approximation to $\det(\Phi(t))$, and using the first three terms in the Binomial expansion, or the Geometric series, find the quadratic in $t$ approximation to $1/\det(\Phi(t))$ valid for small $|t|$.
   (e) Find a quadratic in $t$ approximation to $\Phi(t)^{-1}$.
   (f) Formally take the derivative of both sides of the equation $I_{n\times n} = \Phi(t) \cdot \Phi(t)^{-1}$ and use Eq. (7.57) to find a similar equation for $d\Phi(t)^{-1}/dt$.
   (g) Verify the equation for $d\Phi(t)^{-1}/dt$ up to linear terms using the quadratic in $t$ approximations obtained.

2. Let $\|\cdot\|$ be the matrix norm and suppose $\exists \lambda > 0$ and $T > 0$ so that $\|A(t)\| \leq \lambda$ for $|t| \leq T$. Prove that the iteration scheme in Eq. (7.63) generates a Cauchy sequence, under sufficient conditions, as follows:

   (a) Use the properties in Eq. (4.62), applied to Expression (7.59), to show that $\|\Phi(t)\| \leq \exp[\lambda \cdot |t|]$ for $|t| \leq T$.
   (b) Explain why $\|I_{n\times n} - \Phi(t)\| \leq \exp[\lambda \cdot |t|] - 1$.
   (c) Show that $\|\Phi^{-1}(t)\| \leq 1/(2 - \exp[\lambda \cdot |t|])$ by applying the matrix norm properties to the expression $\Phi^{-1}(t) = (I_{n\times n} - \Phi(t)) \cdot \Phi^{-1}(t) + I_{n\times n}$. For this estimate to make sense, what is the condition on $\Delta t$ where $|t| \leq \Delta t$?
   (d) Suppose that $\|\vec{G}_p(t, \vec{0})\| \leq M \cdot e^{\lambda|t|}$ for $|t| \leq \Delta t$. Then estimate the magnitude of the first iterate $\|\vec{y}_1(t)\|$ for $|t| \leq \Delta t$ using the formula in Eq. (7.63).
   (e) Suppose that $\|\vec{G}_p(t, \vec{u}) - \vec{G}_p(t, \vec{v})\| \leq L \cdot e^{\lambda|t|} \cdot \|\vec{u} - \vec{v}\|$ for $\forall \vec{u}, \vec{v} \in \mathbb{R}^n$ and $|t| \leq \Delta t$. Then estimate $\|\vec{y}_{n+1}(t) - \vec{y}_n(t)\|$ in terms of $\sup_{|s|\leq|t|} \|\vec{y}_{n+1}(s) - \vec{y}_n(s)\|$ using Eq. (7.63).
   (f) Express the solution $\vec{y}(t)$ of Eq. (7.55) in the form suggested by Eq. (7.47), and express $\vec{y}_{n+1}(t)$ as a truncated series.
   (g) Given $\epsilon > 0$, find $N_\epsilon$ similar to Eq. (7.51) so that $\{y_n(t)\}$ is Cauchy on the space $\mathcal{C}^0([-\Delta t, \Delta t] \to \mathbb{R}^n)$.

## 7.7 Stability Versus Instability

Suppose a solution to the nonlinear Eq. (7.45) has components that grow unboundedly. Such a solution is considered to be *unstable*. Conversely, a solution with components that approach finite constants over time is considered *stable*. Although there are a range of behaviors between these extremes, solutions where all components remain bounded, but never become constant, are commonly found in nature. Thus if a mathematical model like Eq. (7.45) results in an unbounded solution, its applicability over time becomes problematic. To improve the model requires either an adjustment of the parameters or the inclusion of stabilizing terms.

### 7.7.1 Equilibria of Autonomous Systems

If the rate of change functions $\{F_j : \mathbb{R}^n \to \mathbb{R}\}_{j=1}^n$ in Eq. (7.45) do not depend on $t$ explicitly, the corresponding system is called *autonomous*. In such cases it is possible that the system will have permanent *equilibria*. These are constant values of the variables from which the system does not change. This gives $n$-algebraic equations

$$F_1(\vec{x}_*) = 0, \ F_2(\vec{x}_*) = 0, \ \ldots, F_n(\vec{x}_*) = 0, \tag{7.64}$$

from Eq. (7.45), which are typically nonlinear, leading to possibly several solutions $\{\vec{x}_{*,k}\}_{k \in Z}$ for $Z \in \mathbb{N}_0$. These solutions are then partitioned into three categories:

(1) $Z_+$ unstable zeros of Eq. (7.64);
(2) $Z_0$ metastable states of Eq. (7.64); and
(3) $Z_-$ stable equilibria of Eq. (7.64);

where $Z = Z_+ \cup Z_0 \cup Z_-$. One common challenge is to classify all the solutions to Eq. (7.64) once they are obtained. This is aided by computing the Jacobian matrix,

$$J[\vec{F}](\vec{x}_{*,k}) \equiv J(\vec{x}_{*,k}) \equiv D\vec{F}(\vec{x}_{*,k}) = \left(\frac{\partial \vec{F}(\vec{u})}{\partial \vec{u}}\right)_{\vec{u}=\vec{x}_{*,k}}$$

The eigenvalues of $J[\vec{F}](\vec{x}_{*,k})$ need to be analyzed and the $\Lambda_k$ determined for each equilibrium point $\vec{x}_{*,k}$, to determine stability properties;

$$\begin{aligned} k \in Z_+ \text{ unstable zeros} &\iff \Lambda_k > 0, \\ k \in Z_0 \text{ metastable states} &\iff \Lambda_k = 0, \\ k \in Z_- \text{ stable equilibria} &\iff \Lambda_k < 0. \end{aligned}$$

Suppose the system is two-dimensional. Then $J[\vec{F}](\vec{x}_{*,k}) \equiv J_{*,k} = \begin{pmatrix} a & b \\ c & d \end{pmatrix}$. An analysis is achieved by using the techniques of Example 7.3 where the trace $\mathcal{T} = a + d$ and the determinant $\mathcal{D} = a \cdot d - b \cdot c$ are sufficient to classify the equilibrium points.

### 7.7.2 Instability and Negative Feedback

As models become more complex by including more variables, the act of obtaining an actual solution becomes less important than the understanding of the range and the nature of the model's variability.

**EXAMPLE 7.6**

The family of Edward Lorenz oscillator equations [65],

$$\frac{d}{dt}\begin{pmatrix} x(t) \\ y(t) \\ z(t) \end{pmatrix} = \begin{pmatrix} \alpha \cdot (y(t) - x(t)) \\ x(t) \cdot (\rho - z(t)) - y(t) \\ x(t) \cdot y(t) - \gamma \cdot z(t) \end{pmatrix}, \quad (7.65)$$

are derived from the study of atmospheric convection, where $x, y$, and $z$ are thermodynamic variables, and $t$ is time. This system has three equilibrium points for parameters $\alpha, \gamma > 0$, and $\rho > 1$. All three equilibria are unstable. However, the curves $\vec{r}(t) \equiv (x(t), y(t), z(t))^T$ are confined to a bounded region of $\mathbb{R}^3$ and they trace out a pattern of transitions between the equilibrium points.

Numerical solutions indicate that the paths generated by (7.65) can switch between neighborhoods of the different equilibria without wandering off to infinity. Thus, the instability in this example refers to the lack of permanence for a solution's path to remain near any one equilibrium. However, the paths do revisit the different equilibria at different times. To show that all solutions remain bounded, the method of constructing an *energy functional*, or *Aleksandr Lyapunov function*, $V \in C^0(\mathbb{R}^n \to \mathbb{R}_0^+)$ can be attempted. This must be a nonnegative single-valued *valuation* of the system's state $\vec{r}(t)$. For the system in Eq. (7.65) a common choice is

$$V[x, y, z] \equiv \rho \cdot x^2 + \alpha \cdot y^2 + \alpha \cdot (z - 2\rho)^2. \quad (7.66)$$

Then the rate of change of this energy, along a solution paths $\vec{r}(t)$, is

$$\begin{aligned}\frac{dV[\vec{r}(t)]}{dt} &= \vec{\nabla}V[\vec{r}(t)] \cdot \frac{d\vec{r}(t)}{dt} = (2\rho \cdot x \ 2\alpha \cdot y \ 2\alpha \cdot (z-2\rho))^T \cdot (x' \ y' \ z')^T \\ &= 2\rho \cdot x(t) \cdot x'(t) + 2\alpha \cdot y(t) \cdot y'(t) + 2\alpha \cdot (z(t) - 2\rho) \cdot z'(t) \\ &= -2\alpha \cdot (\rho \cdot x^2 + y^2 + \gamma z \cdot (z - 2\rho)) \\ &\leq -2 \cdot \min\{1, \gamma\} \cdot V + 4\alpha\rho\gamma \cdot \sqrt{V}. \end{aligned} \quad (7.67)$$

This means that the value of $V$ is bounded for solutions of Eq. (7.65). If the initial value was $V_0 \equiv V[\vec{x}_0] \geq (2\alpha\rho\gamma/\min\{1,\gamma\})^2$, then the solution is restricted to a box with sides

$$|x(t)| \leq \sqrt{V_0/\rho}, \ |y(t)| \leq \sqrt{V_0/\alpha}, \ |z(t) - 2\rho| \leq \sqrt{V_0/\alpha},$$

using Eq. (7.66). This gives qualitative control over the behavior of the solutions.

### EXAMPLE 7.7

The following represents an autonomous nonlinear system:

$$\frac{d}{dt}\begin{pmatrix} x(t) \\ y(t) \\ z(t) \end{pmatrix} = \begin{pmatrix} x(t) \cdot z^4(t) + x^3(t) \\ x^2(t) \cdot y(t) + 2 \cdot z^3(t) \\ -y(t) + x^2(t) \cdot z(t) \end{pmatrix} \equiv \begin{pmatrix} f_0(x,y,z) \\ g_0(x,y,z) \\ h_0(x,y,z) \end{pmatrix} \equiv \vec{F}_0[\vec{r}(t)]. \quad (7.68)$$

Define the Lyapunov function $V(x, y, z) \equiv x^2 + y^2 + z^4$, and then consider its rate of change along a solution of Eq. (7.68),

$$\frac{dV[x(t), y(t), z(t)]}{dt} \geq 2 \cdot x^2(t) \cdot V[x(t), y(t), z(t)]. \quad (7.69)$$

Thus, if $|x(t)| \geq \epsilon > 0$, $V$ grows at least exponentially along solutions. So Eq. (7.68) is unstable.

Suppose the system in Eq. (7.68) is to be modified to make it stable. Then the following negative feedback term,

$$-\frac{\lambda}{2} \cdot \vec{\nabla}(V^2) = -2 \cdot \frac{\lambda}{2} \cdot V \cdot \vec{\nabla}V = -\lambda \cdot (x^2 + y^2 + z^4) \begin{pmatrix} 2x \\ 2y \\ 4z^3 \end{pmatrix}, \quad (7.70)$$

can be added to Eq. (7.68). The effect of this addition is to change the outcome of the computation in Eq. (7.67), which will now be

$$\frac{dV[\vec{r}(t)]}{dt} = \vec{\nabla} V[\vec{r}(t)] \cdot \frac{d\vec{r}(t)}{dt} = \vec{\nabla} V \cdot \vec{F}_0[\vec{r}(t)] - \lambda \cdot V[\vec{r}(t)] \cdot \|\vec{\nabla} V[\vec{r}(t)]\|^2. \tag{7.71}$$

A sufficiently large $\lambda > 0$ will often give stability. Indeed, it can be shown that

$$\frac{dV[x(t), y(t), z(t)]}{dt} \leq -2 \cdot (\lambda^4 - 1) \cdot V^2[x(t), y(t), z(t)], \tag{7.72}$$

for the system in Eq. (7.68). In general, if a system is known to have an instability, then a Lyapunov function $V : \mathbb{R}^n \to \mathbb{R}_0^+$ can help to understand its behavior. Furthermore, this provides a mechanism for the modification of the system, using this understanding. Then, as in the example just discussed, a policy parameter $\lambda > 0$ allows for a choice of the sufficient effort required to induce stability.

### ■ Exercises

1. Consider the system of nonlinear, autonomous ordinary differential equations:

   $$x' = -x \cdot (y - 3) = f(x, y), \quad y' = -y \cdot (x - 2) = g(x, y).$$

   (a) Find both equilibrium points $(x_1, y_1)$ and $(x_2, y_2)$ by solving $f = g = 0$.
   (b) Compute the Jacobian matrix of the system: $J = \begin{pmatrix} f_x & f_y \\ g_x & g_y \end{pmatrix} = D \begin{pmatrix} f \\ g \end{pmatrix}$.
   (c) Compute the Jacobian matrix at the equilibria, i.e., find $J(x_1, y_1)$ and $J(x_2, y_2)$.
   (d) For these Jacobian matrices, compute the traces $\mathcal{T}$ and the determinants $\mathcal{D}$.
   (e) Use Table 7.2 to classify the equilibrium points.
   (f) An equilibrium point will be stable if $\mathcal{T} < 0$ and $\mathcal{D} > 0$. Is either equilibrium point stable, and can the species $x$ and $y$ coexist? Briefly explain.

2. Consider the system of nonlinear, autonomous ordinary differential equations:

   $$x' = -x + x \cdot y = f(x, y), \quad y' = 2 \cdot y - y^2 - x \cdot y = g(x, y).$$

   (a) Find all three equilibrium points $\{(x_j, y_j)\}_{j=1}^3$ by solving $f = g = 0$.
   (b) Compute the Jacobian matrix at the equilibrium points $\{J(x_j, y_j)\}_{j=1}^3$.

(c) For each Jacobian matrix, compute the trace $\mathcal{T}$ and the determinant $\mathcal{D}$.
(d) Use Table 7.2 to classify the equilibrium points.
(e) Can the species $x$ and $y$ coexist? Briefly explain.

3. Consider the system $\dfrac{d}{dt}\begin{pmatrix} x(t) \\ y(t) \end{pmatrix} = \begin{pmatrix} x \cdot y^4 - 2 \cdot y^4 \\ x \cdot y + y^3 \end{pmatrix}$.

(a) Let $V(x,y) = x^2 + y^4$ and show that

$$dV/dt \geq 2 \cdot \min\{1, y^4\} \cdot V, \tag{7.73}$$

along a solution path $\vec{r}(t) = (x(t)y(t))^T$.

(b) Suppose $|y(t)| \geq \epsilon > 0$ for all $t \geq 0$ and $V(0) = V_0 > 0$. Integrate Eq. (7.73) to find the growth rate of $V(t)$.

(c) Add the stability term $-(\lambda/2) \cdot \vec{\nabla} V^2$ and find $C_\lambda > 0$ for $\lambda$ sufficiently large so that $\max\{|x|, |y|\} \geq 1$ implies that

$$dV/dt \leq -C_\lambda \cdot V^2. \tag{7.74}$$

(d) If $V_0 > 1$, then integrate Eq. (7.74) and obtain an upper bound on the time it takes the solution to enter the box $\max\{|x|, |y|\} \leq 1$.

## ■ 7.8 System Instability Due to Parameter Changes

Once a mathematical model has been chosen, determining the typical size and range of the parameters must be done by a scientist. The consequence can be significant to the theoretical existence and stability of equilibrium points. Changing one parameter, even slowly, can convert a stable equilibrium point into an unstable one, forcing the system to find new equilibria. During this search, the system may experience large variations or unusual fluctuations just before a transition to a new equilibrium. The *critical* value for a parameter, delineating between stability and instability, is called a *bifurcation point* for the parameter that is being changed. If there are two parameters being changed, then drawing a two-dimensional parameter-space diagram will reveal *bifurcation lines* for the system.

Solutions of difference and differential equations give predictive and postdictive estimations. The long-term behaviors of the solutions can be cataloged. As often occurs, a system will approach an equilibrium point as the dependent variables approach a constant value. This can be used to classify the solutions. However, if a limit does not exist, then the frequencies of oscillation, the centers of oscillation,

## 7.8.1 Phase Plots and Bifurcations

When one parameter is varied, a system can be analyzed for its stability properties. For example, in the population model in Eq. (7.11) the parameter $k$ has a dramatic effect on the solution if it passes through the $k = 0$ value, as indicated in Eq. (7.12). This is also seen in Eq. (7.16) where, if $b$ is held constant, then the sign of $a$ determines the stability of the solutions.

For a general single-dependent-variable *autonomous* system, $y' = g(y)$, suppose $g(y_*) = 0$ but $g'(y_*) \neq 0$. Then $y_*$ is an equilibrium, and for $y(t) \simeq y_*$, we have,

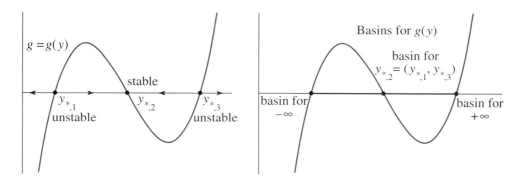

**FIGURE 7.1** ■ Plot of the slope function $g(y)$ and the basins for system $y' = g(y)$.

using a Taylor Expansion of $g(y)$ about $y_*$, the approximation

$$\frac{dy}{dt} = g(y) \simeq g'(y_*) \cdot (y(t) - y_*). \tag{7.75}$$

Thus, a qualitative assessment, based on Eq. (7.11) and Eq. (7.16), gives

$$0 = g(y_*) \implies \begin{cases} g'(y_*) < 0 \implies y_* \text{ is stable} \\ g'(y_*) > 0 \implies y_* \text{ is unstable} \end{cases}, \tag{7.76}$$

which means that the solution "$y$" is *attracted* to the equilibrium point $y_*$ if $g'(y_*) < 0$. Such a "$y$" region is called a *basin for $y_*$* as indicated in Figure 7.1 where "$g$" is plotted against "$y$." If $g'(y_*) = 0$, then more terms of the Taylor Expansion about $y_*$ are needed in Eq. (7.75), i.e., if $g(y_*) = g'(y_*) = 0$ but $g''(y_*) \neq 0$,

$$\frac{dy}{dt} \simeq \frac{g''(y_*)}{2} \cdot (y(t) - y_*)^2, \tag{7.77}$$

in which case $y_*$ is called *semistable* or a *node*. In scientific applications such points are effectively unstable. Regions of unbounded instability are basins of $\pm\infty$.

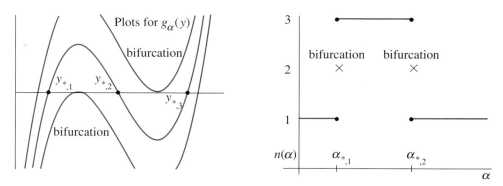

**FIGURE 7.2** ■ Left: Plot of the slope functions $g_\alpha(y)$. Right: The number of equilibria function $n(\alpha)$ for systems $y' = g_\alpha(y)$ with bifurcations $\alpha \in \{\alpha_{*,1}, \alpha_{*,2}\}$.

Now if $g(y)$ is varied according to a single parameter, then the equilibria will change, along with their associated basins. The simplest example is found by creating the family $g_\alpha(y) = g(y) + \alpha$ in terms of the parameter $\alpha \in \mathbb{R}$. Then the number of equilibria $n(\alpha)$ for $y' = g_\alpha(y)$ is a discontinuous function of $\alpha$ as seen in Figure 7.2. If $g_\alpha(y)$ is differentiable, then $\alpha_*$ is a bifurcation point if an equilibrium point $y_*$ is a critical point for $g_\alpha(y)$. This is expressed as the simultaneous equations

$$① \; g_{\alpha_*}(y_*) = 0, \qquad ② \; \frac{\partial g_{\alpha_*}(y_*)}{\partial y} = 0. \tag{7.78}$$

For each $\alpha \in \mathbb{R}$ the equilibria $\{y_{*,j}\}_{j=1}^{n(\alpha)}$ can be found from ①. Then, this can be substituted into ② to see if $\alpha$ is a value where the number $n(\alpha)$ changes.

### ■ 7.8.2 Parameter Space Diagrams

When two parameters of a system are varied, the solutions of the system can be analyzed to search for different characteristic regions. The purpose may be to identify parameters that will lead to desirable or undesirable characteristics. This can be used to set policy with the intent of maintaining desirable characteristics.

## 7.8 System Instability Due to Parameter Changes

Often, parameters are actually variables that change slowly compared to other dependent variables. By fixing the slow parameters, a system can be modeled with simplified equations, and its stability properties assessed. Predictions can be made under different scenarios. If a system is heading toward a catastrophe (i.e., when a variable is becoming unbounded, like current $I$, or vanishing, like a population level $P$), then parameter changes may prevent an unwanted outcome. Conversely, if a parameter is identified as being in transition, then a parameter space diagram could reveal an eventual change from an equilibrium state of the system.

As an example, consider the following as a model for human population levels:

$$\frac{dP(t)}{dt} = k \cdot P^2(t-\tau) \cdot \left(1 - \frac{P(t-\tau)}{\mu}\right), \quad P(2000) \simeq 6 \text{ billion.} \quad (7.79)$$

Now, suppose that the parameter $\mu \simeq 9$ billion corresponds to the carrying capacity of the human population and $\tau \simeq 20$ years corresponds to its reproductive and consumptive maturity. What might be surprising is that as $\tau$ is increased for fixed $\mu$, the system becomes more unstable. Similarly, as $\mu$ is increased for fixed $\tau$, the system becomes unstable. The problem is that the effect of delay $\tau$ on the system allows $P$ to increase above $\mu$. The result of a numerical study is presented in Figure 7.3.

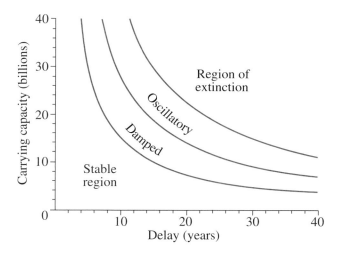

**FIGURE 7.3** ■ Parameter space diagram of the carrying capacity $\mu$ vs. delay $\tau$.

Another concern arises when efforts are made to delay an unwanted, but necessary state of the system. Consequently, when a downturn occurs, it may end up being more severe than allowing the natural stages of evolution of a system to occur. In society, this manifests itself in the phenomena of unintended consequences, and can be modeled by an adaptive cycle [32]. The three main components or critical variables are a society's potential for growth, interconnectedness, and resilience to crises. It is suggested that a dynamic for societal evolution starts with the potential for improvement in one's life. This drives people to become more connected.[3] Correspondingly the resilience of the system increases until a peak of all three variables is reached. However, with no further room to grow, the perceived potential and thus connectedness falls to a point where the resilience can no longer be maintained. A single crisis or multiple disturbances will reduce the society to its lowest levels of potential, connectedness, and resilience. Over time, like near the end of an economic recession, the potential begins to increase again. This model is hard to treat mathematically, and there is much criticism with such approaches [33]. Thus an analysis must be qualitative. The important qualitative feature is that the longer it takes for a system to deal with its loss of resilience, the more devastating the consequence of an unexpected crisis.

## ■ Exercises

1. Consider the autonomous differential equation $y'(x) = y^2 - 5y + 4 = g(y)$.

    (a) Find the equilibrium points of $y' = g(y)$, i.e., solve $0 = g(y)$.
    (b) Plot the function $g = g(y)$ for $y \in [0, 5]$.
    (c) Classify the equilibrium points as stable or unstable.
    (d) Identify the basins on the plot of $g = g(y)$.

2. Consider the family of autonomous differential equations: $y'(x) = y^2 - 4y + \alpha = g_\alpha(y)$.

    (a) Plot the functions $g = g_\alpha(y)$ on $y \in [0, 5]$ for $\alpha \in \{-5, 0, 5\}$.
    (b) What is the number of the equilibrium points of $y' = g_\alpha(y)$ in the three cases $\alpha \in \{-5, 0, 5\}$?
    (c) Use Eq. (7.78) to find the bifurcation value of $\alpha$.
    (d) Plot the number of equilibria vs. $\alpha$.

3. Consider the family $y'(x) = (y-1) \cdot (y-3) \cdot (y-5) \cdot (y-6) + \alpha = h_\alpha(y)$.

    (a) Plot the functions $h = h_\alpha(y)$ on $y \in [0, 8]$ for $\alpha \in \{-20, -10, 0, 10\}$.
    (b) Visually estimate the three bifurcation points and their corresponding equilibria $\{\langle \alpha_{*,k}; \{y_{*,j,k}\}_{j=1}^{n(\alpha_{*,k})} \rangle\}_{k=1}^{3}$ to the nearest integer.
    (c) Write Newton's Method (6.87) to approximate solutions to Eq. (7.78).

---

[3]Theories exist for explaining individual motivation that can be studied scientifically [30].

(d) Use the visual estimate, and one iteration of Newton's Method to improve the estimate of the pairs $\{\langle \alpha_{*,k}; \{y_{*,j,k}\}_{j=1}^{n(\alpha_{*,k})}\rangle\}_{k=1}^{3}$. Make a qualitative statement on the likelihood of the method converging, for each pair.
(e) Plot the number of equilibria vs. $\alpha$ for $\alpha \in [-20, 10]$.

4. Recall the Bernoulli Equation (7.15), $P' = k \cdot P - \beta \cdot P^m$.

   (a) For $m = 2$ and $k > 0$ find the bifurcation point $\beta_*$, not using Eq. (7.78), and describe the number of equilibria function $n(\beta)$.
   (b) Do the same for $m = 3$ and $k > 0$.
   (c) Find the bifurcation point $\beta_*$ for the equation $P' = k \cdot P - \beta \cdot P^2 + \beta$ for fixed $k > 0$, using Eq. (7.78). Also describe $n(\beta)$.

## 7.9 Models of Mass and Electric Charge

In 1687, Sir Isaac Newton presented a mathematical model for the attractive force of gravity [52]. About 100 years later, Charles Augustin de Coulomb published a similar result for electrostatic forces. With gravity, one has the conserved parameter of mass $m > 0$. With electrostatics, the conserved parameter is charge $q \in \mathbb{R}$. If two particles have masses $m_1$ and $m_2$, and charges $q_1$ and $q_2$, then the forces on the second particle, due to the first, are

$$\vec{F}_g = -k_g \cdot \frac{m_1 \cdot m_2}{r^2} \cdot \hat{r}, \quad \vec{F}_e = k_e \cdot \frac{q_1 \cdot q_2}{r^2} \cdot \hat{r}, \tag{7.80}$$

where $\hat{r} \in \mathcal{S}^2$ points from the first particle to the second, and $r \in \mathbb{R}^+$ is the distance between the particles. Both $k_g$ and $k_e$ are positive universal constants.

A natural question is whether the quantities $k_g$, $k_e$, $m_1$, $m_2$, $q_1$, and $q_2$ can really be thought of as constants. A seemingly unrelated question is whether the power dependence on distance $r^{-2}$ is actually a square of a reciprocal. These questions are related. Experiments may establish a range of values for the constants, with a high degree of confidence. However, it can be shown that near $r = 0$,

$$\left( \vec{F} \propto \frac{\hat{r}}{r^2} \right) \iff (\exists \text{ a source at } r = 0). \tag{7.81}$$

Suppose that the first particle is placed at the origin. Then the position of the second particle is at $\vec{r} \in \mathbb{R}^3$, where

$$\vec{r} = r \cdot \hat{r} = \begin{pmatrix} x & y & z \end{pmatrix}^T, r = \|\vec{r}\| = \sqrt{x^2 + y^2 + z^2}. \tag{7.82}$$

The first particle generates a field, experienced by the second particle, which is an entity that remains even as $m_2 \to 0$, $q_2 \to 0$. These are the gravitational fields $\vec{g}(\vec{r})$ and electric fields $\vec{E}(\vec{r})$,

$$\vec{g} \equiv -k_g \cdot \frac{m_1}{r^2} \cdot \hat{r} = -k_g \cdot m_1 \cdot \frac{\vec{r}}{r^3}, \quad \vec{E} \equiv k_e \cdot \frac{q_1}{r^2} \cdot \hat{r} = k_e \cdot q_1 \cdot \frac{\vec{r}}{r^3}. \quad (7.83)$$

It was observed in the mid-1800s by Pierre-Simon Laplace that these can be integrated from $\infty$ to the position $\vec{r}$, which is a distance $r$ from the source, to obtain the gravitational $\phi_g(\vec{r})$ and electric $\phi_g(\vec{r})$ *potential-energy* model functions,

$$\phi_g \equiv \frac{k_g \cdot m_1}{r} = \frac{k_g \cdot m_1}{\sqrt{x^2 + y^2 + z^2}}, \quad \phi_e \equiv \frac{-k_e \cdot q_1}{r} = \frac{-k_e \cdot q_1}{\sqrt{x^2 + y^2 + z^2}}, \quad (7.84)$$

which have units of energy. Corresponding forces $\vec{F}$ can be created from the negative gradient of a potential-energy function $\phi$. Such forces are called *conservative*. Friction is an example of a nonconservative force, and is thus not integrable.

The scalar potential functions in Eq. (7.84) have the further property that $\Delta\phi = 0$, called *harmonicity*. In particular,

$$[(\forall r > 0) \implies (\Delta r^\alpha \equiv \mathrm{div}\,(\mathrm{grad}(r^\alpha)) = 0)] \iff (\alpha \in \{0, -1\}), \quad (7.85)$$

using the definition of $r$ in Eq. (7.82). Thus a general scalar harmonic function of just $r > 0$, in $\mathbb{R}^3$, is of the form

$$\phi(x, y, z) \equiv \phi(\vec{r}) = A + \frac{B}{r^2}, \quad (7.86)$$

where $A$ and $B$ are parameters of the potential. The corresponding force field is

$$\begin{aligned}\vec{F}(\vec{r}) &= -\mathrm{grad}\,(\phi(\vec{r})) = -B \cdot \vec{\nabla}(x^2 + y^2 + z^2)^{-1/2} \\ &= \frac{B}{(x^2 + y^2 + z^2)^{3/2}} \cdot \begin{pmatrix} x \\ y \\ z \end{pmatrix} = \frac{B \cdot \vec{r}}{r^3} = \frac{B \cdot \hat{r}}{r^2}.\end{aligned} \quad (7.87)$$

Thus the constant $A$ has no effect on the force, and so no effect on the physics. However, $A$ can be used as a reference value for the potential-energy function $\phi$.

Now define the function

$$\delta_0(\vec{r}) \equiv -\Delta\phi,$$

## 7.9 Models of Mass and Electric Charge

for $\phi$ in Eq. (7.86). By (7.85) one would expect that $\delta_0(\vec{r}) \equiv 0$, which is true if $\vec{r} \neq \vec{0}$. However, a surprising thing occurs if $\delta_0$ is integrated over a volume $\mathcal{V} \subset \mathbb{R}^3$, with smooth boundary $\partial\mathcal{V}$;

$$\iiint_\mathcal{V} \delta_0(\vec{r}) dx \, dy \, dz = \begin{cases} 0 & \text{if } \vec{0} \notin \mathcal{V} \cup \partial\mathcal{V}, \\ 4\pi \cdot B & \text{if } \vec{0} \in \mathcal{V} - \partial\mathcal{V}. \end{cases} \quad (7.88)$$

In particular, there is no dependence on the size of $\mathcal{V}$ in this result. This means that the source of the potential energy, represented by the delta function $\delta_0(\vec{r})$, which is also the source of the field and the forces they generate, are all locked up at the single point of the origin $\vec{0}$. The parameter $B$ is thus a measure of the strength of that source.

The preceding analysis was fully developed and expanded upon by Emmy Noether in the early 1920s. This lead to an understanding of all conserved quantities like mass, charge, energy, spin, and linear and angular momentum [2], [22].

### ■ Exercises

1. Compute the acceleration due to gravity on the surface of the Earth $g$, in units of m/s$^2$, using the formula in Eq. (7.83). To find $r_E$ use the fact that the circumference of the Earth is 40,000 km and its mass is $m_E \simeq 5.97 \times 10^{24}$ kg. Also, the gravitational constant is $k_g \simeq 6.67 \times 10^{-11}$ m$^3/(kg \cdot s^2)$.

2. Prove the statement in Eq. (7.85).

3. Take the gradients of the potentials in Eq. (7.84) and simplify to obtain Eq. (7.80). In particular, show the details leading to Eq. (7.80).

4. A positive charge $q > 0$ is placed at $(\delta, 0, 0)$ and an equal-strength negative charge $-q < 0$ is placed at $(-\delta, 0, 0)$. This results in two forces on a test charge fixed at a distant point $\vec{r}_0 = (x_0, y_0, z_0)^T$. Let $\vec{r}_\pm = (x_0 \mp \delta, y_0, z_0)^T$.

    (a) Add the two electric fields $\vec{E}_\pm$ associated with $\vec{r}_\pm$ using Eq. (7.83) and let $\delta \to 0$ while requiring the source charges to grow as $q = q_0/\delta$. Obtain an expression in terms of $1/r_0$ for the new (*dipole*) field, denoted as $\vec{D}$.
    (b) Find the unit direction vector for $\vec{D}(\vec{r}_0)$ denoted $\hat{D}$.
    (c) Determine the square of the magnitude of $\vec{D}$ and show that it never vanishes. Also estimate how quickly the magnitude vanishes as $r_0$ increases.
    (d) How does the decay of the dipole field compare with the individual electric fields $\vec{E}_\pm$. Give a short reason why the dipole field decays faster as $r_0 \to \infty$.
    (e) Why does a dipole field not exist in gravitation?

5. To prove Eq. (7.88) it is not necessary to consider the case where $(\mathcal{V} \cup \partial\mathcal{V}) \cap \{\vec{0}\} = \emptyset$ because then $\delta_0(\vec{r}) = 0$, $\forall \vec{r} \in \mathcal{V} \cup \partial\mathcal{V}$. However, since $\tilde{\delta}(\vec{r})$ can only be nonzero at $\vec{r} = \vec{0}$, consider the case where $\exists \epsilon > 0$ so that

$$\{\vec{r} \in \mathbb{R}^3 : \|\vec{r} - \vec{0}\| \leq \epsilon\} \equiv B^3_{\vec{0},\epsilon} \subset \mathcal{V} - \partial\mathcal{V}. \tag{7.89}$$

In this case, since $\delta_0(\vec{r}) = 0$, $\forall \vec{r} \in (\mathcal{V} \cup \partial\mathcal{V}) - B^3_{\vec{0},\epsilon}$, one obtains

$$\iiint_\mathcal{V} \delta_0(\vec{r}) dx\, dy\, dz = \iiint_{B^3_{\vec{0},\epsilon}} \delta_0(\vec{r}) dx\, dy\, dz, \tag{7.90}$$

which, according to Eq. (7.88), is nonzero if the parameter $B \neq 0$. Amazingly, the result will *not* depend on $\epsilon > 0$. To verify this, first define the potential function $f_\alpha(r) \equiv B \cdot \alpha \cdot (\alpha + 1) \cdot r^{\alpha-2}$. Then:

(a) Show that $\Delta r^\alpha = \operatorname{div}(\operatorname{grad}(r^\alpha)) = f_\alpha(r)$ for $\alpha \in (-1, 0)$. See Exercise (2).
(b) Rewrite the function $f_\alpha(r)$ as a first-order derivative of another function that vanishes as $r \to \infty$.
(c) Convert the triple integral to polar coordinates.
(d) Use Fubini's theorem to separate the triple integral.
(e) Integrate in $r$, $\theta$, and $\phi$ to obtain an expression in $\alpha$, $B$, and $\epsilon$.
(f) Let $\alpha \to -1^+$ to finally obtain Eq. (7.88).

## ■ 7.10 Motions in Space

The position of a particle $\vec{r}(t)$ of mass $m_2$ will experience an acceleration $\vec{r}\,''(t)$ in a force field $\vec{F}(t, \vec{r})$ according to the equation

$$\vec{r}\,''(t) \equiv \frac{d^2\vec{r}(t)}{dt^2} = \frac{1}{m_2} \cdot \vec{F}(t, \vec{r}(t)), \quad \text{IC} : \vec{r}(t_0) = \vec{r}_0,\ \vec{r}\,'(t_0) = \vec{v}_0, \tag{7.91}$$

where $\vec{r}_0 \in \mathbb{R}^3$ is the *initial position* and $\vec{v}_0 \in \mathbb{R}^3$ is the *initial velocity*. In the cases of a gravitational or electric field, the path $\vec{r}(t)$ solves an autonomous equation;

$$\frac{d^2\vec{r}_g(t)}{dt^2} = \frac{\vec{F}_g}{m_2} = \left( -k_g \cdot \frac{m_1}{r_g^3(t)} \cdot \vec{r}_g(t) \right), \tag{7.92}$$

$$\frac{d^2\vec{r}_e(t)}{dt^2} = \frac{\vec{F}_e}{m_2} = \left[ \frac{q_2}{m_2} \right] \cdot \left( k_e \cdot \frac{q_1}{r_e^3(t)} \cdot \vec{r}_e(t) \right), \tag{7.93}$$

where the forces were presented in Eq. (7.80). Equation (7.93) demonstrates the effect of the source of charge $q_1$ on the path $\vec{r}_e(t)$. It also demonstrates how the parameters $m_2$ and $q_2$ effect the motion of this particle. However, in the case of gravity, Eq. (7.92) shows no dependence on the parameters of the moving particle. This implies the statement:

*All objects accelerate the same, at the same point, in a gravitational field.*

Consider the following definitions, related to a single particle of mass $m_2$:

**Inertial mass:** The ratio of the applied force to the experienced acceleration;
**Gravitational mass:** The strength of the source of a gravitational field.

Then the canceling of the gravitational and inertial masses in Eq. (7.92) occurs because

$$inertial\ mass = gravitational\ mass.$$

It is not obvious why this should occur, but it has been verified by measurement to a high degree of accuracy [1]. Thus, charged particles of different masses can be distinguished in the presence of an electric field, due to their motion. This is not the case for gravitational fields.

In general the acceleration of a particle $\vec{r}_2''$ is determined by many forces that are first combined through vector addition, and then scaled using a division by the inertial mass $m_2$, as in Eq. (7.91). Equations in physics are often derived in this manner resulting in second-order differential equations whose parameters are masses, charges, initial positions, and initial velocities. These are solved using techniques of mathematics.

### ■ 7.10.1 Quantum and Electromagnetic Fields

At subatomic scales, particles are not well represented by moving paths $\vec{r}(t)$. Rather, one associates a pdf $P(t, \vec{r})$ for the probability of finding the particle at different places throughout space. Furthermore this pdf changes over time. Due to the observation that quantum particles have wave properties resulting in interference effects [73], it is natural to introduce a *wave function*, or *quantum state* $\phi(t, \vec{r})$ that characterizes the particle, so that

$$P(t,\vec{r}) \equiv |\phi(t,\vec{r})|^2, \quad 1 = \int_{\mathbb{R}^3} P(t,\vec{r}) d^3\vec{r} = \int_{\mathbb{R}^3} |\phi(t,\vec{r})|^2 d^3\vec{r}. \tag{7.94}$$

The function $\phi: \mathbb{R} \times \mathbb{R}^3 \to \mathbb{C}$ is associated with an evolution equation that involves partial derivatives. In particular, the Klein-Gordon equation

$$-\partial_t^2 \psi(t, \vec{r}) = -\Delta \psi(t, \vec{r}) + m^2 \cdot \psi(t, \vec{r}), \tag{7.95}$$

governs the behavior of a particle with *inertial mass* $m$ and where the pdf takes the form [72]

$$P \equiv \psi^* \cdot (i\partial_t \psi) - \psi \cdot (i\partial_t \psi^*). \tag{7.96}$$

Now suppose that the particle described by $\psi$ has an electric charge $q$. Then one modifies Eq. (7.95) by making the transformation

$$\vec{\nabla} \to \vec{D}_q \equiv \vec{\nabla} + iq \cdot \vec{A}_b, \quad \Delta = \vec{\nabla} \cdot \vec{\nabla} \to \vec{D}_q \cdot \vec{D}_q. \tag{7.97}$$

Here $\vec{A}_b \in \mathcal{C}^3(\mathbb{R} \times \mathbb{R}^3 \to \mathbb{R}^3)$ is called the *vector potential* [26]. Along with the electric potential $\phi_e$ in Eq. (7.84) one obtains the electric $\vec{E}$ and magnetic $\vec{B}$ fields

$$\vec{E}(t, \vec{r}) = -\vec{\nabla} \phi_e - \partial_t \vec{A}_b, \quad \vec{B}(t, \vec{r}) = \vec{\nabla} \times \vec{A}_b, \tag{7.98}$$

whose energy density[4] $\mathcal{E}(t, \vec{r}) \equiv (\epsilon_0 \cdot \vec{E} \cdot \vec{E} + \vec{B} \cdot \vec{B}/\mu_0)/2$ flows in the direction of $\vec{E} \times \vec{B}$ in free space.

The interaction between $\psi$, $\vec{E}$, and $\vec{B}$ is expressed in terms of a system of seven partial-differential equations in four variables

$$\begin{aligned}
-\partial_t^2 \psi_q(t, \vec{r}) &= -\vec{D}_q \cdot \vec{D}_q \psi_q(t, \vec{r}) + m^2 \cdot \psi_q(t, \vec{r}), & & (7.99) \\
\partial_t \vec{E}(t, \vec{r}) &= c^2 \cdot \text{curl}(\vec{B}(t, \vec{r})) & \text{Ampère's Law}, & (7.100) \\
\partial_t \vec{B}(t, \vec{r}) &= -\text{curl}(\vec{E}(t, \vec{r})) & \text{Faraday's Law.} & (7.101)
\end{aligned}$$

The parameter $c \equiv (\epsilon_0 \cdot \mu_0)^{-1/2} \simeq 3 \times 10^8$ m/s is the *speed of light*, which is a quantity that James Clark Maxwell derived by unifying the separate theories of electric and magnetic fields [2].

### ■ Exercises

1. Consider $\psi(t, \vec{r}) \equiv \exp(i\omega t - i\vec{k} \cdot \vec{r})$ where $\vec{k} \in \mathbb{R}^3$ and $\omega > 0$.

   (a) Show that $\psi$ solves Eq. (7.95) if $\omega$ is an appropriate function of $\vec{k}$ and $m$.
   (b) Compute and simplify $P$ in Eq. (7.96) and show that Eq. (7.94) cannot hold.

---

[4]Here $\epsilon_0$ is the *dielectric constant* and $\mu_0$ the *magnetic permeability* so that $\vec{E} \equiv \vec{D}/\epsilon_0$ and $\vec{B} \equiv \mu_0 \cdot \vec{H}$ where $\vec{D}$ is the *electric displacement field* and $\vec{H}$ is the *magnetizing field* [26].

2. Suppose $\psi_q$ and $\vec{A}_b$ solve Eq. (7.99). Now, for any $f \in C^2(\mathbb{R}^3 \to \mathbb{R})$, define the pair $\psi_{q,f} \equiv e^{-iqf(\vec{r})} \cdot \psi_q$ and $\vec{A}_{b,f} \equiv \vec{A}_q + iq \cdot \mathrm{grad}(f)$.

   (a) For $\vec{D}_{q,f} \equiv \vec{D}_q + iq \cdot \mathrm{grad}(f)$, show that $\vec{D}_{q,f}\psi_{q,f} = e^{-iqf(\vec{r})}\vec{D}_q \cdot \psi_q$.
   (b) Show that $\psi_{q,f}$ solves Eq. (7.99) with $\vec{D}_q$ replaced with $\vec{D}_{q,f}$.
   (c) Show the $P$ is the same for $\psi_q$ and $\psi_{q,f}$.

3. Let $\vec{E}(t,x,y,z) \equiv (0 \ \exp(i\omega \cdot t - ik_1 \cdot x) \ 0)^T$ where $\vec{k} \in \mathbb{R}^3$ and $\omega > 0$.

   (a) Compute $\partial_t \vec{E}$ and $\partial_t^2 \vec{E}$. Express both in terms of $\vec{E}$.
   (b) Compute $\mathrm{curl}(\vec{E})$ and $\mathrm{curl}(\mathrm{curl}\ \vec{E})$. Express $\mathrm{curl}(\mathrm{curl}\ \vec{E})$ in terms of $\vec{E}$.
   (c) Take the curl of Eq. (7.100) and the $\partial_t$ of Eq. (7.101) to obtain an equation just involving $\vec{B}$. This is a *wave equation* for the magnetic field.
   (d) Take the curl of Eq. (7.101) and the $\partial_t$ of Eq. (7.100) to obtain an equation just involving $\vec{E}$. This is a *wave equation* for the electric field.
   (e) Use the wave equation for the electric field and the expression for $\vec{E}(t,x,y,z)$ to find a relation between $\omega$ and $k_1$.

# 8 The Scientific Method

The scientific method will be demonstrated in this chapter by considering a range of problems. The space of influence for each problem will be increased from section to section. Each issue will require different mathematical models and techniques. However, one should look for the similarities between the different approaches with the goal of developing one's own sense of how mathematics is complementary to science.

## Introduction

Sir Isaac Newton performed an amazing achievement with the completion of *Principia* [52]. This work contains the development of calculus and differential equations along with its application to the understanding of the motion of the observable planets. However, what predicated this work was the analysis of astronomical data by Johannes Kepler, which had been carefully gathered by Tycho Brahe. Furthermore, the concept of central gravitating bodies had to be logically deduced by Galileo Galilei due to observations of Jupiter's four largest and brightest moons, using his refined version of the telescope. This, along with the idea of Nicolaus Copernicus that larger objects exert greater forces of attraction, gave a strong indication that the sun must be at the center of the solar system.

This process of shared ideas and development occurred over the rather short period from 1550–1650. There was very little need for statistical analysis at the time because a very clean experiment was being performed. Indeed, observing the points of light on the celestial sphere every evening, which appeared to move in a continuous manner, allowed for the development of theories that could then be tested.

In science theories can be shown to be false by testing some of the logical consequences that they imply. Other theories gain support by the models they suggest. The predictions generated by these proposed models can then be verified through experiment.

The driving force behind both mathematical and scientific discovery is the *deductive-inductive* cycle, suggested by Walter Shewhart and Edward Demming [31].

- Mathematics strives to establish absolute truths using deductive reasoning;
- Science searches for empirical truths using inductive reasoning.

Mathematics is responsible for precision and comprehension, whereas science endeavors to achieve accuracy and dependability. The following is one way of understanding the scientific method:

$$\begin{aligned}
Data &\rightarrow Mathematical\ model \\
&\rightarrow More\ data \rightarrow Compute\ predictive\ power \\
&\rightarrow Improve\ the\ mathematical\ model \rightarrow More\ data \\
&\rightarrow Recompute\ predictive\ power\ for\ new\ model\ldots \quad (8.1)
\end{aligned}$$

This is a version of the *Continuous Never-Ending Improvement* approach to knowledge, based upon feedback, which has been proposed by George Box [5]. One may question the need to keep improving a mathematical model, but this is an issue that is relative to one's need for reliable understanding. A technician or engineer may be satisfied with just a chart, table, schematic, or a simply-expressed relationship between variables, which allows for a quick assessment of the functioning of a system. Such a chart or table starts with an experimental or empirical study. Then a relationship is theorized and tested.

***Experimental-system input*** $=$ data
$\quad \rightarrow$ *Experimental-system output* $=$ statistics

***Theoretical-system input*** $=$ independent variables
$\quad \rightarrow$ *Theoretical-system output* $=$ dependent variables

***Meta-system input*** $=$ parameters
$\quad \rightarrow$ *Meta-system output* $=$ characteristics

The *characteristics* of a system are associated with its behavior or performance. Piecing together subsystems that perform in a predictable manner allows for the development of new complex technologies that confidently behave as they had been designed. However, a research scientist or mathematician may be looking to discover new relationships between variables, leading to subtle corrections of known formulae. This drives the advancement of scientific knowledge, which only much later may be of use to design engineers. For example, the relationship between current $I$ and voltage $V$ in a resistor is typically linear, so that the resistance parameter $R$ can be defined giving the formula $V = I \cdot R$. However, careful experiments reveal that

$$V = V(I) = R \cdot I + a_2 \cdot I^2 + a_3 \cdot I^3 \ldots$$

where the parameters or coefficients $\{a_j\}$ represent higher-order corrections to the model of the relationship between current and voltage. In fact, the coefficients may depend on other unspecified factors like temperature, the surrounding magnetic field strength, or even the age of the component. The question then arises as to the value of greater refinements in a model when the use of a resistor is for a limited range of conditions. An engineer may want to know the temperature dependence of $R$ in order to make a decision on a component's need for ventilation, and so will ask a physicist to perform a series of experiments, and construct a mathematical model for the dependence of the coefficients on temperature, based on the data.

At this point, a distinction should be made with other types of modeling. For example, to explain the scientifically observed phenomena of the diversity of species, Charles Darwin presented the notion of natural selection [10], summarized here by the statement that:

*Evolution, speciation, and extinction occur by a process of adaptive radiation. The constraints of an environment result in a survival of only the fittest. Thus descendants are modified toward a species with the most pertinent characteristics.*

This statement may help us rationalize the complexity of an eye, without the presence of a whole sequence of simpler sensory organs, and the appearance of an animal like a giraffe, without many present-day transitional species. Furthermore, the statement allows one to classify the connection between species, past and present, using a flow chart to explain the process of evolution while emphasizing the feedback component in that model. However, there is no inherent rate of evolution in Darwin's model. Predicting or postdicting evolution is quite inexact and often speculative. This is not a mathematical model in the sense intended in this textbook. As the various topics are studied, it may be useful to keep the following definitions in mind (see, for example, [7]).

---

**Definition 8.1** *Existentialism* is the process of understanding oneself as an entity motivated by *free will*. This "will" endeavors to succeed through a process of individuation toward the understanding of oneself through reflection. *Etymology* is the study of words, their origin, and the changes that occur through history to their meaning. *Epistemology* is the philosophical study of knowledge, and in particular, its limitations. *Ontology* is the study of gathering, expressing, sharing, and incorporating knowledge for the purpose of understanding and decision making. *Phenomenology* is the development of a universal perspective, through consciousness, of the relationships between oneself, others, and the environment, which strives to be independent of oneself.

The smallest scale allowed within present models of space is known as the Planck length $\ell_P \simeq 10^{-35}$ m, and the corresponding shortest time is $t_P \simeq 10^{-43}$ s. These are connected by the speed of light $c \simeq 3 \times 10^8$ m/s, which is the fastest speed allowed in our universe. Conversely, the age of the universe is $T_U \simeq 13.8 \times 10^9$ yr and the farthest observable objects are $R_U \simeq 10^{10}$ km away. It is within these ranges of parameters and variables that scientific exploration must take place, for now.

## ■ 8.1 Quarks and Groups

Leptons (e.g., electrons and neutrinos) are tiny particles ($\leq 10^{-30}$ m) that travel easily through open space. Mesons are larger subatomic particles ($\simeq 10^{-20}$ m) that appear in nuclear reactions and consist of a quark and an anti-quark pair. Baryons (e.g., protons and neutrons, $\simeq 10^{-15}$ m) are found in the nuclei of an atom and are made up of quark triplets. The suggestion for these building blocks of nuclear matter was made by Murray Gell-Mann, George Zweig, and others in 1964, and came about as the result of studying tables of subatomic particle interactions. This lead to what is called the *Standard Model* for hadrons, which are particles that are made up of two quarks (mesons) or three quarks (baryons). In total there are 18 different quarks and 18 anti-quarks, which are cataloged according to six possible flavors for each quark, and three colors within each flavor [44]. The light flavors are $u$, $d$, $s$ and the heavy flavors are $c$, $b$, $t$. The transitions between these types of quarks form the elements of a group, where the group actions can be stimulated by bombarding atoms with high-energy particle beams.

The result of a high-energy interaction can be to transform a nucleus into a new state. This process is represented by components $V_{ij}$ of the Nicola Cabibbo matrix where the proportion of $u$ quarks that transform into $d$ quarks is written $|V_{ud}|^2$. All possible transformations in this system are elements of a group. If energies are too high, the system breaks up into separate particles, and if they are too low, no change of state is observed. Thus to study the group action, physicists must perform experiments within a certain, carefully chosen, energy range. These energies are produced by particle accelerators, which can be tuned for specific types of particles and speeds desired at impact with a target.

## EXAMPLE 8.1

A deuterium nucleus $D$, also known as heavy hydrogen $^2$H (one proton and one neutron), consists of six quarks, each of which can have two possible spin states $\pm 1/2$. The group of possible spin states for six quarks is isomorphic to $(E_2)^6$. However, when confined into a baryon a quark of the same type must have opposite spins. For example, suppose a deuterium nucleus has been prepared with 0 spin, which would mean that the neutron and proton have opposite spins. Suppose a high-energy bombardment on this prepared state results in an even distribution of spins $\{+1, 0, -1\}$. This is observed by passing the products through a magnetic field [69]. Then four group actions, which changed the state, have taken place:

$I$ = no change, $P$ = proton change, $N$ = neutron change, $PN$ = both changed.

From these definitions the internal states of deuterium can be explained.

## ■ Exercises

1. Deuterium $D$ consists of three up quarks $u$, and three down quarks $d$:

   (a) Draw a large ellipse and two large, non-overlapping circles within it, side-by-side.
   (b) Draw three smaller circles within each large circle, placing two circles, side-by-side, forming a base, and then center the third circle above those two circles. In the other large circle, do the same, but inverted.
   (c) In the top three small circles write $u$ and in the bottom three write $d$.
   (d) Label the large $uud$ circle a *proton*, and the smaller $udd$ circle a *neutron*.
   (e) $D$ has only four spin states because $uu$ and $dd$ come in spin-antispin pairs. Draw four ellipses, side-by-side, below the large ellipse, and inside indicate the possible spin states using up and down arrows. Label the four ellipses according to elements in $\{+1, 0, -1\}$.
   (f) Starting with one of the 0 spin states, draw the four group actions $I$, $P$, $N$, and $PN$ using arrows.
   (g) Is the group $\{I, P, N, PN\}$ isomorphic to $E_4$ or $E_2 \times E_2$? Identify a correspondence between the group elements using arrows $\longleftrightarrow$.

2. An $n \times n$ Cabibbo matrix, with real components, is not a probability matrix, so the sum of each row is typically not 1. Rather, it is the sum of the squares of each row that is 1.

   (a) Find the general condition so that the product of two different $2 \times 2$ Cabibbo matrices is another Cabibbo matrix. In particular, compute $\begin{pmatrix} a & b \\ c & d \end{pmatrix} \cdot \begin{pmatrix} e & f \\ g & h \end{pmatrix}$ and set the sum of the squares of the rows to 1.

(b) Show that the determinant of a 2 × 2 Cabibbo matrix is ±1. (*Hint:* Compute the square of the determinant and use the properties to obtain 1.)

(c) Find the condition ensuring that the inverse of a 2 × 2 Cabibbo matrix is another Cabibbo matrix. What does this say about the columns of a 2 × 2 Cabibbo matrix?

(d) By recapping, explain why the set of 2×2 Cabibbo matrices forms a group.

(e) It has been estimated that $V_{ud} \simeq 0.97$ in the Cabibbo submatrix
$$C \equiv \begin{pmatrix} V_{ud} & V_{us} \\ V_{cd} & V_{cs} \end{pmatrix}$$
of a more complicated 3×3 matrix. Find the other three components with the caveats that $\det(C) = 1$ and $V_{us} > 0$.

## ■ 8.2 Measuring Radioactivity

Nuclear matter can be unstable, especially if it contains a large number of hadrons. Radioactive atoms will decay into other material at a rate proportional to the sample size. The reason is that over any period of time, each atom has a probability of decaying. Though this is typically small for small times, in aggregate one can often detect the result of decay in a large sample using a Geiger counter.

### EXAMPLE 8.2

A 1 cm² surface area detector is held 1 m away from a 100 g sample of pure $U^{235}$. Over a 10 second period, 600 clicks are detected. Assuming that each click corresponds to 1 atom's decay, what is the half-life of $U^{235}$?

The basic steps needed to answer the question in Example 8.2 are as follows:

(1) The total number of particles in the sample, at the time of measurement $t$, is
$$N(t) = (100 \text{ g} \times 6.022 \times 10^{23} \text{ particles/mole})/(235 \text{ g/mole}) \simeq 2.56 \times 10^{23} \text{ particles}.$$

Here one uses that one mole[1] of $U^{235}$ has a mass of about 235 g.

(2) The area covered by the Geiger counter is 1 cm² = $(10^{-2} \text{ m})^2 = 10^{-4} \text{ m}^2$, which is a fraction of the sphere of radius 1 m, surrounding the sample. Only the proportion $p = 10^{-4} \text{ m}^2/(4\pi \cdot (1 \text{ m})^2) \simeq 8 \times 10^{-6} = 0.0008\%$ of particles emitted

---

[1] A mole is $6.022 \times 10^{23}$, which is defined to be the number of atoms in 12 grams of Carbon 12.

will be detected. Thus the total number of particles streaming out of the sample, in all directions, per unit time, is

$$N'(t) = 600/(10 \text{ seconds} \times p) \simeq 7.5 \times 10^6 \text{ particles/second}.$$

(3) To determine the rate constant, assume an exponential-decay model, and compute the ratio $N'/N$,

$$k = N'(t)/N(t) = (7.5 \times 10^6 \text{ particles/second})/(2.56 \times 10^{23} \text{ particles}) \simeq 3 \times 10^{-17} \text{ s}^{-1}.$$

(4) The solution for the number of particles that remain in the sample over time is $N(t) = N_0 \cdot \exp(-kt) = N_0 \cdot 2^{-kt/\ln(2)}$, which has half-life

$$t_{1/2} = \ln(2)/k \simeq 2.4 \times 10^{16} \text{ seconds} = 7.5 \times 10^8 \text{ years}.$$

The actual measured value is $7.13 \times 10^8$ years [73].

## ■ Exercises

1. Radioactive radium, $Ra^{226}$, decays at a rate that is modeled by the function $N(t) = N_0 \cdot 2^{-t/1620}$ where $N_0$ is the initial number or concentration of $Ra^{226}$, and $t$ is in years.

    (a) Convert the expression $N(t)$ into $N_0 \cdot \exp(\lambda \cdot t)$. In particular, find $\lambda$.
    (b) Use one of the two expressions of $N(t)$ to find the half-life $t_{1/2}$ of $Ra^{226}$, and explain why you made the choice.
    (c) For a sample to have 25% of its present day amount of $Ra^{226}$, how long must one wait?

2. A 2 kg fossil is found that emits $N = 55 \pm 3$ particles per hour, as detected from a Geiger counter with a surface area of 1 cm² held 1 dm = $10^{-1}$ m away. The amount of carbon is determined (by chemically studying a small sample) to be 62% of the mass of the fossil. Using the method for studying Example 8.2 determine the age of the fossil, assuming that the source of the particles is radioactive carbon 14 ($C^{14}$), which has a half-life of 5730 years. Note that naturally occurring radioactive $C^{14}$ is $10^{-12}$ of the nonradioactive $C^{12}$. Also $C^{13}$ occurs in trace amounts and does not contribute to the number of particles counted.

3. Let $r$ be the distance of a Geiger counter from a radioactive source. It is observed that the number of counts per second $\mathcal{C} \equiv N'(t)$ depends on the

distance from the source $r$ according to the function $\mathcal{C}(r) = 7500/(6+r)^2$. It is not clear why $\mathcal{C}$ does not follow an inverse-squared law for small $r$, however it is proposed that the Geiger counter needs to reset itself before another particle can be detected. Plot $\mathcal{C}(r)$ and $7500/r^2$ for $r \in [0, 5]$ and restricted to the range $[0, 2000]$. Find the *dead time* based on $\mathcal{C}(r)$ and rationalize the computation.

## ■ 8.3 Chemical Equilibrium and Reaction Rates

Controlled chemical reactions typically involve quantities (e.g., reactants $R_j$, products $P_j$, catalysts) in an appropriate environment (e.g., temperature, acidity, pressure). The rate at which a reaction occurs is mainly influenced by the changing concentrations of the quantities and the surrounding temperature. One goal of a chemist is to optimize the yield of the products of a reaction. Therefore, models are needed to explain the mechanisms of a chemical reaction. Over time, a reaction will reach an equilibrium with all quantities present to some extent. In order to optimize the amount of products, one must understand the factors influencing the reaction. This involves experimentally determining parameters, like reaction rates.

### ■ 8.3.1 Rates of Reaction

When chemicals $\{R_1, R_2, \ldots\}$ are first combined in a solution, a reaction will take place resulting in products $\{P_1, P_2, \ldots\}$. At the molecular level, reactants will combine in certain integer ratios $\{\rho_1, \rho_2, \ldots\}$ producing products with ratios $\{\pi_1, \pi_2, \ldots\}$. This reaction is then written

$$\rho_1 \cdot R_1 + \rho_2 \cdot R_2 + \ldots \rightarrow \pi_1 \cdot P_1 + \pi_2 \cdot P_2 + \ldots. \tag{8.2}$$

With a small amount of products, the rate of change in the concentration of reactants depends only on the concentration of reactants. Thus, for each $j$,

$$\frac{d[R_j]}{dt} = -k \cdot [R_1]^{a_1}(t) \cdot [R_2]^{a_2}(t) \ldots, \tag{8.3}$$

where the parameters $\{a_1, a_2, \ldots\}$ define the *order of the reaction*. The reactant that is present in the (proportionally) smallest amount determines the initial evolution of the process, whereas the other reactants remain relatively constant. Suppose that this is the case for $R_1$. Then Eq. (8.3) becomes

$$\frac{d[R_1]}{dt} = -\tilde{k} \cdot [R_1]^{a_1}(t), \; [R_1](0) = [R_1]_0, \tag{8.4}$$

for a reaction-rate constant $\tilde{k} > 0$. This is an $a_1$th-order reaction equation where the rate constant $\tilde{k}$ depends on the concentrations of the other reactants and the environmental conditions. Using integration, Eq. (8.4) can be solved.

### ■ 8.3.2 Chemical Equilibrium

Once a reaction has been allowed to proceed for a sufficient time, the products begin reacting to recreate the reactants. When the two reaction rates agree, an equilibrium has been reached, which is then expressed as

$$\rho_1 \cdot R_1 + \rho_2 \cdot R_2 + \ldots \rightleftharpoons \pi_1 \cdot P_1 + \pi_2 \cdot P_2 + \ldots. \tag{8.5}$$

In this situation $d[P_i]/dt = d[R_j]/dt = 0$ and so Eq. (8.3) is replaced with the condition

$$[P_1]^{\pi_1} \cdot [P_2]^{\pi_2} \ldots \propto [R_1]^{\rho_1} \cdot [R_2]^{\rho_2} \ldots, \tag{8.6}$$

where the concentrations are now their asymptotic values $[P_i](\infty)$, $[R_j](\infty)$.

An example of an equilibrium situation is the dissociation that occurs in water

$$H_2O \rightleftharpoons H^+ + OH^-, \quad [H^+] \cdot [OH^-] \simeq [H^+]^2 \propto [H_2O]. \tag{8.7}$$

Water has the *amphoteric* property that it acts as an acid and a base. The concentration of free hydrogen, written $[H^+]$, is about 1 ion for every 10 million water molecules. This is also expressed as

$$\text{pH} = 7 = -\log_{10}[H^+],$$

which is a short form for the *negative power of the hydrogen ion concentration*. This means that for neutral water, $[H^+] = 10^{-7}$ mol/L. Here L $\equiv$ liter and 1 mole of atoms has a mass of about 18 grams of water. The equilibrium constant of proportionality in Eq. (8.7) is $K_{H_2O\pm} = 10^{-14}$ mol/L.

As the acidity of an aqueous solution increases, the pH decreases. For example, water can dissolve carbon dioxide $CO_2$ leading to an increase of dissociated hydrogen ions

$$CO_2 + H_2O \underset{k_P}{\rightleftharpoons} CO_2 \cdot H_2O \overset{K_H}{\rightleftharpoons} H_2CO_3 \overset{K_{C\pm}}{\rightleftharpoons} H^+ + HCO_3^-. \tag{8.8}$$

The higher the surrounding pressure of $CO_2$, the more $CO_2$ is dissolved, which is represented by the model

$$[CO_2] = \frac{1}{k_P} \cdot \text{pres}(CO_2), \quad k_P \simeq 30 \text{ atm}/(\text{mol/L}), \qquad (8.9)$$

where $\text{pres}(CO_2)$ is the partial pressure of carbon dioxide. Once dissolved, carbonic acid $H_2CO_3$ is produced with equilibrium constant $K_H \simeq 0.0017$. Finally, the dissociation of acid constant is $K_{C\pm} \simeq 0.00025$. This results in the formula

$$[H^+] \simeq \sqrt{10^{-14} + 1.4 \times 10^{-8} \cdot \text{pres}(CO_2)}. \qquad (8.10)$$

If the atmospheric partial pressure of carbon dioxide increases, the acidity of the oceans will increase.

## ■ Exercises

1. Solve Eq. (8.4) for $a_1 = 1 + \alpha$ where $\alpha > 0$. Determine the half-life of the reaction $t_{1/2}$ by solving $[R_1](t_{1/2}) = (1/2) \cdot [R_1]_0$. How long will the reaction last? Use L'Hôpital's Rule to compute $\lim_{\alpha \to 0^+} t_{1/2}$.

2. Two chemicals $R_1$, $R_2$ are mixed together and a 1:1 reaction $R_1 + R_2 \to$ *products* takes place in which the products are quickly removed from the solution (through evaporation or sedimentation). Suppose that the reactants satisfy Eq. (8.3) with $a_1 = 1$, $a_2 = 2$, and rate constant $k = 10^{10}$ L/(moles · min).

   (a) If $[R_2](0) = [R_1](0) = 10^{-5}$ moles/L, set up and solve Eq. (8.3) for $a_1 + a_2 = 3$, corresponding to a third-order reaction. What is the half-life $t_{1/2}^a$ for $[R_1]$?
   
   (b) If $[R_2](0) = 10 \cdot [R_1](0) = 10^{-4}$ moles/L approximate Eq. (8.3) as a first-order reaction for $a_1 = 1$, adjusting $k$ accordingly. Find the half-life $t_{1/2}^b$ for $[R_1]$.
   
   (c) Plot $[R_1](t)$ in the two cases for $t \in [0, \max\{t_{1/2}^a, t_{1/2}^b\}]$ in seconds.

3. Derive the formula in Eq. (8.10) using Eq. (8.9) and Eq. (8.8), in combination with Eq. (8.7). If $\text{pres}(CO_2)$ starts at 0.28 milli-atm and is increased by 35%, use the logarithmic derivative of Eq. (8.10) to determine the percent increase of $[H^+]$ and the percent decrease of the pH (*Hint*: Factor out $10^{-14}$ in Eq. (8.10) and note that $d\ln[H^+] = d[H^+]/[H^+]$.)

## 8.4 A Small Ising Model of Magnetism

A model for understanding magnetism in solid materials was suggested by Ernst Ising in 1924. An energy is defined whereby the alignment of neighboring magnetic moments is more likely than an opposite alignment [40]. The phenomena at high temperatures is that all magnetic orientations are equally likely, approximately. However, at low temperatures a complete alignment is the most probable state.

Most theoretical studies consider large lattice arrays of magnetic moments. However, to demonstrate the actual modeling required, consider a configuration where a fixed magnet is placed at the origin and positioned to point upward, in the $\hat{k}$ direction. In the two-dimensional symmetric case, four other magnetic moments are placed along the positive and negative $x$- and $y$-axes, which point in the $\pm\hat{k}$ directions. Each of these surrounding magnetic moments are assumed to have one of two possible orientation states $\sigma \in E_2 \equiv \{-1, 1\}$. Then a *state of the system* is a four-component string: $\langle \pm 1, \pm 1, \pm 1, \pm 1 \rangle = (E_2)^4$. Clearly, the system can have $2^4 = 16$ possible states. The energy due to each coupling is modeled by a *coupling-strength* parameter $J > 0$, which is specific to the material being studied. Thus the energy assigned to the system, which is in a state $\langle \sigma_1, \sigma_2, \sigma_3, \sigma_4 \rangle$, is evaluated to be

$$E\langle \sigma_1, \sigma_2, \sigma_3, \sigma_4 \rangle = J \cdot \sigma_1 + J \cdot \sigma_2 + J \cdot \sigma_3 + J \cdot \sigma_4. \tag{8.11}$$

The effect of a temperature $T$, from the surrounding material, when the system is in equilibrium, is to *thermally activate* a state transition. The Ising configuration will either absorb (endothermic) or emit (exothermic) energy, depending on its initial condition. The likelihood of absorbing energy increases with the neighboring temperature. In particular, it has been found theoretically, and verified experimentally, that the local energy available to each magnetic moment is $\beta = k \cdot T$. Here $k$ is Boltzmann's *constant of proportionality* between the system energy $E$ and the ambient temperature $T$. Then the probability of the appearance of an energy level $E$ is proportional to the *Boltzmann factor*, also known as the *fugacity* [54],

$$\exp[-E/(kT)] = \exp[-\beta \cdot E\langle \sigma_1, \sigma_2, \sigma_3, \sigma_4 \rangle] = \prod_{j=1}^{4} e^{-J \cdot \beta \sigma_j}. \tag{8.12}$$

Let $Z$ be the *normalization factor* for the distribution, also called the *partition function*. Then, in this example, five energy levels are possible, from Eq. (8.11) and Eq. (8.12),

$$P(E\langle 1,1,1,1\rangle) = e^{-4\cdot J\cdot \beta}/Z, \tag{8.13}$$
$$P(E\langle -1,1,1,1\rangle) = e^{-2\cdot J\cdot \beta}/Z, \tag{8.14}$$
$$P(E\langle -1,-1,1,1\rangle) = 1/Z, \tag{8.15}$$
$$P(E\langle -1,-1,-1,1\rangle) = e^{2\cdot J\cdot \beta}/Z, \tag{8.16}$$
$$P(E\langle -1,-1,-1,-1\rangle) = e^{4\cdot J\cdot \beta}/Z. \tag{8.17}$$

There are four unique configurations equivalent to Eq. (8.14) and Eq. (8.16), whereas Eq. (8.15) has six configurations. Thus, the partition function must satisfy

$$Z(\beta) = e^{-4\cdot J\cdot \beta} + 4\cdot e^{-2\cdot J\cdot \beta} + 6 + 4\cdot e^{2\cdot J\cdot \beta} + e^{4\cdot J\cdot \beta}. \tag{8.18}$$

The physical system of magnetic moments has a probability distribution. There are two extreme states or phases that the system transitions between depending on $T$.

### ■ 8.4.1 High Temperature Limit

As $T \to \infty$, the coefficient $\beta \to 0$, in which case all states have probability $P \to e^0/Z = Z^{-1}$ where $Z \to 16$. The expected energy is

$$\langle E\rangle \equiv \frac{1}{Z}\cdot \sum_{k=1}^{16} E_k \cdot P(E_k) = \frac{-1}{Z}\frac{dZ}{d\beta} = -\frac{d\ln(Z)}{d\beta}. \tag{8.19}$$

This surprising formula, involving the logarithmic derivative, is useful throughout thermodynamics. From Eq. (8.18) one obtains, for $\beta = 0$,

$$\langle E\rangle = \frac{4\cdot J + 8\cdot J + 0 - 8\cdot J - 4\cdot J}{16} = 0. \tag{8.20}$$

This indicates a lack of knowledge by the system that a fixed magnetic moment exists at the origin, oriented in the $\hat{k}$ direction.

### ■ 8.4.2 Low Temperature Limit

As $T \to 0°K$, the coefficient $\beta \to \infty$. Then $Z \simeq e^{4\cdot J\cdot \beta}$ and $\ln(Z) \simeq 4\cdot J\cdot \beta$. Thus $\langle E\rangle \simeq -4J$, which is the lowest energy state of the system. This is also called the *ground-state* energy level.

## ■ Exercises

1. For the configuration described in this section, nearest neighbors are assumed not to interact.

   (a) Suppose the coupling strengths to nearest neighbors is $J$ and compute the 16 energies, the Boltzman factors, and the partition function $Z$. Compute the expected energy in the high-temperature limit, and the ground-state energy level.
   (b) Provide the same analysis as in (a), but suppose that the coupling strengths to nearest neighbors is $-J/2$.

## ■ 8.5 The Shape of DNA

A major contribution to the understanding of the cellular mechanisms in biology was made with the discovery that DNA exists in the nucleus of a cell as a double helix. A strand of DNA, held upward with its center along the $z$-axis, can be modeled by a pair of parameterized spirals for $m_1, m_2 \in \mathbb{Z}$:

$$\vec{r}_1(m_1) = (a_0 \cdot \cos(b_0 \cdot m_1),\ a_0 \cdot \sin(b_0 \cdot m_1),\ c_0 \cdot m_1)^T \qquad (8.21)$$
$$\vec{r}_2(m_2) = (a_0 \cdot \cos(b_0 \cdot m_2),\ a_0 \cdot \sin(b_0 \cdot m_2),\ c_0 \cdot m_2 + d_0)^T.$$

This structure was discovered by the chemist Rosalind Franklin in 1952 [15], not with a microscope, but by properly interpreting diffraction patterns produced by X-ray scattering off a pure form of crystalized DNA. Her particular achievement was in determining the parameters $\{a_0, b_0, c_0, d_0\}$ associated with molecular placement, which are independent of the discrete, independent variables $\{m_1, m_2\}$. The parameters were measured by using the results of X-ray crystallography, and an application of inverse Fourier transforms. The wavelength of the X-rays used was about $\lambda \simeq 10^{-10}$ m = 10 nm. She found that $b_0/(2\pi) = 0.1$ fit the observed spacing between dots on a photographic plate, which meant that 10 molecules, on each helix, completed a single cycle. These were identified to be phosphorous atoms, with spacing 0.71 nm. She also stated that the radius of the helix was about $a_0 = 1$ nm.

The structure of DNA, which is far too small to see directly, can be deduced because X-ray scattering is a nearly linear process. The microscopic double helix is linearly transformed into an X pattern, with periodic dark and light bands, which appear as dots on a large photographic plate. In particular, for a fixed angle of incidence $\theta_{inc}$ the spacing between layers of molecules $\Delta d$ occurs in regular intervals, according to the formula

$$n \cdot \lambda = 2 \cdot \Delta d \cdot \sin(\theta_{inc}), \forall n \in \mathbb{N}, \qquad (8.22)$$

which is known as William Bragg's law [38]. Here the wavelength $\lambda$ of the X-ray beam must be adjusted until a pattern of dots appears. Then $\Delta d$ can be inferred.

### ■ Exercises

1. For an incident beam with angle $\theta_{inc} = 30°$ find the wavelength of the X-ray beam, expressed in meters, when the first dot $n = 1$ appears. Be sure to use Eq. (8.22) and the, now known, spacing between phosphorus atoms in DNA. Also find the frequency in Hz = s$^{-1}$ of the X-ray required, using $c = 3 \times 10^8$ m/s.
2. Draw a single helix as defined in Eq. (8.21) indicating $a_0$, the 10 phosphorous atoms, the spacing between the atoms, and the height of the helix after one cycle $10 \cdot c_0$. Compute the parameter $c_0$ by setting $10 \times 0.71$ nm $= \int_0^{2\pi} \|\vec{r}_1'(t)\| dt$.
3. Suppose that the minimal angle between the lines in the X pattern is $\theta_{DNA} \simeq 40°$ (the complementary angle is 50°). Estimate the shift parameter $d_0$ in Eq. (8.21) by speculating a relationship between $d_0$ and $\theta_{DNA}$.
4. Rewrite Eq. (8.21) using the parameters now determined.

### ■ 8.6 Cellular Homeostasis

A cell can be understood as a large system consisting of many processes. To stay alive a cell must absorb oxygen, water, nutrients, and ATP, which contains the energy needed by the cell. Concurrently, the cell must expel carbon dioxide, biproducts of metabolism, and ADP, which is the exhausted form of ATP. To maintain a balanced metabolic state internal to the cell, known as *homeostasis*, the cell must interact with its highly variable surroundings. The key mechanism by which this is done is known as *negative feedback*. A *receptor* detects an imbalance in a system and sends a signal to the *effector* that must try to maintain or even restore homeostasis.

Models used to understand the functioning of $\beta$-cells in the pancreas must consider the concentrations of calcium ions [Ca$^{2+}$] and potassium ions [K$^+$]. Here the example by Arthur Sherman is considered [66]. The movement of calcium ions into the cell causes a current $I_{Ca}$, which is driven by a voltage difference $v(t)$. This function vanishes at a rate dependent on the capacitance of the cell membrane $C_M$ and its resistance $R_{Ca}$. The relaxation time constants are derived from computations like $\tau_{Ca} = R_{Ca} \cdot C_{Ca}$, etc. The cell is neutralized by the movement of potassium out of the cell, gated by a variable $n(t) \in [0,1]$, causing a current $I_K$. There is also a slow mechanism resulting in changes of the cell's charge that is combined into one variable $s(t) \in [0,1]$ resulting in a current $I_s$. The system of autonomous equations are

$$\begin{aligned} v'(t) &= -(v(t) - v_{Ca})/\tau_{Ca} - n(t) \cdot (v(t) - v_K)/\tau_K - s(t) \cdot (v(t) - v_K)/\tau_s, \\ n'(t) &= -(n(t) - n_\infty(v))/\tau_n, \\ s'(t) &= -(s(t) - s_\infty(v))/\tau_s. \end{aligned} \qquad (8.23)$$

The steady-state parameters $v_{Ca}$ and $v_K$ put limits on the range of voltages

$$v(t) \in [v_K, v_{Ca}] \simeq [-75 \ mV, \ 25 \ mV]. \qquad (8.24)$$

The parameters $n_\infty$ and $s_\infty$ are assumed to depend on the voltage according to the models

$$n_\infty(v) \simeq [1 + e^{-(v+17)/6}]^{-1}, \ s_\infty(v) \simeq [1 + e^{-(v+50)/5}]^{-1}. \qquad (8.25)$$

These gating variables are always changing toward their equilibria, which take their values in the interval $(0, 1)$. In this way, the system in Eq. (8.23) is designed to evolve in a constrained manner, as is observed in a real-life pancreas.

To complete the model, it is suggested in [66] that

$$\tau_{Ca} \simeq 5 \ ms, \ \tau_K \simeq 2 \ ms. \qquad (8.26)$$

The effect of $s(t)$ is typically negligible. However, when $v \simeq v_{Ca} = 25 \ mV$ and $n \simeq 0$, then, from Eq. (8.25) and Eq. (8.23) $s = [1 + e^{-15}]^{-1} \simeq 1$, which will drive the voltage toward the alternate extreme of $v_K = -75 \ mV$. The time constants in Eq. (8.23) also satisfy

$$\tau_n \simeq 20 \ ms \ll 20 \ s \simeq \tau_s, \qquad (8.27)$$

suggesting that $s(t)$ will slowly decay as compared to the behavior of $v(t)$ and $n(t)$. In such situations, an analysis proceeds by considering $s$ to be a parameter, rather than a variable. It is therefore reasonable to consider Eq. (8.23) to be a two-dimensional system, where the different values of $s$ are responsible for causing transitions between the different states of the system.

# Exercises

1. Rewrite the system in Eq. (8.23) for just $v$ and $n$ in the two cases (a) $s = 0$ and (b) $s = 1$. In each case approximate the equilibrium point using one iteration of Newton's Method, using seed $(v_0, n_0) = (-17, 0.5)$. Then determine the stability properties of each $s$ state using Table 7.2.

## 8.7 Epidemics

The Kermack-McKendrick or SIR model for the introduction and spread of a disease in a *population* $P$ identifies the *susceptible* $S$, *infected* $I$, and *recovered* $R$ subpopulations. The vector $(SIR)^T$ represents the *state* of the epidemic system. An epidemic begins with a small number of infected people, $I_0$, being introduced into the larger population, $P$, at some time $t_0$. The infection rate is jointly proportional to the number of infected people $I$ and the number of susceptible people $S$,

$$\frac{dI}{dt} \propto I \cdot S \iff \frac{dS}{dt} \propto -I \cdot S.$$

The constant of proportionality is denoted as $\nu > 0$. The recovery rate depends only on the number of infected people $I$,

$$\frac{dR}{dt} \propto I \iff \frac{dI}{dt} \propto -I.$$

The constant of proportionality is denoted as $\rho > 0$. Collecting these equations gives a nonlinear autonomous system,

$$\frac{d}{dt}\begin{pmatrix} S \\ I \\ R \end{pmatrix} \equiv \begin{pmatrix} S'(t) \\ I'(t) \\ R'(t) \end{pmatrix} = \begin{pmatrix} -\nu \cdot S \cdot I \\ \nu \cdot S \cdot I - \rho \cdot I \\ \rho \cdot I \end{pmatrix}, \begin{pmatrix} S(t_0) \\ I(t_0) \\ R(t_0) \end{pmatrix} = \begin{pmatrix} P - I_0 \\ I_0 \\ 0 \end{pmatrix}, \quad (8.28)$$

where the parameters correspond to

$\nu =$ interaction rate between uninfected $S$ and infected-contagious $I$ people,
$\rho =$ rate that infected-contagious people $I$ recover $R$.

There are several conditions on the model that must hold to make it applicable to the study of epidemics. Clearly $\nu > 0$, $\rho > 0$, and $0 < I_0 < P$ are required of the parameters for a nontrivial solution to exist. In these cases $S, I, R \geq 0$. For a nonlethal illness, the conservation law

$$P(t) = P_0 = S(t) + I(t) + R(t) \geq I_0, \qquad (8.29)$$

holds. From the third equation in system (8.28), the rate of change of $R(t)$ is never negative and depends only on $I(t)$. Thus, by the Fundamental Theorem of Calculus

$$R(t) = \rho \cdot \int_{t_0}^{t} I(s)ds, \text{ for } t \geq t_0.$$

From the second equation in system (8.28) the value of $I(t)$ reaches a maximum when $S(t) = \rho/\nu$. This will never occur if $\rho > \nu \cdot P_0$ due to Eq. (8.29), in which case $S(t)$ steadily decreases while $I(t)$ and $R(t)$ steadily increase.[2]

The main point presented here is that $S(t)$ never actually becomes 0 in this model, which is consistent with what is observed in nature. To use this theoretically, divide the first two equations in system (8.28) to obtain

$$\frac{dI}{dS} = \frac{dI/dt}{dS/dt} = \frac{I'}{S'} = \frac{(\nu \cdot S(t) - \rho) \cdot I(t)}{-\nu \cdot S(t) \cdot I(t)} = \frac{(\nu \cdot S - \rho) \cdot I}{-\nu \cdot S \cdot I},$$

which is a separable ODE, independent of $t$, leading to

$$dI = (-1 + \rho/(\nu \cdot S))\, dS = -dS + (\rho/\nu) \cdot dS/S.$$

Now integration, independent of $t$, from $(I_0, S_0) = (I_0, P_0 - I_0)$ to $(I, S)$, gives

$$I - I_0 = -(S - (P_0 - I_0)) + ((\rho/\nu) \cdot (\ln(S) - \ln(P_0 - I_0))).$$

Setting $I = 0$ corresponds to the end of the epidemic. Then, since $S$ is never negative, elimination of $I_0$ gives

$$0 \leq S = P_0 + (\rho/\nu) \cdot (\ln(S) - \ln(P_0 - I_0)).$$

Rearranging leads to the inequality

$$-(\nu/\rho) \cdot P_0 \leq (\ln(S) - \ln(P_0 - I_0)),$$

---

[2]Lawrence Moore and David Smith developed calculus labs using real data to estimate the epidemic parameters [51]. This is used to model the death rate observed in an actual epidemic.

and taking the exponential of both sides, which preserves the inequality, gives

$$\exp\left(-(\nu/\rho)\cdot P_0\right) \leq S/(P_0 - I_0).$$

**Conclusion:** The proportion of the population $P_0$ that never gets infected is more than $e^{-\nu \cdot P_0/\rho} > 0$. Increasing the parameter $\nu$ corresponds to the infection becoming more transmittable. Decreasing $\rho$ suggests that the illness is lasting longer in the host. Both situations will increase the spread of the disease and decrease the number of people that never get infected.

A common occurrence with epidemics is a second surge, or *recrudescence*, of infections. There are many possible reasons for this, like people becoming less concerned as $I$ decreases. A simple bifurcation can be modeled by supposing that

$$\nu = \nu(I) = \nu_0 - \gamma \cdot I, \gamma \geq 0, \tag{8.30}$$

which supposes that the transmission rate decreases when the number of infections increases, resulting in a cessation of new cases when $I = \nu_0/\gamma$. One then replaces the infection rate equation with

$$S'(t) = -\nu \cdot S \cdot I + \gamma \cdot S \cdot I^2, \; I'(t) = \nu \cdot S \cdot I - \rho \cdot I - \gamma \cdot S \cdot I^2, \tag{8.31}$$

where $\gamma > 0$ is a correction due to secondary exposures of susceptible people to those who are infected and contagious.

## ■ Exercises

1. Replace the $I$ derivative in system (8.28) with that given in Eq. (8.31). For fixed $\rho > 0$ find values of $\nu > 0$ so that some $\gamma > 0$ will result in a critical point for $I$. The answer will depend on the parameters and the other variables $S$ and $I$. What is the consequence of having $\gamma > 0$ on $S$, for critical $I$?

2. Express the iterative solutions of system (8.28) for $\delta t = 1$. Consider the initial conditions $S_0 = 9$, $I_0 = 1$, and $R_0 = 0$ with parameters $\nu = 0.2$ and $\rho_0 = 0.4$. Create two charts for the iterative solutions: one for $\gamma = 0$ and another for $\gamma = 0.1$ (or near this value). Keep only one decimal place. Compute the fifth iterate to see the effect of the modified model of Eq. (8.30), and point it out.

## ■ 8.8 PCP Diagnosis

When doctors prepare to make a determination as to whether a patient has a disease or not, they must consider the responses to questions and the results of tests. This

## 8.8 PCP Diagnosis

information is then collated leading to an estimation of the likelihood for a disease. *Anchoring* occurs when a diagnosis has been made, and from this point forward the primary care physician (PCP) is convinced of their diagnosis [24]. However, since a sequence of data was used, each with its uncertainties, the conclusion must have a level of uncertainty, leaving open the possibility of another cause for the symptoms reported. Such is the case when a Bayesian analysis is used to catalogue a patient's condition, or state of health.

Consider a disease $D$ associated with a quantitative random variable $X$ that has values $x \in [4, 25]$, in appropriate units.[3] To verify that a diagnosis is possible, volunteers are tested for their values of $X$. This is compared with the results of a *gold standard* used to determine whether the subject truly did not have the disease $N$, was in a borderline state $B$, or had the disease $D$. Associated with these physical states are probability distributions, as displayed in Figure 8.1,

$$
\begin{aligned}
P(x|N) &= \text{Exp}(2)[x-4], \\
P(x|B) &= U(6,8)[x], \\
P(x|D) &= Pois(10-5)[\lceil x-5 \rceil].
\end{aligned}
$$

Suppose that a patient, who goes to a PCP complaining about symptoms, has an $X$ reading of 7.5. What is the probability that the person has the disease $D$?

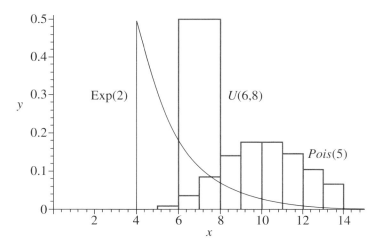

**FIGURE 8.1** ∎ Probability distributions for a disease with three possible states.

---

[3]This example is motivated by diabetes. However, the distributions presented here are hypothetical and do not correspond to any actual study.

Using Bayes Rule, with $\Delta x = 1$ centered at $x = 7.5$, one computes

$$\Delta P_N = \int_7^8 \exp(-[x-4]/2)dx = \frac{e^{-(x-4)/2}}{2}\Big|_7^8 = e^{-2} - e^{-1.5} \simeq 0.0878,$$

$$\Delta P_B = \int_7^8 \frac{1}{8-6}\chi_{[6,8]}(x)dx = \frac{1}{2}x\Big|_7^8 = \frac{1}{2} = 0.5000,$$

$$\Delta P_D = Pois(5)[3] = e^{-5} \cdot \frac{5^3}{3!} \simeq 0.1404,$$

using the definitions in Eq. (3.28), Eq. (3.26), and Eq. (3.33). Assuming a uniform *a priori* distribution for patients with the same *qualitative factors*, or *soft signs*, the probability of this person having the disease is,

$$P(D|x = 7.5) = \frac{\Delta P_N}{\Delta P_N + \Delta P_B + \Delta P_D} \quad (8.32)$$
$$= \frac{0.1404}{0.0878 + 0.5000 + 0.1404} = \frac{0.1404}{0.7282} = .1928 \simeq 19\%.$$

Suppose, conversely, that no initial screening was performed before testing the patient. Then one needs to know that, in the general population, the distribution of disease states $N : B : D$ are $p_N : p_B : p_D = 70 : 20 : 10$ in percentages [58]. The computation of the individual's probability of disease $P(D)$, which was performed in Eq. (8.32), is now modified by Bayes nonuniform *a priori* rule to

$$P(D|x = 7.5) = \frac{\Delta P_N \cdot p_D}{\Delta P_N \cdot p_N + \Delta P_B \cdot p_B + \Delta P_D \cdot p_D} \quad (8.33)$$
$$= \frac{0.1404 \cdot 0.1}{0.0878 \cdot 0.7 + 0.5 \cdot 0.2 + 0.1404 \cdot 0.1}$$
$$= 0.01404/0.1755 = .08 = 8\%.$$

This example is meant to demonstrate the importance of understanding the population, or subpopulation from which a patient or subject has been drawn.

## ■ Exercises

1. Compute the probabilities for the other two health states in the uniform *a priori* case.

2. Compute the probabilities for the other two health states in the 70 : 20 : 10 *a priori* case.

3. Replace the uniform distribution $U(6,8)[x]$ with a Gaussian distribution using the mean and standard deviation from formulas in Section 3.6.2. Then repeat the computations for the uniform *a priori* and the nonuniform *a priori*. Express the results in percentages. Is there a significant difference in the diagnosis using the Gaussian distribution rather than the uniform distribution?

## 8.9 Single-Species Population Models

In the late 1700s Thomas Malthus suggested that human society would experience an eventual catastrophic collapse due to its continued growth. He suggested that food production grew linearly, but population was growing exponentially. However, experience has shown that when society approaches a resource crisis, it changes to resolve the environmental stress. So wars, or price changes, or external aid will restructure the system in a manner that keeps the human population $P$ increasing. The following will explore several population models. A deeper study can be found in the work of Holt and Gomulkiewicz [34].

### 8.9.1 Exponential Growth

The Malthus model assumes that population change is proportional to the population size

$$P' \propto P \iff P' = k \cdot P. \tag{8.34}$$

Given the initial condition that $P(t_0) = P_0$, the exact solution to the model is

$$P(t) = P_0 \cdot e^{k \cdot (t-t_0)}. \tag{8.35}$$

The problem is that this model is wrong, and this can be seen from the data

$$(t_1, P_1) = (1960, 3B), \ (t_2, P_2) = (2000, 6B), \ B \equiv 10^9 \ people, \tag{8.36}$$

from which the growth-rate constant can be computed by substitution into Eq. (8.35), in units of $B$,

$$3 = P_0 \cdot e^{k \cdot (1960 - t_0)}, \tag{8.37}$$
$$6 = P_0 \cdot e^{k \cdot (2000 - t_0)}. \tag{8.38}$$

The model in Eq. (8.34) appears to have three parameters due to the solution in Eq. (8.35). Thus three data points are needed for a unique solution to exist. However, this is an illusion that can be seen by rewriting Eq. (8.35) as

$$P(t) = P_0 \cdot e^{k \cdot t - k \cdot t_0} = P_0 \cdot e^{k \cdot t} \cdot e^{-k \cdot t_0} = A \cdot e^{k \cdot t}. \tag{8.39}$$

This should explain why dividing Eq. (8.38) by Eq. (8.37) gives

$$2 = e^{k \cdot (2000-1960)} \implies \ln(2) = k \cdot 40 \implies k \simeq 0.0173 \text{ yr}^{-1} \simeq 1/(57.7 \text{ yr}).$$

Substituting this into Eq. (8.37), but using the expression in Eq. (8.39) gives

$$3 = A \cdot e^{0.0173 \cdot 1960} \implies A = 3/e^{1960/57.7} = 3/e^{34.0} \simeq 5.3 \times 10^{-15}. \tag{8.40}$$

This completes the Malthus model based on recent data

$$P(t) = 5.3 \times 10^{-15} \cdot e^{t/57.7} B = 5.3 \cdot e^{(t-1993)/57.7} B.$$

Now, a postdiction can be made about the population at the year 0,

$$P(0) = = 5.3 \cdot e^{(0-1993)/57.7} B = 5.3 \cdot e^{-34.0} \times 10^9 \simeq 9.1 \times 10^{-6} \ll 1.$$

The model suggests that it is unlikely that even one person existed 2000 years ago. This is absurd, so the model needs to be changed to better fit the data.

### ■ 8.9.2 Faster-Than-Exponential Growth

In the mid-1960s, Heinz von Foerster suggested that the human population grows quadratically, rather than linearly, as a function of $P$ [76],

$$P' \propto P^2 \iff P' = k_{vF} \cdot P^2. \tag{8.41}$$

The idea is that the growth of the human population is due to two factors:

(1) Fertility less mortality $\implies P' \propto P^1$;
(2) Cooperation and communication $\implies P' \propto P^1$.

Combining these two effects gives $P' \propto P^1 \cdot P^1 = P^{1+1} = P^2$ as given in Eq. (8.41). Plant and animal populations only grow due to factor (1), so the Malthus model is appropriate in the studies of these other species.

Using the initial condition $P(t_0) = P_0$, the exact solution to the von Foerster equation is

$$P(t) = \frac{P_0 \cdot (T - t_0)}{(T - t)}, T > t_0, \tag{8.42}$$

which is a *rational* function, not an *exponential* one! This shows that a more complicated model as in Eq. (8.41) does not necessarily lead to a more sophisticated solution. Using the same data as in Eq. (8.36) results in the equations,

$$3 = \frac{A}{T - 1960}, \tag{8.43}$$

$$6 = \frac{A}{T - 2000}. \tag{8.44}$$

Once again, dividing Eq. (8.44) by Eq. (8.43) gives

$$2 = \frac{T - 1960}{T - 2000} \implies 2 \cdot T - 4000 = T - 1960 \implies T = 4000 - 1960 = 2040.$$

Substituting this into Eq. (8.43) gives $A = 240\ B$ yrs. Thus von Foerster's model is

$$P(t) = 240\ B \text{ yrs}/(2040 \text{ yr} - t). \tag{8.45}$$

For this model the postdiction of the human population at the year 0 is

$$P(0) = \frac{240\ B \text{ yrs}}{2040 \text{ yr} - 0 \text{ yr}} = \frac{240}{2040} \times 10^9 \simeq 1.18 \times 10^8 = 118\ M \text{ people}.$$

This is a more reasonable value for the population 2000 years ago.[4] Adjustments of $A$ and $T$ may allow a better fit if more data is available. Be that as it may, it is now possible to make predictions with some confidence using this model. The main thing to notice from Eq. (8.42) is that $P(t)$ remains positive, but growing as $t \to T^-$, similar to the Malthus model. However, the exponential model never stops growing and is always finite, whereas the von Foerster model predicts an infinite population attained in a finite time. This can be written as $P(T^-) = +\infty$, which again is unrealistic. Still, because this model fits the data so well, something is implied about human behavior. The year $T = 2040$ is referred to as the *doomsday*, and mathematically it means that at some time between now and $T$ human behavior must break out of the simple relationship given in Eq. (8.41). If humanity does not

---

[4] Some figures put the world population at year 0 to be as high as 300 million ($M$) people.

change the parameters, then the rationality of nature will. This may express itself in a more realistic and complicated model that takes into account more subtle aspects of human behavior and its interaction with the environment.

### ■ 8.9.3 Faster-Than-Exponential Logistic Growth

At this point, one may wish to believe that the world population will settle on a *maximally sustainable* level $M$, also known as the *saturation value* of the population, or the *carrying capacity* of the environment. In this case a negative feedback is required to always return the population back to its equilibrium value $M$. A new correcting factor must then appear where:

(3) The population always changes (increases or decreases) toward its maximally allowed level $\implies P' \propto -(P - M)$.

This leads to the logistic von Foerster model:

$$P' \propto P^2 \cdot (M - P) \iff P' = k_{LvF} \cdot P^2 \cdot \left(1 - \frac{P}{M}\right). \tag{8.46}$$

The resulting solutions abruptly go from increasing to nearly constant values, called an *S-shaped growth pattern*. There are two phases of this behaviorial pattern:

(1) ∧ (2) Growth phase $\implies P' \propto P^2$;
(3) Leveling-off phase $\implies P' \propto (M - P)$.

The problem with this model is that $M$ is an unknown so must be estimated using many assumptions about the existence of sustainable resources.

### ■ 8.9.4 Oscillatory Logistic Example

Perfectly smooth logistic growth is not common in nature. Rather, there is often an overshooting of a population size beyond $M$. This leads to an eventual collapse of the population, followed by several oscillations of the population near the equilibrium value $P = M$. This is called a *J-shaped growth pattern*. Factor (3) must be replaced by a factor that facilitates oscillatory behavior. A simple method, often employed, is to *stratify* the total species population by age

$$Y = children, \ A = adults, \ P = Y + A.$$

Then the young portion of the population will only increase if the adult population is below $M$, which now represents the carrying capacity for the most highly consuming part of $P$. Thus, the two equations are

$$Y' = -\alpha \cdot Y + \beta \cdot (M - A), \quad A' = \alpha \cdot Y - \gamma \cdot A, \tag{8.47}$$

where the positive parameters $\alpha$, $\beta$, and $\gamma$ have the obvious meanings. To obtain a qualitative sense of this system, set $\alpha = \beta = \gamma$. Then the equilibrium equations are

$$0 = -\alpha \cdot Y + \alpha \cdot (M - A), \quad 0 = \alpha \cdot Y - \alpha \cdot A, \tag{8.48}$$

which have the only solution $Y = A = M/2$. Defining the new variables

$$Y_* = Y - M/2, \quad A_* = A - M/2,$$

implies that Eq. (8.47) becomes

$$\begin{cases} Y'_* = -\alpha \cdot Y_* - \alpha \cdot A_*, \\ A'_* = \alpha \cdot Y_* - \alpha \cdot A_* \end{cases} \implies \frac{d}{dt}\begin{pmatrix} Y_* \\ A_* \end{pmatrix} = \begin{pmatrix} -\alpha & -\alpha \\ \alpha & -\alpha \end{pmatrix} \cdot \begin{pmatrix} Y_* \\ A_* \end{pmatrix}. \tag{8.49}$$

Since $\mathcal{T} = -2\alpha$ and $\mathcal{D} = 2\alpha^2$, the eigenvalues of the system, $\lambda$, are solutions to the characteristic equation

$$0 = \lambda^2 + 2\alpha \cdot \lambda + 2\alpha^2, \tag{8.50}$$

resulting in the complex solutions $\lambda_\pm = -\alpha \pm i \cdot \alpha$, which leads to associated solution vectors $\langle Y(t), A(t) \rangle$ with *damped oscillations* about the *equilibrium value* $\langle M/2, M/2 \rangle$. Thus, the system will decay at the rate $\alpha$, with angular frequency $\omega = \alpha$.

### ■ 8.9.5 The Lemming Example

Lemmings are small furry mammals that live in the arctic. Their populations are known to fluctuate wildly over a period of several years. Although they have many predators, the chief mechanism for the oscillation in their numbers appears to be their fast reproductive rate coupled with their voracious appetite [48]. This is a species that causes its own instability.[5] To model this population phenomena,

(1) Let $M$ be the carrying capacity for the adult population;
(2) Let $\tau$ be the time-to-reproductive maturity for the young population; and
(3) Measure the adult population $P(t)$ over the period $t \in [t_0 - \tau, t_0]$.

---

[5]A mathematical model is presented here that provides an alternative to the myth that lemmings commit mass suicide at regular time periods.

The differential equation for this model of the species is of the form

$$\frac{dP(t)}{dt} = k \cdot P(t-\tau) \cdot \left(1 - \frac{P(t-\tau)}{M}\right), \quad ICP : P \in \mathcal{C}^0\left([t_0-\tau, t_0] \to \mathbb{R}_0^+\right), \tag{8.51}$$

where an Initial-Condition Path (ICP) must be provided. Expression (8.51) is a type of *delayed-differential equation*, which requires an input path for a unique solution to exist. An iterative approximation to $P(t)$ can be obtained for $t \geq t_0$ using the initial portion of the population provided over an interval of length $\tau$. In particular, define $J \equiv \lceil \tau/\Delta t \rceil \in \mathbb{N}$. Then, the corresponding difference equation is of the form, for $j \geq J$,

$$\begin{aligned} t_{j+1} &= t_j + \Delta t = t_0 + j \cdot \Delta t, \\ P_{j+1} &= P_j + k \cdot \Delta t \cdot P_{j-J} \cdot \left(1 - \frac{P_{j-J}}{M}\right). \end{aligned} \tag{8.52}$$

For lemmings, zoologists have determined a delay parameter of $\tau \simeq 3$ months. Furthermore, ecologists have observed population swings of between 10 *Adults*/ha to nearly 0.1 *Adults*/ha.[6] Thus the carrying-capacity parameter may be reasonably set at $M \simeq 5$ *Adults*/ha. This leaves the growth-rate parameter $k$ to be determined by data. To obtain an accurate value, the growth periods need to be isolated from several cycles of the population's evolution. The result is $k \simeq 1$ month$^{-1}$. Iterative solutions from Eq. (8.52) will depend on the ICP, which should be continuous with values in the range $(0, 10)$ *Adults*/ha. Qualitative results, for a variety of choices of ICP, give oscillatory solutions that have peak-to-peak periods of $T \in [18 \text{ months}, 36 \text{ months}]$.

A solution to the nonlinear delayed-differential Eq. (8.51) has no obvious answer, except for the equilibrium values $P \equiv 0$ and $P \equiv M$. It can be seen that $P = 0$ is unstable, but that $P = M$ is stable. Thus, to linearize Eq. (8.51) define $Q(t) \equiv P(t) - M$, which becomes the alternative delayed-differential equation

$$\begin{aligned} \frac{dQ(t)}{dt} &= k \cdot (Q(t-\tau) + M) \cdot \left(\frac{-Q(t-\tau)}{M}\right) \\ &= -k \cdot Q(t-\tau) - \left(\frac{k}{M}\right) \cdot Q^2(t-\tau). \end{aligned} \tag{8.53}$$

---

[6] A hectar is abbreviated ha and equals $(100 \text{ m})^2 = 10^4 \text{ m}^2$.

In this example, rather than making the mathematical model more complex, a simplification results by assuming that $|Q(t)| \ll M$, which gives the linear delayed-differential equation by dropping the HOT of $Q^2$

$$\frac{dQ(t)}{dt} = -k \cdot Q(t-\tau), \quad \text{ICP}: Q \in \mathcal{C}^0\left([t_0 - \tau, t_0] \to (-M, M)\right). \quad (8.54)$$

There are as many solutions to this equation as there are ICPs. However, a simple harmonic model family of solutions can be guessed,

$$Q(t) = A \cdot \sin(\omega \cdot t). \quad (8.55)$$

This is suggested by the observed behavior of lemming populations. To fit the models in Eq. (8.55) to those in Eq. (8.54), the condition on the parameters becomes

$$A \cdot \omega \cdot \cos(\omega \cdot t) = \frac{dQ(t)}{dt} = -k \cdot Q(t-\tau) = -k \cdot A \cdot \sin(\omega \cdot t - \omega \cdot \tau). \quad (8.56)$$

Using the identity $\sin(\theta - \pi/2 - 2n\pi) = -\cos(\theta)$ for each $n \in \mathbb{N}_0$, gives the two algebraic equations

$$\omega = k, \quad \omega \cdot \tau = \pi/2 + 2n\pi.$$

These *relations on parameters* imply that there are potentially many contributions to the oscillations observed in the lemming population. In particular, in order for the model in Eq. (8.55) to apply, a slower growth rate $k$ requires a longer delay time $\tau$. For lemming populations, the lowest $\omega$, or the *fundamental* frequency, is at $n = 0$

$$\omega_0 = \pi/(2 \cdot \tau) \simeq \pi/(2 \cdot 3 \text{ months}) \simeq (1/2) \text{ month}^{-1}, \quad (8.57)$$

which implies a period of $T = 2\pi/\omega_0 \simeq 12$ months. Although the average period of oscillation is $T \simeq 24$ months, the mathematical model presented in this section gives a qualitative explanation for the *intraspecies* instability. Improvements are possible by noting that $A_0 \simeq 5$ $adults/\text{ha} \simeq M$ is commonly observed for lemmings [77], violating the $|Q(t)| \ll 10$ assumption. This suggests that the nonlinear term in Eq. (8.53) should be taken into account at the population extremes.

### ■ 8.9.6 Logistic Growth with Harvesting

To demonstrate how varying parameters lead to different behaviors, consider a population with (exponential) growth-rate parameter $k$, carrying capacity $M$, and harvest rate $H$. The appropriate mathematical model is

$$\frac{dP(t)}{dt} = k \cdot P(t) \cdot \left(1 - \frac{P(t)}{M}\right) - H, \quad \text{IC}: P(t_0) = P_0 > 0. \tag{8.58}$$

There are three possibilities depending on the size of the harvest rate $H \geq 0$:

(1) $H > k \cdot M/4$. Define parameter $a$ where $a^2 = M \cdot H/k - M^2/4$;
(2) $H = k \cdot M/4 \Longrightarrow a = 0$; and
(3) $H \in [0, k \cdot M/4)$.

A quantitative analysis of case (1), for some fixed $P_0 > 0$, shows that the species population will be positive for a finite time $\Delta t$ from $t_0$, where

$$\Delta t = \frac{1}{a} \cdot \left\{\arctan\left(\frac{M}{2a}\right) - \arctan\left(\left[1 - \frac{2 \cdot P_0}{M}\right] \cdot \left(\frac{M}{2a}\right)\right)\right\}. \tag{8.59}$$

If $P_0 = M/2$, then $\Delta t = (1/a) \cdot \arctan(M/(2a))$, and this approaches $\infty \cdot \pi/2 = \infty$ as $a \to 0^+$. An interpretation is that *extinction is imminent* for $H > k \cdot M/4$. Choosing a harvest level of $H = k \cdot M/4$ gives a semistable equilibrium at $P = M/2$.

### ■ Exercises

1. Consider the data points $\{(0, 0.3B), (1960, 3B)\}$.
   (a) Use the exponential model $P(t)$ in Eq. (8.35) that fits the data and predict $P(2000)$. Compare to $(2000, 6B)$ by computing the relative percentage error.
   (b) Obtain a von Foerster model $P(t)$ as in Eq. (8.42) that fits the data and predict $P(2000)$. Compare to $(2000, 6B)$ by computing the relative percentage error.
   (c) Suppose the population at year 0 was $0.2B$ and recompute the percentages.

2. Set $\beta = 2\alpha$ and $\gamma = 3\alpha$ in Eq. (8.47), find the the equilibrium point, change variables to $(Y_*, A_*)$, write the vector rate of change equation, and find the characteristic equation as in Eq. (8.50). Find the eigenvalues $\lambda_\pm$ and give a brief qualitative analysis of the system.

3. Use the relation in Eq. (8.57) to express $T$ in terms of $\tau$, and find the minimal value of delay $\tau$, in days, that will give the actually observed period $T$.

4. Obtain $\Delta t$ in Eq. (8.59) by integrating Eq. (8.58) in case (1).

5. Find the equilibrium value of $P$ for Eq. (8.58) when $H \in [0, k \cdot M/4)$.

6. The solution to $P''(t) = 0$ is called the *inflection point* for the population $P(t)$.

   (a) Take the derivative of both sides in Eq. (8.58), set $P'' = 0$, and solve for $P$ in terms of $M$.
   (b) Do the same for both sides in Eq. (8.46), solving for $P$ in terms of $M$.
   (c) Comment on the different values of $P$ for inflection, and the consequence for anticipating the approach of $P$ to its carrying capacity $M$.

## ■ 8.10 Interacting Species

An environment can be sustained if its components interact in a manner that ensures survival of all constituent species. Herbivore populations, like deer $H$, will increase in numbers until the region cannot support any larger population levels. This will often motivate migration, which reduces reproductive activity. If an equilibrium is possible (unlike the case of lemmings) then to understand population changes, one can use a logistic model $H' = \eta \cdot H \cdot (1 - H/M)$, where $\eta > 0$ is the growth rate parameter for $H$, and $M$ is its carrying capacity. Carnivores, like wolves $C$, will only increase in numbers if sufficient prey, like deer, are available. Without a sufficient source of sustenance the carnivores will perish with rate $\omega > 0$. This gives a Malthus model with positive-feedback forcing $C' + \omega C = \nu \cdot H \cdot C$. Here $\nu > 0$ is the *predation rate*. This results in a corresponding negative-feedback term for the herbivores, with *reduction rate* $\rho \cdot H \cdot C$. The parameter $\rho > 0$ is a measure of the negative feedback for the prey. Thus one derives a system of equations that couple population sizes of the predator $C$ and prey $H$

$$\frac{d}{dt}\begin{pmatrix} H \\ C \end{pmatrix} = \begin{pmatrix} H' \\ C' \end{pmatrix} = \begin{pmatrix} \eta \cdot H \cdot (1 - H/M) - \rho \cdot H \cdot C \\ -\omega \cdot C + \nu \cdot H \cdot C \end{pmatrix}. \tag{8.60}$$

This is a nonlinear autonomous system also referred to as a *Lotka-Volterra equation*. Numerical methods can be used to obtain approximate solutions for different initial conditions. It often occurs that if $C(t)$ and $H(t)$ do not vanish, they oscillate in size. When $C$ is plotted against $H$ to obtain a *phase diagram* the *orbits* are somewhat triangular, indicating different phases of the species interaction.

### ■ 8.10.1 Equilibria

Typically, population levels will oscillate about an equilibrium point $\langle H_{eq}, C_{eq} \rangle$, which can be identified from a sufficiently broad data set. If the model in Eq. (8.60) is accurate, then the equilibrium can be expressed in terms of the parameters $\{\eta, \rho, \omega, \nu, M\}$ by solving $\vec{0} = (H', C')^T$. At least three data points are needed to reconstruct the equation, and they should be spread by sufficient time intervals to allow a sense of the rate of change. Data should reveal the different dominant subprocesses taking place at different times during the evolution of the equations.

### 8.10.2 Few Carnivores and Herbivores

When $C$ and $H$ are at their lowest values, then $H$ will increase as if it had been isolated, $H' \simeq \eta \cdot H$. If data is collected during this growth period of $H$, then the value of $\eta$ can be estimated.

### 8.10.3 Few Carnivores and Many Herbivores

If $C$ is low, but $H$ is large, the $C$ value will increase according to $C' \simeq \nu \cdot H \cdot C$. Thus the last parameter $\nu$ can be estimated from the data during this period. Also, at its peak, $H \lesssim M$. Thus the decay rate parameter $\rho \simeq -H'/(H \cdot C)$ also follows from the data, if a sufficiently large sample is collected.

### 8.10.4 Few Herbivores and Many Carnivores

Finally, when $C$ is large and $H$ values plummet, the $C$ value will also begin to decay according to the equation $C' \simeq -\omega \cdot C$. Solving this equation gives the value of $\omega$.

### Exercises

1. Set the parameters in Eq. (8.60) to

$$\eta = \rho = \omega = \nu = M \equiv 2.$$

   For a stationary solution, first set $C' = 0$ and find two solutions. Then set $H' = 0$ and consider each case separately. Then find the Jacobian matrix of the right-hand side of Eq. (8.60) and classify the stationary points as either stable or unstable.

2. Plot the following data points on a phase plot of $C$ vs. $H$ and close the orbit by connecting Year 5 to Year 0. Estimate the parameters $\eta$, $\rho$, $\omega$, and $\nu$ by identifying regions (8.10.2), (8.10.3), and (8.10.4) and using the approximations suggested in this section, along with the assumption that $M = max(H)$. From system (8.60) and estimates on the parameters, compute and identify the equilibrium point $\langle H_{eq}, C_{eq} \rangle$ on the phase plot.

| Time | Year 0 | Year 1 | Year 2 | Year 3 | Year 4 | Year 5 |
|---|---|---|---|---|---|---|
| Herbivores | 7 | 12 | 15 | 10 | 7 | 4 |
| Carnivores | 1 | 2 | 3 | 4 | 3.5 | 2.4 |

## 8.11 Thermodynamics of Oceans and Atmosphere

There are several equations of state governing any bounded thermodynamic system. These involve variables that do not depend on the size of the system, known as *intensive* variables, and variables that change proportionally with the size of the system, known as *extensive* variables:

$T$: Temperature (*intensive*)
$P$: Pressure (*intensive*)
$\rho$: Density (*intensive*)
$V$: Volume (*extensive*)
$S$: Entropy (*extensive*)
$U$: Internal Energy (*extensive*)

A closed system, of finite volume $V$, will have many possible states, each with different energies $e_j$. With an increase in internal energy $U$, more states become available to the system. Assuming that each state is equally likely, entropy $S$ is defined to be the quantity

$$S \equiv k \cdot \ln \left( \text{number of states } \langle \{e_j\}_{j=1}^N \rangle \text{ where } \sum_{j=1}^N e_j = U \right). \tag{8.61}$$

Here the constant of proportionality $k \simeq 1.38 \times 10^{-23} J/K$ is Boltzmann's constant.

Given two systems, where $N_1$ and $N_2$ are their respective number of states, the combined system has the number of states $N_{1,2} = N_1 \cdot N_2$. Definition (8.61) implies that the entropy of the combined system is $S_{1,2} = S_1 + S_2$. Thus, entropy is a naturally occurring *extensive* variable of a system [78]. Furthermore, the entity

$$k \cdot \beta \equiv \frac{1}{T} \equiv \left( \frac{\partial S}{\partial U} \right)_V, \tag{8.62}$$

is a measure of how the complexity of a system increases with an increase in energy $U$. Both energy and entropy are extensive quantities. The derivative in Eq. (8.62) defines the *temperature* $T$, where $k$ converts from *Kelvin* units to units of *heat* energy

*Joules*. Heat corresponds to the average of the kinetic energy of the constituent particles making up a substance. Temperature is an *intensive* variable of a system [78] so remains constant throughout its constituent parts, when the system is in equilibrium.

Two different systems with volumes $V_1$ and $V_2$ may have different temperatures $T_1$ and $T_2$. When brought together, the number of states of the combined system increases due to the larger volume $V_1 + V_2$, and so volume is *extensive*. The two systems are in thermal equilibrium when their temperatures agree. However, the process of getting to equilibrium can take time and is often too complex to model in terms of individual particles. Bringing two volumes together, without adding energy, naturally results in the entity

$$P \equiv \frac{1}{k \cdot \beta} \cdot \left(\frac{\partial S}{\partial V}\right)_U = T \cdot \left(\frac{\partial S}{\partial V}\right)_U, \qquad (8.63)$$

defined to be the *pressure* of the combined system. $P$ is an *intensive* variable, so if the two subsystems can exchange particles, then, as with temperature, equilibrium will require the pressures $P_1$ and $P_2$ to agree.

Pressure has units of energy-per-volume, or *energy density*. Thus $P \cdot dV$ is the amount of potential energy in a small volume $dV$ of the volume $V$. To measure this quantity, a barometer can be put on a small portion $dA$ of the surface $\partial V$ of the volume $V$, and the force measured. Thus pressure also has units of force-per-area.

Any four of the variables can be written in terms of the remaining two. Thus there are always four dependent and two independent variables in thermodynamic problems. A small amount of work done on a system $dU$ is a sum of the amount of heat added to the system $\delta Q \equiv T \cdot dS$ and the amount of work done on the system $\delta W = -P \cdot dV$. Thus

$$dU = \delta Q + \delta W = T \cdot dS - P \cdot dV. \qquad (8.64)$$

For example, a solid or liquid may be compressed with little change in their volumes. However, gases are quite compressible, so in an open environment equality of pressures is achieved quickly. The calculations of energy exchanges during an experiment are simplified if a different internal energy function, called *enthalpy*

$$H \equiv U + P \cdot V \implies dH = T \cdot dS + V \cdot dP, \qquad (8.65)$$

is used. $H$ is a *state function* meaning that it depends only on the initial and final states of the system. *Hess's Law* asserts that the changes in enthalpy, from the start of a reaction to its completion, is the same as the change in heat $\Delta H$. This can be determined using a change of temperature measurement $\Delta T$.

## 8.11 Thermodynamics of Oceans and Atmosphere

Finally, note that by Eq. (8.64) if the volume of a system increases, while staying at the same pressure and maintaining its heat content, then its internal energy will decrease. In a slow open-air experiment, pressure is usually the same as the *ambient* value during the entire reaction in which case Eq. (8.65) is preferred.

### ■ 8.11.1 Dissolved $CO_2$

On earth the partial pressure, or *fugacity*, of carbon dioxide in the atmosphere pres($CO_2$) is in equilibrium with the dissolved gas in the oceans. The concentration of gas, dissolved in water, satisfies *Henry's Law*

$$[CO_2 \cdot H_2O] = k(T) \cdot \text{pres}(CO_2). \qquad (8.66)$$

For the same atmospheric gas pressure, the proportionality parameter $k(T)$ is expected to decrease as temperature increases. If the atmospheric temperature of the earth increases, then the effect of increasing $CO_2$ may result in more or less of this gas being dissolved in the oceans.

### ■ Exercises

1. The atmospheric pressure on the surface of the earth is about 1 atmosphere $\equiv$ 101.3 kiloPascals and $CO_2$ is presently at levels of about 385 parts-per-million (ppm). Measurements of the concentrations of dissolved carbon dioxide gives:

   | Temperature $(T^\circ K)$ | 275 | 285 | 295 | 310 |
   |---|---|---|---|---|
   | $[CO_2 \cdot H_2O]$ g/L | 0.03 | 0.02 | 0.015 | 0.01 |

   (a) Use the exponential model $k(T) \simeq k_0 \cdot \exp[-\lambda \cdot T]$ for the constant of proportionality in Henry's Law (8.66) and the preceding data to find $k_0$ and $\lambda$ that best fits, according to regression. Express $\lambda$ in $K^{-1}$.
   (b) Let $P = \text{pres}(CO_2)$ and write the model as a function of the two variables $T$ and $P$. Express $k_0$ in $g/(L \cdot Pa)$.
   (c) For the Henry model in Eq. (8.66), assume that the dissolved carbon dioxide is constant and take the differentials of both sides of Eq. (8.66). Estimate the necessary change in temperature $dT$ required to allow a constant level of aqueous $CO_2$ when the pressure $P \equiv \text{pres}(CO_2)$ increases by 20%, i.e., $(dP/P) = 0.2$. What is this temperature in Celsius, and is this possible?

2. Atmospheric concentrations of carbon dioxide over time $(t_i, [CO_2]_i)$ have been collected by a meteorologist. A summary of the results over a 50-year period, in units of (year, ppm), are:
$\{(1960, 315), (1970, 325), (1980, 340), (1990, 350), (2000, 370), (2010, 385)\}$.

   (a) Plot the data points, and plot a Staircase function between the points using an average of years before and after the collected data point.

   (b) Define $s_i = (t_i - 1960)/10$ and $y_i = ([CO_2]_i - 300)/5$. Then use linear regression to fit the model $\hat{y}_\ell(s) = m \cdot s + b$. Convert back to a model for $[CO_2](t)$ and plot it on the graph from (a).

   (c) By taking the logarithm of the data, and defining the variable $s$ as in (b), fit the model $\hat{y}_e(s) = A \cdot \exp(k \cdot s)$ using linear regression. Keep four significant figures! Convert back to a model for $[CO_2](t)$ and plot it on the graph from (a).

   (d) Comment on the graphical difference between the two models.

3. Ice core data $(\tau_i, [CO_2]i)$ has been collected by a geologist dating back 400,000 years. The maxima and minima during this period are:
$\{(-4, 280), (-3.5, 180), (-3.2, 300), (-2.6, 180), (-2.3, 280),$
$(-1.5, 180), (-1.2, 280), (-0.2, 180), (0.0, 280)\}$ in ($10^5$ years, ppm).

   (a) Plot the data as a Sawtooth function, as defined in Eq. (2.14), assuming that the $[CO_2]$ levels reported are local extreme points.

   (b) Find the mean and standard deviation of the $[CO_2]$ levels.

   (c) Assuming a normal distribution, what is the probability of obtaining 385 ppm or greater? Why is it unreasonable to use control chart techniques for a study of this data, and does this make the true probability $P([CO_2] \geq 385)$ higher or lower?

   (d) From the times $\{\tau_i\}$ compute the average times $\Delta\tau$ between peaks or valleys to estimate the period of these oscillations, and give the 68% confidence interval in years.

   (e) State the estimated natural frequency $\nu$ and angular frequency $\omega$. Then create a model of $[CO_2](t)$ in the form $f(t) = sd_{CO_2} \cdot \cos(\omega \cdot (t - t_0)) + mean_{CO_2}$, by adjusting the shift parameter $t_0$, and plot it on the graph in (a).

4. Two connected thermodynamic systems have variables $\{T_1, P_1, V_1, S_1, U_1\}$ and $\{T_2, P_2, V_2, S_2, U_2\}$. The total change in heat for the combined systems is

$$\delta Q \equiv P_1 \cdot dV_1 + dU_1 + P_2 \cdot dV_2 + dU_2.$$

Designate $\{V_1, U_1, V_2, U_2\}$ to be the independent variables, where the remaining variables are dependent on these.

(a) If the total volume is held constant, how are $dV_1$ and $dV_2$ related?
(b) If the total energy is held constant, how are $dU_1$ and $dU_2$ related?
(c) Using the property of differentials that $d^2 f = 0$, simplify the equation $d\delta Q = 0$ and state the resulting extra condition that must hold.
(d) If $d\delta Q \neq 0$ then show that $\delta Q \cdot d\delta Q = 0$, using the property that $df \cdot df = 0$.
(e) Suppose that subsystem 1 has the pressure function $P_1(U,V) = U/V$, and the system starts in the state $(U_{1,i}, V_{1,i}) = (0,1)$. Now suppose that the final state $(U_{1,f}, V_{1,f}) = (3,2)$ is obtained by either first increasing $V_1$ by 1, then increasing $U_1$ by 3; or first increasing $U_1$ by 3, then increasing $V_1$ by 1. Which path results in system 1 having more heat $Q$?

## ■ 8.12 Climate and Tree Rings

There is a known relationship between the climate conditions over a year and the size of a tree ring. Although there are many factors that can be used to define *climate* in a terrestrial *niche*, only yearly precipitation $P$ and averaged growth-period temperature $T$ are considered here.[7] In this simplified example, the set

$$\mathcal{D}_{Climate} = \{\langle P, T \rangle : 0 \leq P \leq 200 \text{ mm/year}, -50°C \leq T \leq 50°C\} \quad (8.67)$$

corresponds to the possible range of values experienced on the earth's surface. Under ideal conditions a tree ring will be large, corresponding to a good growth year. Conversely, if there is drought, excessive rain, and/or a temperature extreme for a lengthy period, the tree ring will be thin. However, under moderate conditions, a positive-linear relationship has been observed, for many species of trees, between the size of a ring and yearly-averaged precipitation and temperature. To find this relationship, data needed to be extracted from tree trunks using a core sample. This process gives a data triple $\{\langle P_j, T_j, R_j \rangle\}_{j=1}^{N}$. The linear model is

$$R = R(P, T) = a \cdot P + b \cdot T + c, \quad (8.68)$$

and $\{a, b, c\}$ are parameters of the model. Three linearly-independent data points are good enough to obtain a result for the parameters, but it is unlikely to give a

---

[7]The specifics of these definitions are quite subtle. In fact, the climate factors from several previous years often need to be considered in rationalizing growth rates of plants [18].

repeatable result. To obtain confidence in the model in Eq. (8.68), the sample size $N$ must be large and randomized. This then leads to the matrix equation

$$\vec{R} = \begin{pmatrix} \vec{P} & \vec{T} & \vec{1} \end{pmatrix} \cdot \begin{pmatrix} a \\ b \\ c \end{pmatrix}. \tag{8.69}$$

The next step in the data analysis is to consider a control chart for the ring sizes for each tree, represented as the sequence $\{\langle t_j, R_j\rangle\}_{j=1}^N$. Ring widths for conifers are initially large (2–3 mm) but level off after many decades to 1 mm. An exponential model commonly used is

$$R(t) = A \cdot e^{-\lambda \cdot t} + B, \tag{8.70}$$

where the constants $A$, $B$, and $\lambda$ are positive parameters. This is a *heuristic* model, meaning that there is no particular philosophy justifying this form, except that it works well to fit data $\{\langle t_j, R_j\rangle\}_{j=1}^N$. More general functions, with more parameters, may be able to fit the data better, however, this might hide the variations of the trees rings that are due to climate change, rather than normal growth patterns. The model in Eq. (8.70) is nonlinear and not easy to separate. The approach suggested here is to determine $B$ first from the data. Since $R(\infty) = B$, but is not symmetric, the median value of $\{R_j\}_{j=1}^N$ is taken as an estimate for $B$. Then an adjusted variable $S(t) = R(t) - B$ can be modeled by the two-parameter exponential decay function $A \cdot e^{-\lambda \cdot t}$, which is also nonlinear. However, taking logarithms of both expressions gives a linear relation

$$\ln(S(t)) = -\lambda \cdot t + \ln(A), \tag{8.71}$$

allowing regression to be used. One concern should be that $S \leq 0$ is possible. Estimation of the parameters $\lambda$ and $A$ needs to be performed on the early data, where $S$ is the largest. Finally, adjusted data is considered $T(t) = S(t) - A \cdot e^{-\lambda t}$. The transformed data set $\{\langle t_j, T_j\rangle\}_{j=1}^N$ will have variations due mainly to environmental changes, and not due to the stage of the tree's development. If several trees are measured in the same region, then an *Analysis of Variance* (ANOVA) can be performed to separate good growth years from the poor ones.

Scientists who study tree rings prefer to use *ring-width indices*, defined as

$$I(t) \equiv \frac{R(t)}{Ae^{-\lambda t} + B}. \tag{8.72}$$

## 8.12 Climate and Tree Rings

Statistical properties of $I(t)$ are that the mean is close to 1 and the standard deviations of different trees, in the same region, are similar [18].

### Exercises

1. Consider the sizes of tree rings, averaged over 5-year periods:
   $\{(5, 2), (10, 5), (15, 3), (20, 3), (25, 2),$
   $(30, 1.5), (35, 1), (40, 2), (45, 1.5)\}$ in (years, mm).
   Consider three models $R_1(t) = A_1 \cdot t$, $R_2(t) = A_2 \cdot e^{\lambda(t-t_2)}$, and $R_3(t) = B$ to account for the three stages of tree-ring growth: (1) seedling, (2) sapling, and (3) mature.

   (a) Use the first two data points (seedling) to obtain the best fit model $A_1 \cdot t$.
   (b) Take an average of the last five data points $mean_R$ to estimate $B$ for mature trees. Compute the standard deviation $sd_R$. Find the median $R_{med}$ for the full data set. How many standard deviations is $R_{med}$ from $B \equiv mean_R$.
   (c) Using $B$ found, create an *adjusted* data set $S_j = (R_{2,j} - B)$. Identify the time intervals associated with the three stages of growth.
   (d) Using only sapling-growth data that have positive values, create the transformed data set $(t_j, \ln(S_j))$ and find $\lambda$ for the model $(R_2(t) - B)$.
   (e) Express the full model $R(t)$ based on the parameters now estimated by the data. Plot the data and the tree-ring models on the same graph.

2. Data from two species of trees are recorded in the following tables.

| Species 1 Precipitation | Temperature 20°C | 25°C | 30°C | Species 2 Precipitation | Temperature 20°C | 2°5C | 30°C |
|---|---|---|---|---|---|---|---|
| 10 mm | 3 | 2 | 1 | 10 mm | 4 | 3.1 | 2.2 |
| 20 mm | 4 | 3 | 2 | 20 mm | 4.5 | 4.2 | 3.3 |
| 30 mm | 5 | 4 | 3 | 30 mm | 5.7 | 5 | 4.4 |

**TABLE 8.1** ■ Tree rings during adult stage of two species (in mm).

   (a) Choose three points for species 1 and three for species 2.
   (b) Use the three points chosen, for each species, and two linear models
   $\hat{r}_1(T, P) = a_1 \cdot T + b_1 \cdot P + c_1$ and $\hat{r}_2(T, P) = a_2 \cdot T + b_2 \cdot P + c_2$,
   to obtain six equations for the six parameters $\{a_1, b_1, c_1, a_2, b_2, c_2\}$.
   (c) Write the six equations as two $3 \times 3$ matrix equations.

(d) Check that the determinant of both 3 × 3 matrices is nonvanishing. If not, change out a data point with one from Table 8.1, and check the determinant again. (Do not erase work!)
(e) Now, using Gaussian Elimination, solve for the parameters.
(f) Clearly express the two models $\hat{r}_1(T, P)$ and $\hat{r}_2(T, P)$.
(g) Tree rings from the same region and the same year are studied. A single measurement taken for two species gives: $r_1 = 2.3$ mm and $r_2 = 3.5$ mm. Use the models to estimate both temperature $T_*$ and precipitation $P_*$ for the year corresponding to the ring measurement.

## 8.13 The Motion of the Planets—the Royal Science

On a clear night, the stars appear to follow a well-defined path in the sky. The points of light in the sky, called the *luminaries*, were once considered to be elements of a *celestial sphere* encircling the Earth. The positions of these objects could be monitored each night through an alignment with fixed sight-stones or *station stones*, from which it becomes clear that there is an incremental shift in the position of the stars of about 1° each day. After about 365 nights, the stars (and the sun) returned to their original positions, with the exception of the moon and the planets (also known as the *wandering stars*). It is the purity of observation, and the ease of questioning, that resulted in astronomy being dubbed the *Royal Science*.

The lines of longitude $\theta = constant$ and latitude $\phi = constant$ were imagined to be imprinted, from the Earth out onto the celestial sphere, to some unknown distance. In this way the motion of a particular star could be modeled [20]. In particular, consider a star observed on the equator, along the *prime meridian* at some moment in time $t_0$. Then, from Earth, the path of that star at some later time $t$ is defined by the simple parameterization, using the definition in Eq. (2.14),

$$\theta(t) = 360° \cdot Sawtooth\left(\frac{t-t_0}{365.242 \text{ days}} + \frac{t-t_0}{24 \text{ hours}}\right)$$
$$\phi(t) = 0°, \qquad (8.73)$$

where $t$ is a measure of time, in appropriate units. The star has returned to the same position in the sky whenever $\theta(t) = 0°$.

**Definition 8.2** A *day* on Earth is defined to be the time from when the sun just rises above the horizon, until it rises again. An *hour* is defined to be 1/24th of a day, a *minute* is 1/60th of an hour, and a *second* is 1/60th of a minute. Based on these definitions, the time for the earth to rotate once around its axis is 23 hr and 56 min. This is less than 24 hr since the Earth rotates while it is also orbiting around the sun.

The axis of rotation of the Earth points toward *Polaris* in the northern hemisphere ($\phi = 90°$), and the faint star *Sigma Octantis* in the southern hemisphere ($\phi = -90°$). Measuring the angle between the horizon and one of these stars indicates one's latitudinal position $\phi$ on the Earth's surface. To keep track of one's longitudinal position requires measuring other stars, in combination with careful time-keeping.

The perspective just described was sufficient for earthly navigation. However, since the planets did not follow the simple pattern of linear motion with respect to time, there was a scientific problem with the model. An approach for incorporating planetary motion was created by Claudius Ptolemy about 2000 years ago [57]. In particular, it was possible to include correction terms for the increased and decreased speeds observed with the planets. This lead to a change in their postulated orbits. These changes were referred to as *epicycles*, and they accounted for the observed *retrograde* motion of the planets.

## ■ Exercises

1. Use the formula in Eq. (8.73) and a diagram to determine whether the Earth rotates on its axis in the same way that it revolves around the sun, or oppositely.

2. Use the model in Eq. (8.73), with $t_0 = 0$, to compute $t$ in hours (to two decimal places), for $\theta = 2\pi^-$ radians. Then convert to hours and minutes.

3. Earth follows a path determined by the angle $\theta_E(t) = 2\pi \cdot t/365$, with the sun at the origin, whereas Mars follows a path with angle $\theta_M(t) = 2\pi \cdot t/687$.

   (a) Express the paths $P_E(t) = (a \cdot \cos(\theta_E), a \cdot \sin(\theta_E))$ and $P_M(t) = (b \cdot \cos(\theta_M), b \cdot \sin(\theta_M))$ in terms of $t$.
   (b) Compute the position vector from the Earth's perspective, which is $\overrightarrow{P_{EM}}(t) = P_M - P_E \equiv (x_{EM}(t), y_{EM}(t))^T$.
   (c) The angle $\theta_{EM}(t)$ can be determined by solving $\tan(\theta_{EM}) = y_{EM}(t)/x_{EM}(t)$. Just write this expression and then re-express using the notation $c_E \equiv \cos(\theta_E)$, $c_M \equiv \cos(\theta_M)$, etc., along with the knowledge that $b \simeq 1.5 \cdot a = 3a/2$.
   (d) Find the time of retrograde motion of Mars, as observed from Earth, by solving $0 = \dfrac{d\tan(\theta_{EM})}{dt}$. The identity for $\cos(\theta_M - \theta_E)$ will be needed.

4. Venus follows a path with angle $\theta_V(t) = 2\pi \cdot t/225$ radians.

   (a) Compute the paths $P_V(t) = (c \cdot \cos(\theta_V), c \cdot \sin(\theta_V))$, and the position vector from the Earth's perspective, which is $\overrightarrow{P_{EV}}(t) = P_V - P_E$.
   (b) The angle $\theta_{EV}(t)$ can be determined by solving $\tan(\theta_{EV}) = y_{EV}(t)/x_{EV}(t)$. Let $c = \rho \cdot a$ and eliminate $a$ from the expression $y_{EV}/x_{EV}$. Then re-express using the notation $c_E \equiv \cos(\theta_E)$, $c_V \equiv \cos(\theta_V)$, etc.
   (c) The time of retrograde motion for Venus is about 42 days, as observed from Earth. By solving $0 = \dfrac{d \tan(\theta_{EV})}{dt}$ find possible values for $\rho < 1$.
   (d) Which value of $\rho$ is correct, if Venus is never seen much more than 45° above the horizon?
   (e) If the distance from the sun to the Earth is $a \simeq 1.5 \times 10^8$ km, estimate the distance from the sun to Venus using the estimate for $\rho > 0$ obtained.

## 8.14 The Sky Is Blue, but the Universe Is Red

Observable light consists of a small interval in the full width of the electromagnetic spectrum. When James Clerk Maxwell combined the forces of electricity and magnetism in the 1860s, a consequence was the prediction that the unified force field could oscillate, creating waves that travel at $c = 3.00 \times 10^8$ m/sec. This was several decades before Max Planck connected the energy $E$ of a photon,[8] to the natural frequency of radiation $\nu$. The relationship is linear $E = h\nu$ where Planck's constant is $h \simeq 6.63 \times 10^{34}$ Joules · sec. Photons interact with atmospheric molecules. The sky is blue is because of the principle that:

*Higher frequencies tend to scatter more readily than lower frequencies.*

This principle is noticed in large animals who use low frequencies to communicate over long distances, whereas small animals use high frequencies to communicate over short distances. High frequency waves are energetic and tend to interact with material at their finer scales. This results in greater detail when such waves are focused, but their ability to penetrate, without doing damage or getting scattered, is limited. Lord Raleigh theoretically showed in the 1870s that the intensity of light that passes through a material decreases as $\nu^{-4}$. Red light has frequency $\nu_{red} \sim 8 \times 10^{14}$ Hz and blue light has $\nu_{blue} \sim 4 \times 10^{14}$ Hz. Thus the proportion of red light that passes through the atmosphere compared to blue light is about

---

[8] Here, energy is measured in Joules = kg · m² / sec² or the force of 1 Newton over 1 meter.

$(\nu_{red}/\nu_{blue})^{-4} = 2^4 = 16$. The blue light is reflected down toward the Earth, in all directions. When the sun is on the horizon, blue light is scattered away from the field of view, leaving mainly red and yellow light in view.

In the 1840s Christian Doppler showed that if a source of waves is moving toward an observer, then the frequency is higher for the observer than for the emitter. One reason for this is that the observer it detecting not just the radiation from an object, but also the motion of that object. Thus waves from an approaching radiating object are getting compressed from the observer's point of view. Therefore, if the emitter is moving toward a stationary observer with speed $s_e$, while emitting a wave with speed $c$ and frequency $\nu_e$, then the detected frequency $\nu_o$ is

$$\nu_o = \frac{c}{c - s_e} \cdot \nu_e, \quad \Delta\nu \equiv \nu_o - \nu_e = \frac{s_e}{c - s_e} \cdot \nu_e, \quad (8.74)$$

where it is assumed that $c \gg |s_e|$. This gives a blue shift if $s_e > 0$. Observing distant galaxies in the 1920s Edwin Hubble was able to demonstrate that in most cases, $\Delta\nu < 0$, which means that they are moving away in all directions. Furthermore, it has been noted that the farther another galaxy is from the *Milky Way* galaxy, the greater the red shift, thus the faster the galaxy is moving away. The relationship between speed and distance is found to be linear. In particular, if $d_e$ is the distance of the emitting galaxy,

$$|s_e| \propto d_e \implies |s_e| = H_0 \cdot d_e, \quad (8.75)$$

where $H_0$ is known as *Hubble's constant*. The distances are measured using size and luminosity estimates, and contain many sources of uncertainty. Using the fact that typically $|s_e| \ll c$, one obtains the formula for the distances from Earth

$$d_e = |s_e|/H_0 \simeq (c \cdot |\Delta\nu|/\nu_e)/H_0 = (|\Delta\nu|/\nu_e) \cdot c/H_0. \quad (8.76)$$

## ■ Exercises

1. Suppose the distribution of intensities from the sun $I_0(\nu)$, in the light-frequency interval $[red, blue] \equiv [400 \text{ TeraHz}, 800 \text{ TeraHz}]$, is uniform. The total intensity of light reaching the outer atmosphere is $I_0 \simeq 3$ Watt $\cdot$ sec/m$^2$ for each frequency. The intensity that reaches the ground is $I(\nu) = I_0 \cdot \exp[-\delta \cdot \nu^4]$. Suppose that the total peak intensity at noon, on the Spring equinox, at the equator, on the surface of the Earth is $\mathcal{I}_{total} = \int_{red}^{blue} I(\nu) d\nu = 1.0$ kW/m$^2$.

   (a) Find the linear-in-$\delta$ approximation of $I(\nu)$ for $\nu \in [red, blue]$ in TeraHz.
   (b) Use the linear approximation of $I(\nu)$ and $\mathcal{I}_{total}$ to estimate $\delta$, denoted $\delta_0$.

(c) To improve the estimate of $\delta$, factor out $I(400)$ from $I(\nu)$ and use a linear approximation of $I(\nu)/I(400)$ in the integral for $\mathcal{I}_{total}$ to obtain a transcendental equation for $\delta$.

(d) Rearrange so that it becomes a fixed-point problem, and use one iteration to obtain an improved estimate for the parameter $\delta$ denoted $\delta_1$.

(e) Restate the nonuniform distribution using $\delta_3$ and compute the ratio $I(red)/I(blue)$ at the outer atmosphere and on the Earth.

2. The sun emits strong electromagnetic waves at $\nu_S = 1.4204 \times 10^9$ Hz.

(a) The Andromeda Galaxy is our closest neighbor at $2.2 \times 10^6$ light years away,[9] yet the closest spectral line to $\nu_S$ occurs at $1.4211 \times 10^9$ Hz. Determine how fast Andromeda is approaching using Eq. (8.74). When will it hit the Milky Way galaxy, assuming a direct path toward the center?

(b) Combine Eq. (8.74) and Eq. (8.75) to obtain Eq. (8.76). Then use the observed frequency of $1.17 \times 10^9$ Hz corresponding to $\nu_S$ for the object *3C 273*, which is at a distance $d_e = 2.5 \times 10^9$ light years, to estimate $H_0$.

(c) For very fast moving objects, the relativistically adjusted formula $\nu_o = (\sqrt{c+s_e}/\sqrt{c-s_e}) \cdot \nu_e$ must be used rather than Eq. (8.74). The most distance galactic-sized object is not a Quasar. From the *Gamma Ray Burst 090423* a frequency of about $3 \times 10^8$ Hz peak is detected. Assuming that the mechanism for emission of this frequency is the same as that for the sun $\nu_S$, estimate the speed of this source $|s_e| \lesssim c$. What is the minimal size and age of the universe?

## ■ 8.15 Quantum Cosmology and the Big-Bang Theory

In 1916 Albert Einstein published equations for his general theory of relativity. It reconciled the force of gravity with the physical observation that the distance between two space-time points $\mathcal{P}_1 \equiv (t_1, \vec{r}_1)$ and $\mathcal{P}_2 \equiv (t_2, \vec{r}_2)$, called *events*,

$$dist(\mathcal{P}_1, \mathcal{P}_2)^2 = c^2 \cdot (t_1 - t_2)^2 - (x_1 - x_2)^2 - (y_1 - y_2)^2 - (z_1 - z_2)^2, \quad (8.77)$$

is the same to all observers. When events occur nearly simultaneously, then the physical *length*

$$length = dist(\vec{r}_1, \vec{r}_2) = \sqrt{(x_1 - x_2)^2 + (y_1 - y_2)^2 + (z_1 - z_2)^2}, \quad (8.78)$$

---

[9] A unit light year ≡ speed-of-light year = $c \cdot 1$ year $\simeq 9.5 \times 10^{15}$ m.

is an invariant to all observers moving with the same velocity, or comoving. The invariance of *length* is part of common experience, but does not hold true in high-energy particle experiments. The theory of general relativity represents a mathematical model for describing the large-scale structure of the universe while remaining faithful to subatomic particle physics. Verification of Einstein's equations come from observing electromagnetic radiation reaching telescopes on Earth and on satellites. This radiation is in the form of waves and particles, created by nuclear, chemical, and thermodynamic interactions at distant locations. The farther the source, the more ancient the source, since electromagnetic waves move at the finite speed of $c$. Based on Hubble's nearly linear grow-rate model for the expansion of the universe, and assuming that Earth-based knowledge about subatomic particles can be extrapolated, the age of the universe is estimated to be about 13.8 Byrs. The understanding of the past of the universe has been developed by Stephen Hawking [27]. The postdiction goes as follows:

The earliest period is known as the *Planck era* ($t < 10^{-43}$ sec) where space variables cannot be defined. With the present understanding of physics this is the extent of scientific postdiction for all things in the universe. After this time, once the elements are separated sufficiently, three space variables and one time variable appeared as independent entities, whose separation can be measured using the space-time distance formula in Eq. (8.77). This marks the beginning of the *Inflation Era* $t \in (10^{-43}$ sec, $10^{-32}$ sec). The *Radiation Era* is characterized by the predominance of energy over matter $t \in (10^{-32}$ sec, $10^{-12}$ sec), although lepton-antilepton pairs formed an equilibrium with the electromagnetic waves. The first building blocks of matter appeared in the *Quark Era* $t \in (10^{-12}$ sec, $10^{-10}$ sec). The *Weak-force Era* occurred at the late Quark Era $t \sim 10^{-10}$ sec and resulted in the appearance of particles responsible for the exchange of the weak nuclear force. With cooling and expansion, the *Confinement Era* began $t \in (10^{-10}$ sec, $10^{-4}$ sec) leading to the creation of hadrons, where quarks, bound together by gluons, formed pairs (mesons) and triples (baryons). The *Proton Era* oversaw the formation of the main fuel for the fusion process, during $t \in (10^{-4}$ sec, $1$ sec), which now powers the stars. Some neutrons, and thus deuterium and helium also appeared in this era. This occurred most significantly during the *Nucleus Era* $t \in (1$ sec, $100$ sec), where protons and neutrons merge in a plasma to form nuclei, which are positive ions surrounded by a sea of negative ions, mainly in the form of electrons. The strong nuclear force appeared in the *Matter Era* $t \in (10^2$ sec, $10^5$ sec) to bind together many protons and neutrons. The creation of nuclei with multiple baryons resulted in minute differences in density that later lead to the distinction between galaxies and the vacuous regions of space that separate them. Also, the photons decreased in their interaction with matter and settled into the background radiation detected today, corresponding to $2.73°$ Kelvin. The slight differences in temperature on the celestial sphere also indicate the density differences established in this era. The *Molecular Era* $t \in (3$ hrs, $500,000$ yrs) corresponded to sufficient cooling that nuclei could, for the first time, attract the electrons into stable orbits, to form atoms. Thus the electric force field generated by a nucleus was able to overpower the high temperatures of the environment. The high temperatures worked to maintain the *plasma state* of matter corresponding to a uniform mixture of ions. With sufficient cooling, chemistry starts to make sense. This was followed by the *Galactic Era* $t \in (0.5 \text{ Myrs}, 1.5 \text{ Byrs})$, which is associated with the formation of galaxies and stars. Gravitational forces become significant during this period. The process of nuclear fusion resulted in a new source of radiation, at much higher frequencies than the background. Finally, the *Star Era*, which we are in now $t \in (1.5 \text{ Byrs}, 13.8 \text{ Byrs})$, consists of stars of various ages. The differences seen in the luminosity of stars allow estimation of their distances. Shifts of the spectral lines in the radiation observed from stars results in estimations of their speeds. The universe now consists of luminaries moving through empty space in three dimensions, in different directions, attracted by galactic centers and globular clusters.

The preceding chronology gives a simplistic postdiction for the life of the universe. Each one of the eras described has a variety of mathematical models that are used and developed by astronomers, physicists, chemists, and other scientists to help make sense of the main processes that took place. The predictions for the universe suggest continued accelerating expansion and cooling, without end. Gravitational forces between galaxies cannot seem to overcome the flood of new energy and matter into the present universe, and it is not clear when or if this process will cease.

Using a variety of measuring devices, like telescopes and photovoltaic plates, the celestial sphere appears to be radiating uniformly, or *isotropically*, in all directions from the past. Since the cosmological equations are unstable to small mass variations, this observation suggests a uniform or *homogeneous* early universe. This is known as the *cosmological principle* and a consequence is that a spatial model of the universe can ignore angular dimensions, approximately, and just consider its *scale factor*, denoted by $a > 0$, which is a measure of the size of the universe.

Questions about the large-scale aspects of the universe are addressed by an area of study known as *cosmology*. To understand what is happening at the largest scales, one must also consider the smallest scales of the universe, an area of study known as *quantum mechanics*. A model that could grasp the full range of physically observed phenomena would be called a *Grand Unified Theory* or simply a *Theory of Everything*. Such a model will have many parameters, which then must be optimized to best fit all the data observed by humankind. The more parameters that are introduced, the less significance any one parameter will have, and thus the more difficult it will be for any one person to fully grasp. The simplest model that describes the universe can be expressed in terms of the following variables and parameters:

**Let** $t$ be time (independent variable);
**Let** $a$ be the scale factor of the universe (dependent variable);

## 8.15 Quantum Cosmology and the Big-Bang Theory

**Let** $\phi$ be the amplitude of the collective particle state (*inflaton*);
**Let** $H_0$ be the Hubble constant (*variable* parameter);
**Let** $\lambda < 1$ be the rate that new energy and mass enter the universe (parameter); and
**Let** $M$ be the Planck mass ($\sqrt{h \cdot c/(2\pi \cdot G)} \sim 2.18 \times 10^{-8}$ kg).

In 1922 Alexander Friedman obtained energy-conservation equations for the large-scale evolution of the universe, using the Einstein equations created six years earlier. In this way he predicted the expansion of the universe seven years before it was observed by Hubble. These equations make the connection

$$\text{Space-time geometry} \quad \longleftrightarrow \quad \text{Mass-energy-momentum physics.}$$

An elementary cosmological model consists of the three coupled equations,

Friedman Equation: $\quad H_0^2(t) + \dfrac{1}{a^2(t)} = \dfrac{8\pi}{3M^2} \cdot \left[ \dfrac{(\phi'(t))^2}{2} + \dfrac{\lambda}{4} \cdot \phi^4(t) \right],$ (8.79)

Hubble's Law: $\quad a'(t) = H_0(t) \cdot a(t),$ (8.80)

Evolution of State: $\quad \phi''(t) + 3 \cdot H_0(t) \cdot \phi'(t) = -\lambda \cdot \phi^3(t).$ (8.81)

This is a nonlinear, autonomous system between the variables $a$, $H_0$, and $\phi$. Initiating this model is a supposed *quantum fluctuation* of amplitude $\phi_0 \sim M/\sqrt[4]{\lambda}$, which appears in a region of size of about $a \sim 1/H_0 \sim 1/M$. Under the further assumption of temporary stasis,

$$(\phi')^2 \ll (\phi)^4, \quad \phi'' \ll H_0 \cdot \phi', \quad H_0' \ll H_0^2, \qquad (8.82)$$

$\phi(t)$ grows initially to a value much greater than $M$, while $a^2$ increases to contain this state. Then, due to the algebraic relation in Eq. (8.79), $H_0$ must become significant. Eq. (8.80) can be integrated, under conditions in Eq. (8.82), to obtain,

$$a(t) = a_0 \cdot \exp\left[\int_0^t H_0(s)ds\right] \simeq a_0 \cdot \exp\left[\dfrac{\pi}{3 \cdot M^2} \cdot (\phi_0^2 - \phi^2(t))\right]. \qquad (8.83)$$

Coincidentally, the quantum state that solves Eq. (8.79) will behave as,

$$\phi(t) \simeq \phi_0 \cdot \exp\left[-M \cdot \sqrt{\dfrac{\lambda}{6\pi}} \cdot t\right]. \qquad (8.84)$$

Thus $\phi(t)$ decreases dramatically from $M/\sqrt[4]{\lambda}$ down to $M$ while $a(t)$ increases from $a_0 \sim \sqrt{3}/(\sqrt{2\pi} \cdot M)$ up to $a_0 \cdot \exp[\pi/(3 \cdot \sqrt{\lambda})]$.

The cosmological-state solution $(a(t), H_0(t), \phi(t))^T$ is an inflation model of the Big-Bang theory for the creation of the universe. It required the simple Equation (8.81) for the first particle ever created $\phi(t)$. As explained by Andrei Linde [37],

"...the global structure of the universe is formed due to quantum effects."

Hubble's constant $H_0$ is not a constant. The continuing acceleration of the universe is due to the continuing production of new photons, leptons, mesons, and baryons. Their source is still unknown, as is the future of our universe.

## ■ Exercises

1. A flashing rocket, leaving the Earth, emits a signal at $\mathcal{P}_1 = (0,0,0,0)$ and $\mathcal{P}_2 = (1 \text{ min}, 0, 0, 10 \text{ km})$.

   (a) Using Definitions (8.77) and (8.78) compute both the space-time distance, and the spatial distance between the events $\mathcal{P}_1$ and $\mathcal{P}_2$.
   (b) How fast is the rocket moving in km/hr according to an observer on Earth?
   (c) Another rocket was sent up at exactly the same time so that the events, relative to the second rocket, were $\mathcal{P}'_1 = (0,0,0,0)$ and $\mathcal{P}'_2 = (\delta t, 0, 0, 0)$. Compute $\delta t > 0$ using $c = 3 \times 10^8$ m/s and the requirement that Eq. (8.77) is an invariant of our universe.
   (d) Relative to the observers on Earth, are the clocks on the rockets moving faster or slower?

2. Use the following steps to obtain the approximate solutions in Eq. (8.83) and Eq. (8.84) using the equations and assumptions.

   (a) From the assumptions in Eq. (8.82) reduce Eq. (8.79) from four terms to two involving only $H$ and $\phi$, and denote this as ①.
   (b) From the assumptions in Eq. (8.82) reduce Eq. (8.81) from three terms to two, and denote this as ②.
   (c) Divide ① by ② to obtain an expression for $H\,dt$ in terms of $\phi\,d\phi$.
   (d) Using Eq. (8.80) and (c), obtain Eq. (8.83) by defining $a_0 = a(0)$ and $\phi_0 = \phi(0)$.
   (e) Take a derivative of Eq. (8.83) and divide both sides by $a(t)$ to obtain an approximation of $H_0(t)$ due to Eq. (8.80).
   (f) Multiply the new expression for $H_0(t)$ by $3\phi'$ and use ② to obtain an expression only involving $\phi(t)$ and its derivative. Call this Equation ③.
   (g) Finally, simplify ③ into a linear equation, taking the negative root, and solve the differential equation using the initial condition $\phi(0) = \phi_0$.

(h) For $\lambda = 1/6$, $M = \sqrt{\pi}$, $a_0 = 1/e$, and $\phi_0 = \sqrt{\pi} \cdot \sqrt[4]{6}$, draw $a(t)$, $H_0(t)$, and $\phi(t)$ on the same plot over the interval $[0, 6]$.

3. Draw a long horizontal (or diagonal) line labeled *Postdiction of the Universe*. Divide the line *evenly*, one indication for each era. On the (bottom) left start with the Planck Era and indicate all eras up to the Star Era (top right). For each era indicate the approximate value of $log_{10}(t)$ where time $t$ is in seconds.

# References

[1] Adler R., Bazin M., and Schiffer M. *Introduction to General Relativity.* 2nd ed., New York, NY: McGraw-Hill Book Company, 1975.

[2] Arfken G.B. and Weber H.J. *Mathematical Methods for Physicists.* 5th ed. San Diego, CA: Acadmic Press; 2001.

[3] Aster R.C., Borchers B., and Thurber C.H. *Parameter Estimation and Inverse Problems.* Burlington, MA: Elsevier Academic Press; 2005.

[4] Blanes S., Casas F., and Ross J. *Improved high order integrators based on the Magnus expansion.* BIT Numerical Math., Vol. 40, No. 3, pp. 434–450, 2000.

[5] Box G. *Statistics as a Catalyst to Learning by Scientific Method, Part II: A Discussion.* Report No. 172. Center for Quality and Productivity Improvement: University of Wisconsin; 1999.

[6] Cardan, G. *The Rules of Algebra: Ars Magna.* Translated by Witmer T.R., Mineola, NY: Dover Publications; 2007.

[7] Bothamley J. *Dictionary of Theories.* Canton, MI: Visible Ink Press, LLC; 1993.

[8] Coddington E.A. *Introduction to Ordinary Differential Equations.* Englewood Cliffs, NJ: Prentice Hall, Inc.; 1961.

[9] Cohen-Tannoudji C., Diu B., and Laloë F. *Quantum Mechanics.* Paris: Hermann; 1977.

[10] Darwin C. *The Origin of Species.* Notes by Levine G. New York, NY: Barnes & Noble, Inc.; 2004.

[11] Eves H. *An Introduction to the History of Mathematics.* Revised Ed. New York, NY: Holt, Rinehart, and Winston, Inc.; 1964.

[12] Farmer J.D. and Foley D. The Economy Needs Agent-Based Modelling. *Nature.* 2009: 460; 685–686.

[13] Finch S.R. *Mathematics and its Applications.* Cambridge University Press; 2003.

[14] Fowler D. and Robson E. Square Root Approximations in Old Babylonian Mathematics: YBC 7289 in Context. *Historia Mathematica.* 1998: 25; 366–378.

[15] Franklin R.E. and Gosling R.G. Molecular Configuration in Sodium Thymonucleate. *Nature.* 1953: 171; 740–741.

[16] Freedman D., Pisani R., and Purves R. *Statistics.* New York, NY: W.W. Norton & Company, Inc.; 1980.

[17] French A.P. *Vibrations and Waves*. New York, NY: W.W. Norton & Company; 1971.

[18] Fritts H.C. *Tree rings and Climate*. London: Academic Press, Inc., Ltd.; 1976.

[19] Genesereth M.R. and Nilsson N.J. *Logical Foundations of Artificial Intelligence*. Los Altos, CA: Morgan Kaufmann Publishers, Inc.; 1987.

[20] Gingerich O. *Was Ptolemy a Fraud*. Royal Atron. Soc. Quarterly Journal. 1980: 21; 253–266.

[21] Ginsberg M.L. Multivalued logics: a uniform approach to reasoning in artificial intelligence. *Comput Intell*. 1988: 4; 265–316.

[22] Goldstein C.S. *Classical Mechanics*. Reading, MA: Addison-Wesley Pub. Co., Inc.; 1981.

[23] Goode S.W. and Annin S.A. *Differential Equations with Linear Algebra*. 3rd ed. Upper Saddle River, NJ: Pearson Education, Inc.; 2007.

[24] Groopman J. *How Doctors Think*. New York, NY: Houghton Mifflin Co.; 2008.

[25] Havil J. *Gamma*. Princeton University Press; 2003.

[26] Haus H.A. *Waves and Fields in Optoelectronics*. NJ: Prentice-Hall Inc.; 1984.

[27] Hawking S. *A Brief History of Time*. Bantam Books Pub.; 1996.

[28] Herstein I.N. *Abstract Algebra*. 3rd ed. New York, NY: Prentice; 1996.

[29] Hofstadter D.R. *Gödel, Escher, Bach: An Eternal Golden Bond*. New York, NY: Basic Books, Inc.; 1979.

[30] Hofstede G. *The Cultural Relativity of the Quality of Life Concept*. Acad Manage Rev. 1984: 9(3); 389–398.

[31] Hogg R.V. and Tanis E.A. *Probability and Statistical Inference*. Upper Saddle River, NJ: Prentice-Hall, Inc.; 1993.

[32] Homer-Dixon T. *The Upside of Down*. Canada: Alfred A. Knopf; 2006.

[33] Holling L.H., Gunderson D., and Ludwig D. *In Search of a Theory of Adaptive Change*. Ed. Gunderson, L. H. et al. In Panarchy, Washington, D.C.: Island Press; 2002. The Resilliance Alliance, 2009.

[34] Holt R.D. and Gomulkiewicz R. *The Evolution of Species: A Population Dynamic Perspective*. Ed. Othmer H.G., et. al. In *Case Studies in Mathematical Modeling*, Upper Saddle River, NJ: Prentice Hall; 1997.

[35] Jacobson N. *Basic Algebra I*. San Francisco, CA: W.H. Freeman and Company; 1974.

[36] Kaw A. *Exact solutions to cubic equations*. Accessed at http://numericalmethods.eng.usf.edu; 2009.

[37] Kerr R.A. Scientists see greenhouse, semiofficially. *Science*. 1995: 269; 1667.

[38] Kittle C. *Solid State Physics*. New York, NY: John Wiley & Sons; 1976.

[39] Kittel C. and Kroemer H. *Thermal Physics*. 2nd ed. San Francisco, CA: W. H. Freeman and Company; 1980.

[40] Kindermann R. and Snell J.L. *Markov Random Fields and their Applications*. Providence, RI: American Mathematical Society; 1980.

[41] Krasovskii N.N. *Stability of Motion.* Stanford University Press; 1963.
[42] Levitus S., et.al. Global ocean heat content 1955–2008 in light of recently revealed instrumentation problems. *Geophysical Research Letters*, Vol. 36, L07608, 2009.
[43] Linde A.D. Externally existing self-reproducing chaotic inflationary universe. *Physics Letters B.*, 1986: 175(4); 395–400.
[44] Zhong-Qi Ma. *Group Theory for Physicists.* World Scientific Publishers; 2007.
[45] Mahler K. *Lectures on Transcendental Numbers.* Springer-Verlag, Lecture Notes in Mathematics 546; 1976.
[46] Malthus T.R. *An Essay on the Principle of Population.* Oxford World's Classics; 1798.
[47] Magnus J. R. and Neudecker H. *Matrix Differential Calculus.* John Wiley & Sons Ltd.; 2001.
[48] Marsden W. *The Lemming Year.* London: Chatto & Windus; 1964.
[49] Moore D.S. and MaCabe G.P. *Introduction to the Practice of Statistics.* New York: W.H. Freeman and Company; 1999.
[50] Maslow A.H. A Theory of Human Motivation. *Psychologic Rev.* 1943: 50(4); 370–96.
[51] Moore L. and Smith D. *Calculus: Modeling and Application.* Boston: Houghton Mifflin Co.; 1996.
[52] Newton, Sir Isaac. *The Principia: Mathematical principles of natural philosophy.* Berkeley, CA: University of California Press, 1650; 1999.
[53] Nicholson W.K. *Linear Algebra.* Boston: PWS Publishing Co.; 1990.
[54] Pathria R.K. *Statistical Mechanics.* Great Britain: Pergamon Press; 1978.
[55] Press W.H., Flannery B.P., Teukolsky S.A., and Vetterling W.T. *Numerical Recipes.* Cambridge: Cambridge University Press; 1987.
[56] Prugovečki E. *Quantum Mechanics in Hilbert Space.* 2nd ed., New York, NY: Academic Press; 1981.
[57] Ptolemy. *Ptolemy's Almagest (Mathematical Syntaxis).* Translated by: Toomer G.J. Springer Verlag; 1984.
[58] Quinn G. P. and Keough M. J. *Experimental Design and Data Analysis for Biologists.* Cambridge University Press; 2006.
[59] Ross K.A. *Elementary Analysis: The Theory of Calculus.* New York, NY: Springer Science; 1980.
[60] Reed M. and Simon B. *Functional Analysis I.* London: Academic Press, Inc.; 1980.
[61] Ross S. *A first Course in Probability.* Upper Saddle River, NJ: Prentice Hall; 6th ed.; 2002.
[62] Royden H. *Real Analysis.* Englewood Cliffs, NJ: Prentice-Hall, Inc.; 1988.
[63] Russell B. and Whitehead A.N. *Principia Mathematica.* Cambridge University Press; 1930.
[64] Sachs A. M. and Ashforth B. E. *Organizational Socialization.* Journal of Vocational Behavior. 1997: 51; 234–279.

[65] Scheinerman E.R. *Invitation to Dynamical Systems.* Upper Saddle River, NJ: Prentice Hall; 1996.
[66] Sherman A. *Calcium and membrane potential oscillations in pancreatic $\beta$-cells.* Ed. Othmer H.G. et. al. In *Case Studies in Mathematical Modeling*, NJ: Prentice Hall; 1997.
[67] Shifrin T. *Multivariable Mathematics.* Hoboken, NJ: John Wiley & Sons, Inc.; 2005.
[68] Spivak M. *Calculus on Manifolds.* Cambridge, MA: Perseus Publishing, LLC; 1998.
[69] Stancu F. *Group Theory in Subnuclear Physics.* UK: Oxford Studies in Nuclear Physics Vol. 19; 1996.
[70] Strayer J.K. *Elementary Number Theory.* Waveland Press, Inc.; 1994.
[71] Sherwood L. *Human Physiology: From Cells to Systems.* Belmont, CA: Brooks/Cole; 4th ed.; 2001.
[72] Schutz B. *Geometrical Methods of Mathematical Physics.* Cambridge, MA: Cambridge University Press; 1984.
[73] Tipler P.A. *Modern Physics.* New York, NY: Worth Publishers, Inc.; 1978.
[74] Thompson W.H. and Leidlein J.E. *Ethics in City Hall.* Boston, MA: Jones and Bartlett Publishers, LLC; 2009.
[75] Triola M.F. *Elementary Statistics.* Boston, MA: Addison Wesley Longman, Inc.; 2000.
[76] von Foester, H., Mora, P.M., and Amiot, L.W. Doomsday: Friday, November 13, AD 2026. *Science.* 1960: 132; 1291–1295.
[77] Wooding F. *Wild mammals of Canada.* Toronto: McGraw-Hill Ryerson, Ltd.; 1982.
[78] Zemansky M.W. and Dittman R.H. *Heat and Thermodynamics.* 6th ed. New York, NY: McGraw-Hill, Inc.; 1982.

# Glossary

$\wedge$ &: And, but...is also true
$\vee$: Or
$\forall$: For all, for every
$a.a.$: Almost always
$a.c.\ a.s.$: Almost certainly, almost surely
$a.a.c.$: Asymptotically almost certainly
$\simeq$: Approximately equals
$\exists$: There is, there are
$\exists!$: There is one and only one; there is a unique element
$\{A \mid B\}$: The set of all $A$ such that $B$ is true
$\implies$ : Implies; if ... then ...; "...it is true that ..."
$\impliedby$ : Implied by; follows from
$\iff$ : If and only if, *iff*
$\cap$: Intersection
$\cup$: Union
$\subset$: Subset
$\subseteq$: Subset or equal as sets
$\subsetneq$: Proper subset
$\in$: Is an element of
$\ni$: Such that, so that
$\equiv$: Is defined as (used for original definitions or equivalences)
$\geq$: Is greater than or equal
$\leq$: Is less than or equal
$\lceil \cdot \rceil$: Least integer greater than or equal to $\cdot$; least integer function
$\lfloor \cdot \rfloor$: Greatest integer less than or equal to $\cdot$; greatest integer function
$\mathcal{C}^0$: The set of continuous functions
$\mathcal{C}^k$: Functions whose $k$th derivative is continuous
**Contradiction:** $\not\implies$
**Such that:** $\ni, \mid, :,$ s.t.
**CDF:** Cumulative distribution function
**IVP:** Initial Value Problem

**PDF:** Probability distribution function
**PCP:** Primary care physician
□ $QED$: End of proof; quod erat demonstrandum (thus it is shown)

## Greek Alphabet

| $\alpha$ alpha | $\epsilon$ epsilon | $\mu$ mu |
| $\beta$ beta   | $\phi$ phi         | $\nu$ nu |
| $\delta$ delta | $\psi$ psi         | $\zeta$ zeta |
| $\gamma$ gamma | $\rho$ rho         | $\iota$ iota |

# Index

## A

a priori, 79, 102, 115, 442
a.a.c., 131
Abelian, 20
abelian group, 158
abscissa, 98
absurdity, 63
acceleration, 236
acceleration due to gravity, 53, 135, 255, 417
acceleration vector, 322
advection, 377
algebra of functions, 60
algebraic equations, 36
algebraic multiplicity, 173, 395
algebraic numbers, 36
algorithm, 139, 160, 186, 383
almost certainly, 128
almost surely, 128
alternate hypothesis, 116
alternative hypothesis test, 117
ambient temperature, 433
amphoteric property, 431
amplitude, 53
anchoring, 441
angle of rotation, 51
angular frequency, 57, 447, 456
antiderivative, 250, 391
antidifferentiation, 249
Antisymmetric, 6
area, 44
area $A$ of the circle of radius $\rho$, 354
area of the parallelogram spanned by $\vec{x}$ and $\vec{y}$, 313
area under the bell curve $e^{-x^2/2}$, 355
argument, 66
array, 156
art of science, 121
association, 91, 133
associations, 124
associativity, 4, 20
asymptotic population, 389
augmented matrix, 161
automorphism, 69
autonomous, 406, 411
average, 49, 94
Axiom of Choice, 45

## B

bar, 95
barometer, 454
base, 29, 53
basin, 411, 414
basis, 73, 146, 162, 171
basis functions, 180
Bayes formula, 101
Bayes rule, 442
Bayesian analysis, 441
belief, 42, 115, 121
bell shape, 98
bell-shape distribution, 83
bell-shaped, 103, 109
Bernoulli distribution, 79
Bernoulli equation, 415

Bernoulli equations, 390
Bernoulli experiment, 107
best-fit model, 125
best-fit parameter set, 183
bias, 117
bifurcation, 414
bifurcation lines, 410
bifurcation point, 410, 412
Big-Bang theory, 468
bijection, 54, 66, 69
bilinear, 153
bimodal, 94
binomial coefficient, 34, 76
binomial distribution, 79
binomial series, 34
bins, 97
bisection method, 246
bivariate function, 133
Boolean algebra, 43
Borel, 44, 46
Borel set, 81
Borel sets, 42
boundary, 44, 343
boundary of a region in $\mathbb{R}^n$, 343
bounded, 17, 166, 406
box plot, 96
branch, 54, 64
branches, 66

## C
$\mathcal{C}^k(I;\mathbb{R})$, 236
$\mathcal{C}^0(\mathcal{S})$, 193
capacitance, 397
cardinality, 20, 43
carnivores, 451
carrying capacity, 287, 413, 446, 448, 451
case functions, 47
cases, 49
cataloged, 410
catastrophe, 413
categorical, 91, 132

category, 97, 100, 406
Cauchy sequence, 27, 36
Cauchy-Schwarz inequality, 143, 309
causal, 385
causation, 91
ceiling function, 51
celestial sphere, 460, 466
cell, 435, 436
cell division, 384
census, 91, 94
center, 94
center of mass, 269, 270, 350
centroid, 351
certainty, 90
chain rule, 210
chain rule for a function of several variables, 329
chain rule for a vector-valued function of several variables, 335
change of coordinates, 342
change of variables, 252, 269, 353, 354
characteristic, 73, 160, 171, 389, 394, 412
characteristic function, 49
characteristic polynomial, 173
charge, 415, 437
charges, 397
classify, 406, 410
classifying data, 384
climate, 457
closed, 141
closed interval, 44
closed rectangle, 44
closure, 20
closure, 60, 146
codomain, 60
collapse, 443
collecting, 77
column space, 162
column spaces, 140
columns, 157
combinations, 76

common denominator, 10
commutativity, 4
compatible, 141
completing the squares, 10
complex conjugate, 37, 59
complex numbers, 36
components, 156, 161
composition, 63
concavity, 237
conditional probabilities, 84
conditional probability, 75
conductance, 397
cone, 320
confidence, 90, 386
confidence interval, 114, 121
confidence level, 74, 125, 183
confidence region, 182
confidence statistic, 125, 134
conics, 320
conjugate, 114
conjugation by permutations, 178
conservation of charge, 397
constant, 49
constant rule, 143, 166
continuity, 53
continuous, 47, 48, 67, 192
continuous at $p$, 193
continuous on the interval, 193
continuous probability distribution, 81
continuously compounded interest, 227
contraction, 142
control, 408
control chart, 123, 456, 458
controlled environment, 89
convection, 377
convenience bias, 117
converge, 292
convergent sequences, 19
converges, 17, 274
converges in probability, 128
correlation coefficient, 134, 182

correlations, 124
cosecant, 62
coset, 21
cosine, 56
cosine law, 143, 154
cosmological model, 467
cosmological principle, 466
cotangent, 62
count, 91, 123
countable, 43
counted, 111
counting, 77, 82
counts, 100, 184
coupling-strength, 433
cover, 43, 44
criterion for continuity, 67
critical, 410
critical point, 412, 440
critical points, 231, 238
critical question, 115
cross product, 153, 311
cross section, 266, 320
cross-sectional area, 266
cumulative distribution function, 81, 99
curl, 372
current, 397, 425, 436
curve, 322
cyclic, 20
cylinder, 320

# D

damped oscillations, 447
Darwin's model, 425
decay, 428, 447
decimal, 29, 35
decreasing, 63, 242
defective, 172, 173, 177, 396
definite integral, 258
definite integral, properties, 262
degree, 9, 15
degrees of freedom, 94

delay, 413
delay parameter, 448
delayed-differential equation, 448
delta function, 417
dependence, 85
dependent, 90, 121, 134
dependent variable, 47, 66
dependent variables, 89
derivative, 195, 198
derivative table, 233
descriptive statistics, 103
design of the experiment, 121
determinant, 160, 167, 311
deviations, 98
diagnosis, 441
diagonal, 167, 171, 178
diagonal matrix, 167
diagonalize, 393
difference equation, 448
difference quotient, 52
difference variance, 119
differences statistic, 119
differentiable, 198, 323, 326, 334
differential of f, 233
differentiation, 198
diffusion, 377
dimension, 60, 73, 157
dimensional, 122
dipole, 417
direction, 143
direction of maximal ascent, 333
directional derivative, 332
discontinuities, 50
discontinuous, 47, 48
discrete, 43
discrete probability distribution, 78
disease, 440
disjoint, 42, 43
displacement vector, 151
distance, 27, 304, 415
distribution, 89, 92

distribution function, 98
distributions, 90
div($\vec{V}$), 370
diverge, 292
divergence, 367, 370
divergence theorem, 367, 369, 371
diverges, 274
divides, 4, 20
DNA, 435
domain, 47, 48, 59
domain space, 163
doomsday, 445
dot plot, 95
dot product, 142, 153, 305
double application of L'Hopital's rule, 276
double helix, 435
double integrals, 344
double-angle formulas, 57
doubling-time, 65
dual basis, 147
dual space, 148

**E**
$e$, 224
effect, 91
effective, 139
effector, 436
eigenvalue, 172
eigenvalues, 171, 176, 393, 406
eigenvector, 172, 176, 393
electrical circuit, 397
electrostatics, 415
elementary matrices, 169
elementary operations, 161, 165, 169
elementary permutations, 178
ellipsoid, 319
elliptic paraboloid, 319
energy, 408, 433, 453
enthalpy, 454
entire, 48, 71

entropy, 453
epicycles, 461
epidemic, 438
equations of motion, 322
equilibrium, 124, 392, 406, 407, 411, 430, 446, 447
equivalence classes, 43, 45
equivalence relation, 43, 45
error vector, 181
Euclidean space, 304
Euler's approximation, 256
Euler's continuity equation, 377
Euler's formula, 58
Euler's identity, 72
Euler's method, 256, 386
even, 54, 59
event, 74, 464
evidence, 117
evolution, 425
evolution operation, 393
evolution operator, 396
expectation, 100
expected value, 100, 113, 129
experiment, 74, 79, 89, 183
experimental study, 91
experiments, 110, 184
exponent, 52
exponential, 71, 445
exponential distribution, 82, 106
exponential function, 222
exponential growth, 286
exponential model, 458
exponential-decay model, 429
extensive variables, 453
extermination rate, 389
external forcing, 390
extraneous, 14

**F**
factorial, 34, 35
factorial function, 76

factors, 84, 122, 124, 444
feedback, 424, 425
Fibonacci evolution equation, 384
Fick's law of diffusion, 377
field, 22, 141, 416
fields, 60
first derivative test, 243
five-number summary, 96
floor function, 50
flux, 367
flux of $\vec{V}$ across $\partial R$, 369
focus, 320
force of gravity, 415
forced linear-vector equation, 396
Fourier's law of thermal conductivity, 377
fraction, 30
frequency, 82, 98
frequency diagram, 98
Fubini's theorem, 347, 418
fugacity, 433, 455
full rank, 164
function, 47
functions, 41
fundamental frequency, 449
fundamental matrix, 394, 403
fundamental set, 394
Fundamental Theorem of Algebra, 173
Fundamental Theorem of Calculus, 259, 260, 263, 387, 401, 404

**G**
gamma function, 71
gating, 437
Gauss-Jordan (GJ) method, 164
Gaussian, 83
Gaussian elimination, 161
generation, 384
geometric multiplicity, 173, 395
geometric series, 30, 34, 72, 159, 292, 403
global extremum, 238

global maximum, 238
global minimum, 238
gold standard, 441
golden ratio, 10, 33
goodness-of-fit, 180
gradient, 326, 327, 333, 416
Gram-Schmidt, 177
Gram-Schmidt process, 173
graph, 48, 325
gravitational constant, 417
Green's theorem, 362, 364
ground-state, 434
group, 20, 73
group isomorphism, 54, 66

## H
half-life, 53, 65
harmonic function, 416
harmonic oscilator model function, 398
harmonic series, 72
harvest rate, 389, 450
Heat Equation, 372
height of a falling object, 255
helix, 435
Henry's law, 455
Herbivore, 451
Hermitian, 176
Hess's law, 454
Hessian matrix, 374
heuristic model, 458
higher derivatives, 236
higher order term, 60, 195
Hippasus, 5, 23, 32
histogram, 109
histograms, 103
homeostasis, 436
homogeneous, 389, 466
homomorphism, 54, 66
horizontal, 98
horizontal asymptote, 61, 62
horizontal-line test, 63

host, 440
Hubble's constant, 463
human behavior, 445
human population, 413, 443, 444
hyperbolic, 62
hyperbolic cosine, 54
hyperbolic paraboloid, 320
hyperboloid of one sheet, 320
hyperboloid of two sheets, 320
hyperplane in $\mathbb{R}^n$, 316, 317
hypotenuse, 30
hysteresis, 86

## I
ideals, 60
idempotent, 178
idempotent property, 50
identity, 20, 60, 70, 140, 161
identity function, 64
identity matrix, 159
illness, 123
image, 48
imaginary numbers, 24, 36
implicit function theorem for regular surfaces, 337
implicitly, 388
improper integral, 274
increasing, 63, 242
indefinite integral, 250
independent, 75, 90, 121, 132, 134
independent and identically distributed, 182
independent variable, 47, 66, 89, 152
indeterminate form $0/0$ or $\infty/\infty$, 248, 275
index, 98
index of the product, 32
index of the sum, 15
individuals, 91
induce stability, 409
inductance, 397

induction, 27
inertial mass, 420
inference, 84, 101
infinite, 7, 20, 34
infinite series, 292
inflation model, 468
inflection point, 451
initial position, 418
initial state vector, 156
initial value problem, 254, 392
initial velocity, 418
injective, 69
inner product, 142, 146, 176
input, 166
input vector, 161
instability, 394
instantaneous velocity, 236
integers, 5
integrable, 258
integral calculus, 397
integral table, 250
integrals of trigonometric expressions, 279
integrating factor, 390, 396, 404
integration by parts, 277, 392
intensive variables, 453
interference, 419
interior, 44
intermediate value theorem, 193
interpolate, 71
interquartile range, 96
intersection, 43
interval, 92, 95
interval estimates, 114
intervals of increase and of decrease, 243
intraspecies instability, 449
invariant, 54, 173
inverse, 20
inverse, 60, 159, 388
inverse function, 41
inverse function theorem, 336

inverse matrix, 164
invertibility, 63
invertible, 164, 167
investments compounded continuously, 286
irrational, 33, 36, 52, 71
iso-segregation, 377
isobar, 325
isoclines, 377
isomorphic, 21
isomorphism, 54, 66
isopycnal, 377
isotherm, 377
isotropically, 466
iteration scheme, 404
IVP, 254

**J**
J-shaped growth, 446
Jacobian, 406
Jacobian matrix, 335
Jordan canonical form, 172, 178

**K**
kernel, 163
Kirchhoff's law, 397
Kronecker delta function, 147
$k$th continuously differentiable functions, 236

**L**
L'Hopital's rule, 248, 275
L'Hopital's rule, double application, 276
Lagrange multiplier, 378
Laplacian, 372
latitude, 53, 460
law of cosines, 307
law of cosines in $\mathbb{R}^n$, 308
law of large numbers, 115
law of natural growth, 286
least-squares fit, 125

Legendre basis, 74
Leibniz, 234, 259
lemming, 447, 449
length, 44, 142
length of a vector, 305
level curve, 320
level of confidence, 114, 134
level of significance, 116
level surface, 325
Lie groups, 308
likelihood, 441
limit, 8, 27
limit at infinity, 273
limit at infinity of a rational function, 274
limits of integration, 258
line, 149, 152
line integrals, 359
line of least squares, 376
line of regression, 376
line through a point $\vec{p}$ in the direction $\vec{d}$, 323
linear, 180
linear approximation, 199, 403
linear combination, 36, 73, 107, 141, 146, 151
linear density, 267
linear function, 51
linear model, 122
linear nonhomogenous, 392
linear operator, 48, 166, 178
linear regression, 374
linear relationship, 125
linearity, 402
linearity property, 100
linearization, 199, 324
linearization of the inverse, 339
linearize, 448
linearly independent, 73, 145, 180
linearly-independent functions, 60
Lipschitz constant, 401

local extremum, 238
local maximum, 238
local minimum, 238
logarithm, 64
logarithmic derivative, 434
logistic growth, 446
logistic model, 451
logistic population model, 287
logistic von Foerster model, 446
longitude, 53, 460
Lorenz oscillator, 407
Lotka-Volterra equation, 451
lower triangular, 167, 178
LRC circuit, 397
luminaries, 460
lurking, 89
lurking variable, 123
Lyapunov function, 408

## M

Maclaurin series, 298
magnetism, 433
magnitude, 142
Malthus model, 388, 444, 451
mapping, 47
margin of error, 114, 121
mass, 415
mass M of a thin rigid wire, 267
mathematical spaces, 3, 6
matrix algebra, 397
matrix equation, 161, 164, 166
matrix norm, 166
matrix valuation, 166
mature, 459
maturity, 413
maximally sustainable, 446
maximum norm, 166
mean, 78, 82, 94, 98, 183, 271
mean value theorem, 242
measure, 44, 447
measurement, 91, 184, 393, 419, 454, 460

median, 94
meromorphic, 60, 71
metastable, 406
method of partial fractions, 287
method of substitution, 251, 252
mod, 5, 21, 37
mode, 114
model, 90, 121, 134
model function, 180, 384
model parameter, 134
model space, 122
modeling, 227
modulus, 21, 37
moment, 78
monic, 9, 11, 390, 396
multivalued function, 64
multivariable Newton's method, 378
mutually exclusive, 42

## N

natural frequency, 57, 456, 462
natural logarithm, 65
Natural numbers, 5
negative feedback, 408, 436, 446, 451
nested, 184
net area, 261
Newton, 234, 259
Newton's $2^{nd}$ law for gases, 377
Newton's method, 246, 247, 386, 414, 438
niche, 457
node, 412
nominal, 92
nonhomogeneous, 377
nonlinear, 52
norm, 142
normal, 316, 317
normal density function, 269
normal distribution, 83
normalizable, 143
normally distributed, 128

nucleus, 426
null hypothesis, 116
null space, 163

## O

observational study, 91
odd, 54, 59
one-to-one, 69
onto, 44, 69
open cover, 46
open interval, 44
operates, 171
operations, 139
operator, 48
optimization, 238, 374
orbit, 322
orbits, 451
order of a group, 20
order of an element, 20
order of the reaction, 430
ordering, 6
ordinal, 92
ordinate, 98
orthogonal, 152, 176, 177
orthogonal complement, 146
orthogonal projection, 149
orthonormal basis, 172
orthonormal vectors, 150
oscillation, 447
out of control, 124
outcomes, 74, 79, 107
outer product, 153
outlier, 131
outliers, 96, 105
output, 166
output of the system, 390
output vector, 159, 161, 164
over-determined, 181

## P

$P$-value, 117

parallel, 152
parameter, 49, 80, 91, 321, 425
parameter of the model, 121
parameter space, 122
parameter vectors, 155
parameter-space, 410
parameterization, 322
parameterize, 339
parameterized, 152
parameterized regions, 342
parameterized surface, 340
parameterized surfaces, 339
parameters, 7, 81, 98, 109, 161, 171, 180, 182, 385, 394, 412
parametrization, 460
partial derivative, 326, 327
partial fraction decomposition, 287, 289
partial fractions, 286
partial pressure, 455
partial sums, 402
particles, 415
partition, 43, 50, 85, 97, 98, 111, 257, 406
partition function, 434
Pascal's triangle, 34
paths, 407
PCP, 385
perimeter, 33, 44
period, 57, 448
periodic, 57
permutation, 69, 76, 110
perp, 377
perpendicular, 152
phase diagram, 451
phases, 434
piecewise continuously differentiable, 362
Planck length, 426
plane, 149, 151, 316, 319
plasma state, 466
point estimate, 114
points, 150
Poisson distribution, 80, 107

poles, 71
policy, 412
policy parameter, 409
polynomial, 9, 48, 52, 73, 166
polynomial rule, 204
population modeling, 226, 227
population size, 114
population variance, 98
position vector, 322
positive-feedback, 451
postdict, 385
postdiction, 444, 445, 465
postdictive, 410
potential-energy model functions, 416
power function, 52
power of the test, 116
power rule, 197, 208, 228, 229
power set, 78
precipitation, 457
predators, 388
predictive, 410
preimage, 42, 46, 48, 63
pressure, 454
pressure gradient, 377
prey, 388
prime numbers, 5
probability, 5, 42, 92, 128, 419
probability density function, 90, 99
probability distribution, 75, 99, 434, 441
probability function, 74
probability model, 74
probability over [a,b], 268
probability theory, 76, 128
product, 70
product inequalities, 166
product of factors, 32
product rule, 205
prognosis, 385
projection, 148
projection operators, 50
propagation, 384

proportion, 114
proportion of variation, 182
proportions, 100, 116

## Q
qualitative, 91
quantitative, 91, 132
quantum fluctuation, 467
quantum state, 419, 467
quark, 426
quasilinear, 404
quotient rule, 213

## R
radians, 55
radical conjugate, 23, 37
radius of convergence, 293
random variable, 81, 92, 100, 114, 129
range, 48, 97, 407, 410, 425, 437
ranges, 426
rank, 149, 162
ratio, 92
ratio data, 121
ratio test, 294
rational function, 445
rational numbers, 5, 8
reaction, 430
real numbers, 29, 36
receptor, 436
reciprocal, 17, 23, 42, 57, 390, 415
reciprocal rule, 207
reciprocals, 11, 60
recrudescence, 440
reduced row-echelon, 178
reflexive, 6
regression function, 182
regular points, 231
relation, 86
relations, 449
relationship, 388, 424, 457
relative error, 16, 18

relaxation, 437
relaxation parameter, 379
remainder, 21
residuals, 375
resilience, 414
resistance, 397
resistor, 425
resolvent force, 304
resonant frequency, 398
responses, 84, 122
restocking rate, 389
resultant vector, 151
retrograde motion, 461
Riemann, 258
Riemann sum, 258
Riemann surface, 66
Riemann zeta function, 72
right, 31
right triangle, 30
ring, 22, 60, 158
Rolle's theorem, 240
root, 9, 16, 64
root test, 293
roots of unity, 38
row space, 140, 162
row vector, 140
rows, 157
Roy Baker problem, 12, 13
rules of probability, 74
rules of rounding, 89
rules of statistical inference, 113

## S
S-shaped growth pattern, 446
sample, 74, 91, 94, 98
sample means, 101
sample size, 100, 114, 125, 128
sample space, 74, 79, 84
sample standard deviation, 94, 101, 185
sample variance, 94
saturation value, 446

sawtooth, 456, 460
scalar, 142, 153
scalar multiplication, 304
scale factor, 466
scatter plot, 133
schools of thought, 115
scientific hypothesis, 116
scientific knowledge, 424
scientific notation, 30
secant, 62
second derivative, 236
second-order linear differential equation, 397
self-adjoint, 149
semistable, 412
sensitivity, 116
sequence, 16, 17, 19, 21, 24, 29, 33, 146, 186
set theory, 42
sgn, 68
significance level, 183
significant figures, 93
simple harmonic model, 449
simple random sample, 114
sine, 54, 56, 71
sine law, 154
sinusoidal function, 57
size, 157
skew Hermitian, 179
skewed, 94
slope, 51, 62
slope function, 52
soft signs, 442
solution space, 37
solvability, 63
source, 417
space, 157
space of continuous functions, 193
span, 145
spanning set, 73
specificity, 116

spectral representation, 177
spectral values, 175
spectrum, 171
speed of light, 420, 426
sphere, 319
sphere of radius $R$, 317
splines, 70
square parameter matrix, 159
squared deviations, 94, 98, 107
stabilizing terms, 406
stable, 406
stable equilibrium point, 394
staircase function, 456
standard basis, 311
standard deviation, 78, 98
standard error, 114, 117, 119
standard model, 426
standard normal, 269
standard normal density, 301, 356
standard normal distribution, 83, 104
state, 18, 101, 107, 119, 124, 387, 433, 441, 453
state function, 454
state of the system, 91, 117, 384, 386, 396, 401, 411, 413, 437
state variables, 92
state vectors, 155
statement, 3, 42
station stones, 460
statistic, 110, 134
statistical analysis, 118
statistical control, 124
statistical hypothesis, 116
statistical inference, 103
statistical model, 98
statistics, 91, 94, 128
steady-state, 437
stepping matrix, 156
Stokes theorem I, 373
stratification, 117
stratify, 446

student's $t$ distribution, 83, 115, 118
subgroup, 21
subprocesses, 452
subsequence, 6, 7
subspace, 146, 149, 152, 396
substitution, 12, 391
substitution method, 252
sum, 70
sum of squares function, 375
sum of terms, 15
sum of the squared deviations, 98
sum of the squared-deviations of the data, 180
sums, 50
support, 64, 79, 81
surface area of the sphere of radius $\rho$, 359
surface integrals, 357
surjective, 69
survey, 91
symmetric, 177
symmetric matrix, 175
symptoms, 123
system of linear equations, 161

## T

table of derivatives, 233
table of integrals, 250
tail, 104
tangent line, 199, 324
tangent plane, 318, 321
tangent vector, 324
tautology, 63
Taylor expansion, 411
Taylor polynomial, 296
Taylor remainder, 296
Taylor series, 296
temperature, 453, 457
tensor product, 147, 148
term, 10, 11
thermal equilibrium, 454

tolerance, 34
torque, 270
totality, 6
trace, 160
transcendental, 33, 35, 70
transformable, 172
transformation, 11, 14, 41, 104, 148, 175, 178, 335, 426
transformed, 82, 134, 435
transition, 387, 407, 410, 413, 433, 437
transition matrix, 394, 403
transition process, 405
transitive, 6
transmittable, 440
transpose, 140, 162
treatment, 91, 385
tree ring, 457
tree rings, 458
trial, 79, 107
triangle, 166
triangle inequality, 143, 310, 402
triangular matrix, 172
trigonometric substitutions, 284
trigonometry, 54
triple integral, 418
triple integrals, 348
truth value, 3
truth-value, 113, 128
two-sided hypothesis test, 118
two-sided test, 126
two-tailed hypothesis test, 118

## U

uncertainty, 166
uncountable, 43
undefined, 8
underdetermined, 174
uniform, 78
uniform distribution, 81, 82
uniformly distributed, 101
union, 43

unit vector, 143
unstable, 406
upper triangular, 167, 178

## V

valuation, 26, 407
variability, 49, 125, 407
variable, 7, 99, 109, 424
variance, 78, 183
variations, 125
vector field, 367, 370
vector norm, 176
vector space, 60, 73, 141, 158, 304
vector-valued function, 321
velocity vector, 322
vertical, 98
vertical asymptotes, 61, 62
vertical-line, 48
voltage, 397, 425, 436

volume of a sphere of radius $\rho$, 355
volumes by cross-sectional area, 266
volumes of revolution, 266
von Foerster model, 445

## W

well-defined, 69
whisker, 96
whole numbers, 5

## X

X-ray, 435

## Z

Z-score, 104, 118, 126, 134, 269
zero, 33
zero matrix, 158
zero vector, 151
zeros, 9, 16, 173